CONTEMPORARY QUANTITATIVE ECOLOGY AND RELATED ECOMETRICS

STATISTICAL ECOLOGY
Volume 12

a publication from the
satellite program in statistical ecology
international statistical ecology program

Statistical Ecology Series

General Editor: G. P. Patil

Volume 1* SPATIAL PATTERNS AND STATISTICAL DISTRIBUTIONS
edited by G. P. Patil, E. C. Pielou, and W. E. Waters (1971)

Volume 2* SAMPLING AND MODELING BIOLOGICAL POPULATIONS AND POPULATION DYNAMICS
edited by G. P. Patil, E. C. Pielou, and W. E. Waters (1971)

Volume 3* MANY SPECIES POPULATIONS, ECOSYSTEMS, AND SYSTEMS ANALYSIS
edited by G. P. Patil, E. C. Pielou, and W. E. Waters (1971)

Volume 4 STATISTICAL DISTRIBUTIONS IN ECOLOGICAL WORK
edited by J. K. Ord, G. P. Patil, and C. Taillie (1979)

Volume 5 SAMPLING BIOLOGICAL POPULATIONS
edited by R. M. Cormack, G. P. Patil, and D. S. Robson (1979)

Volume 6 ECOLOGICAL DIVERSITY IN THEORY AND PRACTICE
edited by J. F. Grassle, G. P. Patil, W. K. Smith, and C. Taillie (1979)

Volume 7 MULTIVARIATE METHODS IN ECOLOGICAL WORK
edited by L. Orloci, C. R. Rao, and W. M. Stiteler (1979)

Volume 8 SPATIAL AND TEMPORAL ANALYSIS IN ECOLOGY
edited by R. M. Cormack and J. K. Ord (1979)

Volume 9 SYSTEMS ANALYSIS OF ECOSYSTEMS
edited by G. S. Innis and R. V. O'Neill (1979)

Volume 10 COMPARTMENTAL ANALYSIS OF ECOSYSTEM MODELS
edited by J. H. Matis, B. C. Patten, and G. C. White (1979)

Volume 11 ENVIRONMENTAL BIOMONITORING, ASSESSMENT, PREDICTION, AND MANAGEMENT — CERTAIN CASE STUDIES AND RELATED QUANTITATIVE ISSUES
edited by J. Cairns, Jr., G. P. Patil, and W. E. Waters (1979)

Volume 12 CONTEMPORARY QUANTITATIVE ECOLOGY AND RELATED ECOMETRICS
edited by G. P. Patil and M. L. Rosenzweig (1979)

Volume 13 QUANTITATIVE POPULATION DYNAMICS
edited by D. G. Chapman and V. Gallucci (1979)

*For these first three volumes, contact:
The Pennsylvania State University Press
University Park, PA 16802 USA

For all of the remaining volumes, contact:
International Co-operative Publishing House
P.O. Box 245
Burtonsville, MD 20730 USA

CONTEMPORARY QUANTITATIVE ECOLOGY AND RELATED ECOMETRICS

edited by

GANAPATI P. PATIL
Department of Statistics
The Pennsylvania State University
University Park, Pennsylvania

MICHAEL L. ROSENZWEIG
Department of Ecology and Evolutionary Biology
University of Arizona
Tucson, Arizona

International Co-operative Publishing House
Fairland, Maryland USA

P. O. Box 245, Burtonsville, Maryland 20730

Library of Congress Catalog Card Number: 79-89711
Volume ISBN: 0-89974-009-X
Series ISBN: 0-89974-000-6

Printed in the United States of America

Mathematical ecology is moving out of its classical phase carrying with it untold promise for the future, but, as H.A.L. Fisher, the historian, remarks, progress is not a law of nature. Without enlightenment and eternal vigilance on the part of both ecologists and mathematicians there always lurks the danger that mathematical ecology might enter a dark age of barren formalism, fostered by an excessive faith in the magic of mathematics, blind acceptance of methodological dogma and worship of the new electronic gods. It is up to all of us to ensure that this does not happen.

J. G. SKELLAM (1972)
in Mathematical Models in Ecology
edited by J. N. R. Jeffers
Blackwell Scientific Publications
Oxford, England

For top management and general public policy development, monitoring data must be shaped into easy-to-understand indices that aggregate data into understandable forms. I am convinced that much greater effort must be placed on the development of better monitoring systems and indices than we have in the past. Failure to do so will result in sub-optimum achievement of goals at much greater expense.

R. E. TRAIN (1973)
National Conference on Managing the Environment
The United States Environmental Protection Agency
Washington, D.C., USA

International Statistical Ecology Program

ADVISORY BOARD

(Chairman: G. P. Patil)

Berthet, P.
Cairns, J., Jr.
Chapman, D. G.
Cormack, R. M.
Cox, D. R.
Holling, C. S.
Iwao, S.
Matern, B.
Matis, J. H.
Ord, J. K.
Pielou, E. C.
Rao, C. R.
Robson, D. S.
Rossi, O.
Simberloff, D. S.
Taillie, C.
Warren, W. G.
Waters, W. E.

SATELLITE PROGRAM IN STATISTICAL ECOLOGY

(1977-1978)

Director: G. P. Patil   Managing Editor: C. Taillie

COORDINATORS AND EDITORS

Artuz, M. I.
Cairns, J., Jr.
Chapman, D. G.
Cormack, R. M.
Gallucci, V.
Grassle, J. F.
Holling, C. S.
Innis, G. S.
Matis, J. H.
O'Neill, R. V.
Ord, J. K.
Orloci, L.
Patil, G. P.
Patten, B. C.
Rao, C. R.
Robson, D. S.
Rosenzweig, M. L.
Shoemaker, C.
Simberloff, D. S.
Smith, W. K.
Solomon, D. L.
Stiteler, W. M.
Taillie, C.
Usher, M. B.
Waters, W. E.
White, G. C.
Williams, F. M.

HOSTS

Matis, J. H.
Gates, C. E.
Waters, W. E.
Noy-Meir, I.
Rossi, O.
Zanni, R.

ADVISORS

Berthet, P.
Engen, S.
Glass, N.
Hennemuth, R.
Iwao, S.
Knox, G. A.
Matern, B.
Seber, G.
Warren, W. G.

SPONSORS

NATO Advanced Study Institutes Program
NATO Ecosciences Program

The Pennsylvania State University
The Texas A&M University
The University of California at Berkeley

National Marine Fisheries Service, USA
Environmental Protection Agency, USA
Fish and Wildlife Service, USA
Army Research Office, USA

Comunita Economica Europea
Universita degli Studi, Parma, Italy

Consiglio Nazionale delle Ricerche, Italy
Ministero dei Lavori Pubblici,
Affari Esteri, e Pubblica Istruzione, Italy
Societa Italiana di Statistica, Italy
Societa Italiana di Ecologia, Italy

The Participants and
Their Home Institutions and Organizations

PARTICIPANTS

SATELLITE A: COLLEGE STATION AND BERKELEY July 18-August 13, 1977

Andrews, P. L., Montana
Anthony, R. G., Oregon
Artuz, M. I., Turkey
Bagiatis, K., Greece
Bajusz, B. A., Pennsylvania
Baumgaertner, J., California
Bell, E., Washington
Bellefleur, P., Canada
Berthet, P., Belgium
Beyer, J., Denmark
Bingham, R., Texas
Boswell, M. T., Pennsylvania
Braswell, J. H., Georgia
Braumann, C. A., Portugal
Brennan, J. A., Massachusetts
Bruhn, J. N., California
Cairns, J. Jr., Virginia
Callahan, C. A., Oregon
Cancela da Fonseca, J. P., France
Caraco, T. B., Arizona
Chapman, D. G., Washington
Cho, A., Pennsylvania
Colwell, R. California
Cormack, R., Scotland
Coulson, R. N., Texas
DeMars, C. J., California
Dennis, B., Pennsylvania
Derr, J., Iowa
deVries, P. G., Netherlands
Doucet, P. G., Netherlands
Elterman, A. L., California
Engen, S., Norway
Ernsting, G., Netherlands
Fiadeiro, P. M., Portugal
Flores, R. G., Brazil
Flynn, T. S., California
Folse, L. J., Texas
Ford, R. G., California
Gallucci, V. F., Washington
Gates, C. E., Texas
Gautier, C., France
Gerald, K. B., Texas
Giles, R. H., Virginia
Gokhale, D. V., California
Grant, W. G., Texas
Guardans, R. C., Spain
Hart, D., Virginia
Hazard, J. W., Oregon
Hendrickson, J. A., Pennsylvania
Hennemuth, R., Massachusetts
Hogg, D. B., Mississippi
Innis, G. S., Utah
Janardan, K. G., Illinois
Johnson, D., Texas
Johnson, D. H., North Dakota
Jolly, G. M., Scotland
Kester, T., Belgium
Kie, J. G., California
Kobayashi, S., Japan
Kubicek, F., Czechoslovakia
Labovitz, M. L., Pennsylvania
Lamberti, G. A., California
Lasebikan, B. A., Nigeria
Lindahl, K. Q., California
Laurence, G. C., Rhode Island
Livingston, G. P., Texas
Ludwig, J. A., New Mexico
Ma, J. C. W., Texas
Macken, C. A., Minnesota
Marsden, M. A., Montana
Mason, R., Oregon
Matis, J. H., Texas
Matthews, G. A., Texas
Minello, T. J., Texas
Mizell, R. F., Mississippi
Monserud, R. A., Idaho
Myers, C. C., Illinois
Myers, R. A., Canada
Naveh, Z., Israel
Nebeker, T. E., Mississippi
Neyman, J., California
Norick, N. X., California
O'Neill, R. V., Tennessee
Ord, J. K., England
Overton, S., Oregon
Patil, G. P., Pennsylvania
Pennington, M. R., Massachusetts
Poole, R. W., Rhode Island
Pulley, P. E., Texas
Quinn, T. J., Washington
Rawson, C. B., Washington
Reynolds, J. F., North Carolina
Riggs, L. A., California
Robson, D. S., New York
Roman, J. R., New York
Rosenzweig, M. L., Arizona
Roughgarden, J., California
Roux, J. J. J., South Africa
Sanders, F., Tennessee
Sen, A. R., Canada
Serchuk, F. M., Massachusetts
Shoemaker, C., New York
Singh, K. P., India
Smith, W. K., Massachusetts
Solomon, D. L., New York
Southward, G. M., New Mexico
Stafford, S. G., New York
Steinhorst, R. K., Texas
Stenseth, N. C., Norway
Stiteler, W. M., New York
Stout, M. L., California
Stromberg, L. P., California
Taillie, C., Pennsylvania
Tracy, D. S., Canada
Usher, M. B., England
vanBiezen, J. B., Netherlands
Walter, G. G., Wisconsin
Waters, W. E., California
Wensel, L. C., California
West, I. F., New Zealand
Wiegert, R. G., Georgia
Williams, F. M., Pennsylvania
Wright, J. R., Alabama
Wu, Y. C., California
Yandell, B. S., California
Zweifel, J. R., California

SATELLITE B: PARMA July 31-September 5, 1978

Arditi, R., France
Azzarita F., Italy
Balchen, J. G., Norway
Bargmann, R. E., Georgia
Barlow, N. D., England
Baxter, M. B., Australia
Behrens, J., Denmark
Beran, H. G., Austria
Berman, M., Maryland
Berryman, A. A., Washington
Boswell, M. T., Pennsylvania
Brambilla, C., Italy
Braumann, C. A., New York
Breitenecker, M., Austria
Buerk, R., West Germany
Callahan, C. A., Oregon
Cancela da Fonseca, J. P., France
Chieppa, M., Italy
Clark, W. G., Italy
Cobelli, C., Italy
Cooper, C., California
Curry, G. L., Texas
DeMichele, D. W., Texas
Derr, J., Iowa
Diggle, P. J., England
Drakides, C., France
Ebenhöh, W., West Germany
Engen, S., Norway
Feoli, E., Italy
Fiadeiro, P. M., Portugal
Fischlin, A., Switzerland
Framstad, E. B., Norway
Frohberg, K., Austria
Gallucci, V. F., Washington
Garcia-Moya, E., Mexico
Gatto, M., Italy
Geri, C., France
Giavelli, G., Italy
Ginzburg, L. R., New York
Gokhale, D. V., California
Goldstein, R. A., California
Granero Porati, M. I., Italy
Greve, W., West Germany
Grosslein, M. D., Massachusetts
Grümm, H. R., Austria
Gulland, J. A., Italy
Gutierrez, A. P., California
Gydesen, H., Denmark
Hanski, I., England
Hanson, B. J., Utah
Hau, B., West Germany
Helgason, T., Iceland
Hendrickson, J. A., Pennsylvania
Hengeveld, R., Netherlands

Hennemuth, R. C., Massachusetts
Hoff, J. M., Norway
Holling, C. S., Canada
Hotz, M. C. B., Belgium
Jancey, R. C., Canada
Kooijman, S., Netherlands
Lamont, B. B., Australia
Levi, D., Italy
Liu, C. J., Kentucky
Marshall, W., Canada
Martin, F. W., Maryland
Matis, J. H., Texas
Menozzi, P., Italy
Meyer, J. A., France
Mohn, R. K., Canada
Mosimann, J., Maryland
Naveh, Z., Israel
Noy-Meir, I., Israel
Olivieri-Barra, S. T., Belgium
Ord, J. K., England
Orloci, L., Canada
Pacchetti, G., Italy
Pagani, L., Italy
Patil, G. P., Pennsylvania
Patten, B. C., Georgia
Pennington, M. R., Massachusetts
Policello, G. E., Ohio
Pospahala, R. S., Maryland
Purdue, P., Kentucky
Radler, K., West Germany
Ramsey, F. L., Oregon
Reyment, R. A., Sweden
Reyna Robles, R., Mexico
Rinaldi, S., Italy
Robson, D. S., New York
Rossi, O., Italy
Russek, E., Washington
Russo, A. R., Hawaii
Sadasivan, G., India
Schaefer, R., France
Shoemaker, C. A., New York
Show, I. T., California
Shuter, B. J., Canada
Simberloff, D., Florida
Slagstad, D., Norway
Smetacek, V. S., West Germany
Smith, W. K., Massachusetts
Sokal, R. R., New York
Soliani, L., Italy
Solomon, D. L., New York
Spremann, K., West Germany
Steinhorst, R. K., Idaho
Stenseth, N. C., Norway
Stiteler, W. M., New York
Subrahmanyam, C. B., Florida
Szöcs, Z., Hungary
Taillie, C., Pennsylvania
terBraak, C. J. F., Netherlands
Torrez, W. C., New Mexico
Tracy, D. S., Canada
Tursi, A., Italy
Vale, C., Portugal
van Biezen, J. B., Netherlands
Vazzana, C., Italy
Wahl, E., West Germany
Walker, B. H., England
Walter, G. G., Wisconsin
Walters, C. J., Canada
Warren, W. G., Canada
Waters, W. E., California
White, G. C., New Mexico
Wise, M. E., Netherlands
Zanni, R., Italy

SATELLITE C: JERUSALEM September 7-September 15, 1978

Austin, M. P., Australia
Baxter, M. B., Australia
Berthet, P., Belgium
Carleton, T. J., Canada
Eisen, P. A., New York
Galluci, V., Washington
Goldstein, R. A., California
Goodman, D., California
Hanski, I., Finland
Hendrickson, J. A., Pennsylvania
Hengeveld, R., Netherlands
Hennemuth, R. C., Massachusetts
Hirsch, A., Washington, DC
Kempton, R. A., England
Naveh, Z., Israel
Noy-Meir, I., Israel
Odum, H. T., Florida
O'Neill, R. V., Tennessee
Patil, G. P., Pennsylvania
Quinn, T. J., Washington
Resh, V. H., California
Rossi, O., Italy
Rosenzweig, M. L., Arizona
Safriel, U., Israel
Sheshinski, R., Israel
Simberloff, D. S., Florida
Sokal, R. R., New York
Solem, J. O., Norway
Stenseth, N. C., Norway
Subrahmanyam, C. B., Florida
Torrez, W., New Mexico
Waters, W. E., California
Whittaker, R. H., New York

AUTHORS NOT LISTED ABOVE

Adams, J. E., Texas
Barber, M. C., Georgia
Barreto, M., Brazil
Batcheler, C. L., New Zealand
Beuter, K. J., West Germany
Bitz, D. W., Rhode Island
Brown, B. E., Massachusetts
Brown, G. C., Kentucky
Brown-Leger, L. S., Massachusetts
Chaim, S., Israel
Chardy, P. France
Clark, G. M., Ohio
Condra, C., California
Connor, E. F., Florida
Costa, H., Brazil
Dale, M., Australia
De LaSalle, P., France
Duek, J. L., Arizona
Elwood, J. W., Tennessee
Ferris, J. M., Indiana
Ferris, V. R., Indiana
Finn, J. T., Massachusetts
Foltz, J. L., Texas
Gibson, V. R., Massachusetts
Giddings, J. M., Tennessee
Gittins, R., Australia
Godron, M., France
Grassle, J. F., Massachusetts
Green, R., California
Halbach, U., West Germany
Halfon, E., Canada
Helthshe, J. F., Rhode Island
Hildebrand, S. G., Tennessee
Hogeweg, P., Netherlands
Iwao, S., Japan
Jacur, G. R., Italy
Kaesler, R. L., Kansas
Kravitz, D., Massachusetts
Kruczynski, W. L., Florida
Lackey, R., Virginia
Laurec, A., France
Lepschy, A., Italy
Malley, J. D., Maryland
Marcus, A. H., Washington
Matern, B., Sweden
McCune, E. D., Texas
Moncreiff, R., California
Moroni, A., Italy
Nichols, J. D., Maryland
O'Connor, J. S., New York
Paloheimo, J. E., Canada
Perrin, S., California
Pickford, S. G., Washington
Plowright, R. C., Canada
Podani, J., Hungary
Ratnaparkhi, M. V., Pennsylvania
Rescigno, A., Canada
Rickaert, M., France
Rohde, C., Maryland
Rosen, R., Canada
Scott, E., California
Scott, J. M., Hawaii
Seber, G. A. F., New Zealand
Siri, E., Italy
Skalski, J. R., Washington
Stehman, S., Pennsylvania
Steinberger, E. H., Israel
Swartzman, G., Washington
Taylor, C. E., California
Taylor, L. R., England
Tiwari, J. L., California
Watson, R. M., Scotland
Wehrly, T. E., Texas
Wigley, R. L., Massachusetts
Wissel, C., West Germany
Yang, M., Florida

Foreword

The Second International Congress of Ecology was held in Jerusalem during September 1978. In this connection, a Satellite Program in Statistical Ecology was organized by the International Statistical Ecology Program (ISEP) during 1977 and 1978. The emphasis was on research, review, and exposition concerned with the interface between quantitative ecology and relevant quantitative methods. Both theory and application of ecology and ecometrics received attention. The program consisted of instructional coursework, seminar series, thematic research conferences, and collaborative research workshops.

The 1977 and 1978 Satellite Program consisted of NATO Advanced Study Institutes at College Station in Texas, Berkeley in California, and Parma in Italy; NATO Advanced Research Institute at Parma; ISEP Research Conferences, Seminars, and Workshops at College Station, Berkeley, Parma, and Jerusalem; and a Research Conference at Jerusalem.

The Satellite Program has been supported by NATO Advanced Study Institutes Program; NATO Ecosciences Program; National Marine Fisheries Service, USA; Environmental Protection Agency, USA; Fish and Wildlife Service, USA; Army Research Office, USA; The Pennsylvania State University; The Texas A&M University; The University of California at Berkeley; Universita degli Studi, Parma; Consiglio Nazionale delle Ricerche, Italy; Ministero dei Lavori Pubblici, Affari Esteri, e Pubblica Istruzione, Italy; Societa Italiana di Statistica, Italy; Societa Italiana di Ecologia, Italy; Communita Economica Europea; and the participants and their home institutions and organizations.

Research papers and research-review-expositions were specially prepared for the program by concerned experts and expositors. These materials have been refereed and revised, and are now available in a series of ten edited volumes.

HISTORICAL BACKGROUND

The First International Symposium on Statistical Ecology was held in 1969 at Yale University with support from the Ford Foundation and the US Forest Service. The three symposium co-chairmen (G. P. Patil, E. C. Pielou, and W. E. Waters) represented the fields of statistics, theoretical ecology, and applied ecology. The program was well attended, and it provided a broad picture of where statistics and ecology stood relative to each other. While effort was apparent, communication between the two disciplines was inadequate.

It was clear that a focal forum was necessary to discuss and develop a constructive interface between quantifiable problems in ecology and relevant quantitative methods. As a partial solution to fill this need at professional organizations' level, the director of the symposium (G. P. Patil) made certain recommendations to the Presidents of the International Association for Ecology (A. D. Hasler), the International Statistical Institute (W. G. Cochran), and the International Biometric Society (P. Armitage). The International Association for Ecology (INTECOL) took a timely step in creating a section in the organization, namely, the statistical ecology section. The three societies together took a timely step in setting up a liaison committee on statistical ecology. The INTECOL Section and the Liaison Committee together developed the International Statistical Ecology Program. Since its inception in 1970, ISEP (as it has come to be known) has put emphasis on identifying the interdisciplinary needs of statistics and ecology at advanced instructional levels, and also at research conference and workshop levels.

The First Advanced Institute on Statistical Ecology in the United States was organized at Penn State for six weeks in 1972 with support from the US National Science Foundation, the US Forest Service, and the Mathematical Social Sciences Board. The participants of the Institute are all enjoying the benefits of their fruitful participation. With support from the UNESCO program of Man and Biosphere, a six month program was held in Venezuela for participants from Latin America in 1974 under the direction of Jorge Rabinovich, himself a participant in the 1972 Institute. With some initiatives from ISEP, special statistical ecology sessions have been held at the international conferences of the International Statistical Institute and the Biometric Society.

While plans were being made for the Second International Congress of Ecology, the then Secretary General and current President of INTECOL (G. A. Knox), and the ISEP chairman (G. P. Patil) discussed the need and the timeliness of a program in statistical ecology. The Satellite Program in Statistical Ecology took its final shape from this beginning under the care and concern of its director, advisors, coordinators, and sponsors.

SCIENTIFIC BACKGROUND AND PURPOSE

The perceptions of Skellam and Train quoted on page v in this volume recapitulate the cautions, inspirations, and objectives responsible for the Satellite Program. The rigorous formulation of a quantitative scientific concept requires and in a sense creates empirically measurable quantities. Conversely, the scientific validity of the concept is totally dependent upon the measured values of those quantities. This is a capsule version of the feedback process known as the "scientific method." In crude modern terms, we might label the first procedure "modeling" and the second "curve-fitting." For reasons of complexity, historical accident, or whatever, these mutually dependent, complementary components have never become firmly integrated within ecology and its application to environmental studies.

Both procedures involve forms of mathematics: In the "modeling" process, the mathematics is used *relationally* — as a system of logic to ensure rigor and clarity of reasoning. This is in the historical tradition of Volterra and Lotka. Validation is most often based on *qualitative* agreement: the right trend, or the correct shape of the curve. This is as it should be, especially for broad general theories: it is the ideas and understanding that count, not so much the quantitative detail.

In the "curve-fitting" tradition, the mathematics is used *numerically* — as a system for precise measurement and prediction. Validation is most often based on *quantitative* agreement: n-place accuracy, or minimum uncertainty. This is also as it should be, especially for application and management: it is the forecast and ability to act confidently that count, not so much the underlying concept.

It need not be taken as a sign of "physics envy" to assert that as a science matures, these two processes must converge — more quantification must be used in concept validation and more concepts must be incorporated into the methodology of quantification.

The purpose of the Satellite Program is to encourage that convergence within the science of ecology and promote its application in the study of the environment and environmental stress. Monitoring and assessment activities to be meaningful and defensible need: (i) a conceptual and philosophical basis, (ii) a theoretical framework, (iii) methodological support, (iv) a technological toolbox, and (v) administrative management. The ultimate purpose of the program is to help identify and integrate the specifics of these important factors responsible for protecting the environment.

We take as our theme the better melding of fundamental ecological concepts with rigorous empirical quantification. The overall result should be progress toward a stronger body of general ecological theory and practice.

PLANNING AND ORGANIZATION

The realization of any program of this nature and dimension often fails to fully meet the initial expectations and objectives of the organizers. Factors that are both logistic and psychological in nature tend to contribute to this general experience. Logistic difficulties include optimality problems for time and location. Other difficulties which must be attended to involve conflicting attitudes towards the importance of individual contributions to the proceedings.

We tried to cope with these problems by seeking active advice from a number of participants and special advisors. The advice we received was immensely helpful in guiding our selection of the best experts in the field to achieve as representative and balanced a coverage as possible. Simultaneously, the editors together with the referees took a rather critical and constructive attitude from initial to final stages of preparation of papers by offering specific suggestions concerning the suitability, and also the structure, content and size. These efforts of coordination and revision were intensified through editorial sessions at the program itself as a necessary step for the benefit of both the general readership and the participants. It is our pleasure to record with appreciation the spontaneous cooperation of the participants. Everyone went by scientific interests often at the expense of personal preferences. The program atmosphere became truly creative and friendly, and this remarkable development contributed to the maximal cohesion of the program and its proceedings within the limited time period available.

In retrospect, our goals were perhaps ambitious! We had close to 350 lectures and discussions during 50 days in the middle of the summer seasons of 1977 and 1978. For several reasons, we decided that an overworked program was to be preferred to a leisurely one. First of all, gatherings of such dimension are possible only every 5-10 years. Secondly, the previous meetings of this nature occurred some 5-10 years back, and the subject area of statistical ecology had witnessed substantial growth in this time. Thirdly, but most importantly, was the overwhelming response from potential participants, many of whom were to come across the continents!

Satellite A at College Station, Texas, and at Berkeley, California covered a four week period during July 18-August 13, 1977 and had 125 participants. Satellite B at

Parma, Italy had 130 participants spread over a six week period during July 31-September 5, 1978. Satellite C at Jerusalem, Israel had 35 participants. Approximately, one-third of the participants were graduate students, one-half were university faculty, and one-third were agency scientists. Approximately, one-third of the participants had an affiliation with one mathematical science or another, one-half an affiliation with one environmental science or the other, and one-quarter had an affiliation with one environmental management program or another. Thus the group was a good mix of great variety contributing to the effectiveness of the program. Not only what one heard was enlightening, but what one over-heard was equally enlightening!

Professors G. P. Patil, Paul Berthet, J. K. Ord, and Charles Taillie served as scientific directors of the program with Professor Patil assuming the responsibility of its direction from its conception to its conclusion. The inaugural speakers were: Professor H. O. Hartley at College Station, Professor J. Neyman at Berkeley, Professor C. S. Holling at Parma, and Professor G. P. Patil at Jerusalem.

SCIENTIFIC CONTENT AND PUBLICATION

The following summary information on the subjects and corresponding coordinators of the satellite program may be of some interest. The details of each subject and its publication volume are reported elsewhere.

Statistical Distributions in Ecological Work: J. K. Ord, G. P. Patil, and C. Taillie.

Spatial and Temporal Analysis in Ecology: R. M. Cormack and J. K. Ord.

Quantitative Population Dynamics: D. G. Chapman and V. Gallucci.

Sampling Biological Populations: R. M. Cormack, G. P. Patil, and D. S. Robson.

Ecological Diversity in Theory and Practice: J. F. Grassle, G. P. Patil, W. K. Smith, and C. Taillie.

Multivariate Methods in Ecological Work: L. Orloci, C. R. Rao, and W. M. Stiteler.

Systems Analysis of Ecosystems: G. S. Innis and R. V. O'Neill.

Compartmental Analysis of Ecosystem Models: J. H. Matis, B. C. Patten, and G. C. White.

Environmental Biomonitoring, Assessment, Prediction, and Management-Certain Case Studies and Related Quantitative Issues: J. Cairns, Jr., G. P. Patil and W. E. Waters.

Contemporary Quantitative Ecology and Related Ecometrics: G. P. Patil and M. L. Rosenzweig.

Three more subjects were organized in the program.

Scientific Modeling and Quantitative Thinking with Examples in Ecology: G. P. Patil, D. S. Simberloff, and D. L. Solomon.

Conceptual Foundations of Ecological Theory and Applications: M. B. Usher and F. M. Williams.

Optimizations in Ecological Theory and Management: C. S. Holling and C. Shoemaker.

It would be fruitful to reorganize these subjects and add a few more for the next satellite program when it occurs.

It should be mentioned here that the close coordination and cooperation between the coordinators and the authors/speakers of potential contributions to the Proceedings (which are particularly intensive during the satellite program) paid themselves handsomely when the editors were confronted with the technical work related to the publications after the close of the program at Jerusalem. It is therefore very satisfying to report that the edited research papers and research-review-expositions prepared for the program are ready for distribution within 12 months of the conclusion of the program. For purposes of convenience, the contributions are organized in ten volumes in the Statistical Ecology Series published by the International Co-operative Publishing House. Altogether, they consist of an estimated 4,000 pages of research, review, and exposition, in addition to this common foreword in each followed by individual volume introductions. Subject and author indexes are also prepared at the end. Every effort has been made to keep the coverage of the volumes close to their individual titles. May this ten volume set in its own modest way provide an example of synergism.

FUTURE DIRECTIONS

We wish there was no need for a program of this nature and dimension. It would be ideal if the needs of an interdisciplinary program were satisfactorily met in the existing institutions. Unfortunately, universities and governmental agencies have not been able to find effective ways to foster healthy interdisciplinary programs. The individuals attempting to do something in this direction tend to feel disheartened or disillusioned.

The satellite-like-programs help create and sustain enthusiasm, inward strength, and working efficiency of those who desire to meet a contemporary social need in the form of some interdisciplinary work. It should be only proper and rewarding for everyone involved that such programs are planned from time to time.

Plans are being made for a satellite program in conjunction with the next Biennial Conference of the International Statistical Institute and with the next International Congress of Ecology. Care should be exercised that the next program not become a mere replica of the present one, however successful it has been. Instead, the next program should be organized so that it helps further the evolution of statistical ecology as a productive field.

The next program is being discussed in terms of subject area groups. Each subject group is to have a coordinator assisted by small committees, such as a program committee, a research committee, an annual review committee, a journal committee, and an education committee. This approach is expected to respond to the need for a journal on statistical ecology, and also to the need of bringing out well planned annual review volumes. The education committee would formulate plans for timely manuals, modules, and monographs. Interested readers may feel free to communicate their ideas and interests to those involved in planning the next program. With mutual goodwill and support, we shall have met a timely need for today's science, technology, and society.

July 1979 G. P. Patil

Program Acknowledgments

For any program to be successful, mutual understanding and support among all participants are essential in directions ranging from critical to constructive and from cautious to enthusiastic. The present program is grateful to the members of the ISEP Advisory Board, and to the referees, editors, coordinators, advisors, sponsors and the participants for their timely advice and support.

The success of the program was due, in no small measure, to the endeavors of the Local Arrangements Chairmen: J. H. Matis and C. E. Gates at College Station, W. E. Waters at Berkeley, O. Rossi and R. Zanni at Parma, and I. Noy-Meir at Jerusalem. We thank them for their hospitality and support.

And finally those who have assisted with the arduous task of preparing the materials for publication. Barbara Alles has been an ever cheerful and industrious secretary in the face of every adversity. Charles Taillie managed both scientific and non-scientific aspects. Bharat Kapur copyedited and proofread. Bonnie Burris, Bonnie Henninger, and Sandy Rothrock prepared the final versions of the manuscripts. Marllyn Boswell helped with the subject and author indexes. So did Bharat Kapur, Satish Patil, and Rani Venkataramani.

All of these nice people have done a fine job indeed. To all of them, our sincere thanks.

July 1979 G. P. Patil

Reviewers of Manuscripts

With appreciation and gratitude, the program acknowledges the valuable services of the following referees who have served as reviewers of manuscripts submitted to the program for possible publication. The editors thank the reviewers for their critical and constructive reviews.

J. Balchen
University of Trondheim

R. E. Bargmann
University of Georgia

C. Chatfield
University of Bath

R. Colwell
University of California

A. Berryman
Washington State University

J. Beyer
Danish Institute of Fisheries and Marine Research

M. T. Boswell
Pennsylvania State University

C. Braumann
Instituto Universitario de Evora

M. A. Buzas
Smithsonian Institution

J. Cairns, Jr.
Virginia Polytechnic Institute and State University

C. Cobelli
Laboratorio per Ricerche di Dinamica dei Sistemi e di Bioingegneria

R. Cormack
University of St. Andrews

B. Coull
University of South Carolina

M. B. Dale
Commonwealth Scientific and Industrial Research Organization

A. P. Dawid
The City University, London

B. Dennis
Pennsylvania State University

F. Diemer
Office of Naval Research

P. Diggle
University of Newcastle upon Type

P. G. Doucet
Vrije University

J. E. Dunn
University of Arkansas

L. Eberhardt
Battelle Pacific Northwest Laboratories

S. Engen
University of Trondheim

R. G. Flores
Fundacao Instituto Brasileiro de Geografia e Estatistica

E. D. Ford
Institute of Terrestrial Ecology

C. E. Gates
Texas A&M University

K. B. Gerald
Rockwell International

L. Ginzburg
State University of New York

J. C. Gittins
University of Oxford

N. Glass
Environmental Protection Agency

D. V. Gokhale
University of California

R. Goldstein
Electric Power Research Institute

M. Granero-Porati
Istituto di Fisica, Parma

J. F. Grassle
Woods Hole Oceanographic Institution

J. C. Griffiths
Pennsylvania State University

I. Hanski
University of Oxford

W. Harkness
Pennsylvania State University

T. Helgason
University of Iceland

J. Hendrickson
Academy of Natural Sciences

R. C. Hennemuth
National Marine Fisheries Service

D. Hildebrand
University of Pennsylvania

A. Hirsch
Fish and Wildlife Service

P. Holgate
Birkbeck College

H. Horn
Princeton University

G. Innis
Utah State University

I. James
Commonwealth Scientific and Industrial Research Organization

K. G. Janardan
Sangamon State University

G. M. Jolly
University of Edinburgh

C. Jones
Woods Hole Oceanographic Institution

R. Kaesler
University of Kansas

R. A. Kempton
Plant Breeding Institute, Cambridge

J. Kirkley
National Marine Fisheries Service

G. Knott
National Institutes of Health

S. Kotz
Temple University

A. M. Kshirsagar
University of Michigan

S. Kullback
The George Washington University

R. C. Lewontin
Harvard University

B. F. J. Manly
University of Otago

A. H. Marcus
Washington State University

F. Martin
Patuxent Wildlife Research Center

J. H. Matis
Texas A&M University

J. R. McBride
University of California

D. Mollison
Heriot-Wett University

R. V. O'Neill
Oak Ridge National Laboratory

J. Newton
University of St. Andrews

J. K. Ord
University of Warwick

L. Orloci
University of Western Ontario

G. P. Patil
Pennsylvania State University

B. C. Patten
University of Georgia

M. Pennington
National Marine Fisheries Service

S. Pimm
Texas Tech University

K. H. Pollock
University of Reading

R. W. Poole
Brown University

R. S. Pospahala
Patuxent Wildlife Research Center

E. Preston
Environmental Protection Agency

F. Preston
Preston Laboratories

P. Purdue
University of Kentucky

F. L. Ramsey
Oregon State University

C. R. Rao
Indian Statistical Institute

P. A. Rauch
University of California

E. Renshaw
University of Edinburgh

R. Reyment
Uppsala University

D. S. Robson
Cornell University

M. L. Rosenzweig
University of Arizona

O. Rossi
University of Parma

W. E. Schaaf
National Marine Fisheries Service

T. Schopf
University of Chicago

H. T. Schreuder
Forest Service

G. A. F. Seber
University of Auckland

J. Sepkoski
University of Rochester

I. T. Show, Jr.
Science Applications, Inc.

D. Simberloff
Florida State University

D. B. Siniff
University of Minnesota

W. K. Smith
Woods Hole Oceanographic Institution

R. R. Sokal
State University of New York

G. M. Southward
New Mexico State University

S. Stehman
Pennsylvania State University

R. K. Steinhorst
University of Idaho

W. M. Stiteler
Syracuse University

P. Switzer
Stanford University

C. Taillie
International Statistical Ecology Program

C. E. Taylor
University of California

C. Tsokos
University of South Florida

E. Ursin
Danish Institute of Fisheries and Marine Research

D. Vaughan
Oak Ridge National Laboratory

G. G. Walter
University of Wisconsin

W. G. Warren
Western Forest Products Laboratory

W. E. Waters
University of California

S. D. Webb
University of Florida

G. C. White
Los Alamos Scientific Laboratory

R. H. Whittaker
Cornell University

M. E. Wise
Leiden University

S. Zahl
University of Connecticut

J. Zweifel
National Marine Fisheries Service

Contents of Edited Volumes

STATISTICAL DISTRIBUTIONS IN ECOLOGICAL WORK
J. K. Ord, G. P. Patil, and C. Taillie (editors) **400 pp. approx.**
Chance Mechanisms Underlying Statistical Distributions: M. T. BOSWELL, J. K. ORD, and G. P. PATIL, Chance Mechanisms Underlying Univariate Distributions. C. TAILLIE, J. K. ORD, J. MOSIMANN, and G. P. PATIL, Chance Mechanisms Underlying Multivariate Distributions.

Ecological Advances Involving Statistical Distributions: M. T. BOSWELL and B. DENNIS, Ecological Contributions Involving Statistical Distributions. W. G. WARREN, Some Recent Developments Relating to Statistical Distributions in Forestry and Forest Products Research.

Abundance Models: J. A. HENDRICKSON, The Biological Motivation for Abundance Models. R.HENGEVELD, On the Use of Abundance and Species-Abundance Curves. S. ENGEN and C. TAILLIE, A Basic Development of Abundance Models: Community Description. S. ENGEN, Abundance Models: Sampling and Estimation. R. HENGEVELD, S. KOOIJMAN, and C. TAILLIE, A Spatial Model Explaining Species-Abundance Curves. S. KOBAYASHI, Species-Area Curves.

Research Papers: T. CARACO, Ecological Response of Animal Group Size Frequencies. J. DERR and J. K. ORD, Feeding Preferences of a Tropical Seedfeeding Insect: An Analysis of the Distribution of Insects and Seeds in the Field. R. A. KEMPTON, The Statistical Analysis of Frequency Data Obtained from Sampling an Insect Population Grouped by Stages. M. PENNINGTON, Fitting a Growth Curve to Field Data. M. YANG, Some Results on Size and Utilization Distribution of Home Range.

SAMPLING BIOLOGICAL POPULATIONS
R. M. Cormack, G. P. Patil, and D. S. Robson (editors) **425 pp. approx.**
P. G. DEVRIES, Line Intersect Sampling — Statistical Theory, Applications and Suggestions for Extended Use in Ecological Inventory. C. E. GATES, Line Transect and Related Issues. F. RAMSEY and J. M. SCOTT, Estimating Population Densities from Variable Circular Plot Surveys. G. A. F. SEBER, Transects of Random Length. G. M. JOLLY, Sampling of Large Objects. R. M. CORMACK, Models for Capture-Recapture. D. S. ROBSON, Approximations to Some Mark-Recapture Sampling Distributions. G. M. JOLLY, A Unified Approach to Mark-Recapture Stochastic Models, Exemplified by a Constant Survival Rate Model. J. R. SKALSKI and D. S. ROBSON, Tests of Homogeneity and Goodness-of-Fit to a Truncated Geometric Model for Removal Sampling. A. R. SEN, Sampling Theory on Repeated Occasions with Ecological Applications. W. G. WARREN, Trends in the Sampling of Forest Populations. G. M. JOLLY and R. M. WATSON, Aerial Sample Survey Methods in the Quantitative Assessment of Ecological Resources. W. K. SMITH, An Oil Spill Sampling Strategy. C. ROHDE, Batch, Bulk, and Composite Sampling.

ECOLOGICAL DIVERSITY IN THEORY AND PRACTICE
J. F. Grassle, G. P. Patil, W. K. Smith, and C. Taillie (editors) **350 pp. approx.**
Species Diversity Measures: General Definitions and Theoretical Setting: G. P. PATIL and C. TAILLIE, An Overview of Diversity. D. SOLOMON, A Comparative Approach to Species Diversity. S. ENGEN, Some Basic Concepts of Ecological Equitability. C. TAILLIE, Species Equitability: A Comparative Approach C. TAILLIE and G. P. PATIL, Diversity, Entropy, and Hardy-Weinberg Equilibrium.

Ecological Theory And Species Diversity: R. COLWELL, Toward an Unified Approach to the Study of Species Diversity. B. DENNIS and G. P. PATIL, Species Abundance, Diversity, and Environmental Predictability.

Estimating Diversity From Field Samples: J. E. ADAMS and E. D. MCCUNE, Application of the Generalized Jackknife to Shannon's Measure of Information as a Measure of Species Diversity. J. F. HELTHSHE and D. W. BITZ, Comparing and Contrasting Diversity Measures in Censused and Sampled Communities. J. HENDRICKSON, Estimates of Species Diversity from Multiple Samples. D. SIMBERLOFF, Rarefaction as a Distribution-Free Method of Expressing and Estimating Diversity. W. K. SMITH, J. F. GRASSLE, and D. KRAVITZ, Measures of Diversity with Unbiased Estimates.

Application of Species Diversity Measures: A Critical Evaluation: R. FLORES, M. BARRETO, and H. COSTA, On the Relationship Between Ecological and Physico-Chemical Water Quality Parameters: A Case Study at the Guanabara Bay, Rio de Janeiro. R. W. MONCREIFF and S. PERRIN, Effects of Increased Water Temperature on the Diversity of Periphyton. Z. NAVEH and R. H. WHITTAKER,

Determination of Plant Species Diversity in Mediterranean Shrub and Woodland Along Environmental Gradients. V. RESH, Biomonitoring, Species Diversity Indices, and Taxonomy. J. SOLEM, A Comparison of Species Diversity Indices in Trichoptera Communities. W. K. SMITH, V. R. GIBSON, L. S. BROWN-LEGER, and J. F. GRASSLE, Diversity as an Indicator of Pollution: Cautionary Results from Microcosm Experiments. C. B. SUBRAHMANYAM and W. L. KRUCZYNSKI, Colonization of Polychaetous Annelids in the Intertidal Zone of a Dredged Material Island in North Florida. C. E. TAYLOR and C. CONDRA, Competitor Diversity and Chromosomal Variation in Drosophia Pseudoobscura. B. DENNIS and O. ROSSI, Community Composition and Diversity Analysis in a Marine Zooplankton Survey.

Bibliography: B. DENNIS, G. P. PATIL, O. ROSSI, S. STEHMAN, and C. TAILLIE, A Bibliography of Literature on Ecological Diversity and Related Methodology.

MULTIVARIATE METHODS IN ECOLOGICAL WORK
L. Orloci, C. R. Rao, and W. M. Stiteler (editors) **400 pp. approx.**
R. BARGMANN, Structural Analysis of Singular Matrices Using Union Intersection Statistics with Applications in Ecology. M. DALE, On Linguistic Approaches to Ecosystems and Their Classification. D. V. GOKHALE, Analysis of Ecological Frequency Data: Certain Case Studies. J. HENDRICKSON, Examples of Discrete Multivariate Methods in Ecological Work. R. HENGEVELD and P. HOGEWEG, Cluster Analysis of the Distribution Patterns of Dutch Carabid Species. R. JANCEY, Species Weighting. A. LAUREC, P. CHARDY, P. DE LASALLE, and M. RICKAERT, Use of Dual Structures in Inertia Analysis: Ecological Implications. J. MOSIMANN and J. D. MALLEY, Size and Shape Analysis. L. ORLOCI, Non-Linear Data Structure and Their Description. J. PODANI, A Generalized Strategy of Homogeneity-Optimizing Hierarchical Classificatory Methods. R. REYMENT, Multivariate Analysis in Statistical Paleoecology. E. SCOTT, Spurious Correlation. W. K. SMITH, D. KRAVITZ, and J. F. GRASSLE, Confidence Intervals for Similarity Measures Using the Two Sample Jackknife. R. K. STEINHORST, Analysis of Niche Overlap. W. M. STITELER, Multivariate Statistics with Applications in Statistical Ecology. Z. SZOCS, New Computer Oriented Methods for Structural Investigation of Natural and Simulated Vegetation Patterns. B. LAMONT and K. J. GRANT, A Comparison of Twenty Measures of Site Dissimilarity. R. GITTINS, Ecological Applications of Canonical Analysis.

SPATIAL AND TEMPORAL ANALYSIS IN ECOLOGY
R. M. Cormack and J. K. Ord (editors) **400 pp. approx.**
J. K. ORD, Time-Series and Spatial Patterns in Ecology. P. J. DIGGLE, Statistical Methods for Spatial Point Patterns in Ecology. R. M. CORMACK, Spatial Aspects of Competition Between Individuals. R. W. POOLE, The Statistical Prediction of the Fluctuations in Abundance in Nicholson's Sheep Blowfly Experiments. W. G. WARREN and C. L. BATCHELER, The Density of Spatial Patterns: Robust Estimation Through Distance Methods. B. MATERN, The Analysis of Ecological Maps as Mosaics. J. A. LUDWIG, A Test of Different Quadrat Variance Methods for the Analysis of Spatial Pattern. S. A. L. M. KOOIJMAN, The Description of Point Patterns. R. HENGEVELD, The Analysis of Spatial Patterns of Some Ground Beetles (Col. Carabidae).

SYSTEMS ANALYSIS OF ECOSYSTEMS
G. S. Innis and R. V. O'Neill (editors) **425 pp. approx.**
R. K. STEINHORST, Stochastic Difference Equation Models of Biological Systems. R. V. O'NEILL, Natural Variability as a Source of Error in Model Predictions. R. K. STEINHORST, Parameter Identifiability, Validation and Sensitivity Analysis of Large System Models. R. V. O'NEILL, Transmutation Across Hierarchical Levels. R. V. O'NEILL, J. W. ELWOOD, and S. G. HILDEBRAND, Theoretical Implications of Spatial Heterogeneity in Stream Ecosystems. R. V. O'NEILL and J. M. GIDDINGS, Population Interactions and Ecosystem Function: Plankton Competition and Community Production. J. P. CANCELA DA FONSECA, Species Colonization Models of Temporary Ecosystems Habitats. E. HALFON, Computer-Based Development of Large Scale Ecological Models: Problems and Prospects. G. S. INNIS, A Spiral Approach to Ecosystem Simulation.

COMPARTMENTAL ANALYSIS OF ECOSYSTEM MODELS
J. H. Matis, B. C. Patten, and G. C. White (editors) **400 pp. approx.**
Applications of Compartmental Analysis to Ecosystem Modeling: R. V. O'NEILL, A Review of Linear Compartmental Analysis in Ecosystem Science. G. G. WALTER, A Compartmental Model of a Marine Ecosystem. M. C. BARBER, B. C. PATTEN, and J. T. FINN, Review and Evaluation of Input-Output Flow Analysis for Ecological Applications. I. T. SHOW, JR., An Application of Compartmental Models to Meso-scale Marine Ecosystems.

Identifiability and Statistical Estimation of Parameters in Compartmental Models: C. COBELLI, A. LEPSCHY, G. R. JACUR, Identification Experiments and Identifiability Criteria for Compartmental Systems. M. BERMAN, Simulation, Data Analysis, and Modeling with the SAAM Computer Program. G. C. WHITE and G. M. CLARK. Estimation of Parameters for Stochastic Compartment Models. R. E. BARGMANN, Statistical Estimation and Computational Algorithms in Compartmental Analysis for Incomplete Sets of Observations.

Stochastic Approaches to the Compartmental Modeling of Ecosystems: J. L. TIWARI, A Modeling Approach Based on Stochastic Differential Equations, the Principle of Maximum Entropy, and Bayesian Inference for Parameters. J. H. MATIS and T. E. WEHRLY, An Approach to a Compartmental Model with Multiple Sources of Stochasticity for Modeling Ecological Systems. P. PURDUE, Stochastic Compartmental Models: A Review of the Mathematical Theory with Ecological Applications. A. H. MARCUS, Semi-Markov Compartmental Models in Ecology and Environmental Health. M. E. WISE, The Need for Rethinking on Both Compartments and Modeling.

Mathematical Analysis of Compartmental Structures: G. G. WALTER, Compartmental Models, Digraphs, and Markov Chains. K. B. GERALD and J. H. MATIS, On the Cumulants of Some Stochastic Compartmental Models Applied to Ecological Systems. A. RESCIGNO, The Two-variable Operational Calculus in the Construction of Compartmental Ecological Models.

ENVIRONMENTAL BIOMONITORING, ASSESSMENT, PREDICTION, AND MANAGEMENT — CERTAIN CASE STUDIES AND RELATED QUANTITATIVE ISSUES
J. Cairns, Jr., G. P. Patil, and W. E. Waters (editors) 450 pp. approx.
Biomonitoring: J. CAIRNS, JR., Biological Monitoring — Concept and Scope. Z. NAVEH, E. H. STEINBERGER, and S. CHAIM, Use of Bio-Indicators for Monitoring of Air Pollution by Fluor, Ozone and Sulfur Dioxide.

Environmental assessment and prediction: G. P. PATIL, C. TAILLIE, and R. L. WIGLEY, Transect Sampling Methods and Their Application to the Deep-Sea Red Crab. W. E. WATERS, Biomonitoring, Assessment, and Prediction in Forest Pest Management Systems. C. A. CALLAHAN, V. R. FERRIS, and J. M. FERRIS, The Ordination of Aquatic Nematode Communities as Affected by Stream Water Quality. R. A. GOLDSTEIN, Development and Implementation of a Research Program on Ecological Assessment of the Impact of Thermal Power Plant Cooling Systems on Aquatic Environments.

Environmental Management: A. HIRSCH, Ecological Information and Technology Transfer, R. H. GILES, JR., Modeling Decisions or Ecological Systems. R. LACKEY, Appliction of Renewable Natural Resource Modeling in Public Decision-Making Process. F. MARTIN, R. S. POSPAHALA, and J. D. NICHOLS, Assessment and Population Management of North American Migratory Birds. J. E. PALOHEIMO and R. C. PLOWRIGHT, Bioenergetics, Population Growth and Fisheries Management. Z. NAVEH, A Model of Multiple-Use Management Strategies of Marginal and Untillable Mediterranean Upland Ecosystems.

Case Studies and Quantitative Issues: M. D. GROSSLEIN, R. C. HENNEMUTH, and B. E. BROWN, Research, Assessment, and Management of a Marine Ecosystem in the Northwest Atlantic — A Case Study. J. NEYMAN, Two Interesting Ecological Problems Demanding Statistical Treatment. B. DENNIS, G. P. PATIL, and O. ROSSI, The Sensitivity of Ecological Diversity Indices to the Presence of Pollutants in Aquatic Communities. D. SIMBERLOFF, Constraints on Community Structure During Colonization.

CONTEMPORARY QUANTITATIVE ECOLOGY AND RELATED ECOMETRICS
G. P. Patil and M. L. Rosenzweig (editors) 725 pp. approx.
Community Structure and Diversity: R. A. KEMPTON and L. R. TAYLOR, Some Observations on the Yearly Variability of Species Abundance at a Site and the Consistency of Measures of Diversity. G. P. PATIL and C. TAILLIE, A Study of Diversity Profiles and Orderings for a Bird Community in the Vicinity of Colstrip, Montana. M. L. ROSENZWEIG, Three Probable Evolutionary Causes for Habitat Selection. O. ROSSI, G. GIAVELLI, A. MORONI, and E. SIRI, Statistical Analysis of the Zooplankton Species Diversity of Lakes Placed Along a Gradient. S. KOBAYASHI, Another Model of the Species Rank-Abundance Relation for a Delimited Community. M. L. ROSENZWEIG and J. L. DUEK, Species Diversity and Turnover in an Ordovician Marine Invertebrate Assemblage.

Patterns and Interpretations: D. S. SIMBERLOFF and E. F. CONNOR, Q-Mode and R-Mode Analyses of Biogeographic Distributions: Null Hypotheses Based on Random Colonization. D. GOODMAN,

Applications of Eigenvector Analysis in the Resolution of Spectral Pattern in Spatial and Temporal Ecological Sequences. R. H. WHITTAKER and Z. NAVEH, Analysis of Two-Phase Patterns. R. R. SOKAL, Ecological Parameters Infered From Spatial Correlograms. M. P. AUSTIN, Current Approaches to the Non-Linearity Problem in Vegetation Analysis. M. GODRON, A Probabilistic Computation for the Research of "Optimal Cuts" in Vegetation Studies. S. IWAO, The m*-m Method for Analyzing Distribution Patterns of Single- and Mixed-Species Populations.

Modeling and Ecosystems Modeling: D. L. SOLOMON, On a Paradigm for Mathematical Modeling. R. W. POOLE, Ecological Models and Stochastic-Deterministic Question. R. ROSEN, On the Role of Time and Interaction in Ecosystem Modelling. E. HALFON, On the Parameter Structure of a Large Scale Ecological Model. G. SWARTZMAN, Evaluation of Ecological Simulation Models. R. WIEGERT, Modeling Coastal, Estuarine and Marsh Ecosystems: State-of-the-Art.

Statistical Methodology and Sampling: J. DERR and J. K. ORD, Field Estimates of Insect Colonization, II. J. A. HENDRICKSON, JR., Analyses of Species Occurrences in Community, Continuum and Biomonitoring Studies. R. SHESHINSKI, Interpolation in the Plane. The Robustness to Misspecified Correlation Models and Different Trend Functions. W. C. TORREZ, The Effect of Random Selective Intensities on Fixation Probabilities. R. GREEN, A Graph Theoretical Test to Detect Interference in Selecting Nest Sites. I. NOY-MEIR, Graphical Models and Methods in Ecology. T. J. QUINN, The Effects of School Structure on Line Transect Estimators of Abundance. J. W. HAZARD and S. G. PICKFORD, Line Intersect Sampling of Forest Residue.

Applied Statistical Ecology: R. C. HENNEMUTH, Man as Predator. V. F. GALLUCCI, On Assessing Population Characteristics of Migratory Marine Animals. P. A. EISEN and J. S. O'CONNOR, MESA Contributions to Sampling in Marine Environments. W. E. WATERS and V. H. RESH, Ecological and Statistical Features of Sampling Insect Populations in Forest and Aquatic Environments. R. L. KAESLER, Statistical Paleoecology: Problems and Perspectives. S. A. L. M. KOOIJMAN and R. HENGEVELD, The Description of Non-Linear Relationship Between Some Carabid Beetles and Environmental Factors. P. E. PULLEY, R. N. COULSON, and J. L. FOLTZ, Sampling Bark Beetle Populations for Abundance.

A Bibliography: B. DENNIS, G. P. PATIL, M. V. RATNAPARKHI, and S. STEHMAN, A Bibliography of Selected Books on Quantitative Ecology and Related Ecometrics.

QUANTITATIVE POPULATION DYNAMICS
D. G. Chapman and V. F. Gallucci, editors **300 pp. approx.**
D. G. CHAPMAN and V. F. GALLUCCI, Population Dynamics Models and Applications. J. G. BALCHEN, Mathematical and Numerical Modeling of Physical and Biological Processes in the Barents Sea. A. BERRYMAN and G. C. BROWN, The Habitat Equation: A Fundamental Concept in Population Dynamics. C. A. BRAUMANN, Population Adaptation to a "Noisy" Environment: Stochastic Analogs of Some Deterministic Models. L. GINZBERG, Genetic Adaptation and Models of Population Dynamics. M. I. GRANERO-PORATI, Stability of Model Systems Describing Prey-predator Communities. K. J. BEUTER, C. WISSEL, and U. HALBACH, Correlation and Spectral Analysis of Population Dynamics in the Rotifer *Brachionus Calyciflorus Pallas*. G. G. WALTER, Surplus Yield Models of Fisheries Management. SOME MORE PAPERS IN PREPARATION.

Contributors to This Volume

Austin, M. P.
Division of Land Use Research, CSIRO
Canberra City, Australia

Connor, Edward F.
Department of Biological Science
Florida State University

Coulson, Robert N.
Department of Entomology
Texas A&M University

Derr, Janice A.
Department of Zoology
University of Iowa

Duek, J. Lee
Department of Ecology and Evolutionary Biology
University of Arizona

Eisen, Paul A.
NOAA Environmental Research Laboratories
State University of New York

Foltz, John L.
Department of Entomology and Nematology
University of Florida

Gallucci, Vincent F.
College of Fisheries
University of Washington

Giavelli, G.
Institute of Ecology
University of Parma

Godron, M.
Centre National De La Recherche Scientifique
Centre D Etudes Phytosociologiques Et Ecologiques Louis Emberger

Goodman, Daniel
Scripps Institution of Oceanography
La Jolla, California

Green, Richard F.
Department of Statistics
University of California, Riverside

Halfon, Efraim
National Water Research Institute
Canada Centre for Inland Waters

Hazard, John W.
Pacific Northwest Forest and Range Experiment Station
Portland, Oregon

Hendrickson, John A., Jr.
Division of Limnology and Ecology
Academy of Natural Sciences of Philadelphia

Hengeveld, R.
Department of Geobotany
Catholic University, Nijmegen

Hennemuth, Richard C.
Northeast Fisheries Center
National Marine Fisheries Service

Iwao, Syn'iti
Entomological Laboratory
Kyoto University

Kaesler, Roger L.
Department of Geology
The University of Kansas

Kempton, R. A.
Plant Breeding Institute
Cambridge, United Kingdom

Kobayashi, Shiro
Faculty of Agriculture
Yamagata University

Kooijman, S. A. L. M.
Department of Biology
Division of Technology for Society
Delft, The Netherlands

Moroni, A.
Institute of Ecology
University of Parma

Naveh, Z.
Faculty of Agricultural Engineering
Israel Institute of Technology

Noy-Meir, I.
Department of Botany
Hebrew University

O'Connor, Joel S.
NOAA Environmental Research Laboratories
State University of New York

Ord, J. Keith
Department of Statistics
University of Warwick

Patil, G. P.
Department of Statistics
The Pennsylvania State University

Pickford, Stewart G.
College of Forest Resources
University of Washington

Poole, Robert W.
Division of Biology and Medicine
Brown University

Pulley, Paul E.
Data Processing Center
Texas A&M University

Quinn, Terrance J., II.
Center for Quantitative Science in Forestry, Fisheries, and Wildlife
University of Washington

Resh, Vincent H.
Department of Entomological Sciences
University of California

Rosen, Robert
Department of Physiology and Biophysics
Dalhousie University

Rosenzweig, Michael L.
Department of Ecology and Evolutionary Biology
University of Arizona

Rossi, O.
Institute of Ecology
University of Parma

Sheshinski, Ruth
Department of Statistics
The Hebrew University of Jerusalem

Simberloff, Daniel S.
Department of Biological Science
Florida State University

Siri, E.
Institute of Ecology
University of Parma

Sokal, Robert R.
Department of Ecology and Evolution
State University of New York

Solomon, D. L.
Biometrics Unit
Cornell University

Swartzman, Gordon
Center for Quantitative Science
University of Washington

Taillie, C.
Department of Statistics
The Pennsylvania State University

Taylor, L. R.
Rothamsted Experimental Station
Hertfordshire, United Kingdom

Torrez, William C.
Department of Mathematical Sciences
New Mexico State University

Waters, William E.
Department of Entomological Sciences
University of California, Berkeley

Whittaker, R. H.
Ecology and Systematics
Cornell University

Wiegert, Richard G.
Department of Zoology
University of Georgia

PREFACE TO THE VOLUME

Ecology remains a nascent science. Its purview is still growing. Its grand generalities hardly see publication before being embarrassed out of existence. The predictions of its theories lack precision. Moreover, even they often go untested because of the childlike sense of wonder which we ecologists still retain and which directs our attention to observing more and more about the basic phenomena of ecology rather than to the more mature occupation of trying to make sense out of what we have seen. Yet, this volume demonstrates that there is a widespread, multifaceted concern among us: let ecology at least face puberty.

Like the adolescent aborning, we are probably all sure as individuals that our comprehension runs deep. One value of a volume like this is to show that this cannot be so. Collected together, our efforts are bound to reveal a certain awkwardness and rawness.

But collected together, these papers also exhibit an intense communal vitality. They bear witness to a level of excitement which we honestly believe is unsurpassed in science today.

Most symposia focus narrowly upon a single topic. Not this one. It obviously could not if it was to reveal successfully the breadth of the ongoing ferment in ecology. Furthermore, the development of sophistication in science may be compared to the development of diversity on an island. When this is done, you will see that we are not at a stage where a narrow focus would be of very great importance.

A defaunated island goes through three stages in achieving maturity. At first, there is a rush of colonization. Then, although species continue to colonize, the ones already there grow rapidly in population; this is often termed the r phase. Finally, there is a sorting out process, the K of the mature state. Now one must not stretch the analogy too far, but the historical development of any branch of science bears much resemblance to this process.

The empty islands, their resources waiting for utilization, are like sets of unprocessed observations. Colonists, in the guise of novel hypotheses and theories, arrive to feed upon the data, to grow and proliferate thereby. Those that succeed to any extent soon find themselves in conflict with others. The interactions generated lead to a sorting out of ideas--rejection of some, coordination of others. As in biogeography however, the island is never static; it remains forever vulnerable to invasion by just the right new conceptual twist; its mature phase is a deceptive intellectual peace awaiting a radical revolution.

Spearheaded by G. E. Hutchinson, his students, and others intrigued by the opportunities for a pioneer existence on the intellectual frontier, the colonization of ecology occupied at least the decade from 1955 to 1965. It has not stopped, but many of the novel concepts and questions that were introduced then have grown, and many have already disappeared. We are somewhere in the r phase.

Perhaps this volume should initiate the K phase? We think not. We believe that these three phases are not sharply distinct and that some sorting out is already taking place. But we see no grand designs on the horizon, nothing like a unified field theory of ecology, let alone two of them to compete with each other. Ecology is best seen as being in an r phase with plenty of colonization still going on. And should it be progress we are after, the study of natural selection teaches us to value variation. At this stage, we are likely to succeed in direct measure to the diversity of approaches we try out.

It is in that spirit that we have consciously avoided judging the appropriateness of topics or approaches included here. We wanted an overview. And we wished it to be provided by research contributions instead of summaries or reviews. These papers provide that overview. And because they are research papers, they address issues in which investigators have recently seen fit to invest their time. Thus we address this book to those searching for the species of quantitative ecological question most likely to generate excitement in the next several years.

Despite the diversity of these papers, they do possess a dominant theme. Over and over they raise the question of method. This is yet another indication of the youthfulness of ecology. We really have no factory methods which many of us agree should be applied in system after system. There is no equivalent of DNA hybridization or density gradient centrifugation. Radioactive tracers seem to have brought us little return. We have not even got the equivalent of starth-gel electrophoresis.

Because field work is so difficult, we have been casting about trying to tailor our methods to each special problem in each special environment. As the papers herein illustrate, even the basic need to census requires new approaches and is hardly standardized.

Our theories too require methodology. In the past, these have come closest to standardization. Linearization and neighborhood stability analysis are examples. Computer simulation is another. But here again there is more ferment than anything else. Should theories be stochastic? How do we deal with truly large systems? What sorts of logical systems will yield interesting ecological results (words? algorithms? graphs?)? What techniques established in other sciences will prove useful in ecology (optimization? compartment analysis?)?

One criticism often levelled at ecological theory is lack of testability. But this is almost never true; the theories are testable, but only with great difficulty or perhaps with impractical amounts of time and money. Einstein once advised some wag who criticized his general theory for lack of testability to return in twenty years (it actually took less). The message is clear: theories are not usually born in testable condition. Perhaps they need to be worked on in order to make them practical to test. More often a theory should inspire the empiricist to develop methods to obtain the data of interest. Making theories testable is a worthy endeavor in itself and is usually the province of the empiricist, *not* the theoretician. A give-away sign of the immaturity of ecology is that we not only allow a theoretician to test his own theory, we virtually insist upon it. This volume contains examples of that practice as well as quite a few instances of investigators' striving to develop methods for making theories more testable.

A topic that continues to fire the imagination of the ecologist is species diversity. Quite a few of these contributions stress it. We do not know why such an obsession should afflict ecologists although we both confess to it. But diversity is clearly a rich vein of ore and it will not be played out for a long time to come.

A more novel topic is evolutionary ecology. We believe it to be represented herein roughly in proportion to its commonness in the journals. That is to say, most of the papers ignore it. Yet it is hard to imagine a usefully predictive ecology without an evolutionary component. Hence we are happy to include several papers in which evolution figures prominently.

This leads naturally to the question of utility. Most scientific symposia contain either basic or applied papers. This one has some of both. There is an urgent need for application of our results, which, in the light of their immaturity, is frightening. But the need does exist. Since "applied" or "basic" are only attitudes in the mind of the investigator, it may be possible to benefit both applied and basic research by fostering their combination. We see several examples of that among the works included here. We thus hope and trust that the reader will find this volume on contemporary quantitative ecology and related ecometrics of some value and use to him in carrying out his research and teaching.

June 1979

G. P. Patil
M. L. Rosenzweig

ACKNOWLEDGMENTS

For permission to reproduce materials in this volume, thanks are due to: the Editors of the Journal of the Fisheries Research Board of Canada for Table 3 on page 306; Springer-Verlag New York Inc. for Figure 2(b) on page 302 and Table 4 on page 311; and the University of Tokyo Press for Figures 1 on page 218 and 2 on page 219.

TABLE OF CONTENTS

CONTEMPORARY QUANTITATIVE ECOLOGY AND RELATED ECOMETRICS

SECTION I

COMMUNITY STRUCTURE AND DIVERSITY

G. P. Patil and M. Rosenzweig, (eds.),
Contemporary Quantitative Ecology and Related Ecometrics, pp. 3-22.

SOME OBSERVATIONS ON THE YEARLY VARIABILITY OF SPECIES ABUNDANCE AT A SITE AND THE CONSISTENCY OF MEASURES OF DIVERSITY

R. A. KEMPTON AND L. R. TAYLOR

Plant Breeding Institute, Cambridge, U.K. and
Rothamsted Experimental Station, Harpenden,
Hertfordshire, U.K.

SUMMARY. The yearly fluctuations in multi-species populations of *Macrolepidoptera* are studied at 14 environmentally stable sites from the Rothamsted Insect Survey. The site discrimination and consistency of site ordering is investigated for diversity measures based on the expected number of species in subsamples of specified size. Species stability, defined as the reciprocal of the pooled variance of species log-abundance over years, appears to be positively correlated with diversity. A model for the reordering of species abundances from year to year is studied.

KEY WORDS. diversity indices, expected species number, site discrimination, log-series model, species stability.

1. INTRODUCTION

A series of investigations was initiated in 1959 by one of us (L.R.T.) into the characteristics of single- and multi-species adult populations. Its object was to seek dynamic parameters for the aggregatory and migratory behavior of individuals that determine the changing distribution and abundance of populations (Taylor, 1961). It was prompted by dissatisfaction with the largely theoretical and temporal approach to population dynamics at that time, and with the unverified assumptions of diversity/stability relations. Since then, many other workers have taken up these problems but not usually in the same way. Both subjects lacked sound field data on an adequate scale of space and time and one consequence of this was that the empirical behavior of statistics adopted to describe observed data patterns was never properly investigated. A long-term program was planned to make good these deficiencies.

The collection of data was started in 1960 by establishing the Rothamsted Insect Survey which, by 1978, is producing daily samples of about 850 species of insects at some 200 sites distributed over the island of Great Britain (Taylor, 1968, 1974). Some of these insects, especially aphids, are of primarily agricultural importance (Taylor, 1973, 1977a), but a major part of the investigation is concerned with a fundamental study of the abundance of moth species sampled by light traps (Taylor and Brown, 1972; Taylor and French, 1974).

A strong initial stimulus was given to the spatial aspect of the investigation by the discovery of a power function relating spatial mean and variance (Taylor, 1961) which suggested that single-species populations could best be treated as density surfaces, dynamic in space and time (Taylor, 1965), whose characteristic parameters of mean density and surface roughness are highly specific (Taylor, 1971), persistent (Taylor, 1977b), universal amongst all organisms (Taylor, Woiwod, and Perry, 1977), and can be attributed to intrinsic behavioral motion (Taylor and Taylor, 1977, 1978, 1979). The length of time needed to accumulate sufficiently long temporal sequences of samples has delayed the application of the same analytical approach to temporal dynamics but this is now in hand.

The study of multi-species populations was hindered especially by the confusion of untested diversity statistics that had developed in the meantime, and we have been especially engaged in verifying the ecological, as distinct from the theoretical, value of these statistics (Kempton, 1979; Kempton and Taylor, 1974, 1976, 1978; Kempton and Wedderburn, 1978; Taylor, Kempton, and Woiwod, 1976) before proceeding further with the analysis (Taylor and Woiwod, 1975; Taylor, French, and Woiwod, 1978; Woiwod, 1979).

One problem has become increasingly clear during this parameter-testing exercise; some statistics in common use, such as the negative binomial k, the information statistic H, and the Simpson-Yule statistic Y are often inappropriate as ecological parameters in population dynamics (Kempton and Wedderburn, 1978; Taylor, Woiwod and Perry, 1979). A recent review (Taylor, 1978) summarized our work in this field up to 1977. The present paper extends the investigations to include a preliminary examination of species ordering, and a suggestion that the variance/mean power function may have potential, in addition to its spatial and temporal role for single species, in linking the dynamic stability approach to multi-species populations.

2. CURRENT INVESTIGATIONS

For the purpose of these investigations we have selected from the Survey 14 sites which have remained environmentally stable during the study period and are well dispersed throughout the island of Great Britain (Figure 1). Samples of moths, comprising the accumulated total of 364 nightly subsamples, are available for each site and replicated over at least seven years. The species used are those nocturnal Macrolepidoptera listed in South (1907), excluding the genus *Eupithecia* for taxonomic reasons. The mean yearly number of individuals and species caught for the period 1968-74, when all traps were fully operational, is shown in Table 1.

In this paper we are chiefly concerned with the variation in the population from year to year at different sites. We shall first follow the line of our previous work (Kempton and Taylor, 1974, 1976; Taylor *et al.*, 1976; Kempton and Wedderburn, 1978) and compare the stability of different measures of diversity for successive yearly samples at a site, an approach prompted by the problem of investigating the effects of environmental change on the diversity of the insect fauna. We have been critical of many existing diversity measures, whose use had too often been justified purely on quasi-theoretical grounds or by their behavior in relation to artificial and highly unrealistic data sets, but which do not perform well in practice (Taylor, 1978).

We shall here be particularly concerned with the family of measures obtained as the expected number of species in a subsample of given size, and shall consider the inconsistency problem (Patil and Taillie, 1979a,b), whereby different diversity orderings of the sites are obtained using different measures. We go on to look at the variability in the abundances of named species at a site. Firstly we investigate the empirical relationship between the temporal variance and mean abundance for all species at a site. This suggests a method for investigating the relationship between species diversity and stability. Secondly we consider a possible theoretical model for the rearrangement in species ordering from year to year and see how closely it is followed in practice.

3. THE STABILITY AND CONSISTENCY OF MEASURES OF DIVERSITY

We shall here be concerned with changes in the multi-species population exemplified by the distribution of species abundances. This is of interest because it allows us to investigate the species balance in a population independently of which named species may be present. In choosing statistics that characterize this distribution to measure the diversity of the population, we

FIG. 1: Geographical location in Great Britain of the 14 *sites chosen for the study.*

TABLE 1: Mean yearly numbers of individuals, N, *and species,* S, *caught at* 14 *sites from the Rothamsted Insect Survey,* 1968-1974. S_m *is the expected number of species when all samples are reduced to the size of the smallest sample* (m=345). α *is the index of diversity from the log-series.*

Site number and name		N	S	S_m	α
16	Stratfield Mortimer	2903	193	98	46.9
22	Rothamsted (Geescroft)	6869	186	81	35.6
29	Rannoch	4815	119	66	22.1
45	Malham	2074	92	55	19.9
46	Alice Holt	2904	238	118	61.5
47	Dundee	777	94	69	28.2
57	Ardross	3689	134	72	27.2
58	Elgin	2267	140	78	33.3
67	Slapton Ley	2056	148	80	37.1
68	Harrogate	879	112	84	35.3
72	Castletown	1480	98	63	23.9
78	Ringwood	2589	209	114	50.6
79	Shardlow	3060	141	76	30.8
92	Nettlecombe	4169	197	90	43.7

have emphasized that any statistic should be independent of sample size and show small variability in replicate yearly samples. This last property will allow good discrimination between different sites.

The most commonly used index of diversity is simply the number of species in the sample, S. Table 1 gives the mean number of species obtained from a year's trapping at the 14 stable sites for the seven years 1968-74. The standard error of these means is around 4-5 species, so discrimination between sites is good. However some of the observed differences in S from site to site are due to differences in sample size which may be caused by differences in population density or sampling efficiency. One way of removing the effect of density is to estimate the number of species when all samples are standardized to have the same number of individuals (Smith and Grassle, 1977).

If a sample of size N contains S species with abundances $N_1, N_2, \cdots, N_s$, $\Sigma N_i = N$, then the expected number of species in a random subsample of size $m \leq N$ is given by

$$S_m = \sum_{i=1}^{S} [1 - C(N-N_i, m)/C(N,m)],$$

where $C(a,b) = a!/[b!(a-b)!]$ for $a \geq b$, and zero otherwise. The maximum value for m is determined by the smallest of all our site x year samples, the 345 individuals caught at Harrogate in 1972. Table 1 gives the expected number of species in samples of size 345 for each site, averaged for the years 1968-74. The ordering of sites is clearly different from that obtained using the number of species uncorrected for differences in sample size. In particular, the trap at Rothamsted caught nearly 70% more species than that at Harrogate, but Harrogate apparently had a slightly higher diversity when measured by S_m with $m = 345$.

3.1 Choice of Subsample Size m *for Site Comparison.* For any single sample of size N we may evaluate S_m for any value of m from 1 to N. Figure 2 shows S_m, averaged for the years 1968-74, plotted against m for two sites at Rannoch and Dundee. It will be observed that the ordering of sites depends on whether the subsample size m is greater than or less than some figure close to 200. Site comparison for the whole range of values of m is not possible from Figure 2 because the variability in S_m greatly increases with m. A log-scale for S_m is not wholly effective in equalizing the variances.

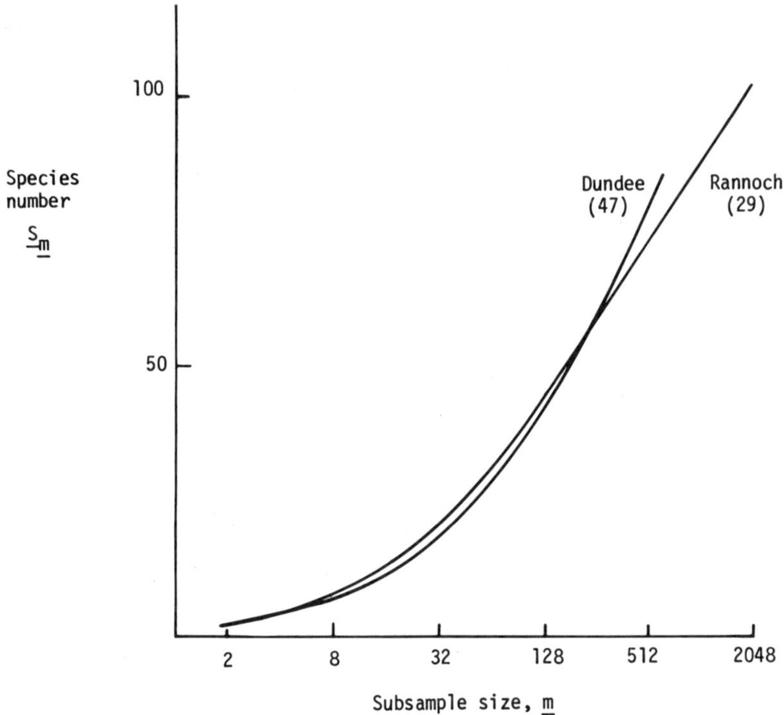

FIG. 2: Expected species numbers, S_m*, averaged over years* 1968-74 *for subsamples of different size,* m*, from sites* 29 *(Rannoch) and* 47 *(Dundee).*

To study the behavior of S_m over its whole range we standardize our mean site values of S_m for each subsample size m by subtracting the mean over all sites and dividing by the standard error of site means, obtained from the within-site root mean square. Figure 3 shows the standardized values of S_m for the 14 sites plotted against m, where m ranges from 2 to 605. The standard error of difference, which is by definition constant for all m, is given to assist site comparison. The mean value of S_m for Harrogate when $m > 345$ is obtained by fitting constants to the sites x years table with one missing value. Extending the range of m above 605 was not thought to be desirable as an increasing number of samples would then be omitted.

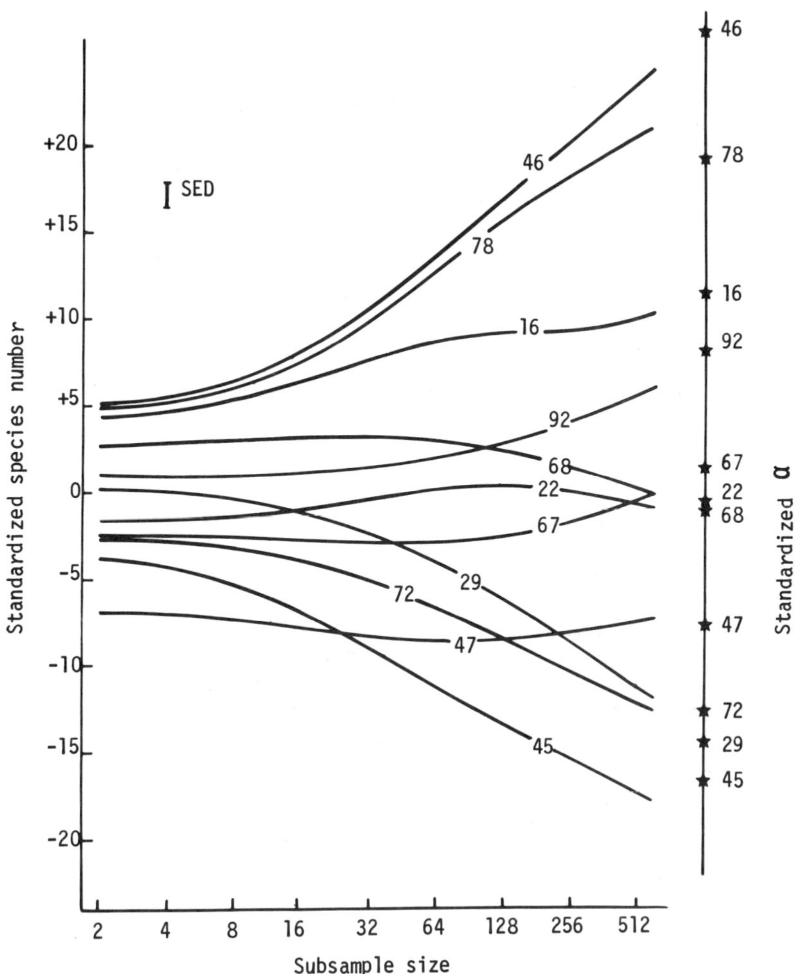

FIG. 3: Standardized species numbers for study sites for sample sizes from 2 *to* 605. SED *represents the magnitude of common standard error of difference of site values for all sample sizes.* * *represents the standardized value of the estimated log-series diversity parameter,* α. *(The site numbers refer to sites listed in Table* 1. *Sites* 57, 58, *and* 79 *are omitted for clarity.)*

The order of the three sites with highest diversity, namely sites 46, 78, and 16, is observed to be fixed for all m within the specified range, but the ordering of all other sites changes with m. This is due to the different emphasis members of S_m give to different parts of the species abundance distribution. For small m, S_m is very much dominated by the commonest species in the sample, but becomes increasingly sensitive to the moderately abundant and rare species as m increases. Thus in many cases there will not be a unique diversity ordering of sites on S_m for all m. Solomon (1975, 1979) and Patil and Taillie (1979a,b) have shown that a necessary and sufficient condition for a unique diversity ordering to exist is that the curves of cumulative species proportional abundance plotted against species rank should not intersect.

Figure 3 also shows that the range of standardized species numbers over sites increases with m, particularly for $m \geq 16$. Clearly if maximum discrimination between sites is the chief criterion we would choose the value of S_m for which m takes its maximum possible value, i.e. that of the smallest sample.

3.2 The Use of the Log-Series Parameter α *as a Diversity Measure.* The method of taking account of differences in sample size by reducing all samples to the size of the smallest could involve considerable loss of information when samples vary greatly in size. For example, in our particular investigation, sample sizes ranged from 9544 individuals obtained at Rothamsted in 1969 down to 345 individuals at Harrogate in 1972.

One way of surmounting this problem is to assume a distributional model for species abundance. The parameters of the model can then be separated into those relating to sample size, N, and those describing the distribution of the proportional abundances, p_i, of species in the population. For our material we have found that the distribution of species abundances in samples is generally well fitted by a logarithmic series, where the expected number of species with abundance r is $e_r = \alpha X^r/r$ (Figure 4). The parameter α is theoretically independent of sample size and we have shown empirically that estimates of α obtained for a series of years at an environmentally stable site are remarkably consistent and that any variability in the species abundance distribution over years is largely due to differences in the sample size parameter X (Taylor *et al.*, 1976).

Assuming a log-series distribution for species abundances we may show that the expected number of species in a sample of size

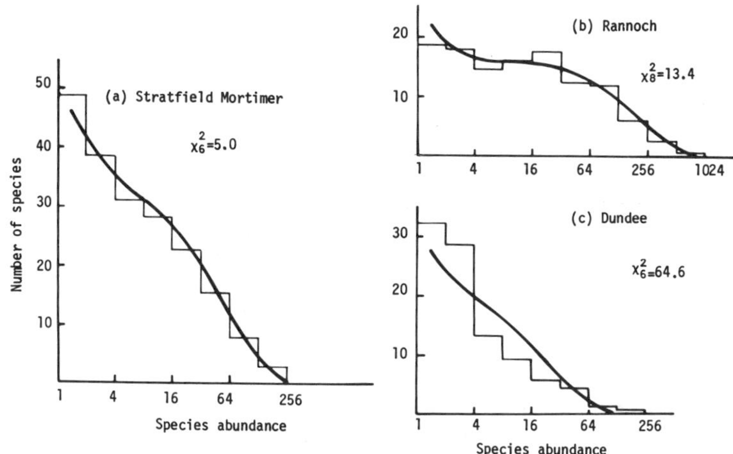

FIG. 4: Mean yearly number of species in each abundance class for 3 sites, with fitted log-series and chi-squared goodness of fit statistic. Sites (a) *and* (b) *show the good fit typical for our material, but site* (c) *has an unusually skewed distribution with species of medium abundance under-represented.*

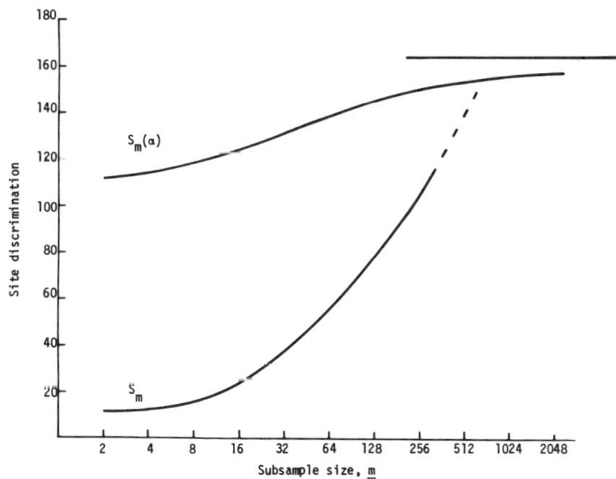

FIG 5: Site discriminant ability of expected species number, S_m, *for different subsample size* m, *compared with the discriminant ability of* $S_m(\alpha)$ *when* S_m *is estimated from parameter of fitted log-series. Broken line indicates the region where one missing sample value was estimated for* S_m. *Horizontal line indicates the asymptotic discrimination achieved by* $S_m(\alpha)$ *when* m *is large.*

m is (see Appendix)

$$S_m(\alpha) = \sum_{i=1}^{m} \frac{\alpha}{\alpha+i-1} .$$

It is easy to show that, for all m, the ordering of sites based on $S_m(\alpha)$ is the same as that based on α, and hence if the log-series holds for all samples, the diversity ordering is unique. Table 1 shows that the ordering of sites on S_m for large m is almost identical to the ordering on α. Another look at Figure 4 suggests that the change in ordering with m shown in Figures 2 and 3 for sites 29(Rannoch) and 47(Dundee) can be explained in terms of deviations from the log-series. Thus while the distribution of abundance for samples from Rannoch fit the log-series moderately well, the distribution for Dundee is too skew with much greater variation in species abundance than predicted for the log-series. Hence subsamples of small or moderate size will consist largely of the few commonest species, since those of medium abundance are under-represented compared with an equivalent population distributed more in accord with the log-series.

As well as helping us in the interpretation of inconsistent orderings, the log-series can have an important role in the efficient estimation of S_m. We have already shown that for small m, S_m varies considerably between years and consequently gives poor discrimination between sites. This behavior is due to the dependence of the statistic on the small number of abundant species, rather than the full complement of species in the population. In fact $S_2 = 2 - \Sigma p_i^2$ is a direct complement of Simpson's index which we have already criticized on just such grounds (Taylor *et al.*, 1976).

Replacing S_m by its estimate $S_m(\alpha)$ from the log-series model enables all the information in the sample to be used, with the consequent reduction in year to year variability and increase in site discrimination. An appropriate measure for the discriminant power of a statistic is given by the ratio of the between-site to the within-site (year to year) variability. Figure 5 plots the discriminant ability of S_m, evaluated directly from the data, and of $S_m(\alpha)$, from the fitted log-series models, for a range of values of m. As expected from Figure 3 the discriminant ability of S_m increases with m, particularly for $m \geq 16$. The improvement in discrimination from using $S_m(\alpha)$, rather than S_m, is sizeable when m is small but becomes less important as m

approaches its maximum value. The discriminant ability of S_m for $345 < m < 605$ has been obtained by again estimating the single missing value, but the behavior for higher m requires further study using a set of sites where catch sizes are more equal. The asymptotic discriminant value for $S_m(\alpha)$ is the discriminant value for α itself since for very large m,

$S_m(\alpha) \simeq \alpha \log(1 + m/\alpha) \simeq \alpha \log m.$

In concluding this Section we would argue that the attraction of using the family of diversity measures S_m, which progressively emphasizes the abundant, moderately abundant, and rare species as m increases, as cited by Sanders (1968) and Smith and Grassle (1977), is to a certain extent illusory as the variability of S_m for small m is so large as to obscure most site differences. However the high efficiency of the statistic for large m, when compared with the fully efficient estimator for the log-series model, is reassuring as it suggests that, for such values, S_m could be a good site discriminator for material where the log-series does not fit and the parameter α can no longer be used with confidence.

4. THE VARIABILITY IN SPECIES ABUNDANCES OVER YEARS

4.1 The Relationship between Variance and Mean. It is commonly found that the relationship between variance and mean abundance from a series of replicate samples for a single species is linear on a logarithmic scale (Taylor, 1961). Plots of log-variance against log-mean abundance over years for all species are similarly found to be linear for each of the 14 sites (Figure 6), though with more scatter about the regression lines than usually observed for single species plots. The slopes of the regression lines ranged from 1.68 to 2.05 with a mean of 1.89, and were generally close enough to 2 to assume that the variance in log-abundance is reasonably constant for all species at a site.

Adopting this assumption we can use the common variance of log-abundance of species over years as an inverse measure of site stability. In theoretical studies stability is often related to the probability of local extinctions of species in a closed population. Clearly populations in which the variability in abundances of species is high will be more prone to such extinctions though in practice their numbers will be renewed by immigration.

Figure 7 shows the reciprocal of the pooled variance of species log-abundances plotted against diversity (measured by α)

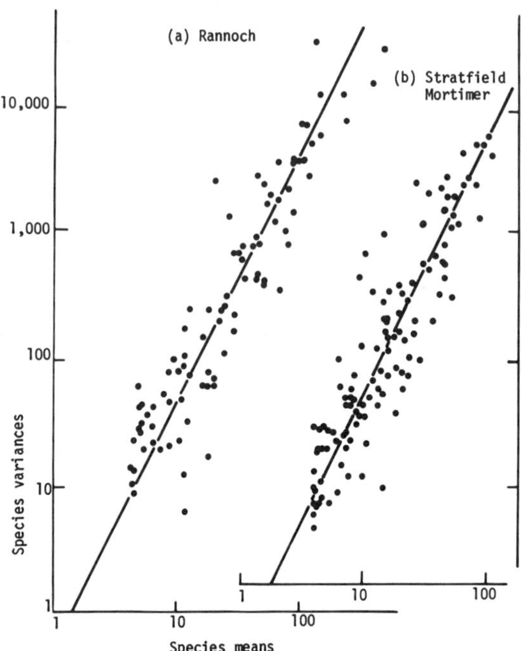

FIG. 6: Relationship between variance and mean abundance of species over years, omitting species with mean abundance of less than 4 individuals/year. Scales are logarithmic and regression lines are constrained to have a slope of 2.

for the 14 sites. There is a suggestion of a positive relationship but more data need to be examined before this can be confirmed.

The variability in abundance, N_i, of a species over years may be expressed in terms of two components: the variability in the size, N, of the total population of all species, and the variability in the relative proportion, p_i, of the population taken up by the particular species. In the absence of correlation, $\mathrm{Var}(\log N_i) = \mathrm{Var}(\log N) + \mathrm{Var}(\log p_i)$, but N and p_i may be correlated if certain species are particularly numerous, compared with other species, in years when conditions are particularly favorable and total population sizes are large.

Figure 8 shows these two components of stability plotted against α and suggests that any relationship between population

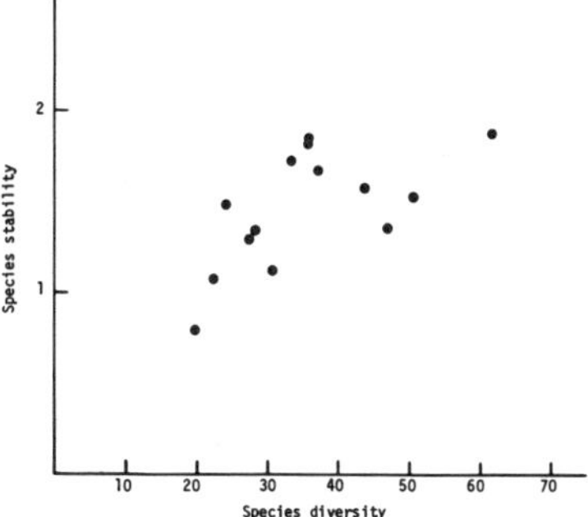

FIG. 7: Plot of species' stability, measured by reciprocal of the variance of species log abundances, against species diversity, measured by α, *for* 14 *sites.*

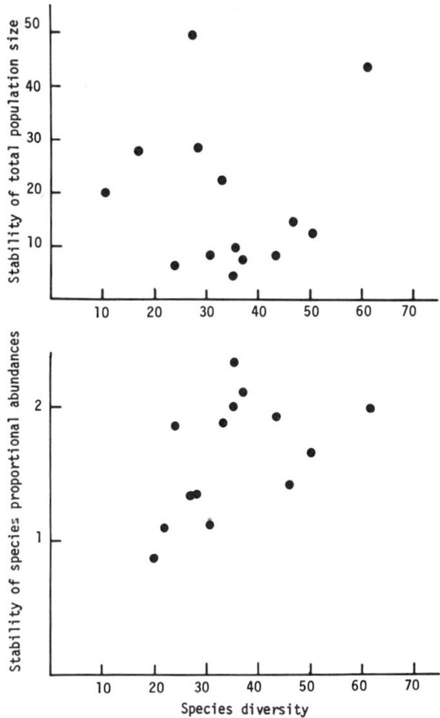

FIG. 8: The two components of species' stability plotted against diversity. (a) *stability of total population size,* 1/Var(log N). (b) *stability of species proportional abundances,* 1/Var(log p).

stability and diversity is due not to the variation in the size of the population as a whole (Figure 8a), but rather to the variation in the species ordering within the population from year to year (Figure 8b).

4.2 Engen's Model for the Yearly Reordering of Species. When considering models of species abundance, it is convenient to specify the hypothetical population from which the samples are drawn. We therefore suppose there exists a fixed number T of species available for capture each year, with abundances specified

A popular description of the pattern of species abundances at a site in any year, is to treat the abundances λ_i $(i = 1 \cdots T)$ as a set of independent identically distributed gamma variates. The log-series is of course a special case of this model. If the species relative abundances, p_i $(i = 1 \cdots T)$, are represented by segments in a partition of the unit interval $(\Sigma p_i = 1)$, the distribution of segment lengths $(u_1, u_2, \cdots, u_T)$ over years is the Dirichlet distribution $f(u) = [\Gamma(kT)/\Gamma(k)^T]\ (u_1 u_2 \cdots u_T)^{k-1}$, where k is the index of the gamma distribution. Note that in any year the set of species relative abundances, $(p_1, p_2, \cdots, p_T)$ is a simple permutation of $(u_1, u_2, \cdots, u_T)$. If we are concerned to describe the change in relative abundance of a named species from year to year, it remains to define the method of allocating segments among different species. If, for example, the segments are allocated each year at random, independent of their length, then the T species will have identical distributions of abundance over years. At the other extreme we could allocate the segments strictly according to their length, so that each year the species have the same ordering with respect to their abundances. Engen (1977), however, considered a possibly more realistic intermediate situation whereby the segments are allocated stochastically with probabilities proportional to their length. Thus p_1, the relative abundance of species 1, is determined by allocating the species to segment i with probability u_i. Having equated p_1 with one of the u_i, we determine p_2 by choosing one of the remaining segments, with probability again proportional to segment length, and so on for the whole set of species.

Engen shows that, with this allocation, the relative abundance of the *ith* ordered species in the limiting log-series situation has expectation $E(p_i) = \alpha^i/(\alpha + 1)^{i-1}$. To test whether our material conforms with this model, it is convenient to express each species abundance successively as a proportion of the total abundance of all species excluding those with higher average abundance than itself. Thus we define $\tilde{p}_1 = p_1$, $\tilde{p}_i = p_i/(1 - \Sigma_1^{i-1} p_j)$ $(i = 2,3 \cdots T)$. The $\tilde{p}_i$ should now possess identical beta distributions with expectations $1/(\alpha + 1)$.

Figure 9 shows the mean values of the relative species abundances, $\tilde{p}_i$, plotted against rank order, i, for two sites, Stratfield Mortimer (1965-76) and Rannoch (1966-76). It is apparent that the commonest species in the sample have consistently higher abundances than predicted by Engen's model. This cannot be explained by deviations of abundances from the log-series as the abundance distributions at both sites closely followed the log-series in all years (Figure 4), but could be due to the high serial correlations between years, particularly evident among the commonest species.

It is of interest to consider the expected relationship between variance and mean species abundance for Engen's model. We may show that for $\alpha^2 \gg 1$,

$$\mathrm{Var}(p_i) \simeq \alpha^{2i-1}[\alpha^2 + 2i(\alpha+1)/(\alpha+2)]/[(\alpha+2)(\alpha+1)^{2(i-1)}]$$

and hence for $i \ll \alpha^2$, $\mathrm{Var}(p_i) \simeq [\alpha/(\alpha+2)]E(p_i)^2$. Hence the variance of a single species over years is linearly related to the square of its mean abundance, as observed to be approximately true in practice. However, for Engen's model the common multi-species variance, $\mathrm{Var}(\log p_i) = \alpha/(\alpha+2)$, is independent of α, except when α is particularly small.

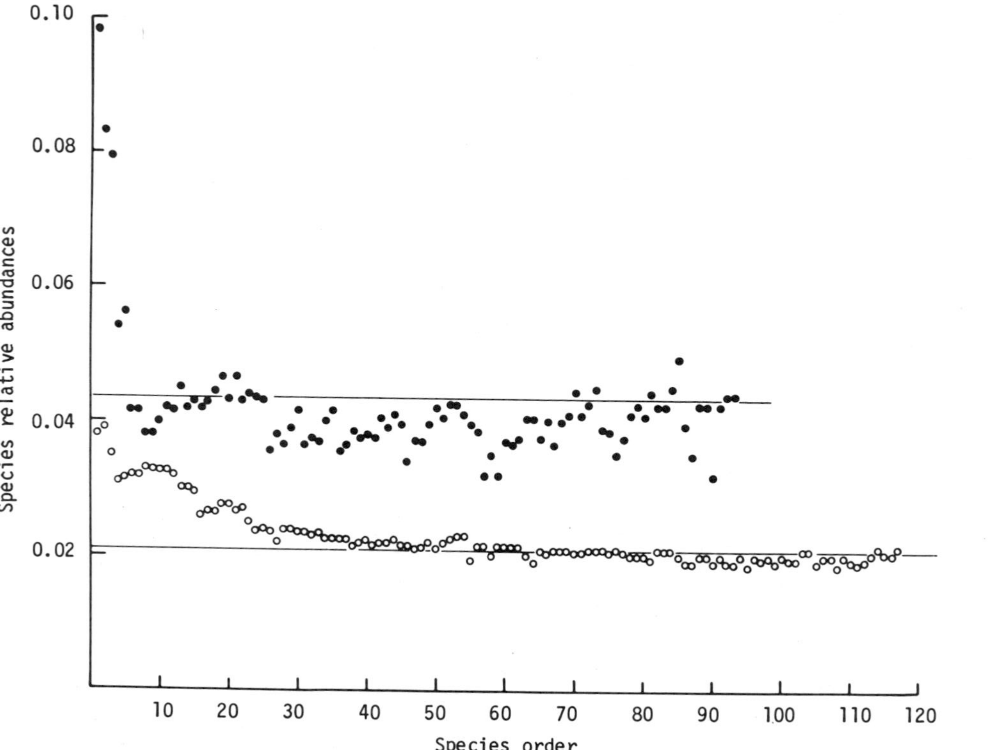

FIG. 9: Species relative abundance, $\tilde{p}$, (for derivation see text) plotted against rank order of absolute abundance for site 29, Rannoch, ●, and site 16, Stratfield Mortimer, o. The horizontal lines represent the expectations from Engen's model.

REFERENCES

Engen, S. (1977). Comments on two different approaches to the analysis of species frequency data. *Biometrics*, 33, 205-213.

Engen, S. (1978). *Stochastic Abundance Models*. Chapman and Hall, London.

Kempton, R. A. (1979). The structure of species abundance and the measurement of diversity. *Biometrics*, 35 (in press).

Kempton, R. A. and Taylor, L. R. (1974). Log-series and log-normal parameters as diversity discriminants for the Lepidoptera. *Journal of Animal Ecology*, 43, 381-399.

Kempton, R. A. and Taylor, L. R. (1976). Models and statistics for species diversity. *Nature*, 262, 818-820.

Kempton, R. A. and Taylor, L. R. (1978). The Q-statistic and the diversity of floras. *Nature*, 275, 252-253.

Kempton, R. A. and Wedderburn, R. W. M. (1978). The measurement of species diversity. *Biometrics*, 34, 25-37.

Patil, G. P. and Taillie, C. (1979a). Diversity as a concept and its measurement. *Journal of American Statistical Association* (in revision).

Patil, G. P. and Taillie, C. (1979b). An overview of diversity. In *Ecological Diversity in Theory and Practice*, J. F. Grassle, G. P. Patil, W. K. Smith, and C. Taillie, eds. Satellite Program in Statistical Ecology, International Co-operative Publishing House, Fairland, Maryland

Sanders, H. L. (1968). Marine benthic diversity: a comparative study. *American Naturalist*, 102, 243-282.

Smith, W. and Grassle, J. F. (1977). Sampling properties of a family of diversity indices. *Biometrics*, 33, 283-292.

Solomon, D. L. (1975). A comparative approach to species diversity. Paper number BU-573-M in the mimeograph series, Biometrics Unit, Cornell University, Ithaca, New York.

Solomon, D. L. (1979). A comparative approach to species diversity. In *Ecological Diversity in Theory and Practice*, J. F. Grassle, G. P. Patil, W. K. Smith, and C. Taillie, eds. Satellite Program in Statistical Ecology, International Co-operative Publishing House, Fairland, Maryland.

South, R. (1907). *The Moths of the British Isles.* Series 1 and 2. New ed. 1961, Frederick Warne, London.

Taylor, L. R. (1961). Aggregation, variance and the mean. *Nature,* 189, 732-735.

Taylor, L. R. (1965). A natural law for the spatial disposition of insects. In *Proceedings of the 12*th *International Congress of Entomology, London,* 1964, 396-397.

Taylor, L .R. (1968). The Rothamsted Insect Survey. *Natural Science in Schools,* 6, 2-9.

Taylor, L. R. (1971). Aggregation as a species characteristic. In *Statistical Ecology, Vol. 1,* G. P. Patil, E. C. Pielou, and W. E. Waters, eds. The Pennsylvania State University Press, University Park. 357-377.

Taylor, L .R. (1973). Monitor surveying for migrant insect pests. *Outlook on Agriculture,* 7, 109-116.

Taylor, L. R. (1974). Monitoring change in the distribution and abundance of insects. *Report of the Rothamsted Experimental Station for 1973, Part 2,* 202-239.

Taylor, L. R. (1977a). Aphid forecasting and the Rothamsted Insect Survey. *Journal of the Royal Agricultural Society of England,* 138, 75-97.

Taylor, L. R. (1977b). Migration and the spatial dynamics of an aphid, *Myzus persicae. Journal of Animal Ecology,* 46, 411-423.

Taylor, L. R. (1978). Bates, Williams, Hutchinson - a variety of diversities. In *The Diversity of Insect Faunas,* L. A. Mound and N. Waloff, eds. Royal Entomological Society of London. 1-18.

Taylor, L. R. and Brown, E. S. (1972). Effects of light trap design and illumination on samples of moths in the Kenya highlands. *Bulletin of Entomological Research,* 62, 91-112.

Taylor, L. R. and French, R. A. (1974). Effects of light trap design and illumination on samples of moths in an English woodland. *Bulletin of Entomological Research,* 63, 583-594.

Taylor, L. R., French, R. A., and Woiwod, I. P. (1978). The Rothamsted Insect Survey and the urbanization of land in Great Britain. In *Perspectives in Urban Entomology,* G. W. Frankie and C. S. Koehler, eds. Academic Press, New York . 31-65 .

Taylor, L. R., Kempton, R. A., and Woiwod, I. P. (1976). Diversity statistics and the log-series model. *Journal of Animal Ecology*, 45, 255-272.

Taylor, L. R. and Woiwod, I. P. (1975). Competition and species abundance. *Nature*, 257, 160-161.

Taylor, L. R., Woiwod, I. P., and Perry, J. N. (1977). The density dependence of spatial behaviour and the rarity of randomness. *Journal of Animal Ecology*, 46, 383-406.

Taylor, L. R., Woiwod, I. P. and Perry, J. N. (1979). The negative binomial as a dynamic ecological model and the density-dependence of k. *Journal of Animal Ecology*, 48, 289-304.

Taylor, L. R. and Taylor, R. A. J. (1977). Aggregation, migration, and population mechanics. *Nature*, 265, 415-421.

Taylor, L. R. and Taylor, R. A. J. (1978). The dynamics of spatial behaviour. In *Population Control by Social Behaviour*, F. J. Ebling and D. M. Stoddart, eds. Institute of Biology Symposium, 1977. 181-212.

Taylor, R. A. J. and Taylor, L. R. (1979). A behavioural model for the evaluation of spatial dynamics. In *Population Dynamics*, R. M. Anderson, B. E. Turner, and L. R. Taylor, eds. Blackwell, Oxford. 1-27.

Woiwod, I. P. (1979). The role of spatial analysis in the Rothamsted Insect Survey. In *Statistical Applications in the Spatial Sciences*, N. Wrigley, ed. Pion, London (in press).

[*Received July* 1978. *Revised November* 1978]

APPENDIX

We here derive an expression for S_m, the expected number of species in a sample of size m where the population abundances of species (permuted on each occasion if necessary) at the time of sampling can be treated as constituting T independent observations from the same gamma distribution with index k.

Suppose we have obtained a sample of size N consisting of S species by sequentially sampling individuals from the population. Kempton and Wedderburn (1978) show that, under the above assumptions, the probability that the $(N+1)th$ individual sampled belongs to the jth species, given that N_j members of this species have been previously caught, is $(k+N_j)/(\alpha+N)$, and the probability that the $(N+1)th$ individual belongs to a new species is $(\alpha-Sk)/(\alpha+N)$.

Let $h_m(z)$ be the probability generating function of the number of species in a subsample of size m, $h_m(z) = \Sigma_s u_m(s)z^s$, where $u_m(s)$ is the probability of a sample of size m containing s species. Now if $q_m(s)$ is the probability of the $(m+1)th$ individual belonging to a new species given s species previously caught,

$$u_{m+1}(s+1) = q_m(s)u_m(s) + [1-q_m(s+1)]u_m(s+1)$$
$$= \frac{\alpha-ks}{\alpha+m} u_m(s) + \frac{m+k(s+1)}{\alpha+m} u_m(s+1).$$

Multiplying both sides by z^{s+1} and summing over $s=1\cdots m+1$,

$$h_{m+1}(z) = \frac{\alpha z+m}{\alpha+m} h_m(z) + \frac{kz(z-1)}{\alpha+m} h_m'(z).$$

Differentiating and setting $z=1$, the expected number of species in a sample of size m, $E_m(s)$ is given by the recurrence relationship

$$E_{m+1}(s) = \frac{\alpha}{\alpha+m} + (1 - \frac{k}{\alpha+m})E_m(s).$$

Hence

$$S_m = E_m(s) = \frac{\alpha}{k} [1 - \frac{\Gamma(\alpha)\Gamma(m+\alpha-k)}{\Gamma(\alpha+m)\Gamma(\alpha-k)}],$$

and for the special limiting case $k\to 0$ obtained for the log-series

$$S_m = \sum_{i=1}^{m} \frac{\alpha}{\alpha+m-i}.$$

Note: A referee has pointed out that an alternative derivation of the above result using the general notation of structural distributions is given by Engen (1978, p. 48).

G. P. Patil and M. Rosenzweig, (eds.),
Contemporary Quantitative Ecology and Related Ecometrics, pp. 23-48.

A STUDY OF DIVERSITY PROFILES AND ORDERINGS FOR A BIRD COMMUNITY IN THE VICINITY OF COLSTRIP, MONTANA

G. P. PATIL AND C. TAILLIE

Department of Statistics
The Pennsylvania State University
University Park, PA 16802 USA

SUMMARY. In this paper, the intrinsic diversity ordering of two communities is portrayed graphically by plotting the right tail sums of their ranked relative abundance vectors against the rank. These plots are called the intrinsic diversity profiles of the two communities. The profile of an intrinisically more diverse community is everywhere above that of a less diverse community. The two profiles may intersect one or more times, in which case no intrinsic ordering of the two communities is possible. In practice, one needs to examine if an intersection is attributable to sampling fluctuation. This paper looks upon isotonic diversity indices primarily as test statistics. The multitude of indices then becomes an advantage, enabling one to develop a variety of tests powerful against different possible changes in community structure. The statistical methods involved are illustrated with the analysis of data collected from a bird community during a three-year period.

KEY WORDS. diversity, diversity ordering, intrinsic diversity ordering, diversity profile, site diversity, Western Meadowlark, bird community, jackknife estimator.

1. INTRODUCTION

Diversity has been an important concept in the study of community structure. Several publications have appeared in which diversity-related issues are discussed. See, for example, Dennis *et al.* (1979). Gradually, there has been emphasis on matters relating to clearer conceptualization, measurement, and comparison.

This paper is an attempt to clarify the notions involved and illustrate the methods using data on a bird community in the vicinity of Colstrip, Montana, collected during the three-year period, 1975-1977.

2. BACKGROUND

As reported in the United States Environmental Protection Agency report (1978, EPA-600/3-78-021), bird populations in the vicinity of Colstrip, Montana were surveyed using methods similar to the North American Breeding Bird Survey (Robbins and Van Velzen, 1970). As stated in the report, the observer starts one-half hour before local sunrise and makes 60 three-minute stops at one-half mile intervals along a predetermined route (Figure 1).

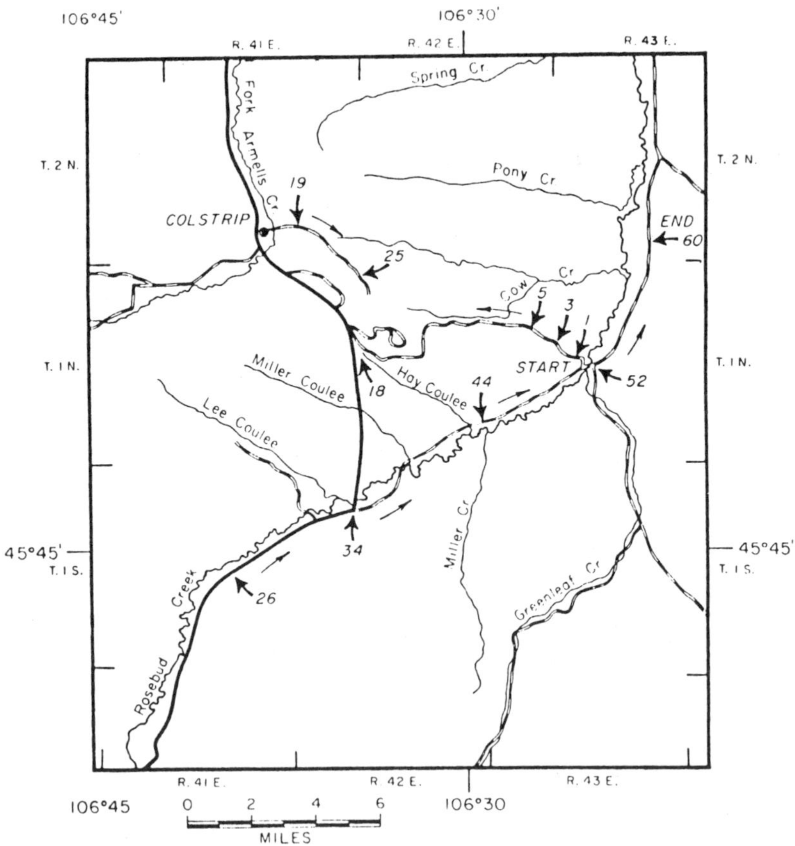

FIG. 1: Map of the survey route (reproduced from the EPA Report, EPA-600/3-78-021, p. 260).

At each stop is recorded the number of birds of each species seen within a 400-meter radius or heard regardless of distance. The route was selected to provide sampling stations at various distances from the power plants along an anticipated gradient of pollution impact. Individual stations span a wide range of habitats including open grassland, streams, rolling hills with ponderosa pine and juniper coverage, cliffs, and habitats affected by a wide range of human impacts.

The 1975 breeding season was considered the baseline period since the power plants were not operational at that time. Between May and September, the route was surveyed on nine dates in 1975, on twenty-one dates in 1976, and on twelve dates in 1977. The summary data for each of these three years is presented in Table 1 as provided to us by Dr. Eric Preston, Environmental Research Laboratory, EPA, Corvallis, Oregon, USA.

To assist in the assessment of environmental impact in the Colstrip region, one may study temporal trends in various characteristics of bird community structure. Diversity is one such characteristic, and the analysis of annual trends in avian diversity is carried out in this paper for the data presented in Table 1. Also, one may study trends in the spatial pattern of the community as a whole or of some of its constituent species. In the latter case, sample size is maximized by studying the spatial pattern of the dominant species, which is Western Meadowlark for the Colstrip region. Table 2 provides some spatial data on the Western Meadowlark during the 1975 breeding season.

3. INTRINSIC DIVERSITY ORDERING

The term diversity has traditionally been associated with its indices, the most popular of which are the (species) richness, the Shannon index, and the Simpson index. But it is widely known that usage of different indices can lead one to different orderings of communities (e.g. Hurlbert, 1971). This raises questions of what diversity means, how it can be measured, and how it can be used.

Broadly speaking, diversity is an attribute of the community which incorporates both its evenness and its richness. This means that diversity should increase with a transfer of abundance from a more to a less abundant species, and it should also increase if a new species enters the community. Thus, without reference to indices, we may say that community C_2 is more diverse than community C_1 if it is possible to pass from C_1 to C_2 by finitely many such changes.

TABLE 1: Species Observed in the Colstrip Bird Survey.*

Species		Proportional abundance (× 10^4)		
Common Name	Scientific Name	1975	1976	1977
Canada Goose	*Branta canadensis*		2	22
Mallard	*Anas platyrhynchos*	7	12	9
Blue Winged Teal	*A. discors*			67
Turkey Vulture	*Cathartes aura*			2
Cooper's Hawk	*Accipiter cooperi*			3
Red-Tailed Hawk	*Buteo jamaicensis*	24	12	7
Rough-Legged Hawk	*B. lagopus*		1	
Golden Eagle	*Aquila chrysactos*	5	6	
Marsh Hawk	*Circus cyaneus*	12	19	14
Prairie Falcon	*Falco mexicanus*	7	2	
Sparrow Hawk	*F. sparverius*	140	118	93
Ruffed Grouse	*Bonasa umbellus*		1	2
Sharp-Tailed Grouse	*Pedioecetes phasianellus*	24	27	68
Ring-Necked Pheasant	*Phasianus colchicus*	339	299	291
American Coot	*Fulica americana*		2	1
Killdeer	*Charadrius vociferus*	15	11	38
Common Snipe	*Capella gallinago*	2		
Upland Sandpiper	*Bartramia longicauda*		4	5
Solitary Sandpiper	*Trianga solitaria*			1
Northern Phalarope	*Lobipes lobatus*			5
Mourning Dove	*Zenaidura macroura*	787	775	802
Black-Billed Cuckoo	*Coccyzus crythropthalmus*		2	6
Great Horned Owl	*Bubo virginianus*		1	
Poor-Will	*Phalaenoptilus nuttallii*	2	2	12
Common Nighthawk	*Chordeiles minor*	10	9	17
Chimney Swift	*Chaetura peligica*	29	3	
Belted Kingfisher	*Megaceryle alcyon*	2	1	
Common Flicker	*Colaptes auratus*	92	143	69
Red-Headed Woodpecker	*Melanerpes erythrocephalus*	7	2	1
Hairy Woodpecker	*Dendrocopos villosus*		1	48
Eastern Kingbird	*Tyrannus tyrannus*	123	155	223
Western Kingbird	*T. verticalis*	131	87	127
Cassin's Kingbird	*T. vociferans*	10	8	
Say's Phoebe	*Sayornis saya*	22	14	17
Western Wood Pee Wee	*Contopus sordidulus*		30	9
Horned Lark	*Eremophila alpestris*	2	2	
Violet Green Swallow	*Tachycineta thalassina*		20	61
Barn Swallow	*Hirundo rustica*	107	121	153
Cliff Swallow	*Petrochelidon pyrrhonota*	228	681	320
Black Billed Magpie	*Pica pica*	17	16	37
Common Raven	*Corvus corax*	5		
Common Crow	*C. brachyrhynchos*	77	85	87
Pinon Jay	*Gymnorhinus cyanocephala*	249	162	45
Black-Capped Chikadee	*Parus atricapillus*	114	70	16
White-Breasted Nuthatch	*Sitta carolinensis*	2	1	

TABLE 1: (Continued)

Species		Proportional abundance (× 10^4)		
Common Name	Scientific Name	1975	1976	1977
House Wren	*Troglodytes aedon*	5	2	45
Winter Wren	*T. troglodytes*	7	1	
Rock Wren	*Salpinctes obsoletus*		1	22
Gray Catbird	*Dumetella carolinensis*		1	6
Brown Thrasher	*Toxostoma rufum*	10	13	8
Sage Thrasher	*Oreoscoptes montanus*			3
American Robin	*Turdus migratorius*	196	176	185
Thrush		5	1	7
Mountain Bluebird	*Sialia currucoides*	39	15	14
Cedar Waxwing	*Bombycilla cedrorum*	5		
Loggerhead Shrike	*Lanius ludovicianus*	17	10	2
Starling	*Sturnus vulgaris*	19	39	43
Yellow Warbler	*Dendroica petechia*	56	131	204
Ovenbird	*Seiurus aurocapillus*	2		
Common Yellowthroat	*Geothlypis trichas*	10		1
Yellow-Breasted Chat	*Icteria virens*	17	26	37
American Redstart	*Setophaga ruticilla*	5		
Western Meadowlark	*Sturnella neglecta*	3448	3457	4339
Red-Winged Blackbird	*Agelaius tricolor*	361	239	119
Northern Oriole	*Icterus galbula*	2	15	5
Brewer's Blackbird	*Euphagus cyanocephalus*	605	428	359
Common Grackle	*Quiscalus quiscula*	12	14	7
Brown-Headed Cowbird	*Molothrus ater*	140	178	103
Black-Headed Grosbeak	*Pheucticus melanocephalus*		4	
Luzuli Bunting	*Passerina amoena*	22	7	2
American Goldfinch	*Spinus tristis*	56	187	173
Red Crossbill	*Loxia curvirostra*	36		22
Rufous-Sided Towhee	*Pipilo erythrophthalmus*	36	54	91
Lark Bunting	*Calamospiza melanocorys*	1165	669	382
Savannah Sparrow	*Passerculus sandwichensis*	157	180	59
Vesper Sparrow	*Pooecetes gramineus*	479	755	324
Lark Sparrow	*Chondestes grammacus*	465	475	578
Dark-Eyed Junco	*Junco hyemalis*		1	
Chipping Sparrow	*Spizella passerina*	31	14	104
Clay-Colored Sparrow	*S. pallida*	5		72
White-Crowned Sparrow	*Zonotrichia leucophrys*	5		
		S 61	66	63
		N 4130	16060	8655

*Our computer programs, as written, required absolute abundances of the various species. These were obtained by multiplying the tabulated proportional abundances by N and rounding to the nearest integer.

TABLE 2: Spatial distribution of Western Meadowlark at 60 *sites in the vicinity of Colstrip, Montana during* 1975.

May 6		July 14		August 26		September 8	
Number of		Number of		Number of		Number of	
Birds	Sites	Birds	Sites	Birds	Sites	Birds	Sites
0	6	0	13	0	34	0	41
1	3	1	2	1	4	1	1
2	4	2	11	2	9	2	3
3	8	3	11	4	2	3	2
4	8	4	7	5	2	4	1
5	10	5	7	6	2	5	4
6	9	6	5	8	1	6	4
7	3	7	1	10	1	8	2
8	2	9	2	11	1	9	1
9	3	13	1	14	1	16	1
10	4			18	1		
				22	1		
				25	1		
N = 275		N = 188		N = 160		N = 102	
S = 54		S = 47		S = 26		S = 19	

To be specific, let the community consist of s species and let π_i, $i = 1,2,\cdots,s$, be the relative abundance of the *ith* species. A transfer of abundance from species i to species j means that π_i is decreased by some amount h and simultaneously π_j is increased by the same amount. Clearly, such a transfer of abundance increases the evenness of the abundance vector $\underset{\sim}{\pi} = (\pi_1,\pi_2,\cdots,\pi_s)$ provided $\pi_i > \pi_j$. Similarly, if the part h of π_i is allotted to a new species, the abundance vector $(\pi_1,\pi_2,\cdots,\pi_s)$ becomes $(\pi_1,\pi_2,\cdots,\pi_i-h,\cdots,\pi_s,\pi_{s+1}=h)$ with increased richness. Finally, species identity does not contribute to diversity so that a rearrangement of the components of $\underset{\sim}{\pi}$ should not change diversity. Thus, the notion of diversity may be formalized without indices in the following way.

Definition: The community C_2 is said to be *intrinsically more diverse* than community C_1 if it is possible to obtain the abundance vector of C_2 from that of C_1 by a finite sequence of changes due to transferring abundance, introducing species, and/or permuting species.

Given two communities, it may be difficult to determine if such a sequence exists. Therefore, a convenient working definition is needed. Several such are available (Taillie, 1977; Patil and Taillie, 1976, 1979), but the following, in terms of the ranked abundance vector, is generally the easiest to apply in practice.

Working definition: Community $\underset{\sim}{\nu}$ is *intrinsically more diverse* than community $\underset{\sim}{\pi}$ if and only if every (right) tail-sum of $\underset{\sim}{\nu}$ is greater than or equal to the corresponding tail-sum of $\underset{\sim}{\pi}$. That is,

$$\sum_{j>i} \nu_j^* \geq \sum_{j>i} \pi_j^* , \quad i = 1,2,3,\cdots,$$

where $\nu_1^* \geq \nu_2^* \geq \nu_3^* \geq \cdots$ and $\pi_1^* \geq \pi_2^* \geq \pi_3^* \cdots$ are the components of $\underset{\sim}{\nu}$ and $\underset{\sim}{\pi}$ arranged in descending order of magnitude.

It may be of interest to observe that the *ith* tail-sum $T_i(\underset{\sim}{\nu}) = \sum_{j>i} \nu_j^*$ is the combined abundance of those species which are rarer than the *ith* ranked species. Thus, in a rough sense, greater diversity means a greater amount of rarity.

So far, we have been speaking of species diversity. However, the concept of diversity can be applied to any situation in which a collection of individuals is grouped into categories. In the Meadowlark spatial data of Table 2, for example, the categories are sites rather than species, and richness s is the number of occupied sites. Transfers of abundance have a very clear physical meaning in this context of site diversity.

4. INTRINSIC DIVERSITY PROFILES

The intrinsic diversity ordering may be portrayed graphically by plotting the tail sums $T_i(\underset{\sim}{\nu})$ and $T_i(\underset{\sim}{\pi})$ against the rank i. These plots are called the *intrinsic diversity profiles* of the two communities. The profile of an intrinsically more diverse community is everywhere above that of a less diverse community.

However, this need not occur for every pair of communities under consideration; the two profiles may intersect one or more times, in which case no intrinsic ordering of the two communities is possible. Since not every pair of communities is comparable, the intrinsic diversity ordering is a partial, rather than a linear, ordering.

We illustrate these ideas with the Meadowlark spatial data of Table 2. To begin with, in Figure 2, the ranked relative abundances π_i^* are plotted against the site-rank i for each of the four selected survey dates. Note that a given rank i may refer to different sites on different dates. Examination of these rank-abundance histograms reveals a steady decrease in the number of occupied sites (richness). There is also a general downward trend in the evenness of the spatial distribution. However, if the first rank, $i = 1$, is ignored, there appears to have been a small increase in evenness between August 22 and September 8. Thus, because of the decreasing richness coupled with increasing evenness, the site diversity on these two dates may not be comparable in the intrinsic sense. This is borne out by an intersection in the intrinsic site diversity profiles of August 22 and September 8 (Figure 3). Although it cannot be seen graphically, these two profiles have a second intersection between ranks $i = 1$ and $i = 2$.

Except for these intersections, the four profiles reveal a steady decrease in intrinsic site diversity, apparently due to territorial nesting behavior of Meadowlark early in the season, later changing to flocking behavior.

The discussion so far has disregarded the sampling aspect of the observed data. One should now examine if the intersection in the two profiles is attributable to sampling fluctuation. For this purpose, statistics are required to test for significance of the distance between the profiles on each side of the intersection. The obvious test statistics are the arithmetic differences between corresponding tail sums of the two communities. However, the sampling distributions of these differences involve ordered multinomial distributions. At present, not enough is known about such distributions to carry out the needed significance tests.

An alternative procedure employs diversity indices as test statistics. At least two indices, say Δ and Δ', are required to establish significance of an intersection. The method is to consider, and presumably reject, the pair of null hypotheses

$$H_o : \Delta(C_1) \geq \Delta(C_2)$$

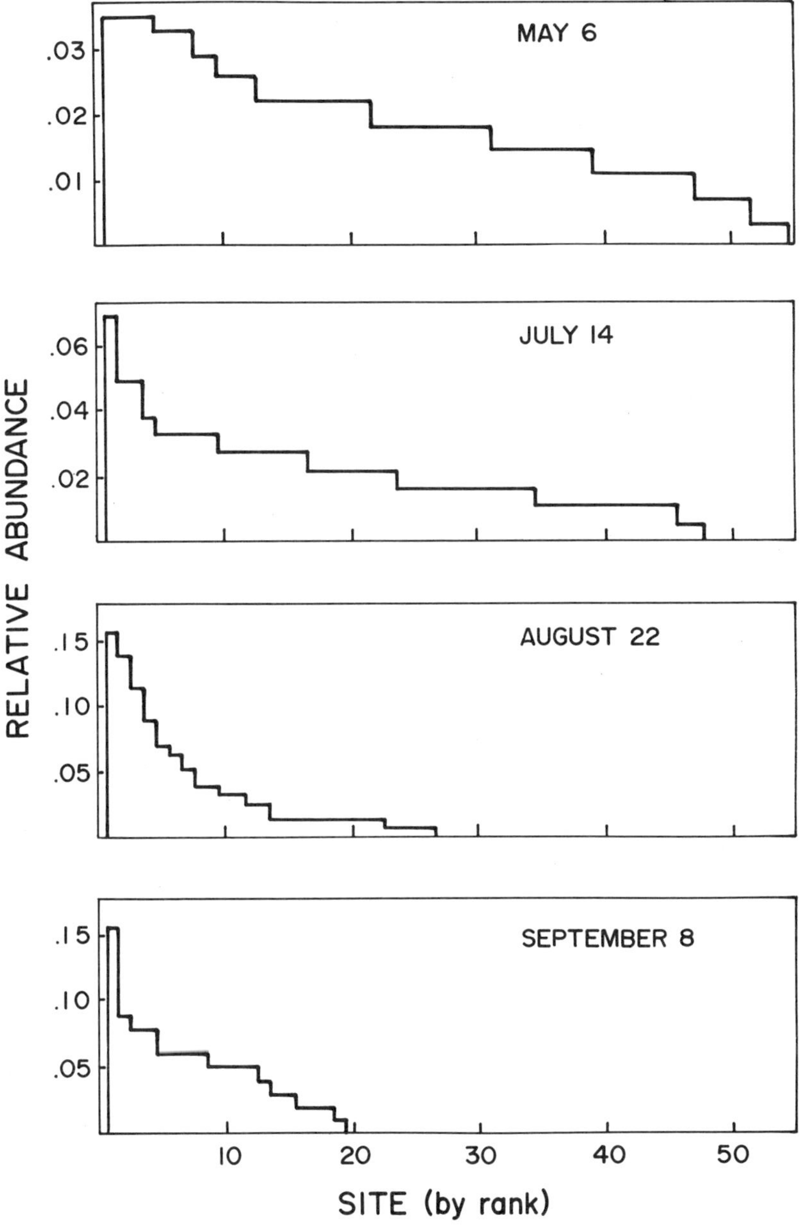

FIG. 2: Relative abundance in 1975 *of Western Meadowlark at* 60 *sites in the vicinity of Colstrip. (Sites arranged in the order of decreasing meadowlark abundance.)*

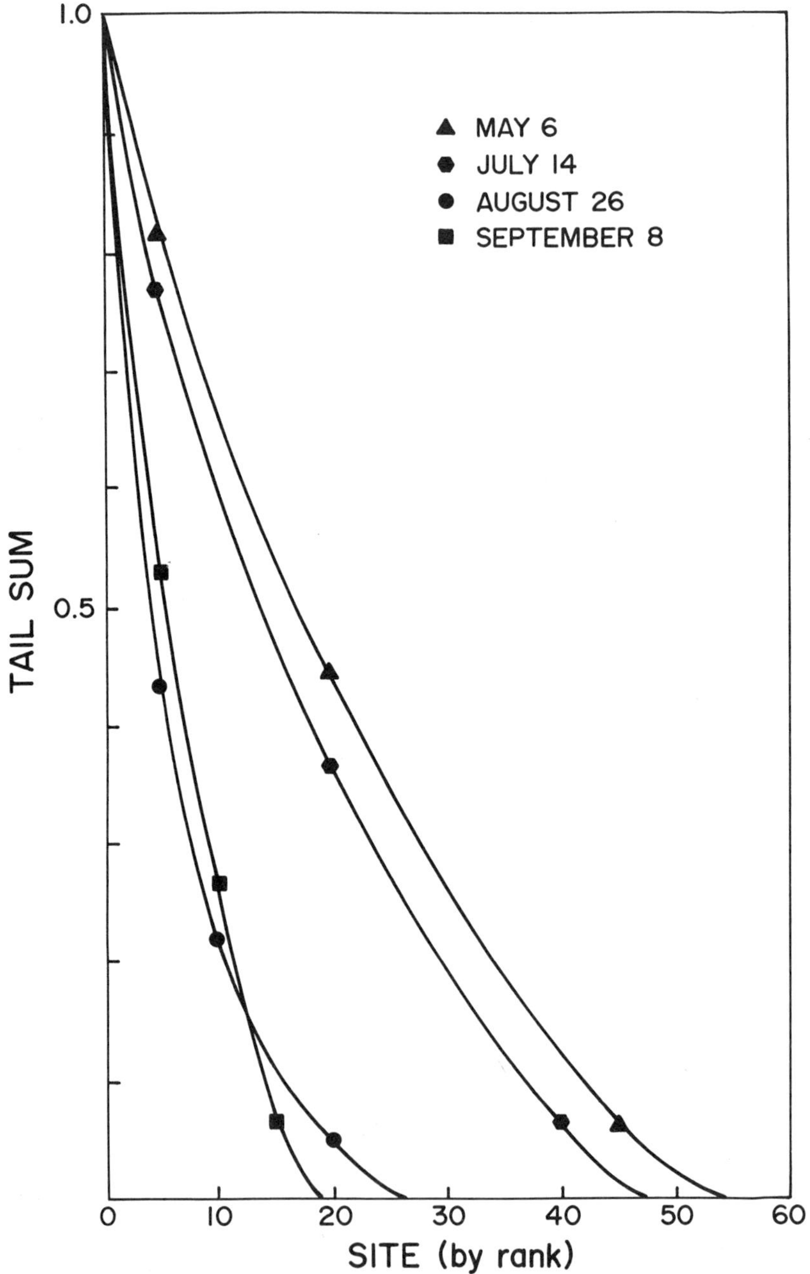

FIG. 3: Intrinsic site diversity profiles of Western Meadowlark on four selected survey dates during 1975.

and

$$H_o : \Delta'(C_1) \leq \Delta'(C_2).$$

These hypotheses are only apparently contradictory since they involve different indices and, hence, different aspects of community structure. Simultaneous rejection of both hypotheses would then establish, at a specified level of significance, that $\Delta(C_1) < \Delta(C_2)$ and $\Delta'(C_1) > \Delta'(C_2)$, implying that the communities are not intrinsically comparable and that their intrinsic diversity profiles must intersect.

From this point of view, diversity indices are primarily test statistics and the multitude of indices becomes an advantage, enabling one to develop a variety of tests powerful against different possible changes in community structure.

In order that the choice of Δ and Δ' not be a hit or miss process, we suggest using the data as an aid in the selection of appropriate test statistics. The procedure, to be described more fully in later sections, is briefly:

(i) Choose a family Δ_θ of diversity indices parametrized by a real parameter θ.

(ii) Plot estimates of $\Delta_\theta(C_1)$ and $\Delta_\theta(C_2)$ against θ. Such plots may be called diversity profiles of the Δ_θ family.

(iii) Match up intersections of these profiles with intersections of the intrinsic diversity profiles.

(iv) For each intersection, select as test statistics Δ_θ and $\Delta_{\theta'}$, where θ and θ' lie on opposite sides of the intersection.

As in many applications of statistics, implementation of such a procedure raises difficult and unresolved questions of simultaneous inference.

5. FAMILIES OF DIVERSITY INDICES AND THEIR PROFILES

As a minimal requirement, a diversity index Δ should be isotonic with respect to the intrinsic diversity ordering. That is, $\Delta(C_2) \geq \Delta(C_1)$ whenever community C_2 is intrinsically more diverse than community C_1. With this as the general definition of a diversity index, the following theorem has been established (Patil and Taillie, 1976, 1979).

Theorem: Community C_2 is intrinsically more diverse than C_1 if and only if $\Delta(C_2) \geq \Delta(C_1)$ for all diversity indices Δ.

An implication of the theorem is that if two communities are not comparable, that is, if their intrinsic diversity profiles intersect, then there exist a pair of diversity indices which order the communities oppositely. This fact has been the motivation for the testing framework described in the last section.

In recent years, several families of indices, satisfying the isotonic requirement, have been proposed. These include the Δ_β family discussed by Patil and Taillie (1976, 1979); Hill's (1973) S_β family; and the "expected number of species" family $s(m)$ of Hurlbert (1971) and Smith and Grassle (1977).

5.1 The Δ_β *Family.* The Δ_β family of diversity indices is defined by

$$\Delta_\beta(\underset{\sim}{\pi}) = (1-\Sigma\pi_i^{\beta+1})/\beta, \ -1 \leq \beta < \infty.$$

The parameter restriction $\beta \geq -1$ is imposed in order that Δ_β be isotonic. For positive integral values of β, $\beta\Delta_\beta$ is the probability of observing more than one species in a hypothetical random sample of size $\beta+1$. The family includes as special cases the Simpson index $1-\Sigma\pi_i^2$ for $\beta=1$, the Shannon index $-\Sigma\pi_i \ \ell n \ \pi_i$ for $\beta=0$, and the species richness $s-1$ for $\beta=-1$.

An important property of any family of indices is its variable sensitivity to rare and abundant species. A precise definition of sensitivity is given by Patil and Taillie (1979) who show that, for large values of β, Δ_β is sensitive to abundant species, whereas it is sensitive to rare species for smaller values of β. In practical terms, this means that, when Δ_β is used as a test statistic, the discriminatory power against changes among abundant species increases with β.

The profiles of the Δ_β family are decreasing and convex. However, they decrease to zero rapidly with increasing β and appear to be of limited value for large β. Figure 4 gives the

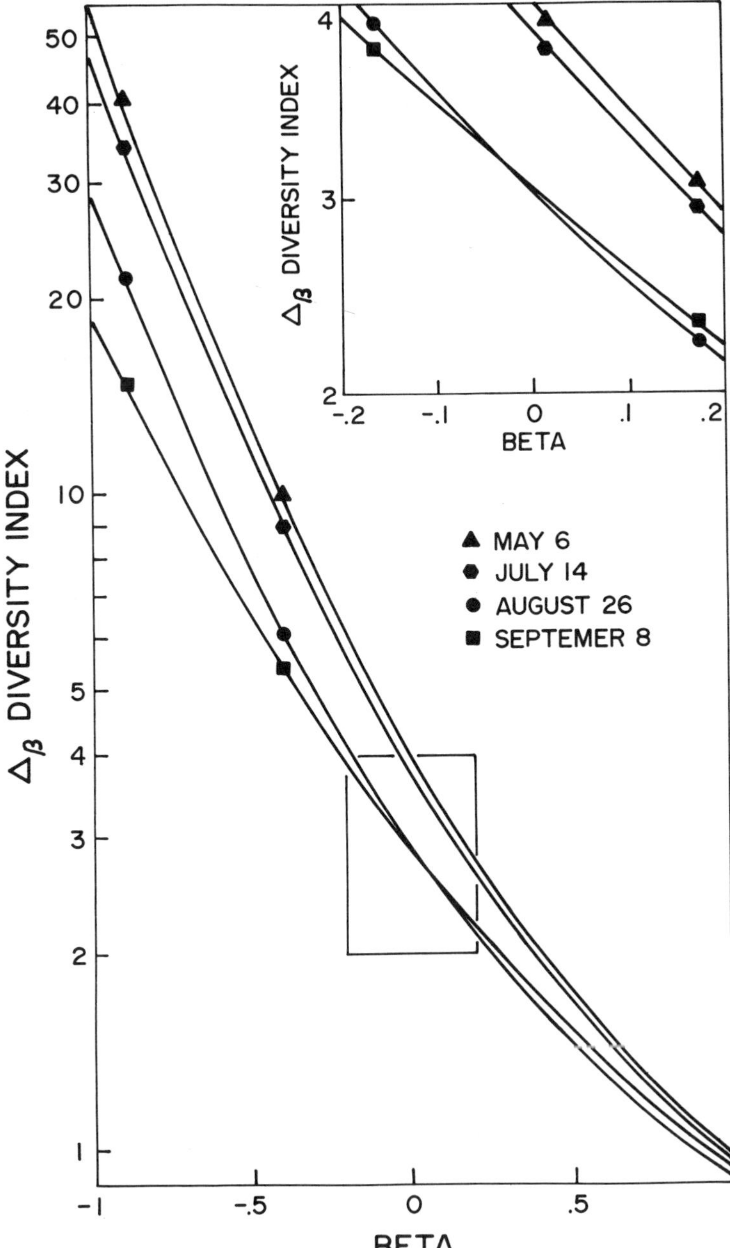

FIG. 4: Jackknifed estimates of the Δ_β site diversity profiles of Western Meadowlark on four selected survey dates during 1975.

Δ_β profiles for the Meadowlark spatial data. These profiles are estimates obtained by a jackknife procedure described in the next section. These Δ_β profiles reflect the qualitative features of the intrinsic diversity profiles of Figure 3. In particular, there is an intersection in the August and September profiles. The major difference is that the August Δ_β profile crosses that of September from above, not from below as in the intrinsic profile. The reason for this is that small β correspond to rare 'species' and, hence, large ranks.

It is interesting to observe that the crossing point occurs approximately at the Shannon index $(\beta=0)$. Thus, *for this data*, the Shannon index has little power for detecting changes in diversity.

5.2 The S_β *Family.* The S_β family of diversity indices is defined by

$$S_\beta(\underset{\sim}{\pi}) = (\Sigma\pi_i^{\beta+1})^{-1/\beta}, \quad -1 \leq \beta < \infty.$$

Some special cases are the species count $\mathcal{s}$ for $\beta = -1$, and the reciprocal of the abundance of the most dominant species, $1/\pi_1^*$, when $\beta=\infty$. The index $S_\beta(\underset{\sim}{\pi})$ has a simple interpretation: it is the number of species that a completely even community would need to have for its Δ_β diversity to be $\Delta_\beta(\underset{\sim}{\pi})$. For this reason, S_β is sometimes called the *equivalent number of species* for Δ_β. Also, S_β is a mathematical function of Δ_β given by

$$S_\beta = (1-\beta\Delta\beta)^{-1/\beta},$$

and has the same variable sensitivity as Δ_β: sensitive to rare species for small β and sensitive to abundant species for large β.

The S_β profiles are decreasing; but unlike those of Δ_β, they do not converge to zero with increasing β. The estimated S_β profiles for the Meadowlark spatial data appear in Figure 5, and have the same characteristics as the Δ_β profiles, except that it is possible to plot a wider range of β values. This permits display of an additional intersection in the August and September profiles corresponding to the hidden intersection of the intrinsic diversity profiles.

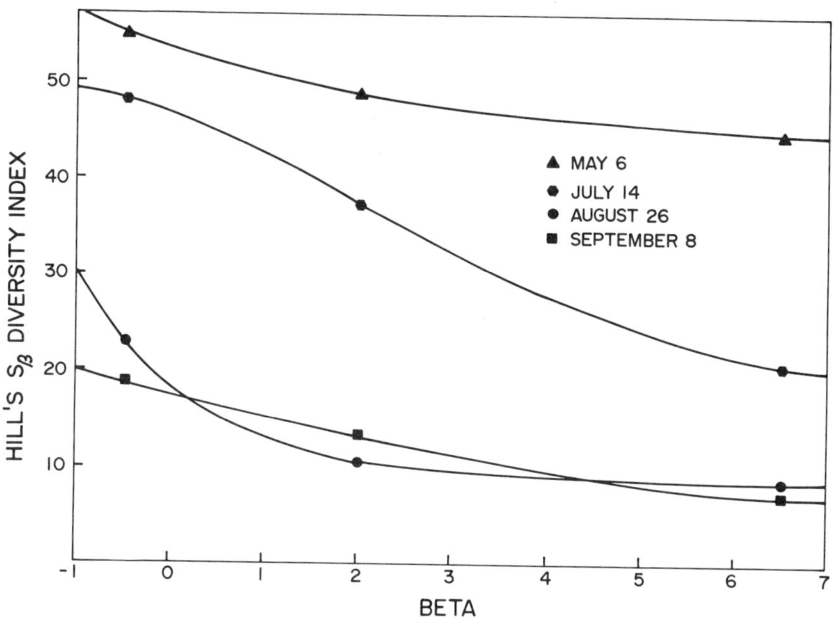

FIG. 5: Jackknifed estimates of the S_β *site diversity profiles of Western Meadowlark on four selected survey dates during* 1975.

5.3 The s(m) *Family.* The s(m) family of diversity indices is defined by

$$s(m) = \Sigma[1-(1-\pi_i)^m], \quad 1 \leq m < \infty.$$

Some special cases are (1 + Simpson index) when $m = 2$, and the species count s for $m = \infty$. When m is a positive integer, s(m) is the expected number of species to be found in a hypothetical random sample of size m. Its pattern of sensitivity is opposite to that of Δ_β : sensitive to abundant species for small m and to rare species for large m. However, as compared with S_β, the s(m) family does not have as wide a range of sensitivity to abundant species.

The s(m) profiles are concave, increasing, and are essentially species-area curves with m replacing area as the measure of sampling effort. The estimated s(m) profiles for the Meadowlark spatial data are given in Figure 6. These have the same characteristics as the intrinsic diversity profiles, except that the August and September s(m) profiles have only one intersection. This intersection, which occurred near the Shannon index for the

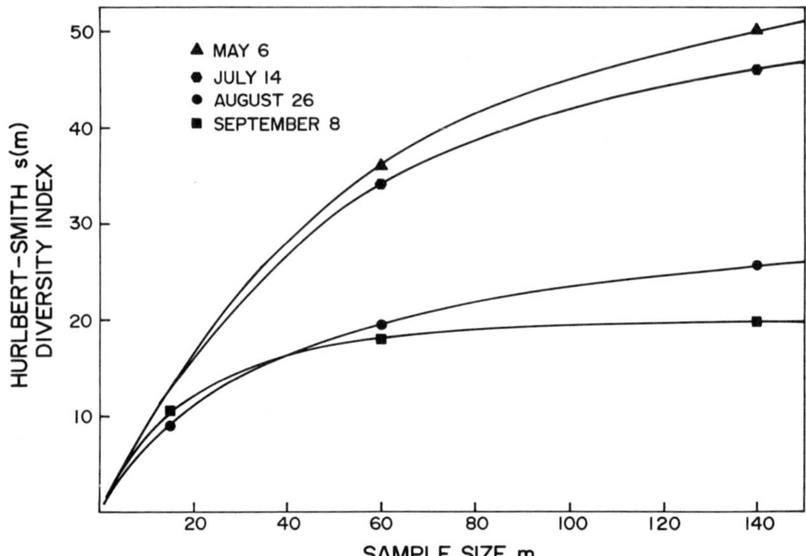

FIG. 6: Jackknifed estimates of the s(m) *site diversity profiles of Western Meadowlark on four selected survey dates during* 1975.

Δ_β profiles, here occurs at about $m = 40$. Generally speaking, there appears to be little information in the Shannon index that is not contained in the $s(m)$ family. However, the value of m for which $s(m)$ is similar to the Shannon index varies considerably with the communities and no universally applicable numerical correspondence is possible.

An index closely related to the $s(m)$ family is the slope of the $s(m)$ profile at $m = 1$ given by

$$s'(1) = -\Sigma(1-\pi_i)\ \ln(1-\pi_i).$$

This index, introduced by Patil and Taillie (1976) in another context, provides the maximal sensitivity to abundant species that is possible within the $s(m)$ family. Even so, $s'(1)$ is unable to detect the hidden intersection in the intrinsic profiles of Figure 2.

6. ESTIMATION OF DIVERSITY INDICES AND SIGNIFICANCE TESTING

This section takes up the problem of estimating and comparing community diversity based on field samples. We will be concerned with bias and its removal and with determining the precision of our estimates.

The question of bias is important because most diversity indices, when evaluated on a sample, consistently underestimate community diversity and by an amount varying with sample size. Thus, for example, we must ask if the apparent decline in Meadowlark site diversity between May and July is due to the smallness of the July sample. Similarly, the differing number of species observed from year to year during the Colstrip Bird Survey (Table 1) may be due entirely to changes in sample size.

Most diversity indices do not have unbiased estimators so that only approximate bias correction is possible. This is true of the Shannon index and of S_β for all values of β. On the other hand, Smith, Grassle, and Kravitz (1979) point out that Δ_β and $s(m)$ do have unbiased estimates, but only when m and β are positive integers. They have shown how to obtain the minimum variance unbiased estimators as well as the sampling variances. However, their formulas are complicated and, for large samples, computationally difficult. Hwere we consider two approximate methods, maximum likelihood and jackknifing, suitable for large samples.

The method of maximum likelihood assumes that one has a random sample from the multinomial population $\underset{\sim}{\pi}$. Let N_i be the number of individuals observed from the *ith* species and $N = \Sigma N_i$ the total number of individuals observed. The likelihood estimate of $\Delta(\underset{\sim}{\pi})$ is then formed by replacing population proportions π_i by sample proportions N_i/N in the expression for Δ. Approximate expressions for the large sample bias and the variance of the estimate can be obtained using statistical differentials (Johnson and Kotz, 1969, p. 29). The resulting formulas are given in Table 3 for the Δ_β, S_β, and $s(m)$ families. These formulas are derived from the assumption of a simple random sample - an assumption seldom met under actual field conditions.

A more robust approach is to use the jackknife estimates of Quenouille (1956) and Tukey (1958). The method as applied to the estimation of diversity can be described as follows. Let Δ be the true value of the index to be estimated and suppose there is available, for each sample size N, an estimator Δ_N for which

$$E[\Delta_N] = \Delta + a/N + 0(1/N^2),$$

where a is an unknown constant. The regression of Δ_N upon $1/N$

TABLE 3: Large sample bias and variance of Δ_β, S_β *and* $s(m)$. $(K_\beta = \Sigma\, \pi_i^{\beta+1})$.

Index	Bias	Variance
Δ_β	$-(\beta+1)(K_{\beta-1} - K_\beta)/(2N)$	$(1+\beta^{-1})^2\ (K_{2\beta} - K_\beta)^2/N$
S_β	$(S_\beta/K_\beta)[\text{Bias}(\Delta_\beta) + (\beta+1)\ \text{Variance}(\Delta_\beta)/(2K_\beta)]$	$(S_\beta/K_\beta)^2\ \text{Variance}(\Delta_\beta)$
$s(m)$	$-m(m-1)[\Sigma\ \pi_i(1-\pi_i)^{m-1}]/(2N)$	$m^2\{\Sigma\ \pi_i\ (1-\pi_i)^{2m-2} - [\Sigma\ \pi_i(1-\pi_i)^{m-1}]^2\}/N$

is then approximately linear for large samples. If there were available two samples of sizes $N > M$, then Δ could be estimated as the intercept of the line passing through $(1/N, \Delta_N)$ and $(1/M, \Delta_M)$, that is, as

$$(N\Delta_N - M\Delta_M)/(N-M) .$$

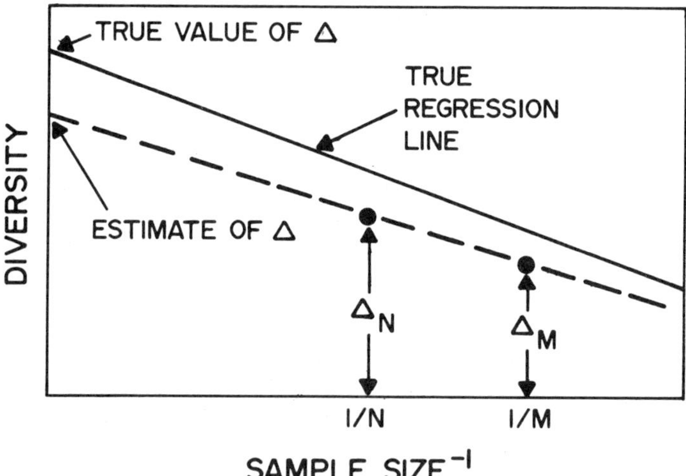

Jackknifing obtains the second sample by omitting an individual from an initial sample of size N yielding the estimate

$$\Psi_N^{(t)} = N\Delta_N - (N-1)\ \Delta_{N-1}^{(t)}$$

when the t*th* individual is omitted. The $\Psi_N^{(t)}$ are called pseudo values. The jackknife estimate $\hat{\Delta}$ of Δ is the average of the pseudo values and the estimated variance of $\hat{\Delta}$ is the ordinary sampling variance. Thus

$$\hat{\Delta} = N\,\Delta_N - (N-1)\,\overline{\Delta^{(\cdot)}_{N-1}}\ ,$$

$$\hat{\text{Var}}\,(\hat{\Delta}) = \frac{1}{N(N-1)} \sum_{t=1}^{N} [\Psi_N^{(t)} - \overline{\Psi_N^{(\cdot)}}]^2$$

$$= \frac{N-1}{N} \sum_{t=1}^{N} [\Delta_{N-1}^{(t)} - \overline{\Delta_{N-1}^{(\cdot)}}\,]^2 .$$

A principal advantage of jackknifing is that, in addition to reducing bias, it provides estimated variances without assuming a theoretical sampling model.

The Meadowlark profiles described in Section 5 were obtained by jackknifing the likelihood estimates of Δ_β, S_β, and $s(m)$. Making use of the estimated variances, the Z-statistic may be calculated to test for significant differences on each side of the intersections. Selected z-values for the August minus September site diversity appear in Table 4. All three indices show significance on one side, but not the other, of the intersection corresponding to $\beta \simeq 0$ and $m \simeq 40$. The hidden intersection is detected only by the S_β family which shows non-significance on both sides. The conclusion is that the decline in richness between August and September is significant but the rise in evenness is not. The data is thus consistent with the hypothesis that the spatial pattern in September is intrinsically less diverse than in August.

7. ANNUAL CHANGES IN THE SPECIES DIVERSITY OF THE COLSTRIP BIRD COMMUNITY

In this section the methodology developed above is applied to the data of the Colstrip Bird Survey (Table 1) to examine for possible trends in avian diversity. Ideally, one would simultaneously jackknife across sites as described by Zahl (1977). However, the authors were not in possession of the needed breakdown of the data by sites for all three years, and have therefore jackknifed across individuals as in Section 6.

The intrinsic diversity profiles for the three years are given in Figure 7. The profile for 1976 closely follows the 1975 baseline profile for the first 20 ranks and then drops below, indicating a decline of diversity among species of intermediate

TABLE 4: Z-values for differences between August and September site diversity.

Beta	Δ_β	S_β	m	s(m)
-1.0	4.54	4.54	2	-1.13
- .8	3.97	3.99	5	-1.16
- .5	2.45	2.50	10	-1.14
- .3	1.40	1.46	20	- .90
0	.19	.21	30	- .45
.3	- .54	- .60	40	.27
.5	- .84	- .92	80	1.86
1.0	-1.13	-1.24	100	2.49
2.0	- .98	- .97	150	3.55
3.0		- .53	200	4.13
5.0		+ .00		
10.0		.33		
20.0		.28		

abundance. Beyond the 55*th* rank, the 1976 profile rises above the baseline, but this apparent rise in diversity is due in part to the very large sample size during 1976 and may well be spurious.

The 1977 profile, as compared with the baseline, is low for the first several ranks. This is largely due to a substantial increase in Meadowlark dominance. This species comprised 34% of the bird population in 1975 and 43% in 1977. Between ranks 5 and 35 the 1977 profile lies above the baseline and thereafter closely follows the baseline, indicating an increase in the diversity among abundant to intermediate-abundant species.

The intrinsic profiles point to some possible changes in diversity, but no consistent temporal trend emerges. Further, without specific knowledge of the prior pattern of fluctuation in Colstrip avian diversity, it is not possible to attribute even the indicated changes to the power plants becoming operational.

In order to assess the significance of the indicated changes, the s(m) profiles together with their 95% confidence bands are given in Figure 8. These show that between 1975 and 1976, there occurred a significant decrease in diversity among species of intermediate abundance. Between 1975 and 1977, diversity significantly decreased among abundant species and significantly increased among abundant to intermediate-abundant species. However, it must be emphasized that statistically significant changes need not be biologically important. Given the large sample sizes, even minor changes in community structure will be found statistically significant.

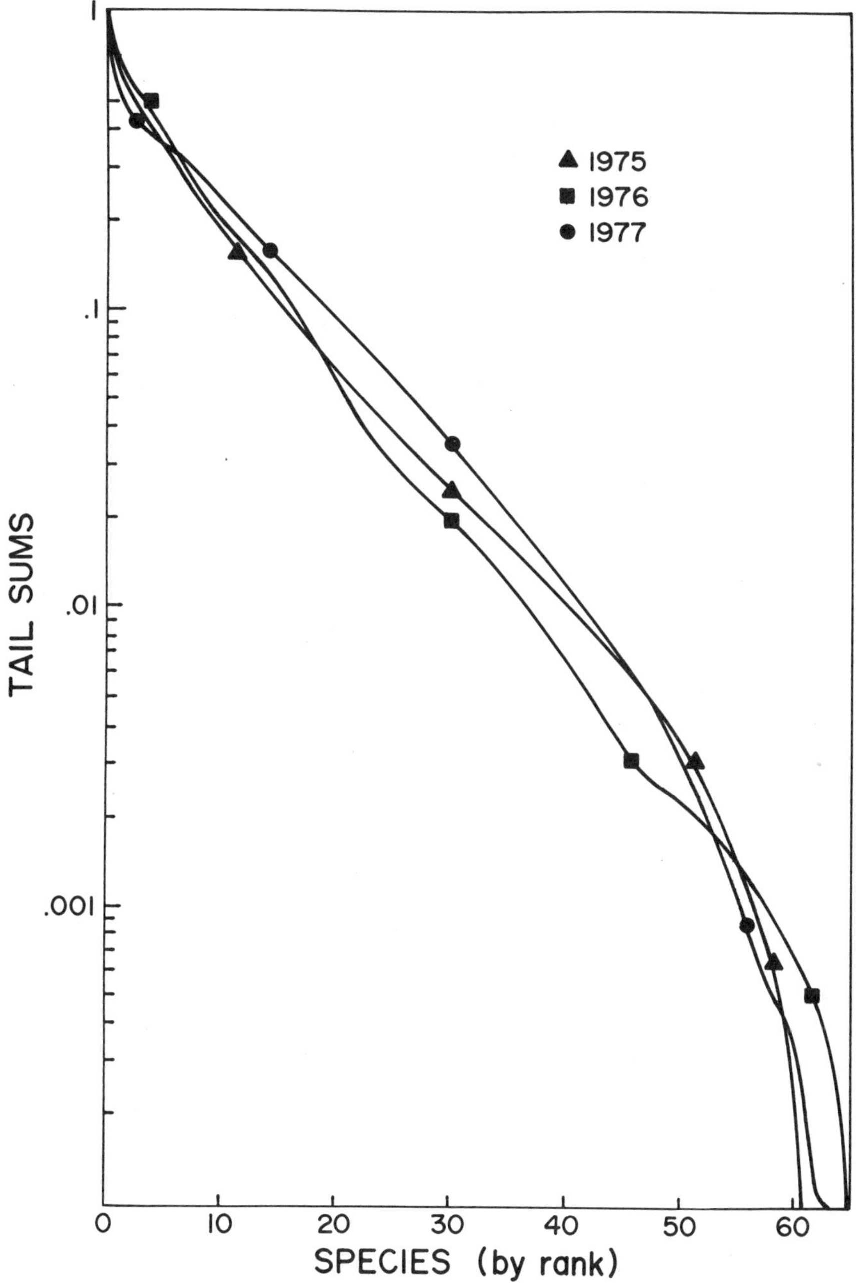

FIG. 7: Intrinsic species diversity profiles of the bird community in the vicinity of Colstrip during 1975, 1976, *and* 1977.

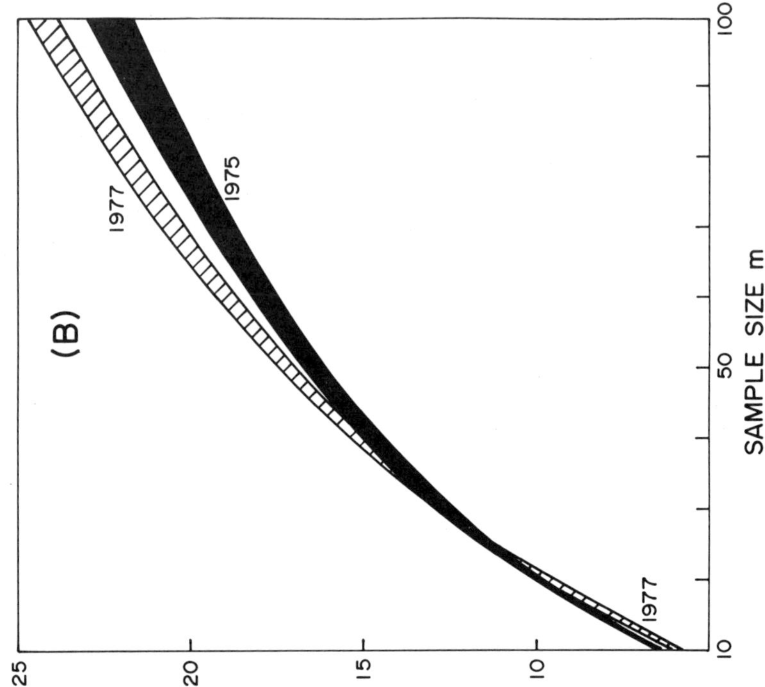
(B)
1977
1975
1977
SAMPLE SIZE m
10
50
100
25
20
15
10

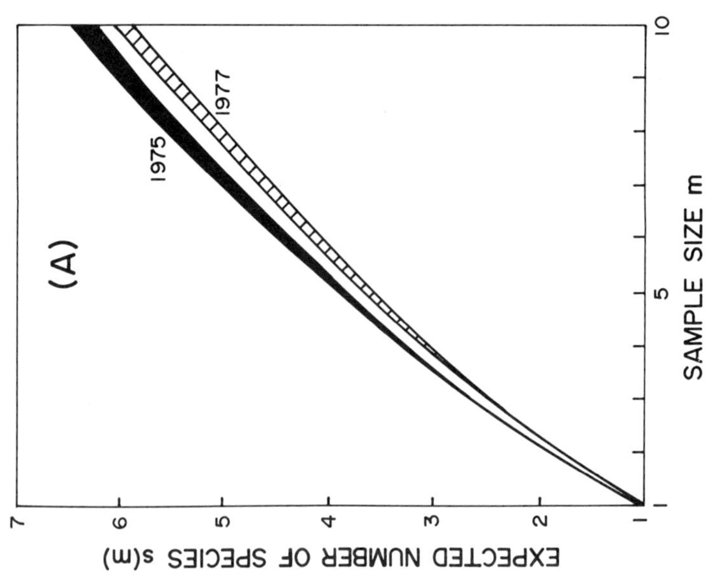
(A)
1975
1977
SAMPLE SIZE m
1
5
10
EXPECTED NUMBER OF SPECIES s(m)
7
6
5
4
3
2
1

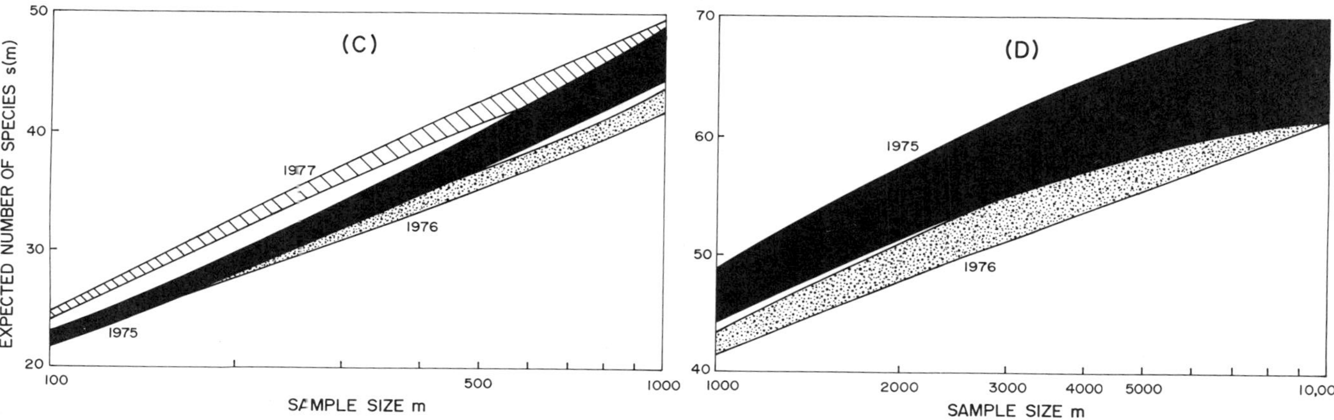

FIG. 8: Ninety-five percent confidence bands for the jackknifed estimates of the avian s(m)-*diversity profiles. A.* $1 \leq m \leq 10$; *B.* $10 \leq m \leq 100$; *C.* $100 \leq m \leq 1000$; *D.* $1000 \leq m \leq 10,000$.

ACKNOWLEDGMENTS

The research reported herein was carried out with support from United States Environmental Protection Agency under contract number 68-03-1308 while the authors were away from the University.

REFERENCES

Dennis, B., Patil, G. P., Rossi, O., Stehman, S., and Taillie, C. (1979). A bibliography of literature on ecological diversity and related methodology. In *Ecological Diversity in Theory and Practice*. J. F. Grassle, G. P. Patil, W. K. Smith, and C. Taillie, eds. Satellite Program in Statistical Ecology, International Co-operative Publishing House, Fairland, Maryland.

Hill, M. O. (1973). Diversity and evenness: A unifying notation and its consequences. *Ecology*, 54, 427-432.

Hurlbert, S. H. (1971). The nonconcept of species diversity: A critique and alternative parameters. *Ecology*, 52, 577-586.

Johnson, N. L. and Kotz, S. (1969). *Discrete Distributions*. Wiley, New York.

Patil, G. P. and Taillie, C. (1976). Ecological diversity: concepts, indices and applications. *Proceedings of 9th International Biometric Conference*, 383-411.

Patil, G. P. and Taillie, C. (1979). Diversity as a concept and its measurement. *Journal of the American Statistical Association* (under revision).

Quenouille, M. (1956). Notes on bias in estimation. *Biometrika*, 43, 353-360.

Robbins, C. S. and Van Velzen, W. T. (1970). Progress reports on the North American Breeding Bird Survey. In *Bird Census Work and Environmental Monitoring*, S. Svenson, ed. Bulletin of Ecological Research, Communication No. 9. Lund, Sweden. 22-30.

Smith, W. K. and Grassle, J. F. (1977). Sampling properties of a family of diversity measures. *Biometrics*, 33, 283-292.

Smith, W. K., Grassle, J. F., and Kravitz, D. (1979). Measures of diversity with unbiased estimators. In *Ecological Diversity in Theory and Practice*, J. F. Grassle, G. P. Patil, W. K. Smith, and C. Taillie eds. Satellite Program in Statistical Ecology, International Co-operative Publishing House, Fairland, Maryland.

Taillie, C. (1977). *The mathematical statistics of diversity and abundance*. Ph.D. thesis, The Pennsylvania State University.

Tukey, J. (1958). Bias and confidence in not quite large samples (abstract). *Annals of Mathematical Statistics*, 29, 614.

Zahl, S. (1977). Jackknifing an index of diversity. *Ecology*, 58, 907-913.

[*Received August* 1978. *Revised July* 1979]

G. P. Patil and M. Rosenzweig, (eds.),
Contemporary Quantitative Ecology and Related Ecometrics, pp. 49-60.

THREE PROBABLE EVOLUTIONARY CAUSES FOR HABITAT SELECTION

MICHAEL L. ROSENZWEIG

Ecology and Evolutionary Biology
University of Arizona
Tucson, Arizona 85721 USA

SUMMARY. Habitat selection has sometimes been rigidly attributed to the trade-off principle: improvement in one function leads to diminished performance of another. While some instances of habitat selection are no doubt due to this principle, others are not. Rendezvous habitat selection occurs for the purpose of species recognition during mating. Dominance habitat selection occurs when one species (the subordinate) is driven out of its preferred habitat either by overt aggression or resource depletion by the dominant species. All three mechanisms may lead to the same stable field situation: no niche overlap and competitive alphas of zero.

Science is the attempt to make the chaotic diversity of our sense-experience correspond to a logically uniform system of thought... The scientific way of forming concepts differs from that which we use in our daily life, not basically, but merely in the more precise definition of concepts and conclusions; more painstaking and systematic choice of experimental material; and greater logical economy. By this last we mean the effort to reduce all concepts and correlations to as few as possible logically independent basic concepts and axioms.

Albert Einstein, Considerations concerning the fundamentals of theoretical physics. Essays in Science. Philosophical Library, New York.

1. STAMPEDING TOWARD SIMPLICITY

For about a decade, many ecologists were striving to please Professor Einstein, departed though he was. We hoped to reduce ecology to a small number of general propositions (e.g. Watt, 1973). While this was no doubt an admirable reaction to what had preceded it, in our zeal we appear to have gone too far. The conjectural generality became a siren song: the thrice-observed pattern an immutable law, deeply etched on our stony brains.

As a graduate student, I remember Robert MacArthur's dismay at the ready acceptance of his stability-diversity theory (MacArthur, 1955). I also remember L. B. Slobodkin's gentle but firm attempt to put this in its place when he wrote his famous book (Slobodkin, 1961): "Unfortunately, there is no immediate evidence that the quantity defined as stability by MacArthur's formulation has any relation to any common-sense definition of the concept of stability."

Did this do any good? Not really. Ecologists had a unifying principle; they could finally hold up their heads in the company of chemists, physicists, and yes, even molecular geneticists. Later and most discouraging was the chatter in The Hague during the first Intecol after May (1972) had introduced many to a counterexample, a case where diversity reduced stability. "Say," I heard, "what did you think of May's Law?" (Anon.)

I could cite other examples. But I should like to dwell on one, because it is rich in side examples and because I have been as guilty as many and more guilty than most in sanctifying it in my classroom.

As ecologists sought rules to organize and explain the phenomena they observe, they turned naturally to natural selection. They wondered how it could help them understand competitive specializations. An early suggestion was called the trade-off principle. The idea is simple: sharpened performance of one ecological function ought to require diminished ability to perform another. The idea can certainly have merit. For example, a carnivore's body size may restrict its optimal choice of victim size, since there will be victims too large for it to kill very profitably, and others too small (Rosenzweig, 1966).

Because of the adage "Jack of all trades is master of none," the trade-off principle is often known as the Jack-of-all-trades principle. Figure 1 is an example of a tool which acts as good public relations for the principle. Anyone ever trying to use such a tool for anything knows it is good for nothing. Other such examples, similarly more sophisticated than mere body size, abound in the world of machinery. Theoretical applications of

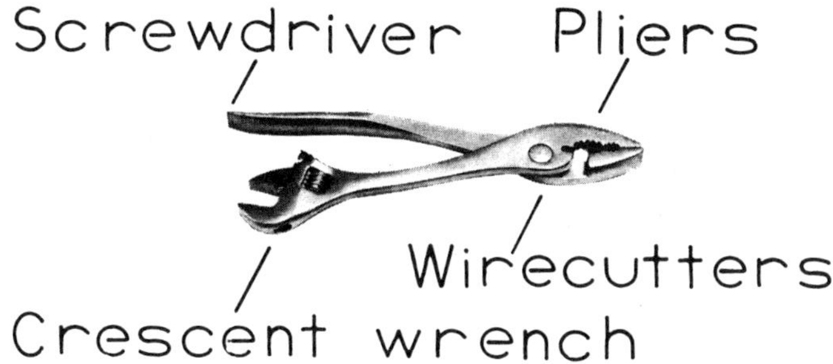

FIG. 1: A multipurpose tool. Such ingenious/ingenuous contraptions are sometimes actually manufactured. They are the best advertising imaginable for the jack-of-all-trades principle.

the principle to evolutionary ecology begin with well-known papers of Levins (1962) and MacArthur and Levins (1964), and include less well-known efforts by Charnov (1976), Schaffer and Gadgil (1975), myself (Rosenzweig, 1974) and many others.

How is the principle applied to real life in the field? The ecologist finds a set of potential competitors, assumes the operation of the principle, and searches for a habitat or resource separation which can be explained by tradeoff. Or, having found a separation, the ecologist searches for an explanation within the confines of the tradeoff axiom: things living mostly in rocks just might not be able to be very adept at living on sand, etc. If he fails to find a satisfactory explanation, the ecologist attributes this to his negligence or stupidity.

Alas, he is too harsh on himself. He is neither negligent nor stupid, but merely rigid. He has donned blinders which prevent him from seeing what is obvious. I shall pursue this idea in respect to habitat separation.

I believe there are at least three distinct mechanisms which can lead to the evolution of ecological segregation by habitat separation. Only one of these is clearly due to the operation of the trade-off principle, and one has, just as clearly, nothing whatever to do with it. Yet we shall see that all three can lead to the same field situation. These three may be termed habitat allocation: 1) for species recognition; 2) owing to classical tradeoff; 3) because of ecological domination of a preferred habitat.

The latter is or is not a case of trade-off, depending on who is consulted. I shall discuss these in order.

2. RENDEZVOUS HABITAT SELECTION

Several years ago, I was enjoying a visit to the laboratory of Dr. Arthur Shapiro of the University of California at Davis. Dr. Shapiro, aware of my interest in habitat separation, brought out a pair of photographs to show me. They were two species of butterfly of the genus *Callophrys* which coexist in New Jersey. Both use trees for larval food. One, *C. hesseli,* feeds only upon *Chamaecyparis;* the other, *C. gryneus,* only upon *Juniperus.* Since the latter cover the dry uplands of the sandy soils of the state, and the former are swamp lovers, this leads to a marked habitat segregation between the butterflies.

There is no use by including a figure of these two species. All you would see is a pair of mottled, non-descript looking butterflies of identical size and shape. There is no obvious way to distinguish the species by external appearance. *C. hesseli* in fact was not discovered until the middle of this century, although it had been captured and even placed in collections long before.

Using the trade-off principle, I immediately assumed that the chemistry of eating *Chamaecyparis* must be considerably different from that of eating *Juniperus.* But I was wrong. So were far more distinguished observers of Lepidoptera who had preceded me in my error by 20 years (Remington and Pease, 1955). Instead, experiments have proved that each butterfly species may be reared almost as successfully on the others' foodplant as on its own (Remington and Pease, 1955; Johnson, personal communication).

I then guessed that the butterflies have just as much trouble identifying each other as we do, and because they simply cannot mate successfully with aught but a member of their own species, they use distinct habitats as a device for species recognition. The trees are, in a manner of speaking, the songs and dances of the butterflies. Sure enough, I was informed that these butterflies, unlike most, pair and breed on the larval foodplant (Shapiro, personal communication). Moreover, I was able to build a model (unpublished) which showed that it was indeed possible for a species pair to evolve to complete habitat segregation for recognition purposes.

I shall not describe the model here; it is too complex to justify occupying your time, and its conclusions are all too obvious anyhow. The most important conclusion is that the evolution of this phenomenon ought to be severely frequency-dependent. The advantage garnered by forsaking a resource depends on the penalty paid for not doing so. And this penalty is great only if one's chances of finding an appropriate mate on the resource are slim because most conspecifics are already elsewhere. If the penalty is great enough to encourage resource abandonment, then the penalty actually grows (as the resource is abandoned) until all of the members of a species have left.

The example of the butterfly is not unique. Dr. Rob Colwell of the University of California at Berkeley has discovered that hummingbird mites of different species specialize on different flowers despite the fact that he can detect no experimental differences in their abilities to use the flowers they shun and those they seek (Colwell, 1973). Moreover, other investigators have independently arrived at the species recognition idea. Dr. Hugh Rowell (personal communication) of the University of California at Berkeley thought of it after detailed investigations of a tropical grasshopper family. Each of these species courts and mates only after landing upon a specific plant in a tropical clearing. One plant species may have many species of grasshopper specific to it, but each grasshopper species is of a different genus. Congeners always choose different species of host plant. Rowell has not been able to discover any physiological basis for the preferences. He has concluded that the plants are simply easier to find in a tropical forest than are potential mates blundering about at random.

As far as I know, the first to discover this mechanism was Professor Helmut Zwolfer (1974). Working with trypetid flies, Dr. Zwolfer (of Bayreuth University) called the host plants "rendezvous points." Following him, we may term the entire phenomenon "rendezvous habitat selection." We do not yet know how widespread or important it is. But even one case acting as counterexample is sufficient to prove that all habitat segregation will not be explained on the basis of the trade-off principle.

3. TRADE-OFF HABITAT SELECTION

Because the penalty for habitat overlap increases as the amount of overlap declines, the end point of the evolution of rendezvous habitat selection should be zero overlap. Herein, perhaps, lies a discernible difference between the outcomes of the tradeoff and rendezvous mechanisms. Many theoretical investigations have concerned themselves with limiting similarity and predicted that a certain stable amount of niche overlap greater than zero should exist between competing species in a mature community (May and MacArthur, 1972; Roughgarden, 1976).

Recently, I have been able to devise a theoretical method which allows for the simultaneous examination of population dynamics and optimal habitat choice in a pair of competitors. Using this method has taught me that habitat selecting competitors may not be able to overlap at all, even if they satisfy the classical trade-off principle.

Let us assume that there are two species, 'mauve' and 'niger,' whose names and densities are symbolized by the letters, M, N. Let us further assume that there are two rewarding types of habitat, m and n, embedded in a spatial matrix useless to both species. M is better at m, and N is better at n, but both habitats can be profitably used by both species.

There are many ways to proceed from here. Let us take the most recently travelled path (Rosenzweig, 1979, in press; Pimm and Rosenzweig, in press) by assuming both density dependence and a cost to habitat selection. By the former I mean that the benefit derived from a patch depends inversely on the density of organisms using it. By the latter I mean simply that an individual pays a price for habitat selecting: it must traverse (at least briefly) the patches it rejects.

Now one erects a state space with coordinates M and N and divides it into different regions of optimal behavior. This is accomplished by drawing *isolegs* for each species. Isolegs (from the Greek *iso* - equal; *lego* - choose) are merely lines on a state space that separate points at which the optimal behavior is strict habitat selection from points at which it is not.

Figure 2 is an example. The M isoleg has positive slope because larger densities of its competitor so reduce resources in patch n that patch n becomes less productive for M. The N isoleg has positive slope for analogous reasons. (The details of proof are in Rosenzweig, in press.)

As a result of the isolegs, we are able to partition the graph into three pieces in Figure 2. In the middle region, Roman Numeral I, both species are restricting their usage of habitats to their own better patch, and so do not overlap ecologically.

Adding isoclines to such a space accounts for the population dynamics. In the region of no overlap, the isocline of each species parallels the axis of the other. Should the isoclines cross in the no-overlap region, and Dr. Stuart Pimm and I have recently developed a theory that predicts they usually will (in press), then a stable node is achieved at densities which preclude ecological overlap (Figure 3). This is the same result as with rendezvous habitat selection. Lawlor and Maynard Smith (1976) reached the same conclusion assuming density dependence but no cost. So did I (Rosenzweig, 1974) while assuming density independence (with or without cost).

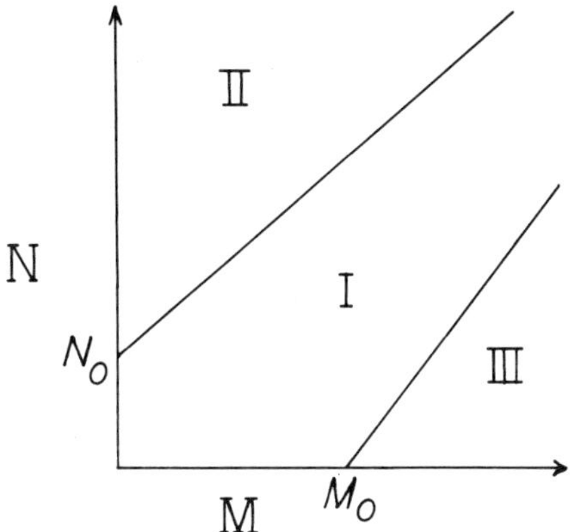

FIG. 2: Regions of optimal behavior. Optimal habitat selection is influenced by the population densities, M *and* N, *of the two competing species. In region* I, *both species should use only their own special patches for foraging, resulting in their complete habitat separation. In* II, M *should specialize whereas* N *should not. In* III, *the species should reverse their roles. The lines separating the regions are called isolegs.*

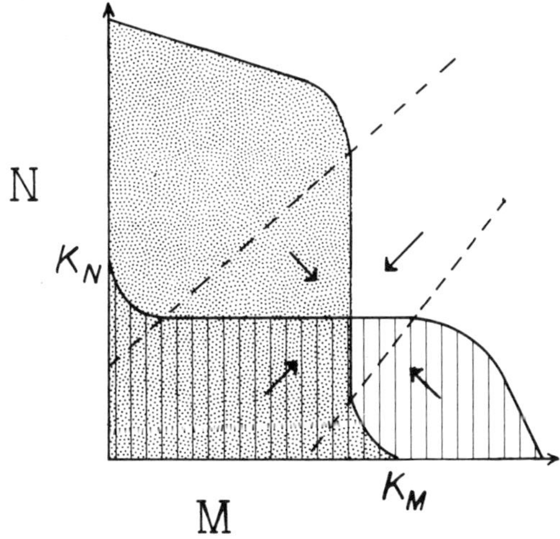

FIG. 3: Population dynamics if individuals optimize their behavior. The K's *are carrying capacities of the species. Dashed lines are the isolegs of Figure 2. Solid lines are the systems zero isoclines.* $\dot{M} > 0$ *in the stippled region;* $\dot{N} > 0$ *in the shaded region. Vectors show the net direction of population change. The equilibrium is a stable node.*

4. DOMINANCE HABITAT SELECTION

Field examples of trade-off habitat selection are not common. Schroder and Rosenzweig (1975) found an example in a pair of kangaroo rats in central New Mexico. And Inouye has found one in a pair of bumblebees in the Rocky Mountains.

Yet this apparent scarcity does not mean that habitat segregation is usually only partial. It means instead that in most field studies, classical trade-off has not been apparent. Instead, it appears that one of the species dominates a habitat which is best for all. For example, Schaffer and Schaffer (in press) have found three species of bees with temporal habitat segregation: honeybees (*Apis*) forage during the times of day when nectar is most abundant. Bumblebees (*Bombus*) work during hours of moderate nectar abundance, and carpenter bees (*Xylocopa*) during the sparsest hours. Experiments have shown that it is not the time of day but the nectar abundance which is the causal variable. Yet they have also shown that neither *Xylocopa* nor *Bombus* needs to be deprived.

Similar experiments on hummingbirds by Pimm (1978) have demonstrated that one large species, the blue-throated hummingbird, excludes a smaller one, the black-chinned hummingbird, from the richest sources of sucrose solution, leaving the poorer places undefended. In a different case, Bovbjerg (1970) has found habitat selection between two species of freshwater crayfish which is based on oxygen concentration. The oxygen-rich waters are good habitat for both species, but the more sluggish species which is also successful in stagnant water is not found there. It is excluded aggressively from the oxygen-rich water by the more active species.

There is also the classic case of barnacles, worked out by Connell (1961). *Balanus*, the barnacle in the richer part of the intertidal, cannot survive in the higher *Chthamalus* zone and actively excludes *Chthamalus* from the rich zone. Another particularly startling case concerns cactophilic drosophila. Fellows and Heed (1972) demonstrated that the dominant species, *Drosophila nigrospiracula*, actually reproduced more poorly in the better cactus than *D. mojavensis*, the species it excludes from the better cactus. There are many other examples in the literature.

Habitat selection because of ecological dominance can also be modelled with isolegs. Figure 4 is an example. Notice that here again the isolegs can cause the isoclines to cross in a manner which precludes niche overlap (Figure 5).

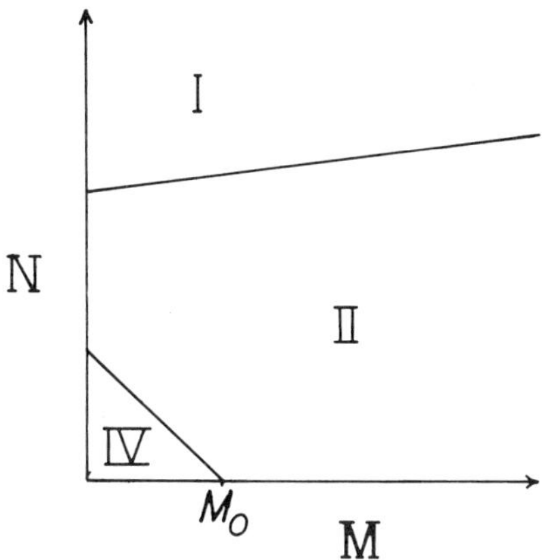

FIG. 4: Regions of optimal behavior when one patch is better for both species but is dominated by N. *Regions* I *and* II *as in Figure 2 except that in* I, *species* M *should specialize on the patches that it should not use at all if* N *is absent. In region* IV, *both species should avoid using this poorer patch type.*

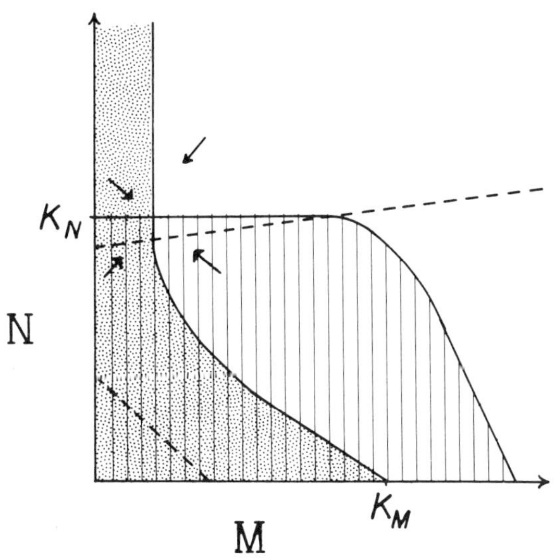

FIG. 5: Possible population dynamics if individuals engaged in dominance competition optimize their behavior. Symbols as in Figure 3. N *is the subordinate species.*

5. CONCLUSION

In all of the foregoing mechanisms, the end result may be zero niche overlap. This is the same as $\alpha = 0$. Limiting similarity theory fails to predict such alphas because it stresses what is tolerable rather than what natural selection is constrained to produce. Hence the trade-off principle must not be dogmatically assumed to underlie the evolutionary resolution of all competitive coexistences.

Where the trade-off principle can be used to focus our research and make it more efficient, let it continue to be used. But when it fails, let us not be too disappointed. And let us certainly not insist that *its* failure is ours. Although Professor Einstein might well have been happier with a single explanation for the evolution of habitat selection, he most certainly would have preferred a complex set of truths to a unitary dogma. Besides, he expected complexity from biology. He might even have congratulated us all for organizing a bewildering chaos into the rich rococco tapestry that is increasingly what ecological science is becoming.

ACKNOWLEDGEMENTS

Drs. R. K. Colwell, C. Remington, H. Rowell, A. Shapiro, and H. Zwolfer provided valuable discussion.

REFERENCES

Bovbjerg, R. V. (1970). Ecological isolation and competitive exclusion in two crayfish (*Orconectes virilis* and *Orconectes immunis*). *Ecology*, 51, 225-236.

Charnov, E. L. (1976). Optimal foraging. The marginal value theorem. *Theoretical Population Biology*, 9, 129-136.

Colwell, R. K. (1973). Competition and coexistence in a simple tropical community. *American Naturalist*, 107, 737-760.

Connell, J. H. (1961). The influence of interspecific competition and other factors on the distribution of the barnacle *Chtamalus stellatus*. *Ecology*, 42, 710-723.

Fellows, D. P. and Heed, W. B. (1972). Factors affecting host plant selection in desert adapted cactiphilic *Drosophila*. *Ecology*, 53, 850-858.

Lawlor, L. and Maynard Smith, J. (1976). The coevolution and stability of competing species. *American Naturalist*, 110, 79-99.

Levins, R. (1962). Theory of fitness in a heterogeneous environment. I. The fitness set and adaptive function. *American Naturalist*, 96, 361-373.

MacArthur, R. H. (1955). Fluctuations of animal population and a measure of community stability. *Ecology*, 36, 533-536.

MacArthur, R. H. and Levins, R. (1964). Competition, habitat selection and character displacement in a patchy environment. *Proceedings of the National Academy of Sciences, USA*, 51, 1207-1210.

May, R. M. (1972). Will a large complex system be stable? *Nature*, 238, 413-414.

May, R. M. and MacArthur, R. H. (1972). Niche overlap as a function of environmental variability. *Proceedings of the National Academy of Sciences, USA*, 69, 1109-1113.

Pimm, S. L. (1978). An experimental approach to the effects of predictability on community structure. *American Zoologist*, 18, 797-808.

Pimm, S. L. and Rosenzweig, M. L. (in press). Competitors: how they should compete.

Remington, C. L. and Pease, R. W., Jr. (1955). Studies in foodplant specificity. I. The suitability of swamp white cedar for *Mitoura gryneus* (Lycaenidae). *Lepidoptera News*, 9, 4-6.

Rosenzweig, M. L. (1966). Community structure in sympatric carnivora. *Journal of Mammalogy*, 47, 602-612.

Rosenzweig, M. L. (1974). On the evolution of habitat selection. *Proceedings of the First International Congress of Ecology*, 401-404.

Rosenzweig, M. L. (1979). Optimal habitat selection in two species competitive systems. *Fortschr. Zool.* (in press).

Rosenzweig, M. L. (in press). A theory of habitat selection.

Roughgarden, J. (1976). Resource partitioning among competing species -- a coevolutionary approach. *Theoretical Population Biology*, 9, 388-424.

Schaffer, W. M. and Gadgil, M. D. (1975). Selection for optimal life histories in plants. In *Ecology and Evolution of Communities*, M. Cody and J. Diamond, eds. Belknap Press of Harvard University Press, Cambridge, Massachusetts. 142-215.

Schaffer, W. M. and Schaffer, M. V. (in press). The adaptive significance of variations in reproductive habit in the agavaceae. II. Pollinator foraging behavior and selection for increased reproductive expenditure. *Ecology*.

Schroder, G. D. and Rosenzweig, M. L. (1975). Perturbation analysis of competition and overlap in habitat utilization between *Dipodomys ordii* and *Dipodomys merriami*. *Oecologia*, 19, 9-28.

Slobodkin, L. B. (1961). *Growth and Regulation of Animal Populations*. Holt, Rinehart, and Winston, New York.

Watt, K. E. F. (1973). *Principles of Environmental Science*. McGraw Hill, New York.

Zwolfer, H. (1974). Das treffpunkt-prinzip als kommunikationsstrategie und isolationmechanismus bei bohrfliegen (Diptera: Trypetidae). *Entom. German*, 1, 11-20.

[*Received* *July* 1977. *Revised* *April* 1979]

G. P. Patil and M. Rosenzweig, (eds.),
Contemporary Quantitative Ecology and Related Ecometrics, pp. 61-88.

STATISTICAL ANALYSIS OF THE ZOOPLANKTON SPECIES DIVERSITY OF LAKES PLACED ALONG A GRADIENT

O. ROSSI, G. GIAVELLI, A. MORONI, and E. SIRI

Institute of Ecology
University of Parma
Parma, Italy

SUMMARY. The main geomorphological and biological characteristics of 250 Italian lakes are taken into account in the present study. Each lake is defined by a vector whose 28 elements represent the main geomorphological characteristics of the lake while the ecological characteristics are represented by the zooplankton species diversity and its components (species richness and species evenness). A principal component analysis carried out on the geomorphological characteristics divides the lakes into three subgroups whose trophism is tendentially different. A statistical analysis carried out on the diversity indices confirms the existence of three subgroups of lakes so characterized: a first subgroup, of tendentially oligotrophic lakes located in high mountains, which exhibit low values of diversity because of the severe environmental conditions; a second subgroup, of tendentially eutrophic lakes located on the plain, which exhibit low diversity because of the effect of human influence; a third subgroup of tendentially mesotrophic lakes, present a greater diversity than the other two subgroups because they escape the influence of both climate and human activity. These results seem to confirm the usefulness of species diversity as a bioindicator.

KEY WORDS. zooplanktonic diversity, environmental gradient, trophic state, principal component analysis.

1. INTRODUCTION

Species diversity is a measurable characteristic of a natural community. The indices of diversity proposed by ecologists, since the first, the Fisher-Williams' index (Fisher, Corbet, and Williams, 1943), have become very numerous. Patil and Taillie (1976) made clear the common conceptual and mathematical foundations of the more frequently used diversity indices: the species richness index, the Shannon-Weaver index, and the Simpson index.

Not all of the ecologists interested in biomonitoring agree about the importance of species diversity as a bioindicator. Two kinds of problems seem relevant to this subject: 1) the high cost of species classification; 2) the degree of sensitivity of species diversity as an environmental indicator.

The first problem is of a practical nature, but the science and technology of pattern recognition can given decisive help in recognizing automatically most of the species present in a given environment.

The second problem encompasses an important and general concept. Some ecologists utilize the presence or the absence of a particular species as an indicator in biomonitoring. The presence of an individual species in an environment implies that the minimal conditions for survival of that species are satisfied, but not much more. But really, we need an alarm signal. The absence of an individual species seems more helpful but it presents some difficulty of interpretation because the absence may be due to reasons other than the presence of pollutants in the environment. Often those species which are less abundant in the community are also more sensitive to pollutants. One of these sensitive but rare species can be useful indicator but its absence from the sample may be due to the failure in collecting the species because of its relative rarity.

The simultaneous absence of many common species in an environment is more informative because it is the expression of a great alteration of community structure not only in terms of absence of those species but, especially, in terms of alteration of the abundances of the other species which can survive. Species diversity indices seem to be a useful means to describe community structure and its alteration concerning both the species richness and the abundances of individual species. From a general point of view an index is not only a helpful scientific means of observing trends and analyzing programs but also a good way of informing people of important events in a very simple way. A set of careful measurements of some important environmental parameters provides a decision maker with a large amount of valuable data. But be useful for evaluation and for decision, these data must aggregated in a meaningful way because the decision maker must arrive at a practical decision in a very short time. Dennis and Patil (1976) proposed the diversity index as an alarm test in water biomonitoring.

After the alarm, the ecologist can investigate further in order to identify the real causes of the alarm.

Is the species diversity index, as an alarm test, sensitive to the environmental changes due to pollutants? This is the first and real question which we must examine.

The feasibility of utilizing zooplankton in biological monitoring has received little attention. In the past most studies have concentrated on diatoms and macroinvertebrates. But Kochsiek, Wilhm and Morrison (1971) considered the possibility of using zooplankton species as an indicator of water pollution. The zooplankton data analyzed in the present study are from a large sample of lakes located along a gradient. The gradient has two important (and related) variables: altitude, and the degree of intensity of nearby agricultural exploitation. High mountainous areas are sparsely populated. Lower plains regions, on the other hand, are crowded areas characterized by modern, intensive, and productive agriculture in Italy.

The aims of the present analysis are: 1) to study the behavior of zooplanktonic diversity along this type of gradient; 2) to evaluate the sensitivity of zooplanktonic diversity and especially its components (species richness and species evenness) as a means of detecting changes along the gradient.

2. THE DATA COLLECTION

The data utilized in the present study come from a sample of 250 small lakes scattered in the mountains and plains in northern Italy. The lakes are comparable because they belong to the same geographical area along the north side of the Apennine Mountains. The specific region concerned extends from west to south-east for about 350 km, from the Tirreno Sea to the Adriatic. The north side of the Appenine Mountains is furrowed by narrow valleys in which the studied lakes are located.

The 250 small lakes analyzed are only a part of a larger sample of lakes in the same geographical area which were studied from 1965 to 1971. Unfortunately, for several reasons, not all the data collected then are available for study and comparison now.

The lakes located in high mountains were visited only a few times and always in the summer season. For each lake studied, the main geomorphological characteristics were surveyed. Many plankton samples were collected from the lakes during the year, and for each sampling the zooplankton species composition was recorded.

In the present study only zooplankton data collected from the 250 lakes during the summer are used. The zooplankton samples were collected between July 10 and August 20. The lakes in high mountains are included in this sample.

It is known that the seasonal fluctuation of species richness in all the Apennine Mountain lakes reaches its maximum value in July. The maximum value is maintained until September (Rossi 1971; Rossi and Baroni 1972; Rossi and Moroni, 1975).

In each lake the zooplankton samples were collected from a point corresponding to the maximum depth of the lake. The zooplankton species present in the sample of 250 lakes analyzed in the present study are 148 and they belong to three taxonomic groups: Rotifera, Copepoda, Cladocera.

One can find many other details about the data collection in these lakes in Moroni, Ghetti and Rossi (1971).

3. PRELIMINARY STATISTICAL ANALYSES OF THE GEOMORPHOLOGICAL FEATURES AND OF SOME SEMIQUANTITATIVE BIOLOGICAL FEATURES OF THE LAKES

The main geomorphological features of the 250 lakes are shown in Table 1. In the same table some simple descriptive or semi-quantitative biological characteristics of the lakes are also given: presence/absence of fish, presence/absence of macrophytes and their concentration (or intensity).

The quantitative measures of the geomorphological and biological characteristics of the same lake i are included in a vector of 28 elements:

$$[x_{i1}, x_{i2}, \ldots, x_{i28}].$$

The variables from x_{i1} to x_{i5} (Table 1) are continuous while those from x_{i6} to x_{i27} can assume only two values: the value (score) 1 when the lake presents a prefixed characteristic and the value (score) 0 when this characteristic is absent. The variable x_{i28}, presence/absence of macrophytes and degree of this presence, has been semi-quantified in the following way:

- for the absence of macrophytes: score 0;
- for the presence of macrophytes only on the lake shores: score 1;

TABLE 1: Eigenvector of the second principal component extracted from the correlation matrix.

	Lake Variables	Eigenvector
(x_1)	Lake altitude	-0.31
(x_2)	Lake length	-0.21
(x_3)	Lake width	-0.20
(x_4)	Lake perimeter	-0.22
(x_5)	Lake mean depth	-0.21
(x_6)	Lake is of tettonic-glacial origin	-0.28
(x_7)	Origin of the lake is by landslide	0.15
(x_8)	Lake is man-made	0.14
(x_9)	Basin drainage is rocky	0.09
(x_{10})	Basin drainage is woody	-0.17
(x_{11})	Basin drainage is with grassland	-0.06
(x_{12})	Basin drainage is with tillage	0.31
(x_{13})	Lake shores are rocky	-0.19
(x_{14})	Lake shores are woody	-0.12
(x_{15})	Lake shores are with grassland	0.03
(x_{16})	Lake shores are with tillage	0.27
(x_{17})	Lake is placed on gravelly of abnormal ground	0.22
(x_{18})	Lake is placed on moth flysch	0.13
(x_{19})	Lake is placed on sandstone	-0.33
(x_{20})	Presence of a perennial effluent	-0.18
(x_{21})	Presence of a seasonal effluent	0.12
(x_{22})	Presence of more than one effluent	0.07
(x_{23})	Presence of a perrennial effluent	-0.16
(x_{24})	Presence of a seasonal effluent	0.10
(x_{25})	Presence of more than one effluent	0.05
(x_{26})	Presence of an ice coverlet from October to April	-0.19
(x_{27})	Presence of fishes	-0.09
(x_{28})	Abundance of macrophites	0.14

- for the presence of macrophytes on the lake shores and, also, by sparse mottles on the lake surface: score 2;

- for the presence of macrophytes that panel the greatest part of the lake surface: score 3.

These scores were assigned to all 250 lakes by the same experimenter.

In this way a large sample of $n = 250$ independent observation vectors is available and each vector is composed of $p = 28$ elements. The vectors can be assembled in the matrix:

$$\underset{\sim}{X} = \begin{bmatrix} x_{1,1} & \cdots & x_{1,p} \\ \vdots & & \\ x_{n,1} & \cdots & x_{n,p} \end{bmatrix}$$

After transformation of the original measurements x_{ij} to standard scores $z_{ij} = (x_{ij} - \bar{x}_j)/s_j$ ($\bar{x}_j$ and s_j represent the sample measures of, respectively, the mean and the standard deviation of the variable j), a principal component analysis was carried out on the correlation matrix. This multivariate statistical method is a useful descriptive technique for examining the structure of the correlation matrix and for revealing the probable presence of some subgroups within the group of the 250 lakes. Principal component analysis instead of cluster analysis was used because the former takes into account the correlations among the geomorphological characteristics of the lakes in an explicit way. Some of the main results of this statistical analysis are summarized in Table 1.

The 250 lakes are plotted in Figure 1 according to their scores along the first y_1 and the second y_2 principal components. Inspection of the scatter of points (lakes) along y_1 does not suggest the presence of subgroups. On the other hand the scattering along y_2 does suggest the presence of some clusters of points (lakes). In fact if one collects the number of lakes falling in a fixed class interval along y_2 one obtains a histogram of frequencies which clearly reveals the presence of three lake subgroups (Figure 2). In order to interpret this graphic result it is necessary to examine carefully the algebraic sign and the magnitude of the coefficients of y_2 (Table 1).

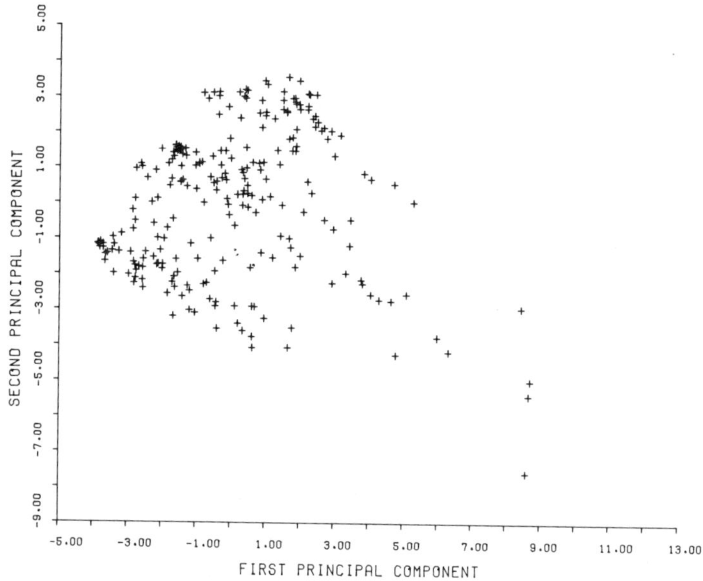

FIG. 1: Map of the 250 lakes. The two coordinates are the first and second principal components.

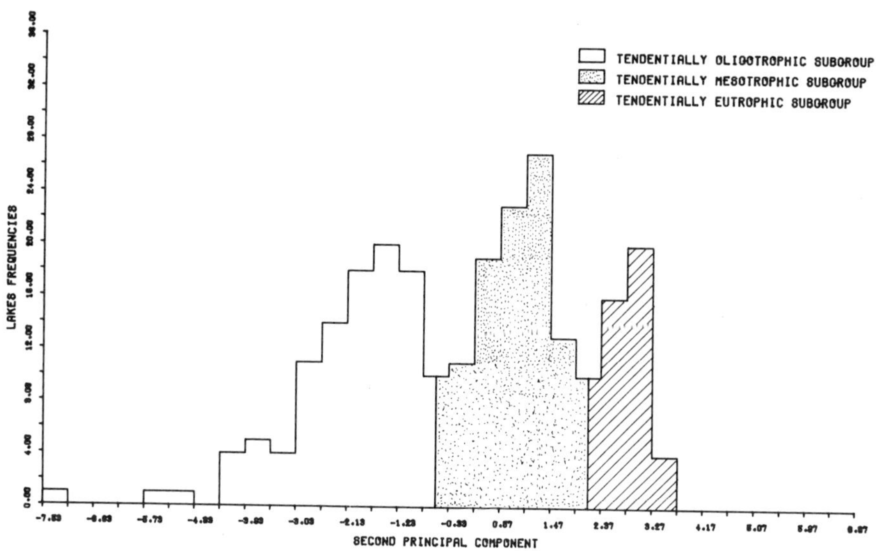

FIG 2: Distribution of 250 lakes along the second principal component.

The subgroup of lakes to the left in Figure 2 whose y_2 scores run from approximately -7.5 to -0.8 shows the influence of the negative coefficients of y_2 more. In other words, the lakes to the left in Figure 2 tend to be those from high mountainous regions (x_1). Their mean altitude is 1306 m. These lakes also tend to be deeper than others (x_5), having a mean depth of 7.27 m. The mountain lakes are of glacial-tettonic origin (x_6). They have drainage basins full of rich woods and grasslands (x_{10}, x_{11}); and have rocky shores (x_{13}). The subgroup of the lakes on the right of Figure 2, whose y_2 scores are positive (from approximately 2.4 to 3.7), show the influence of the positive coefficients of y_2 more. Consequently, they exhibit features which are opposed to those of the subgroup on the left of Figure 2. In other words this second subgroup of lakes has a mean altitude of 315 m. and a mean depth of 3.02 m. They are of artificial origin (x_8), and drain not from woods but from lands of rich ploughed land (x_{12}). Tillage may occur quite near the shore (x_{16}). These lakes show a greater abundance of macrophytes than do the lakes of the subgroups on the right of Figure 2 (mean values: 2.21 vs. 1.50). The interpretation suggested by these observations is as follows: the y_2 scores are, in some sense, correlated with the trophic state of the 250 lakes. The lakes on the left of Figure 2 present, very probably, a lesser degree of trophism than those on the right of the same figure. According to their y_2 scores, the 250 lakes can be divided in three subgroups, whose trophism is tendentially different. It is also necessary to assign to a subgroup the lakes whose y_2 scores are contained in the class intervals (-0.8, -0.3) and (1.9, 2.4).

Each of these lakes has been associated with a number from a random number table: if this random number is even, the lake is assigned to a prefixed trophic group; if the random number is odd, the lake is assigned to the other group. The number of the lakes assigned in this way is small: 10 for each of the above two class intervals. Thus the sample of $n = 250$ lakes has been subdivided into three subgroups whose size n_i are as follows:

1. Tendentially oligotrophic subgroup n_1 (size) = 102

   $(-7.5 \leq Y_2 \leq -0.5)$

2. Tendentially mesotrophic subgroup n_2 (size) = 103

$(-0.5 \leq Y_2 \leq 2.1)$

3. Tendentially eutrophic subgroup n_3 (size) = 45.

$(2.1 \leq Y_2 \leq 3.7)$

It seems obvious to note that this trophic classification is strictly relative to the 250 lakes examined. It seems also important to observe that the method of recording some of the binary data gives rise to a functional dependence among the correspondent elements of a data vector. Might the three clusters observed in Figure 2 be the consequence of this functional dependence? First of all the three clusters are interpretable easily *a posteriori* by the examination of the principal component coefficients. The substantial realism of the interpretation is demonstrated also in light of the subsequent results (zooplankton species diversity analysis). Secondly the binary variables which give rise to the functional dependence have been left out and a principal component analysis has been performed on the remaining variables. Using these eight variables only (29% of all the variables) a clear presence of three clusters along the *2nd* principal component scores has been obtained. These clusters are very similar to those of Figure 2: the lakes on the right tend to be eutrophic ($x_{28} = 0.29$).

4. STATISTICAL COMPARISON OF THE ZOOPLANKTONIC DIVERSITY IN THE SUBGROUPS OF LAKES.

The hypothesis suggested by the results of the principal component analysis of the preceding section can, in a sense, be tested by the zooplanktonic data collected in the same 250 lakes.

It is expected that the values of zooplanktonic diversity are not equal in the three trophic classes. The results of two independent species countings relative to two independent samplings collected from the same lake and on the same day have combined. In this way a single and more realistic measure of the zooplanktonic diversity has been obtained for each lake. The Shannon-Weaver diversity index has been used. According to this index the zooplanktonic diversity is

$$H = - \sum_{i=1}^{s} p_i \log_2 p_i,$$

where s represents the number of zooplanktonic species and p_i represents the proportion of the total individuals belonging to species i. This index is of importance when one intends to compare the species diversity of two or more ecological situations. When the sample size is greater than about 400, the variance of the index is small and practically constant (Kochsiek, Wilhm, and Morrison, 1971). This condition is always satisfied by the samplings used in the present paper. The frequency histograms of the diversity index values relative to the three trophic classes of the lakes are presented in Figure 3. It is interesting to observe that the three histograms seem close enough to the hypothesis of normal density distribution of the index. Moreover the three mean values of the diversity indices do not appear to be equal for the three trophic classes. To test this hypothesis the analysis of covariance model was used:

$$H_{ij} = \mu + \alpha_i + \beta(\alpha_{ij} - \bar{d}) + \varepsilon_{ij}, \tag{1}$$

where i $= 1,2,3,$ (trophic class);

j $= 1,2,\cdots,n_i$ (number of lakes: $n_1 = 102$; $n_2 = 103$; $n_3 = 45$);

H_{ij} : species diversity value of lake j in trophic class i;

μ : grand-mean of the H_{ij};

α_i : effect of trophic class i measured as $(\mu_i - \mu)$ where μ_i is the mean value of the class i;

β : coefficient of regression of the H_{ij} values on the covariate d_{ij} (depth of the sampling) calculated from the 'within trophic classes' of the analysis of covariance;

d_{ij} : depth of the sampling in lake j of trophic class i (covariate);

$\bar{d}$: mean value of covariate d_{ij};

ε_{ij} : random normal deviate with zero mean and variance σ^2.

If $H_{i,adj\cdot}$ represents the mean value of the diversity in the class after adjustment by the covariance analysis, the null hypothesis was tested:

$$H_0 : \mu_{1,adj\cdot} = \mu_{2,adj\cdot} = \mu_{3,adj\cdot}$$

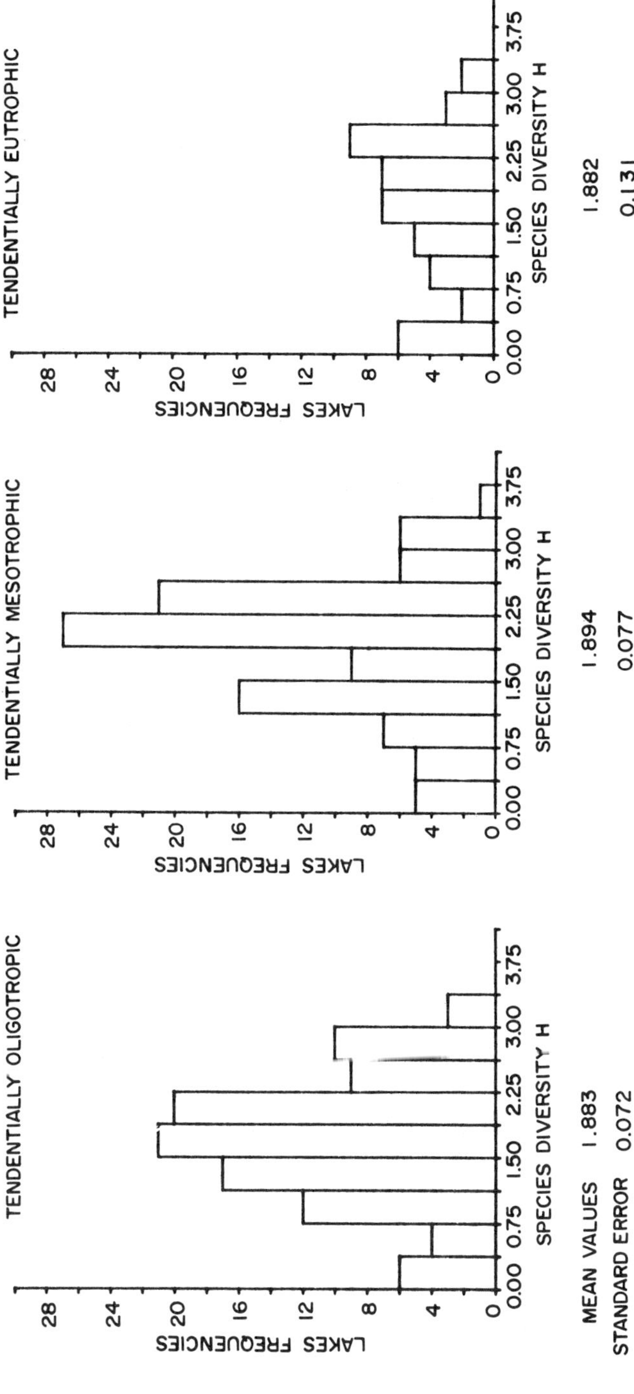

FIG. 3: Distribution cf 250 lakes according to their species diversity H *indices in three tropic subgroups.*

It is assumed that the three trophic classes are not a random sample from a population of trophic class but represent the particular trophic classes which are of interest (Model 1 of the covariance analysis). It is assumed also that from the *ith* trophic class a random sample of size n_i of lakes has been drawn from which we obtained n_i diversity index values. According to the covariance analysis model the regression coefficient β is the same for the three trophic classes (parallelism hypothesis). This hypothesis will be tested on the sample of data (Table 3). The results of covariance analysis are summarized in Table 2 and in Table 3. In Table 3 the intermediate steps necessary for the parallelism test are shown. The results of this test are that the three sample coefficients relative to the trophic subgroups can be considered three independent estimates of the same coefficient β. After this result it is possible to accept and to interpret more confidently the results of Table 2. These results are very interesting because the F value is very close to the significant level. Practically the three mean values of the zooplanktonic diversity in the three trophic classes are statistically different. The species diversity is able to detect that something is changed along the trophic gradient. In particular we observe that the oligotrophic and eutrophic lakes have practically the same mean value of zooplanktonic diversity, but their mean values are statistically lower than that of the mesotrophic lakes. The 'correction' due to the covariate has a small effect ($t = 1.79$; $0.05 < P < 0.10$). The adjusted means are close to the non-adjusted ones.

5. FURTHER ANALYSIS OF THE ZOOPLANKTONIC DIVERSITY: THE RICHNESS AND EVENNESS COMPONENTS

The species diversity of the three trophic classes has been further investigated by a separate analysis of the components of diversity: species richness and species evenness.

5.1 Species Richness. The value of species richness is $s-1$, where s represents the number of species. In practice, and especially for comparative purposes, it is necessary to take into account that $(s-1)$ is affected by the sample size. When the sample size increases, very often the value of $(s-1)$ also increases so that a useful transformation may be $(s-1)/\log_e n$, where n represents the sample size. In this way the relation between $(s-1)$ and $\log_e n$ becomes, roughly, linear. The frequency histograms of zooplanktonic species richness are presented in Figure 4.

TABLE 2: Analysis of covariance carried out on the zooplanktonic species diversity of the lakes. The covariate is the depth of the sampling from which the diversity has been calculated.

Sources of Variation	d.f.	Sum of Squares	Sum of Squares (due)	Sum of Squares (about)	Mean Squares
Between subgroups	2	2.70			
Within subgroups	247	150.32	1.94	148.38	0.60
Total	249	153.02	1.56	151.46	

Null hypothesis: No difference among mean diversities after adjustment with the covariate:

$F_{(2,246)} = 2.56 \quad (0.05 < P < 0.10)$.

	Coefficients of regression on the covariate	t	P
Between subgroups	-0.35		
Within subgroups	0.038	1.79	>0.05
Total	0.034	1.59	>0.05

Subgroup	Non-adjusted Mean	Adjusted mean	St. error adjusted
Oligotrophic	1.68	1.67	0.08
Mesotrophic	1.89	1.90	0.08
Eutrophic	1.68	1.69	0.11

TABLE 3: Parallelism test on the three 'within' regressions of the analysis of covariance (see Table 2).

	Deviation from regression		
	d.f.	Sum of Squares	Mean Square
Oligotrophic subgroup	100	53.10	0.53
Mesotrophic subgroup	101	61.06	0.60
Eutrophic subgroup	43	32.78	0.76
Sum of the single regressions	244	146.94	0.60
Pooled regression	246	148.38	0.60

Comparison for slopes: $F = \dfrac{(148.38 - 146.95)/2}{0.60} = 1.18 \quad (P > 0.05)$

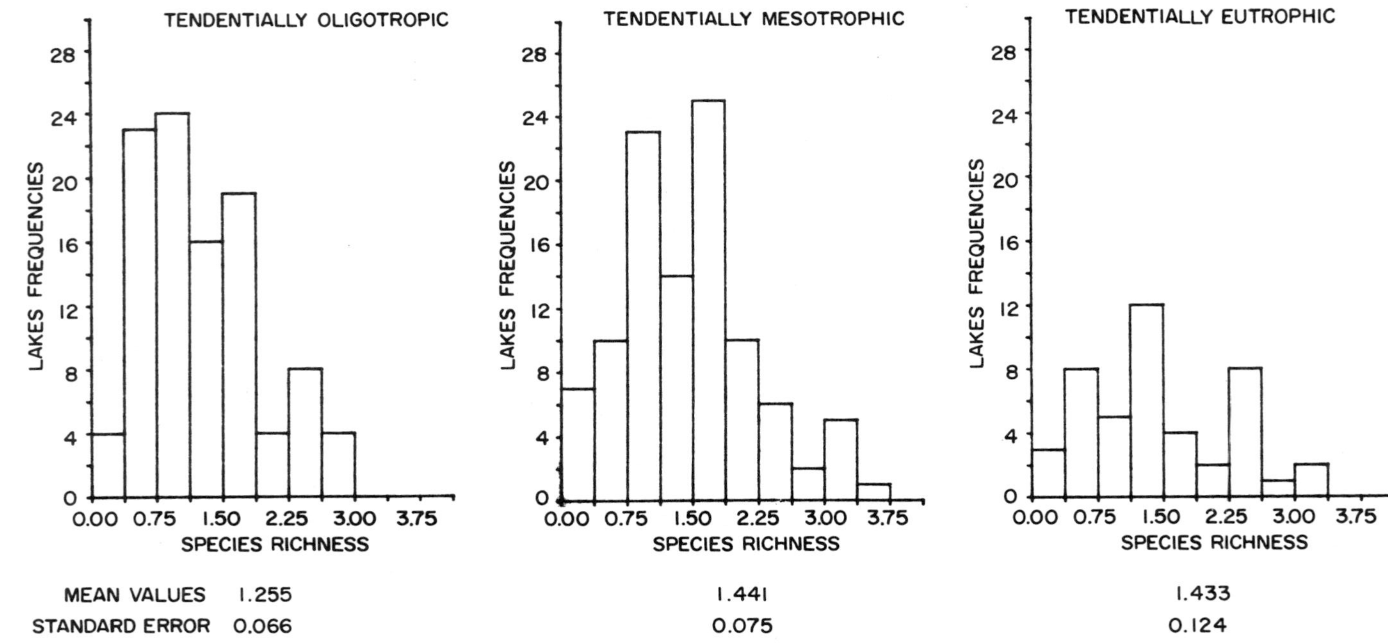

FIG. 4: Distribution of 250 lakes according to their species richness in three tropic subgroups.

The statistical model to test the hypothesis that the species richness is the same for the three trophic groups is a simple development of equation (1):

$$(s\text{-}1)_{ij} = \gamma + w_i + \beta\ (d_{ij} - \bar{d}) + \tau(\log_e n^*_{ij} - \overline{\log_e n^*}) + \varepsilon^*_{ij} \qquad (2)$$

Equation (2) represents a multiple covariance model where:

$i = 1,2,3$ (trophic class); $j = 1,2,\cdots,n_i$ (number of lakes of the *ith* class;

γ : grand-mean of the $(s\text{-}1)_{ij}$;

w_i : effect of the *ith* class measured as $w_i = (\gamma_i - \gamma)$ where γ_i represents the mean value of the richness $s\text{-}1$ of the *ith* class;

β : regression coefficient of the $(s\text{-}1)_{ij}$ on the depth d_{ij}; the depth is the first covariate of the model;

d_{ij} : depth of the sampling relative to the richness value $(s\text{-}1)_{ij}$;

$\bar{d}$: mean value of the d_{ij};

τ : regression coefficient of the $(s\text{-}1)_{ij}$ on the natural logarithm of sample size, $\log_e n^*_{ij}$;

$\log_e n^*_{ij}$: $\log_e$-size of the sample relative to the richness value $(s\text{-}1)_{ij}$;

$\overline{\log_e n^*}$: mean value of the $\log_e n^*_{ij}$;

ε^*_{ij} : random normal deviate with zero mean and variance σ^{*2}.

The null hypothesis tested is:

$$H_0 : \gamma_{1,adj\cdot} = \gamma_{2,adj\cdot} = \gamma_{3,adj\cdot},$$

where the symbol adj· represents the adjustment due to the two covariates.

It is assumed that the three trophic classes are the particular classes which are of interest and from which random samples of size n_i^* have been selected. According to the assumptions of the multiple covariance model, the values of the pair of coefficients β and τ are the same for all of the three trophic groups (parallelism test). This hypothesis will be tested on the sample data. The main results of the multiple covariance analysis are presented in Table 4 and Table 5. First of all one observes that (Table 5) the sample data agree well with the parallelism hypothesis. Secondly, the results of Table 4 yield the interesting conclusion that the three mean values of the species richness are not statistically different. In fact the comparison of the mean values adjusted with the non-adjusted ones confirms that statistical conclusion.

5.2 Evenness. Evenness is defined as the ratio of the actual value to the maximum value of the diversity (Pielou, 1969) so that:

$$e = \frac{H}{\log_2 s} \quad . \tag{3}$$

The value of the evenness in each subgroup of the lakes has been obtained by linear regression analysis of diversity H on $\log_2 s$. The regression coefficients represent the evenness.

Clearly when $s = 1$, $\log_2 s = 0$ and also H must be 0. However both the regressions, with and without the intercept, have been carried out. From the results summarized in Tables 6, 7, and 8, it appears that using a two parameter regression does not improve the fit.

In Figure 5 are represented the regression lines of H on $\log_2 s$ for the subgroups of lakes. The slope of a line represents the evenness of a subgroup. The evenness values have been compared by the simple model

$$e = e_3 + fx_1 + gx_2, \tag{4}$$

where x_1, x_2 are dummy variables which can assume only two values: 1 and 0. More precisely,

$$x_1 = \begin{cases} 1, & \text{for each } e \text{ value belonging to the mesotrophic subgroup.} \\ 0, & \text{otherwise.} \end{cases}$$

TABLE 4: Analysis of multiple covariance carried out on the zooplankton richness of the lakes. Covariates are depth of the sampling and log of the number of zooplankton animals caught.

Source of Variation	d.f.	Sum of Squares	Sum of Squares (due)	Sum of Squares (about)	Mean Squares
Between subgroups	2	45.66			
Within subgroups	247	3288.80	383.07	2905.73	11.86
Total	249	3334.46	379.53	2954.94	

Null hypothesis: No difference among mean diversities after adjustment with the two covariates:

$F_{(2.245)} = 2.074$ $(0.10 < P < 0.20)$

		Coefficient of regression on the covariates	t	P
Between	depth	-1.86		
	log sample size	-1.16		
Within	depth	0.05	<1	>0.05
	log sample size	0.72	5.67	<0.01
Total	depth	0.03	<1	>0.05
	log sample size	0.72	5.63	<0.01

Subgroup	Non-adjusted Mean	Adjusted Mean	St. error adjusted
Oligotrophic	5.98	5.95	0.34
Mesotrophic	6.78	6.88	0.34
Eutrophic	6.98	6.81	0.51

TABLE 5: Parallelism test on the three 'within' multiple regressions of the analysis of multiple covariance (see Table 4).

	Deviations from multiple regression		
	d.f.	Sum of Squares	Mean Squares
Oligotrophic subgroup	99	1006.25	10.16
Mesotrophic subgroup	100	1286.92	12.87
Eutrophic subgroup	42	602.20	14.34
Sum of the single multiple regressions	241	2895.37	12.01
Pooled multiple regression	245	2905.73	11.86

Comparison for slopes: $F = \dfrac{(2905.73 - 2895.37)/4}{11.86} < 1$ $(P > 0.05)$

$$x_2 = \begin{cases} 1, & \text{for each } e \text{ value belonging to the oligotrophic subgroup} \\ 0, & \text{otherwise.} \end{cases}$$

When $x_1 = x_2 = 0$ one obtains e_3 which represents the evenness of the eutrophic subgroup (see also Table 8). When $x_1 = 1$ and $x_2 = 0$ one obtains that f represents the difference between the evenness values of the mesotrophic and the eutrophic subgroups. Finally, when $x_1 = 0$, $x_2 = 1$ one obtains that g represents the difference between the evenness values of the oligotrophic and the eutrophic subgroups.

The estimates of e_3, f, and g have been obtained by multiple regression of the e values on x_1, x_2. The results are in Table 9. It results that the species evenness of the mesotrophic subgroup is statistically greater than that of the eutrophic subgroup (the t value is very close to the significance level), while the species evenness values of the eutrophic and oligotrophic subgroups are not statistically different.

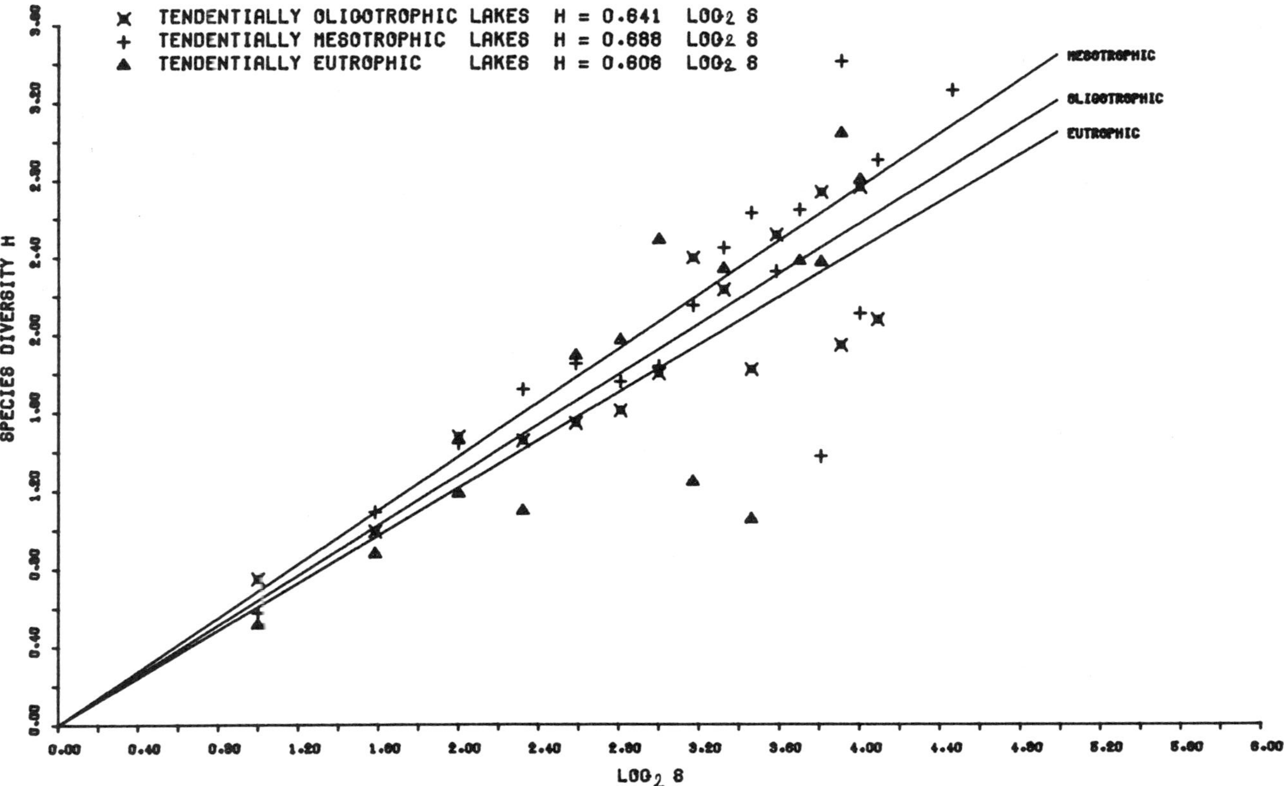

FIG. 5: Linear regression of species diversity H *on* log_2 *of the number* S *of species in the three subgroups of lakes. The slopes represent the evenesss values.*

TABLE 6: Results of the analysis of the regression of H *on* $\log_2$ s *in the oligotrophic subgroup of lakes.*

First trial: $H = e_1 \log_2 s + a_1$ (with intercept)

	Coefficient		t ratio	P
e_1 (evenness)	0.602	0.067	9.09	<0.01
a_1 (intercept)	0.112	0.210	0.88	>0.05

Analysis of Variance

	d.f.	Sum of Squares	Mean Squares
Due to regression	1	24.121	24.121
Residual	100	29.522	0.295

Second trial: $H = e_1 \log_2 s$ (without intercept)

	Coefficient		t ratio	P
e_1 (evenness)		0.019	32.66	<0.01

Analysis of Variance

	d.f.	Sum of Squares	Mean Squares
Due to regression	1	313.016	313.016
Residual	101	29.634	0.293

Comparison of the two types of regression:

$$F_{(1,000)} = \frac{(29.634 - 29.522)}{0.295} < 1 \quad (P > 0.05)$$

TABLE 7: Results of the analysis of the regression of H *on* $\log_2 s$ *in the mesotrophic subgroup of lakes.*

First trial: $H = e_2 \log_2 s + e_2$ (with intercept)

	Coefficients		t ratio	P
e_2 (evenness)	0.693	0.062	11.19	<0.01
a_2 (intercept)	-0.017	0.221	0.77	0.05

Analysis of Variance

	d.f.	Sum of Squares	Mean Squares
Due to regression	1	34.552	34.552
Residual	101	27.865	0.275

Second trial: $H = e_2 \log_2 s$ (without intercept)

	Coefficient		t ratio	P
e_2 (evenness)	0.688	0.017	38.45	<0.01

Analysis of Variance

	d.f.	Sum of Squares	Mean Squares
Due to regression	1	404.131	404.131
Residual	102	27.868	0.273

Comparison of the two types of regression:

$$F_{(1,101)} = \frac{(27.868 - 27.865)}{0.275} < 1 \quad (P > 0.05)$$

TABLE 8: Results of the analysis of the regression of H *on* $\log_2 s$ *in the eutrophic subgroup of lakes.*

First trial: $H = e_3 \log_2 s + a_3$ (with intercept)

	Coefficients		t ratio	P
e_3 (evenness)	0.658	0.118	5.58	<0.01
a_3 (intercept)	-0.152	0.342	0.36	>0.05

Analysis of Variance

	d.f.	Sum of Squares	Mean Squares
Due to regression	1	14.381	14.381
Residual	43	19.875	0.462

Second trial: $H = e_3 \log_2 s$ (without intercept)

	Coefficient	S.D.	t ratio	P
e_3 (evenness)	0.068	0.034	17.67	<0.01

Analysis of Variance

	d.f.	Sum of Squares	Mean Squares
Due to regression	1	141.669	141.669
Residual	44	19.965	0.453

Comparison of the two types of regressions:

$$F_{(1,44)} = \frac{(19.965 - 19.875)}{0.462} < 1 \quad (P > 0.05)$$

TABLE 9: Results of the analysis of multiple regression $H = e_3 + fx_1 + gx_2$ *in order to compare the evenness values of the three subgroups of lakes.*

Coefficient	Value	S.D.	t ratio	
e_3	0.608			
f	0.0802	0.0422	1.91	$0.05 < P < 0.10$
g	0.0330	0.0423	<1	$P > 0.05$

6. DISCUSSION OF THE RESULTS

6.1 The Trophic State Concept and the Multivariate Approach.

The subdivision of the original 250 lakes and the trophic interpretation of the three subgroups so obtained are based on the results of a multivariate statistical analysis. In effect, the trophic state of a lake is a multidimensional concept because it depends on the functional interrelationships among many abiotic and biotic characteristics of the lake. All these lake characteristics are, principally, related to: 1) the quality and quantity of nutrients received from the lake drainage basin; 2) the geological origin and of the lake; and 3) its morphometrics (Odum, 1971; p. 312).

Lakes are transitory features of the earth's surface. All lakes, regardless of their origin, pass through the process of ecological succession and result in terrestrial ecosystems. In the first stage of this process a lake is a water body, deep, with low concentrations of nutrients and generally low productivity. At this first successional stage the lake is called oligotrophic. Owing to the importation of allochthonous materials the water becomes richer in nutrients which stimulate the production of organic substances. This greater production of organic materials increases the sedimentation rate, thus accelerating the ecological succession. In this way, a lake becomes, gradually, eutrophic. Even though there is good agreement concerning the oligotrophic-eutrophic successional scheme, the problem of the measurement of the trophic state of a lake at a given point of time has to be considered attentively. Several authors (Brezonik and Shannon, 1971; Piwoni and Lee, 1975; Vallentyne, 1974; Boland, 1976) have discussed and summarized the principal indicators of the trophic state of a lake at a given point in time. A more recent review and a general discussion of all the principal indicators of the trophic state of a lake has been made by Rest and Lee (1978). Brezonik and Shannon (1971) were the first to use multivariate statistical techniques in order to quantify the trophic state of a lake. In all these quoted papers there is an explicit aim to propose a trophic index for general use so that one can compare, for instance, an American lake with an European one.

The second principal component, by which the trophic groups have been individualized within the original group of 250 lakes, is a linear combination of many lake characteristics closely related to the trophic state. By this linear combination it is possible to summarize in a single score the trophic state of a lake.

The quality and the quantity of the nutrients (especially nitrogen compounds and phosphates) are not present explicitly in the second component because these chemical data are not available for all the 250 lakes. However, the variables from x_9 to x_{16} and x_{28} (Table 1) are, without doubt, related closely to the quality and quantity of nutrients in the lakes.

6.2 Sensitivity of the Species Diversity Index Along the Trophic Gradient. From a general point of view it is expected that, during the natural succession, species diversity increases at first and then decreases toward the final stage of succession (Margalef, 1967; Odum, 1971).

The results of the statistical analysis of the zooplankton diversity agree with the trophic classification based on the principal component. However, the diversity values of these 250 lakes are low if they are compared with those of other Italian lakes. In one of the largest Italian lakes, Lago Maggiore, the estimated zooplankton diversity ranged from $H = 2.6$ to $H = 4.0$ (Margalef, 1963). These Apennine lakes are small and their number of species is small (Figure 4) if compared with Lago Maggiore. The power of a simple numerical index, like the Shannon-Weaver one, in distinguishing among these three different ecological situations seems interesting.

It is important to observe that the results concerning the zooplanktonic diversity obtained in the present study derive from the analysis of a very large sample. These results represent a further suggestion of the feasibility of using the diversity index as an alarm test in biomonitoring. Utilizing the tables published by Cohen (1977), it is also possible to calculate quickly the power $(1 - \beta)$ of the statistical test used in the analysis of covariance carried out on the zooplankton diversity. As a result, the statistical power is $(1 - \beta) \simeq 0.70$ at $\alpha = 0.10$. This result requires comment; under the alternative hypothesis H_1

(so that the three mean species diversities are really different) and using large samples like the present one, 30 times out of 100 these differences of species diversity would not be detected by the statistical significance test at $\alpha = 0.10$. This conclusion stresses the importance of the sample size in evaluating the sensitivity of the species diversity index. This aspect has been neglected by all the authors who tested the sensitivity of the diversity index to water pollutants. If, for instance, a sample of 120 lakes had been used instead of 250, all other things being equal, the value of $(1 - \beta)$ would be $\simeq 0.42$. Consequently, 58 times out of 100 the real differences would not have been detected by the statistical test using this sample size.

The subgroup of tendentially oligotrophic lakes presents a low mean value of zooplanktonic species diversity probably because it is composed of lakes in their early stage of ecological succession. It is likely that the adverse environmental conditions of the mountains prevent them from becoming mature more quickly.

In the subgroup of tendentially eutrophic lakes, the zooplanktonic species diversity is low and not different from the oligotrophic group's one, but it is significantly lower than that of the mesotrophic group. This result is interesting because these 45 lakes are located on low hills and on a plain where there is intensive and very modern agriculture. Very probably the excessive input of nutrients from the agricultural land into these lakes tends to simplify the structure of the zooplanktonic communities.

The 103 tendentially mesotrophic lakes present a zooplanktonic species diversity higher than that of the other two subgroups because they escape both from the adverse conditions of the high mountains and from excessive human influence.

6.3 The Components of Diversity Along the Trophic Gradient. From the analysis of the components of zooplanktonic diversity, it follows that richness is not able to discriminate among the three subgroups of lakes. The three mean values of richness are not statistically significant at the conventional level of $\alpha = 0.05$ but the following observations seem indicative (Figure 4): passing from the oligotrophic lakes of the mountains to the mesotrophic lakes, the mean richness is growing, as expected; passing from the mesotrophic lakes to the lakes of the plain, where the general climatic conditions are even better, the mean richness decreases, probably because these lakes demonstrate the effects of the agricultural lands around them.

The results of the analysis of evenness values agree perfectly with the diversity index H's ones. The interpretation of the low values of evenness in the lakes of the plain seems to be easy. Probably the excess of nutrients which arrives at the plains lakes from their drainage basins stimulates the growing of many zooplanktonic species. Not all the zooplanktonic species respond equally to this stimulus. First of all some substances which arrive together with the nutrients may be toxic or lethal for some species. Secondly each one of the other species responds to the excess of nutrients in different ways according to its rate of increase. Consequently the excess of nutrients is followed by a decrease of the evenness value.

In summary the species evenness index seems very sensitive to the changes along the trophic gradient and it might be a good candidate as an environmental indicator.

ACKNOWLEDGEMENTS

This research was supported by the Programma Finalizzato "Promozione della Qualità dell'Ambiente", N.R.C. (Italy) (Contract number: 7600948-90).

REFERENCES

Boland, D. H. P. (1976). Trophic classification of lakes using LANDSTAT-1 (ERTS-1) Multispectral Scanner data. Ecological Research Series.

Brezonik, P. L. and Shannon, E. E. (1971). Trophic state of lakes in north central Florida. Publication No. 13, Florida Water Resources Research Center. University of Florida, Gainsville.

Carlson, R. E. (1974). A trophic state index for lakes. Contribution No. 14, Limnological Research Center. University of Minnesota, Minneapolis.

Cohen, J. (1977). *Statistical Power Analysis for the Behavioral Sciences*, Revised edition. Academic Press, New York.

Dennis, B. and Patil, G. P. (1977). The use of community diversity for monitoring trends in water pollution impacts. *Tropical Ecology*, 18, 36-51.

Fisher, R. A., Corbet, A. S., and Williams, C. B. (1943). The relation between the number of species and the number of individuals in a random sample of an animal population. *Journal of Animal Ecology*, 12, 42-58.

Kochsiek, K. A., Wilhm, J. L., and Morrison, R. (1971). Species diversity of net zooplankton and physiochemical conditions in Keystone Reservoir, Oklahoma. *Ecology*, 52, 1119-1125.

Margalef, R. (1962). Diversità dello zooplankton nel Lago Maggiore. *Memorie Istituto Italiano di Idrobiologia*, 15, 137-151.

Margalef, R. (1968). *Perspectives in Ecological Theory*. The University of Chicago Press, Chicago.

Moroni, A., Ghetti, P. F., and Rossi, O. (1971). Prospettive del censimento degli ecosistemi lacustri del versante Nord dell'Appennino Settentrionale. *Atti 1° Simposio Nazionale sulla Conservazione della Natura*, 2, 171-180.

Moroni, A., Ferrari, I., and Rossi, O. (1973). Il Lago Santo Parmense: note di fisiografia e dinamica dei popolamenti mesoplantonici. *Bollettino Pesca, Piscicoltura e Idrobiologia*, 28, 5-43.

Odum, E. P. (1971). *Fundamentals of Ecology*, third edition. Saunders, Philadelphia.

Patil, G. P. and Taillie, C. (1976). Ecological diversity: concepts, indices and applications. *Proceedings of the 9th International Biometric Conference*, Boston, August 22-27.

Piwoni, M. D. and Lee, G. F. (1975). Report of nutrient load-eutrophication response of selected South Central Wisconsin impoundments. EPA Report. Environmental Research Laboratory, Corvallis, Oregon.

Rest, W. and Lee, F. G. (1978). Summary analysis of the North American (U.S. portion) O.E.C.D. Eutrophication Project: nutrient loading-lake response relationships and trophic state indices. EPA Report. Environmental Research Laboratory, Corvallis, Oregon.

Rossi, O. (1971). Prime analisi degli aspetti dinamici delle comunità zooplantoniche del Lago Santo Parmense. *Bollettino Zoologia*, 38, 558-559.

Rossi, O. and Baroni, A. (1972). Prime indagini sulla dinamica degli ecosistemi lacustri del versante Nord dell' Appennine Settentrionale. *Bollettino Zoologia,* 39, 50-51.

Rossi, O. and Moroni, A. (1975). Ecologia quantitativa di 343 biotopi lacustri (Appennino Settentrionale). *Bollettino Zoologia,* 42, 20-21.

Vallentyne, J. R. (1974). The algal bowl. Special publication No. 22. Department of the Environment, Fisheries, and Marine Service, Ottawa, Ontario.

[*Received July* 1978. *Revised January* 1979]

G. P. Patil and M. Rosenzweig, (eds.),
Contemporary Quantitative Ecology and Related Ecometrics, pp. 89-108.

ANOTHER MODEL OF THE SPECIES RANK-ABUNDANCE RELATION FOR A DELIMITED COMMUNITY

SHIRO KOBAYASHI

Faculty of Agriculture
Yamagata University
Tsuruoka, Japan

SUMMARY. On the basis of a species-area relation, a mathematical model which describes the species rank-abundance relation for a delimited community is developed as an alternative to that for a community in an open habitat. The model equation includes three parameters each representing the species equitability, the sample size, and the total number of species present in a habitat. The first parameter is related to the specific diversity of the corresponding species-area curve which would be expected if the spatial distribution of individuals is uniform. Numerical simulations show that a species rank-abundance curve given by this model results in different species-area curves according to the spatial distribution of individuals. When the spatial distribution is strongly contagious, the corresponding species-area curve approaches a form expected for a community in an open habitat. This implies that a species richness can be expressed as a function of the species equitability, the spatial distribution, and the sample size. Application of the model to some field data including the birds in Quaker Run Valley (Saunders, 1936) and the moths in a light trap (Williams, 1943) results in satisfactory agreement.

KEY WORDS. species abundance, species diversity, species equitability, species-area curve.

1. INTRODUCTION

In an earlier paper (Kobayashi, 1976) it has been shown that the species-area relation for a delimited community is well described by

$$S = T[1 - (1 + x/E)^{-A}] \tag{1}$$

where S is the number of species encountered in an area x, T is the total number of species present in a habitat, E is a constant called the *elemental area*, and A is a constant which can be written in terms of the *specific diversity* (λ) as

$$A = \ln[T/(T - \lambda)] \tag{2}$$

From the premises of this modeling, equation (1) is expected to be effective in a community which consists of the species belonging to a synusia.

The purpose of this paper is to derive a model equation of the species rank-abundance relation from (1). The modeling procedure used here is virtually identical with that given for a community in an open habitat (Kobayashi, 1977).

2. MODELING

When the spatial distribution of individuals is uniform, the number of species (S) occurring in a sample with N individuals is given by

$$S = T[1 - (1 + h'N)^{-A}] \tag{3}$$

where $$h' = [T/(T - 1)]^{1/A} - 1 \tag{4}$$

(Kobayashi, 1976). (In this paper some constants are denoted by letters with a prime to avoid confusion with those given in Kobayashi, 1976). This gives

$$N = [\{T/(T - S)\}^{1/A} - 1]/h' \tag{5}$$

Let r*th* ($r \leq S$) abundant species be represented by n_r individuals in a sample with N. If the number of individuals of the r*th* species decreases to z' as the sample size is reduced to N' ($< N$) individuals belonging to r species, the following relations are expected:

$$n_r/N \simeq z'/N' \tag{6}$$

and

$$N' = [\{T/(T - r)\}^{1/A} - 1]/h' \quad . \tag{7}$$

Relations (5), (6), and (7) give

$$n_r \simeq z'N/N'$$
$$= z'[\{T/(T-S)\}^{1/A} - 1]/[\{T/(T-r)\}^{1/A} - 1] \quad . \quad (8)$$

Here writing

$$\phi' = z'[\{T/(T - S)\}^{1/A} - 1] \quad ,$$

we have

$$N = \sum_{i=1}^{S} n_i \simeq \sum_{i=1}^{S} [\phi'/\{(T/(T - i))^{1/A} - 1\}] \quad ,$$

and hence

$$\phi' = N/\sum_{i=1}^{S} [\{T/(T - i)\}^{1/A} - 1]^{-1} \quad . \quad (9)$$

Then equation (8) becomes

$$n_r \simeq \phi'/[\{T/(T - r)\}^{1/A} - 1]$$
$$= N/[\{T/(T-r)\}^{1/A} - 1] \sum_{i=1}^{S} [\{T/(T-i)\}^{1/A} - 1]^{-1} \quad (10)$$

which is a model of r-n_r relation. It is noted that the r-n_r equation depends only on T, N, and S because A is given by (3). As will be shown later, however, the value of A in an r-n_r curve is larger than that in the corresponding S-N curve even if the spatial distribution is uniform. Therefore the following model will be effective in general.

When the spatial distribution of the individuals of each species is not uniform, suppose the observed species-area relation can be described by (1). Suppose also that if the spatial distribution were uniform, the same community could be described by

$$S(U) = T[1 - \{1 + x/E(U)\}^{-A(U)}] \quad (11)$$

where S(U) is the number of species which would be encountered in an area x if individuals were distributed uniformly, A(U) and E(U) are respectively the values of A and E expected in the uniform distribution. Putting S(U)=1 in (11), we have a specific area (i.e., an area in which one species occurs on an average) $h'(U)E(U)$, where

$$h'(U) = [T/(T - 1)]^{1/A(U)} - 1 \quad . \tag{12}$$

Since $h'(U)E(U)=x/N$, equation (11) becomes

$$S(U) = T[1 - \{1 + h'(U)N\}^{-A(U)}] \tag{13}$$

whence

$$N = [\{T/(T - S(U))\}^{1/A(U)} - 1]/h'(U)$$

$$\text{and} \qquad N' = [\{T/(T - r)\}^{1/A(U)} - 1]/h'(U) \tag{14}$$

where N' is the same as in (6). Relations (6) and (14) give

$$\begin{aligned} n_r &\simeq z'N/N' \\ &= z'[\{T/(T-S(U))\}^{1/A(U)} - 1]/[\{T/(T-r)\}^{1/A(U)} - 1] \quad . \end{aligned}$$

Writing

$$\phi'(U) = z'[\{T/(T - S(U))\}^{1/A(U)} - 1]$$

we have

$$N = \sum_{i=1}^{S(U)} n_i \simeq \sum_{i=1}^{S(U)} [\phi'(U)/\{(T/(T-i))^{1/A(U)} - 1\}]$$

whence

$$\phi'(U) = N/\sum_{i=1}^{S(U)} [\{T/(T - i)\}^{1/A(U)} - 1]^{-1} \quad . \tag{15}$$

Then the expected abundance is given by

$$\begin{aligned} n_r &\simeq \phi'(U)/[\{T/(T - r)\}^{1/A(U)} - 1] \\ &= N/[\{T/(T-r)\}^{1/A(U)} - 1] \sum_{i=1}^{S(U)} [\{T/(T-i)\}^{1/A(U)} - 1]^{-1} \end{aligned} \tag{16}$$

which is a model equation expected to be effective generally for a delimited community. It is noted that the r-n_r curve is determined by T, N, and $A(U)$ because $S(U)$ is given by (13).

If we write $h'E/h'(U)E(U) = G'$, equation (1) which fits an observed species-area curve is rewritten as

$$S = T[1 - (1 + h'N/G')^{-A}] \tag{17}$$

because $h'E=G'x/N$. Similarly, the substitution of S for $S(U)$ in (11) yields

$$S = T[1 - \{1 + h'(U)N/g'\}^{-A(U)}] \quad , \tag{18}$$

where $g' = [\{T/(T-S(U))\}^{1/A(U)} - 1]/[\{T/(T-S)\}^{1/A(U)} - 1]$.

If $G' = 1$, the spatial distribution of individuals must be uniform. In this case, $A=A(U)$ and hence (17) corresponds with (3).

3. RELATION BETWEEN AN r-n_r CURVE AND THE CORRESPONDING S-x CURVE

Kobayashi (1977) showed that when the values of T, $A(U)$, and $\phi'(U)$ are known, the corresponding species-area curve can be determined from a rank-abundance curve by

$$S = \sum_{r=1}^{T} F_r \quad , \tag{19}$$

where F_r is the probability that r*th* species occurs in an area x and is given by one of the following expressions:

(i) If the individuals are uniformly distributed,

$$F_r = \begin{cases} 1 & (\rho_r x \geq 1) \\ \rho_r x & (\rho_r x < 1) \quad . \end{cases} \tag{20}$$

(ii) If the individuals are distributed at random,

$$F_r = 1 - \exp(-\rho_r x) \quad . \tag{21}$$

(iii) If the number of individuals in x is a negative binomial variate with parameters $\rho_r x$ and k,

$$F_r = 1 - (1 + \rho_r x/k)^{-k} \quad . \tag{22}$$

(iv) The number of individuals in x is a negative binomial variate with $\rho_r x$ and $k=0.1$ for $r=1\sim5$, $k=0.5$ for $r=6\sim10$, $k=1.0$ for $r=11\sim20$, or $k=\infty$ for $r \geq 21$. (This shows that the individuals of high ranked species are more contagiously distributed.)

(v) The number of individuals in x is a negative binomial variate with $\rho_r x$ and $k=\infty$ for $r=1\sim5$, $k=1.0$ for $r=6\sim10$, $k=0.5$ for $r=11\sim20$, or $k=0.1$ for $r \geq 21$. (This shows that the individuals of high ranked species are less contagiously distributed.)

Now ρ_r is the number of individuals of the *rth* species in a unit area and is given by

$$\rho_r = \phi'(U)/[\{T/(T - r)\}^{1/A(U)} - 1] \quad . \tag{23}$$

Let $T=100$, $A(U)=0.5$, and $\phi'(U)=0.5358$ (i.e., $N=100$ at $x=1$). As an approximation of (19), the values of $\sum_{r=1}^{99} F_r$ are calculated for various x in each of the above expressions and plotted in Figure 1. Since equation (12) gives

$$h'(U) = [100/(100 - 1)]^{1/0.5} - 1 = 0.0203 \quad ,$$

the corresponding S-N equation is

$$S = 100[1 - (1 + 0.0203N)^{-0.5}] \tag{24}$$

(Figure 1, thick solid line).

Simulation, by (20) and (23), results in a curve slightly inconsistent with (24) (Figure 1, dotted line). This curve approximately fits (18) with $T=100$, $A(U)=0.44$, and $g'=1.35$. This suggests that (10) is not effective even if in a uniform distribution. Therefore equation (16) may be in better agreement with an observation regardless of spatial distribution. If T is small, the deviation from (3) is so small that (10) may approximately fit a rank-abundance curve.

The result of (21) and (23) is also inconsistent with (24) (Figure 1, thin solid line with $k=\infty$), while that of (22) and (23) shows that the more contagiously the individuals of each species are distributed, the more gently the corresponding S-log x curve slopes. It should be noted that as k value decreases the S-N curve approaches a form given by

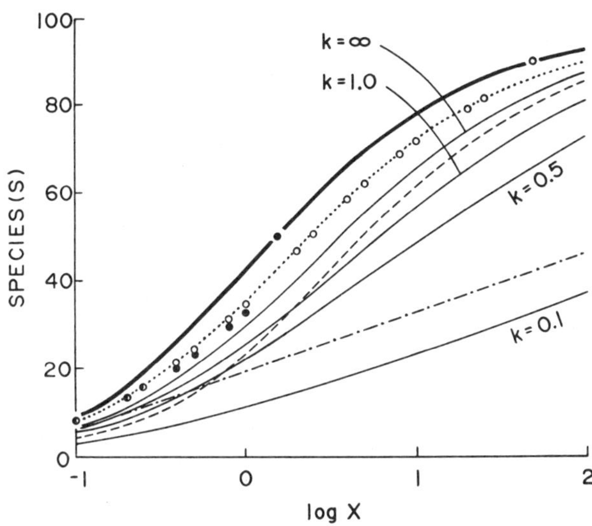

FIG. 1: Species-area curves of a model community which conforms to (23) *with* T=100, A(U)=0.5 *and* $\phi'(U)=0.5358$. *Thick solid line:* $S=100[1-(1+x/0.4925)^{-0.5}]$, *dotted line: uniform distribution of individuals, thin solid line: random or contagious distribution, broken line: more contagious distribution in high ranked species, chain line: less contagious distribution in high ranked species. Open and solid circles show* S *determined by the rarefaction method.*

$$S = \lambda \ln(1 + hN) \quad , \tag{25}$$

where $h=\exp(1/\lambda) - 1$ (see Kobayashi, 1977).

If the individuals of high ranked species are more contagiously distributed, the corresponding S-log x curve becomes more steep [i.e., $A(U) < A$] (Figure 1, broken line), while if the individuals of high ranked species are less contagiously distributed, the corresponding S-log x curve becomes more gently sloping [i.e., $A(U) > A$] (Figure 1, chain line).

Sanders (1968) showed that an S-N relation can be determined from a rank-abundance list by using the rarefaction method. Kobayashi (1977) pointed out that this method may be effective for a community in an open habitat only if the spatial distribution of individuals is uniform. When we use the rarefaction method on the basis of a rank-abundance list with N_0 individuals and S_0 species which conforms to (23) with $T=100$ and $A(U)=0.5$, the resulted S-N relation agrees with the dotted line in Figure 1 if $S_0=90$ [i.e., $\phi'(U)=25.1174$]; while it is inconsistent with the dotted line if $S_0=50$ [i.e., $\phi'(U)=0.7802$]. It may be noted that a point for N_0 and S_0 lies naturally on the thick solid line, and hence it deviates from the dotted line even though S_0 is large. If the spatial distribution is random or contagious, a point for N_0 and S_0 lies on one of the thin solid lines in Figure 1, so that it lies below the dotted line. Then we can roughly estimate the spatial distribution of individuals according to whether a point for N_0 and S_0 lies above or below the S-N curve fitted to the result of rarefaction method.

In conclusion, the following can be noted:

(i) Even if the spatial distribution is uniform, equation (10) is not effective.

(ii) To a first approximation, (16) may be applicable, because the curve resulted from (19), (20), and (23) takes a form similar to that of (24). In this case, the value of $A(U)$ is slightly larger than the value expected in a uniform distribution.

(iii) When the spatial distribution of individuals is strongly contagious, the S-N curve approaches a form given by (25). Therefore, even if an $r\text{-}n_r$ curve conforms to (16), the corresponding S-N curve tends to conform to (25).

(iv) Since an S-N curve determined by the rarefaction method disagrees with the expectation by (3) even if in a uniform distribution, the rarefaction method is not available to estimate $A(U)$ though it is useful in estimating T.

4. ESTIMATION OF PARAMETERS

To apply (16) to field data, we should estimate the values of T, $A(U)$, and $\phi'(U)$. The procedure is as follows:

(i) Unless additional information is available the value of T should be determined from a rank-abundance list. An S-N curve is determined by the rarefaction method. The resulted curve must approximately fit (18) unless the spatial distribution of individuals is strongly contagious.

(ii) Equation (18) can be rewritten as

$$\log[T/(T - S)] = A(U)\log[1 + h'(U)N/g'] \quad . \qquad (26)$$

Since 1 is negligible compared with $h'(U)N/g'$ if $N >> g'/h'(U)$, it follows that

$$\log[T/(T - S)] \simeq A(U)[\log N + \log\{h'(U)/g'\}] \quad , \qquad (27)$$

which shows that the relation between $y=\log[T/(T - S)]$ and $\log N$ is almost linear for larger N. By using the values of S determined by the rarefaction method, we can calculate y. The correlation coefficient between y and $\log N$ for some plots of larger N must show a maximum if a proper value is assigned to T. It is noted that such a small sample as $S \leq T/2$ should be tested by (31) given later, not (16), because it is impossible to estimate T in such a sample unless otherwise given.

(iii) We find T value according to the above criterion and obtain a curve which conforms to (26) by plotting y against $\log N$. The slope of the linear part of this curve gives $A(U)$ of (18) which may be slightly smaller than that of (16). Extension of this linear part cuts the abscissa (i.e., the axis of log N) at $\log[g'/h'(U)]$ (Figure 2A).

(iv) Equation (16) can be rewritten as

$$\log[T/(T - r)] = A(U)\log[1 + \phi'(U)/n_r] \quad . \qquad (28)$$

Since 1 is negligible compared with $\phi'(U)/n_r$ if $n_r << \phi'(U)$, it follows

$$\log[T/(T - r)] \simeq A(U)[\log \phi'(U) - \log n_r] \qquad (29)$$

which shows that the relation between $y'=\log[T/(T - r)]$ and $\log n_r$ is almost linear for small n_r (i.e., large r). Then by plotting y' against $\log n_r$, we find $A(U)$ value from the slope of the linear part of this curve and $\log \phi'(U)$ from the intersection of the extension of this linear part and the abscissa (i.e., the axis of $\log n_r$) (Figure 2B).

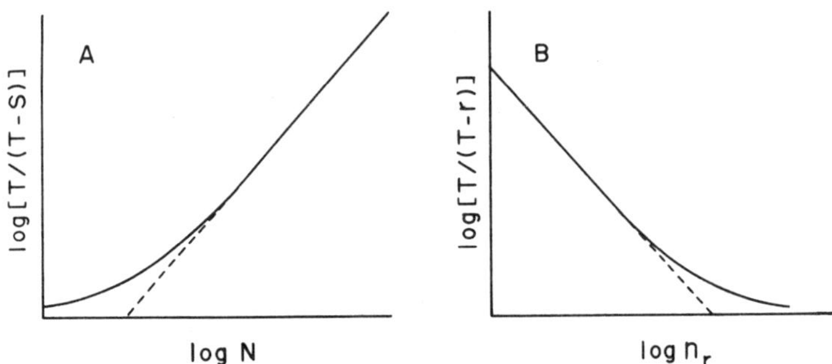

FIG. 2: A: Curve given by (26). *B: Curve given by* (28).

5. APPLICATION OF THE MODEL

5.1 Ostrea communities in the Misaki cliff. Motomura (1935) presented the rank-abundance list of *Ostrea* communities in the Misaki cliff. This list has shown the mean number of individuals of each species per square meter in each of the nine horizontal strips stratified above the low tide. Figure 3 shows the r-n_r relations in the strips IV (1.08 m above the low tide), V (1.04 m), and VI (1.00 m) where *Ostrea spinosa* dominated.

The value of T is estimated by eye to be $S+1$ in each of these strips. Since the estimated T is small, (10) may approximately fit these observations. The value of A is found by (3) with the known values of S, N, and T (=$S+1$). Then the value of ϕ' is obtained by (9). The fitted equations are

$$\text{Strip IV:}\quad n_r = 7.32/[\{14/(14 - r)\}^{1/0.90} - 1] \quad ,$$

$$\text{Strip V:}\quad n_r = 9.05/[\{14/(14 - r)\}^{1/0.85} - 1] \quad ,$$

$$\text{Strip VI:}\quad n_r = 4.56/[\{17/(17 - r)\}^{1/1.11} - 1] \quad .$$

The good agreement between the expection by (10) and the observation suggests that the spatial distribution of individuals is not far from uniform.

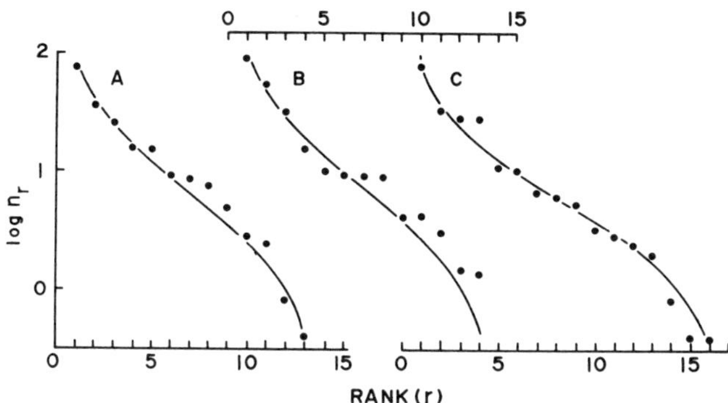

FIG. 3: Rank-abundance curves of Ostrea *communities (data from Motomura,* 1935*). Solid lines show the expectation by* (10). *A: Strip IV, B: Strip V, C: Strip VI.*

5.2 Nonpredatory mite community in soil. The rank-abundance list of a nonpredatory mite community in soil was presented by Hairston (1969). The value of T has been estimated to be 33 by Kobayashi (1976). The values of other parameters are determined to be $A(U)=0.39$ and $\phi'(U)=32$ by plotting $\log[33/(33 - r)]$ against $\log n_r$. The fitted equation is

$$n_r = 32/[\{33/(33 - r)\}^{1/0.39} - 1]$$

(Figure 4B). The S-N relation determined by the rarefaction method fits

$$S = 33[1 - (1 + 0.0867N/1.10)^{-0.37}] \quad ,$$

while the observed S-N relation fits

$$S = 33[1 - (1 + 0.0867N/1.79)^{-0.37}]$$

(Figure 4A). The observed S-x curve was given as

$$S = 33[1 - (1 + x/10)^{-0.37}]$$

(Kobayashi, 1976). The point corresponding to $N=902$ and $S=25$ lies below the curve fitted to the result of rarefaction method. This suggests that the individuals of each species are contagiously distributed.

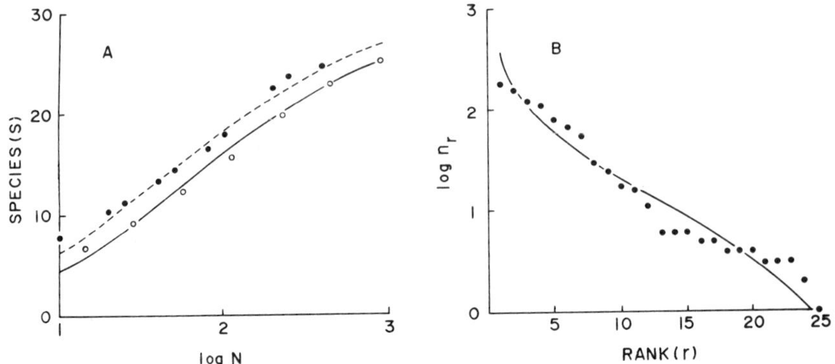

FIG. 4: A: S-N curve of a nonpredatory mite community in soil (solid line) and that determined by the rarefaction method (broken line). B: Rank-abundance curve (data from Hairston, 1969*).*

5.3 Carabid and staphylinid beetles in Finland. From the data presented by Kontkanen (1957), Community II is used for exemplifying the application of (16). The value of T was estimated to be 30 by Kobayashi (1976). By the method mentioned in the preceding section, we find $A(U)=3.00$ and $\phi'(U)=2$ graphically. The fitted equation is

$$n_r = 2/[\{30/(30 - r)\}^{1/3} - 1]$$

(Figure 5B). The S-N curve determined by the rarefaction method fits

$$S = 30[1 - (1 + 0.0287N)^{-1.2}] \quad ,$$

while the observed one fits

$$S = 30[1 - (1 + 0.0229N/2.37)^{-1.5}]$$

(Figure 5A). The observed S-x curve was

$$S = 30[1 - (1 + x/200)^{-1.5}]$$

on the assumption that a sample corresponds to $x=100$ (Kobayashi, 1976). Since the point corresponding to $N=519$ and $S=28$ lies below the curve fitted to the result of rarefaction method, the spatial distribution of individuals may be contagious.

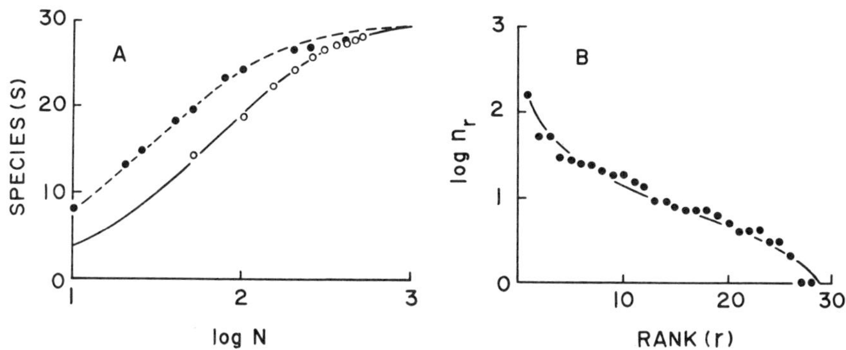

FIG. 5: A: S-N curve of a carabid and staphylinid beetle community in Finland (solid line) and that determined by the rarefaction method (broken line). B: Rank-abundance curve (data from Kontkanen, 1957*).*

5.4 The birds of Quaker Run Valley. The data presented by Saunders (1936) has been tested the fitness of a truncated lognormal (Preston, 1948) and a truncated negative binomial (Brian, 1953). If the community is delimited in terms of breeding species, the value of T may be given by the number of species which bred or have bred in the Quaker Run Valley (i.e., 90 species). Plotting the values of $\log[90/(90 - r)]$ against $\log n_r$ we find $A(U)=0.40$ and $\phi'(U)= 160$. The expected r-n_r equation is

$$n_r = 160/[\{90/(90 - r)\}^{1/0.40} - 1]$$

(Figure 6B). The S-N curve determined by the rarefaction method fits

$$S = 90[1 - (1 + 0.0283N)^{-0.40}]$$

(Figure 6A). The spatial distribution of the breeding pairs may be contagious because the point corresponding to $N=14353$ and $S=80$ lies below the expectation.

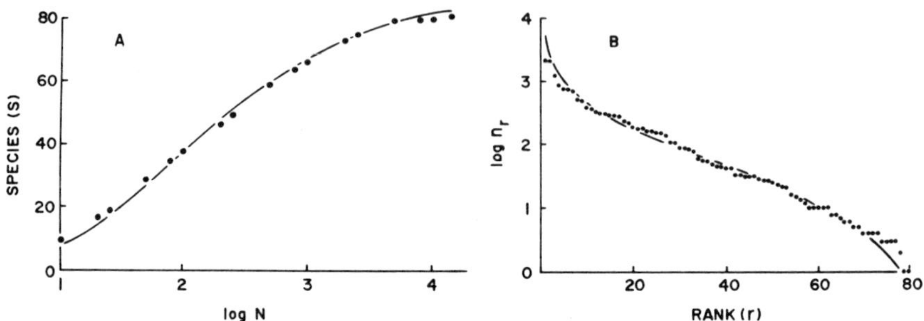

FIG. 6: A: S-N curve of the birds in Quaker Run Valley determined by the rarefaction method. B: Rank-abundance curve (data from Saunders, 1936*).*

5.5 Moths in a light trap at Rothamsted. The rank-abundance list of Macrolepidoptera captured in a light trap at Rothamsted during the four years from 1933 to 1936 was presented by Williams (1943). This was originally tested the fitness of a log series distribution (Fisher *et al.*, 1943).

Since no information is available as to the delimitation of this community, the value of T is estimated by the method given in the preceding section. When $T=276$, the correlation coefficient between $\log[276/(276 - S)]$ and $\log N$ calculated from 8 plots for $N \geq 2000$ shows a maximum by using S determined by the rarefaction method. This value is close on 273 species estimated from a truncated lognormal distribution by Bliss (1965) and Preston (1948). Then $A(U)=0.42$ is determined from the slope of the linear part of the curve obtained by plotting $\log[276/(276 - r)]$ against $\log n_r$, and $\phi'(U)=36$ is read from the intersection of the extension of this linear part and the abscissa. The fitted equation is

$$n_r = 36/[\{276/(276 - r)\}^{1/0.42} - 1]$$

(Figure 7). The S-N curve determined by the rarefaction method fits

$$S = 276[1 - (1 + 0.0091N/1.30)^{-0.40}]$$

(Figure 8). The monthly or yearly catches of each species may be distributed uniformly because the point corresponding to $N=15609$ and $S=240$ lies above the expectation.

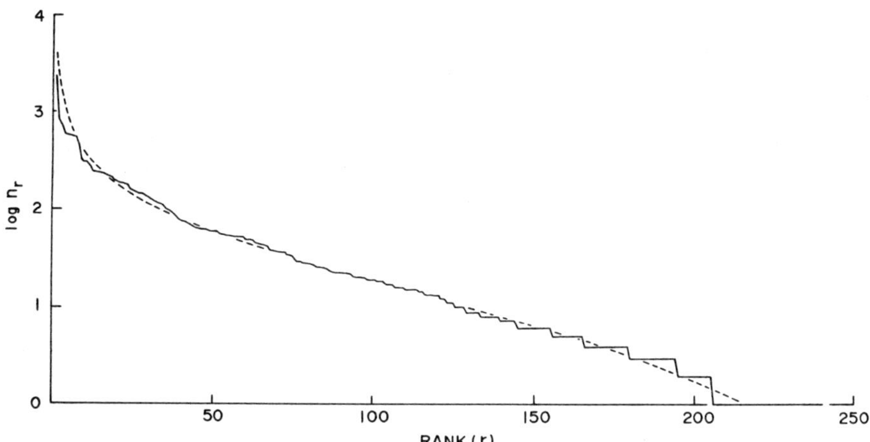

FIG. 7: Rank-abundance curve of Macrolepidoptera captured in a light trap for 4 years (data from Williams, 1943).

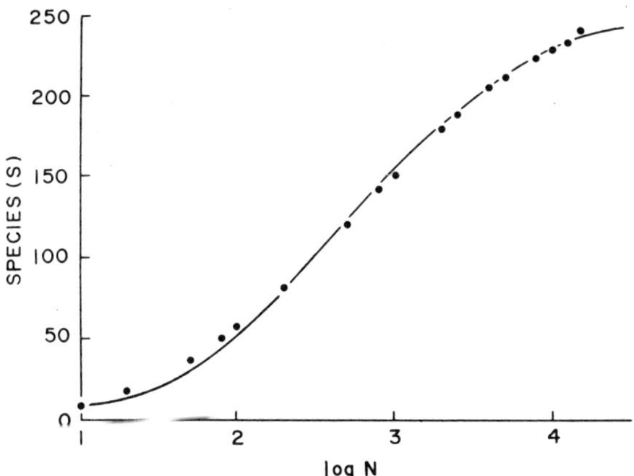

FIG. 8: S-N curve of the Macrolepidoptera determined by the rarefaction method.

6. DISCUSSION

The specific diversity λ or $\lambda(U)$ is obtained by (2) as

$$\lambda = T[1 - \exp(-A)]$$

or

$$\lambda(U) = T[1 - \exp\{-A(U)\}] \quad . \tag{30}$$

The value of $A(U)$ in an S-N curve determined by the rarefaction method is equal to the value expected in a uniform distribution unless the observed number of species is much smaller than T. Therefore $\lambda(U)$ can be estimated from the result of rarefaction method. For some communities exemplified in the preceding section, the values of $\lambda(U)$ are shown in Table 1.

On the assumption of uniform distribution, the result of numerical simulation of a community which conforms to

$$n_r = \phi(U)/[\exp\{r/\lambda(U)\} - 1] \tag{31}$$

was in satisfactory agreement with the corresponding S-N equation (25) (Kobayashi, 1977), whereas that of a community which conforms to (16) results in disagreement with the expectation by (3) as Figure 1 shows. This disagreement may mainly be due to the assumption that $n_r = \rho_r x$ in equations (20)-(22) despite the fact that the relative abundance (n_r/N) of *rth* species depends on sample size. Equations (9) and (15) show that ϕ' and $\phi'(U)$ are not directly proportional to N because

$$N/\phi' = \sum_{i=1}^{S} [\{T/(T - i)\}^{1/A} - 1]^{-1}$$

and

$$N/\phi'(U) = \sum_{i=1}^{S(U)} [\{T/(T - i)\}^{1/A(U)} - 1]^{-1}$$

are clearly dependent on sample size. From this it follows that n_r/N changes with sample size. Table 2 shows the values of n_r/N (r=1, 10, and 50) in different sizes of samples drawn from a model community which conforms to (16) with $T=100$ and $A(U)=0.5$. This inconstancy of n_r/N shows that (6) is an approximate relation so that $A(U)$ in (16) becomes larger than that in (11), though this discrepancy can be corrected by small allowances.

TABLE 1: $\lambda(U)$ values calculated from T *and* A(U) *in the S-N curve determined by the rarefaction method.*

Community	T	A(U)	λ(U)
Nonpredatory mites	33	0.37	10.21
Carabid and staphylinid beetles	30	1.20	20.96
Birds in Quaker Run Valley	90	0.40	29.67
Macrolepidoptera	276	0.40	90.99

TABLE 2: Relative abundance of rth *species (*r=1, 10, *and* 50*) in different sample sizes drawn from a model community having* T=100 *and* A(U)=0.5.

S	N	n_1/N	n_{10}/N	n_{50}/N
10	11.553	0.3543	0.0307	-
20	27.704	0.2982	0.0258	-
50	147.754	0.2601	0.0225	0.0018
80	1182.030	0.2540	0.0220	0.0017
90	4875.874	0.2537	0.0220	0.0017

As another result of the inconstancy of n_r/N, the value of A(U) in (16) is not equal to that in the S-N curve determined by the rarefaction method. Although A(U) in (16) thus differs from that expected in a uniform distribution, it may still be regarded as a parameter representing the species equitability. Since the specific diversity λ has been defined in terms of S encountered in an area (e - 1)E, λ(U) should be determined by the value of A(U) in the equation fitted to the result of rarefaction method.

The value of S can be estimated by (18) if T, A(U), N, and g' are known. Namely, the components of species diversity must be enumerated as

(i) the total number of species present in a habitat, T,

(ii) the species equitability, A(U),

(III) the species richness, S,

(iv) the sample size, N, and

(v) the departure from a uniform distribution, g'.

Each of these five items may theoretically be estimated from the others.

Kobayashi (1977) showed that z' is equal to the relative abundance of the most abundant species (n_1/N) which has been termed "junritsu" (the rate of purity) by Motomura (1943). Putting r=1 in (8), we have

$$n_1 \simeq z'[\{T/(T-S)\}^{1/A} - 1]/[\{T/(T-1)\}^{1/A} - 1] = z'N,$$

whence $z' \simeq n_1/N$. May (1975) has stated that n_1/N is superior to other dominance measures because it is simple and is not much affected by sample size.

Preston (1962) proposed that a species-area relation for isolate universes can be described by

$$S = 2.07(\rho/m)^{0.262}x^{0.262}$$

where ρ is the density of individuals per unit area and m is the minimum number of individuals which allow a species to be viable. Kobayashi (1979) points out that the above relation may be expected between T and x for isolate universes characterized by a common value of E: Putting S=T-1 in (1) we have

$$T \simeq E^{-A}x^{A} \quad .$$

If the value of A is about 0.262 in any universe, a possible explanation is that the more the distribution of abundance among species is equitable (i.e., the A(U) value in its r-n_r curve increases), the more the individuals of each species are aggregative (i.e., the A value in its S-x curve decreases). A balance of these two tendencies may stabilize the value of A.

REFERENCES

Bliss, C. I. (1965). An analysis of some insect trap records. In *Classical and Contagious Discrete Distributions*, G. P. Patil, ed. Statistical Publishing Society, Calcutta. 385-397.

Brian, M. V. (1953). Species frequencies in random samples from animal populations. *Journal of Animal Ecology*, 22, 57-64.

Fisher, R. A., Corbet, A. S., and Williams, C. B. (1943). The relation between the number of species and the number of individuals in a random sample of an animal population. *Journal of Animal Ecology*, 12, 42-58.

Hairston, N. G. (1969). On the relative abundance of species. *Ecology*, 50, 1091-1094.

Kobayashi, S. (1976). The species-area relation. III. A third model for a delimited community. *Researchers on Population Ecology*, 17, 243-254.

Kobayashi, S. (1977). A model of the species rank-abundance relation for a community in an open habitat. Paper presented at the Advanced Study Institute in Statistical Ecology, Berkeley.

Kobayashi, S. (1979). Species-area curves. In *Statistical Distributions in Ecological Work*, J. K. Ord, G. P. Patil, and C. Taillie, eds. Satellite Program in Statistical Ecology, International Co-operative Publishing House, Fairland, Maryland.

Kontkanen, P. (1957). On the delimitation of communities in research on animal biocoenotics. *Cold Spring Harbor Symposium on Quantitative Biology*, 22, 373-375.

May, R. M. (1975). Patterns of species abundance and diversity. In *Ecology and Evolution of Communities*, M. L. Cody and J. M. Diamond, eds. Belknap Press, Cambridge. 81-120.

Motomura, I. (1935). *Ostrea* communitics in the Misaki cliff. *Ecological Review* (in Japanese), 1, 55-62.

Motomura, I. (1943). A statistical method in animal synecology (Continued Report). *Ecological Review* (in Japanese), 9, 117-119.

Preston, F. W. (1948). The commonness, and rarity, of species. *Ecology*, 29, 254-283.

Preston, F. W. (1962). The canonical distribution of commonness and rarity. *Ecology*, 43, 185-215, 410-432.

Sanders, H. L. (1968). Marine benthic diversity: A comparative study. *American Naturalist,* 102, 243-282.

Saunders, A. A. (1936). Ecology of the birds of Quaker Run Valley, Allegany State Park, New York. *New York State Museum Handbook,* 16. Albany, New York. (Cited from Preston, 1948).

Williams, C. B. (1943). (See Fisher, Corbet, and Williams, 1943).

[*Received July* 1978]

G. P. Patil and M. Rosenzweig, (eds.),
Contemporary Quantitative Ecology and Related Ecometrics, pp. 109-119.

SPECIES DIVERSITY AND TURNOVER IN AN ORDOVICIAN MARINE INVERTEBRATE ASSEMBLAGE

MICHAEL L. ROSENZWEIG

J. LEE DUEK

Department of Ecology & Evolutionary Biology
University of Arizona
Tucson, Arizona 85721 USA

SUMMARY. Data on 79 species from a marine invertebrate province in the Upper Ordovician of North America are analyzed. The number of species increased to an equilibrium or quasi-equilibrium in *circa* one million years. It maintained this state for about 3 million years despite considerable turnover averaging about 15% per my. Over the 3.5my period of the equilibrium, half the steady-state diversity turned over. Probabilistic formulae are developed to correct for inaccuracies in the estimation of species' life spans, and to estimate the number of fossilizable species so far missed by the paleontologists working in this province. Approximately 20% of the fauna remains to be found, but these species should all have brief lifespans. This adds 6% to 9% to the turnover estimate and causes it to match very closely a previous estimate on the recent marine invertebrate fauna by Stanley.

KEY WORDS. paleoecology, species diversity, extinction, marine invertebrates.

1. INTRODUCTION

There are two practical ways of testing theories in evolutionary ecology. One is to cause the theories to yield predictions about patterns in time, and the other, predictions about patterns which are instantaneous (more or less). Most sciences are rather even-handed about these methods or prefer tests in time. That is because processes in physics and chemistry are so quick compared to an investigator's lifetime. The extreme case is the experimental physicist whose objects of study are sometimes processes so brief that he must expend great sums of money and develop considerable expertise to learn how to record them for analysis. Somewhere near the opposite extreme is the student of biological evolution who can hope to live long enough to see the course of evolutionary change in only rare instances. Consequently, evolutionary ecology has emphasized the study of community structure and of geographical patterns.

The lesson of paleontology has been that evolutionists do not have to restrict themselves to the human timescale. But because most evolutionary ecology treats life at the species level or below, whereas most paleontology is usually considered secure only at the genus level or above, this lesson has mostly been ignored. That is too bad, because the repertoire of testable predictions would be greatly extended if evolutionary ecology could use fossils.

Fortunately, in the past decade paleontologists have turned their attention to sophisticated analyses of fossils at the specific level. They have generated sets of data which fairly beg for study and restudy. It is time for the evolutionary ecologists to take notice.

One such data set was developed by Peter and Sara Bretsky (1975) of the State University of New York at Stony Brook. Working in Quebec, the Bretskys carefully analyzed a sequence of 231 bedding planes from a great inland ocean of the Upper Ordovician, some 450 million years ago. In the bedding planes are fossils of all sorts of marine invertegrates: sponges, trilobites, brachiopods, mollusks and others. The entire sequence spans about 5 million years during Ordovician time. Toward the end of this time, the basin filled up with sediment.

These data and others should be useful for investigating many questions about speciation and species diversity. As you shall see, they also invite the application of sophisticated statistical techniques. Since we are by no means statisticians we confidently predict that most or all of the statistics presented will be found to be too crude, and require repair with more sophisticated assumptions and methods by more competent investigators. However, someone has to be willing to initiate the series of mistakes that eventually leads to scientific understanding.

We should like to use the data to discuss the problem of the equilibrial nature of species diversity (Rosenzweig, 1975). The proposition that species assemblages attain such equilibria is really twofold. First, that they achieve equilibria in situations where new species most commonly enter by immigration: these situations are called islands. Second, that they do so where most originations are by some process of speciation: these situations are called provinces. We shall make no attempt to discriminate islands from provinces. However, internal evidence, as suggested elsewhere (ms.), is consistent with the hypothesis that what we are about to show you is the record of some unknown fraction of a province. We shall refer to it as a province hereafter.

2. TECHNICAL NOTES

The samples from each stratum are very, very roughly equal in size, and none was accepted by the Bretskys unless it contained more than a minimum number of individuals. But the samples come only from one small area of the province and the Bretskys present compelling evidence that the environmental characteristics of this area changed over the 5 million year period (1975). Therefore, what we shall see is definitely not one community. The Bretskys (1975) estimate that there are 4 or perhaps a few more recurring ones. Since each type of environment lasts for quite a few sample intervals, we are faced with the statistical problem of serial dependence. Believing that this is likely to be a second order problem, we have ignored it in this treatment and have instead assumed that each sample is drawn independently from the total pool of species available at any one time. Since there are 231 samples from about 4 environments, each of which appears several times in the record, we do not expect this assumption has led to gross errors. But future, improved statistical analysis may not support that.

Species that could be identified are recorded as alive in the province during the entire time between their first and last appearances in the record. Of course they may also have lived before and after that, but escaped notice so far. Later we shall explain how a statistical approach may be used to correct for these errors.

3. PRELIMINARY RESULTS

Figure 1 shows the record of the number of species in the province. Notice that this record might be interpreted in several ways:

1) An equilibrium or quasi-equilibrium is attained and maintained by the province during the middle 3.5 million years or so.

2) The equilibrium appearance is false. Species have such long lifespans that 5 million years is insufficient to observe

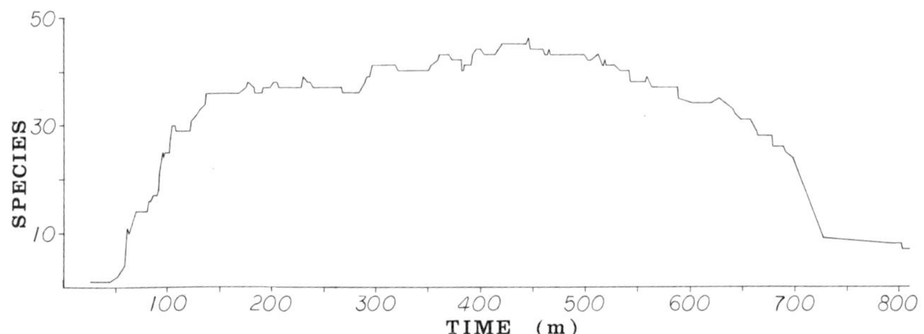

FIG. 1: The known number of species as a function of time (represented by the elevation of the sample in meters).

dynamics. In that case, what we are seeing would be simply one fixed assemblage with an insignificant number of additions and subtractions to reveal the true dynamical trend.

3) There is no equilibrium appearance. The province continues to diversify, but we are reading our biases into the picture. What one actually would be seeing in this picture would derive from the limitations of the paleontological samples and methods. Initially we see an increase, but that would be caused by increasing knowledge of the fauna. The subsequent decline would be caused by the fact that since observations terminate at a certain point in time, the probability of final appearances in the fossil record increases as that time is approached. These two processes would combine to yield a gentle arch which may be misinterpreted as a plateau.

How may one discriminate among these three interpretations? We have tried several ways. They illustrate the sort of statistical analyses one can perform on such data and so we shall present them in some detail. Along the way, we shall have to try to answer several other biological questions, such as:

Can we correct the lifespans at least to some extent, to account for the time before and after appearance in the fossil record? Can we determine the turnover rate for the fauna? Can we predict how many more fossil species will be discovered in this province?

4. WAS THERE AN EQUILIBRIUM?

Let us begin by comparing the first and third interpretations. Under the third interpretation, diversity is really arch-shaped and the record is inadequate to show the true shape of the diversity curve. Old species are still being discovered in the middle and even the latter part of the record. Given this interpretation, a true picture of this 5 million year interval would not emerge until more samples, probably over a much greater interval, were examined.

Since adding samples is more difficult than subtracting them, we examined the converse of interpretation 3: fewer samples should lower the arch and make it more pronounced. We deleted the first and last 10% of the samples. This is a net reduction of 20% of the information. In a second trial, we deleted the first and last 20% of samples for a reduction of 40%. Figure 2 shows the results.

Notice that the loss of 20% of the information had no effect on the shape of the arch. It merely diminished the breadth of the plateau. Otherwise, the 4 million year record is a close match to the 5 million year record and is identical to it over much of the plateau period.

By contrast, the 3 million year record has a very slightly reduced peak and is a bit less flat on top. We can conclude that the samples taken over 4 million years were sufficient to reveal the steady-state. The 5*th* million years was valuable for other reasons including confirmation of the adequacy of the sample, but did not add to the picture of species diversity.

Yet another indication that the number of species is actually about constant comes from the original presentation by the Bretskys (1976). They discovered that the number of species per sample bedding plane was fairly steady. This is especially so during the period of the plateau.

Let us turn our attention now to the second interpretation, that the steady number of species is a result of the short time span, a time span during which few speciations and extinctions might be expected. An obvious way to test this interpretation is to enumerate originations and extinctions, and so achieve a crude estimate of turnover.

For purposes of turnover calculation, we do the following. First, eliminate the beginning and ending parts of the record. During these times, many interpretations of cause are possible, and so their data may not be due to speciations and extinctions. The middle of the record is much more secure in this respect. Second, notice that during a steady state, extinction balance originations. Thus one needs to measure only one or the other to estimate turnover. This leads directly to the procedure.

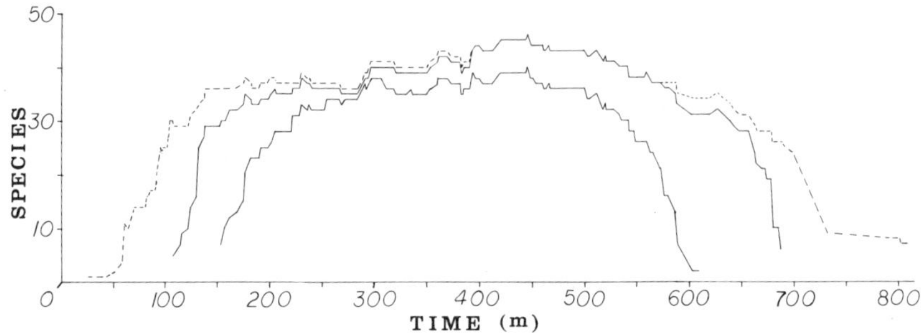

FIG. 2: The truncation experiments. The uppermost record is repeated from Figure 1 as a reference. The intermediate record is the known number of species given that we ignore the first and last half million years of the record. This record is identical to that of Figure 1 for a large fraction of its duration. The lower curve is the known number of species given that we ignore the first and last million years. Axes as in Figure 1.

From the arbitrary cutoff point for the end of the beginning at 137m, to the peak diversity of 46 species at 465m, occupied about 2.1 my (million years). During this 2.1my, the fauna appeared to undergo 28 originations. However, the diversity at 137m was ten species lower than the peak, so ten of the originations cannot be ascribed to turnover. The remaining eighteen can be, yielding a turnover estimate of 18 species/46 species/2.1my. This is 18.7%/my. In 5my, that is going to mean a substantial fraction of the community will need renewal. Notice also that the exclusion of ten of the originations is conservative; it biases the result in favor of finding out that turnover is insignificant in 5 my.

The post-peak period also saw a substantial turnover, although probably not quite as much. In the 1.4my from 465m to the arbitrary cutoff point of the plateau period at 688m, we counted 25 extinctions. However, only 26 species appeared to remain at 688m. Therefore 20 of the extinctions may be non-turnover cases. Again, this exclusion is conservative. It leaves only 5 extinctions/46 sp./1.4my for an estimated turnover of 7 2/3%/my.

Low as this figure is, it is still significant in a 5my time span. Of a cohort of one hundred species that enter a 5my period with such a turnover rate, only 2/3 will survive to its end. In a

3my period, fully 25% would suffer extinction. Moreover, this figure is the lowest one. The actual turnover for the 3.5my plateau period is 23 species, or 50% of the diversity. Clearly, a plateau of several million years could not result from these species merely hanging on as a mostly unaltered community or province.

5. CORRECTING FOR INCOMPLETENESS IN THE FOSSIL RECORDS

The calculations of turnover which have been described so far are quite crude. We may have missed species which have been fossilized, but remain undiscovered. One would like to know how many. And one might also like to know their 'lifespans.' One might further like a more accurate picture of the lifespans of the species which were observed. These questions turn out to be related, and can be answered in one series of calculations.

What is wanted is the vector x_j, the number of species which actually survived for j strata (no more and no less) where j takes the set of values 1 to 231 in this series of observations. The vector appears in the set of linear equations for S_i, the number of species which *appear* to have survived for i strata (again, no more and no less):

$$S_i = \sum_{j=1} p(i,j)x_j \quad , \tag{1}$$

where $p(i,j)$ is the proportion of those species living through j strata which appear to have lived for i strata. There is an equation like this for each of the n apparent lifespans. The set of S_i comes from the data itself. Hence, if one could determine the set of probabilities, he would be left with a set of n linear equations in n unknowns. Such a system is readily solved for the x_j vector.

The probabilities $p(i,j)$ may in fact be estimated from the detailed data. The key statistic is p, the probability that an extant species will fail to appear in a given bedding plane.

A species which actually survives for j strata has a chance $(1-p)$ of being recorded in its first, and $(1-p)$ of being recorded during its last stratum. So

$$p(i,j) = (1-p)^2 \quad , \quad j>1.$$

To be counted as surviving for $(j-1)$ strata, the species must appear in its first and next to last, but not its last; or it must appear in its second and its last, but not its first. So

$$p(j-1,j) = 2(1-p)^2p \quad , \quad (j-1)>1 \quad .$$

Similarly,

$$p(j-2,j) = 3(1-p)^2p^2 \quad , \quad (j-2)>1 \quad .$$

In general,

$$p(j-k,j) = (k+1)(1-p)^2p^k \text{ for all integers } (j-k)>1 \quad .$$

It is easy to evaluate $p(j-j,j) = p^j$.

To complete the reckoning, we need $p(1,j)$. To be counted once, a species must not be counted in all its strata but one. There are j different ways this might happen, one for each stratum it might appear in. Hence,

$$p(1,j) = j(1-p)p^{j-1} \quad .$$

Here is a sample calculation based upon $j = 8$, $p = 0.8$:

	0	1	2	3	4	5	6	7	8
p(i,8)	.17	.34	.07	.08	.08	.08	.08	.06	.04

Notice that half the species appear only once or not at all.

Now we turn to the data for our estimate of p . To avoid biasing the statistic to high probabilities for short-lived species, we don't count either first or last occurrence in the sample. We use only the strata between first and last occurrence. Thus a species which appeared to occur for 5 strata contributes only its interior 3 strata to the estimate. Suppose it occurred once in those three. Then it contributes one actual occurrence and 3 potential occurrences.

The 65 species which could contribute were all those appearing to have survived at least three strata. These occurred 1437 times and could have occurred 7306 times. Hence $(1-p) = .2$ and $p = .8$. These species were also examined individually, and their values of $(1-p)$ were not significantly different from a Poisson distribution (Table 1).

TABLE 1: Distribution of the % Appearance. (% Appearance is the % of strata in between a species' first and last appearance, for which it was actually recorded in the fossil record. % Appearance = 100 (1-p).)

% Appearance	Number of Species	% Appearance	Number of Species
0 - 9	22	50-59	0
10-19	17	60-69	3
20-29	11	70-79	0
30-39	5	90-100	1
40-49	7		

Because the entire community consisted of only 79 species, we could not use the simplest version of equation (1). We instead lumped lifespans into logarithmic intervals and used a slightly modified formula. The results of the calculations do not show us what to expect of each species, but do show us how the entire provincial picture ought to appear after an infinite amount of paleontological work is done at the current level of technical expertise (Table 2). We predict that the entire set of fossilizable species will turn out to be 17 to 21 species larger than it seems at present. We also predict that most of these 17 to 21 will turn out to have very short lifespans. Adjustments elsewhere in the table are minor, so we expect that little change in the steady state will be discovered. However, all those short-lived species should add considerably to the turnover, because each had to originate and to become extinct. This results in a revised turnover estimate for the 3.5my equilibrium period of 21.7%/my to 23.5%/my.

Recently, Stanley (1975), using Durham's work, has estimated that the median lifespan of marine invertebrate species in the immediate past has been 6my. Our figures of turnover yield a median lifespan estimate of from 5.9my to 6.4my, agreeing very well with Stanley.

TABLE 2: Distribution of species' lifespans by octave.

Octave	Lifespan (strata)	Number of Species: Data	Number of Species: Most Probable Reconstruction	Number of Species: Second Probable Reconstruction
1	1	12	5	0
2	2	2	25	25
3	3-4	1	0	0
4	5-8	2	7	7
5	9-16	2	0	0
6	17-32	4	7	8
7	33-64	4	1	1
8	65-128	23	23	23
9	129-231[a]	29	32	32
Total species:		79	100	96

[a]The number 231 is the number of bedding planes examined.

6. FINAL REMARKS

The controversy regarding equilibrium species diversity has been alive for 15 years and is not going to be settled by this single crude analysis of one province. However, several years ago one of us published a quantitative theory which predicts that such equilibria ought to be present in natural systems (Rosenzweig, 1975). The literature also contains some biogeographical patterns which support the notion that at least for birds, these equilibria are, in fact, similar to the actual diversities (Cody, 1975). And now we have a temporal record which supports it for some fossilizable marine invertegrates. The only serious evidence against it has been shattered by Raup (1976a,b). This is not to say that the equilibrium is attained in every taxon. Nor is it meant to suggest that the value of the equilibrium is fixed; other paleontological evidence shows us that it fluctuates greatly. It *is* meant to say that equilibria of species diversity may be much more commonly and rapidly achieved than had been formerly supposed.

REFERENCES

Bretsky, P. W. and Bretsky, S. S. (1975). Succession and repetition in Late Ordovician fossil assemblages from the Nicolet River Valley, Quebec. *Paleobiology*, 1, 225-237.

Bretsky, P. W. and Bretsky, S. S. (1976). The maintenance of evolutionary equilibrium in Late Ordovician benthic marine invertebrate faunas. *Lethaia*, 9, 223-233.

Cody, M. L. (1975). Towards a theory of continental species diversity. In *Ecology and Evolution of Communities*, M. L. Cody and J. M. Diamond, eds. Belknap Press of Harvard University, Cambridge, Massachusetts. 214-257.

Raup, D. M. (1976a). Species diversity in the Phanerozoic: a tabulation. *Paleobiology*, 2, 279-288.

Raup, D. M. (1976b). Species diversity in the Phanerozoic: an interpretation. *Paleobiology*, 2, 289-297.

Rosenzweig, M. L. (1975). On continental steady states of species diversity. In *Ecology and Evolution of Communities*, M. L. Cody and J. M. Diamond, eds. Belknap Press of Harvard University, Cambridge, Massachusetts. 212-140.

Stanley, S. M. (1975). A theory of evolution above the species level. *Proceedings of the National Academy of Sciences, USA*, 72, 646-650.

[*Received September* 1978. *Revised March* 1979]

SECTION II

PATTERNS AND INTERPRETATIONS

G. P. Patil and M. Rosenzweig, (eds.),
Contemporary Quantitative Ecology and Related Ecometrics, pp. 123-138.

Q-MODE AND R-MODE ANALYSES OF BIOGEOGRAPHIC DISTRIBUTIONS: NULL HYPOTHESES BASED ON RANDOM COLONIZATION

DANIEL S. SIMBERLOFF AND EDWARD F. CONNOR

Department of Biological Science
Florida State University
Tallahassee, Florida 32306 USA

SUMMARY. Consider an $r \times c$ matrix with the r rows representing species, the c columns representing locations (usually islands), and the entries 1 or 0 connoting presence or absence, respectively. Such a matrix has been examined analogously to Q-mode and R-mode analysis in numerical taxonomy. In the analog to Q-mode analysis, pairwise similarities between sites have been calculated from the columns by computing various statistics (e.g., Jaccard's index, the Sørensen quotient) based on matching 1's and o's or mismatches. Classically, these similarities are then ordinated and inferences about biogeographic, ecological, or evolutionary affinities between a given pair of sites are drawn from the rank of their similarity in the ordination. In the analog to R-mode analysis, on the other hand, the number of sites shared between pairs or larger sets of species has been used to draw inferences about interactions (especially competition) between species, as well as to define evolutionary or ecological groups of species.

In both modes, however, the statistics have generally been *ad hoc* constructs with no underlying null hypothesis, no resultant null distribution, and no associated confidence limits. Deductions made from such analyses are therefore suspect. In an initial attempt to remedy this problem, we assume for both modes a null hypothesis of random distribution of species, subject to their observed numbers of occurrences in some archipelago of interest. For the Q-mode analysis (site- or island-comparison) of Galapagos plants, Florida Keys insects, and West Indies birds, the matrix was filled by maintaining column sums (island species 'richnesses') and randomly drawing species with probabilities proportional to their observed numbers of occurrences. For the R-mode analysis (species-comparisons) of birds of the New Hebrides and West Indies and bats of the West Indies, both row sums (species occurrences) and column sums were maintained.

For all three biotas subjected to Q-mode analysis, number of island pairs sharing more species than expected was much greater than the number sharing fewer than expected, suggesting that there exist determinsitic forces not accounted for in the null hypothesis. But most values of expected number of species shared are quite close to those observed, so one may reasonably infer that the colonization process is mostly stochastic. Use of classical *ad hoc* similarity analyses has led to the opposite conclusion, that most colonization is deterministic. The R-mode analysis, on the other hand, showed for all three faunas a remarkably good match between observed and expected number of species pairs sharing i islands, for all i; in particular, the number of exclusive pairs of species $(i = 0)$ is not much more than what one would expect under the null hypothesis of random distribution. Previous claims of competitive exclusion based on selected exclusive species pairs are therefore invalid.

KEY WORDS. biogeographic distributions, Q-mode analysis, R-mode analysis.

1. INTRODUCTION

Consider an $r \times c$ matrix M in which the r rows represent species, the c columns represent locations, and each matrix entry m_{ij} is 1 or 0 depending on whether species i is present or absent, respectively, in location j. Such a matrix can be examined in either of two modes, analogously to Q-mode and R-mode analysis in numerical taxonomy. In a taxonomic context Q-mode analysis leads to a classification of taxa (or 'OTU's') by calculation of similarities among columns, while R-mode analysis groups characters through examination of similarities among rows (Sneath and Sokal, 1973). Analogously in biogeography, Q-mode analysis classifies sites while R-mode analysis groups species.

As in numerical taxonomy, Q-mode analysis in biogeography has a richer and older tradition than R-mode analysis. Most uses have been phytogeographical; Simpson (1960) provides a review of the early literature. The first explicit statistic used in such analyses, aside from simply number of species shared, was Jaccard's index (1908), $J = N_c/(N_1 + N_2 - N_c)$, where N_c is number of shared species, N_1 is number of species at site 1, and N_2 is number of species at site 2. Another old and commonly used similarity statistic is that proposed independently by Dice (1945) and Sorensen (1948): $2N_c/(N_1 + N_2)$. In fact, there are a plethora of *ad hoc* similarity indices consisting of the quantities N_c, N_1, N_2, A (number of species absent from both site 1 and 2 but present elsewhere), and N_t (number of species present in all sites together) combined in various combinations and permutations. Cheatham and Hazel (1969) give nineteen such expressions which have been used in one or more biogeographic or taxonomic analyses of matrices of the form of M. These indices share with others in the literature, such as that derived by Hagmeier and Stults (1964) from Preston's 'resemblance equation' (1962) or one based on both map distance and species shared between sites (Tobler, Mielke, and Detwyler, 1970), the related properties that they are arbitrary (there is no *a priori* reason to prefer one of them over the others as a more suitable expression of biotic similarity between sites) and do not directly test hypotheses. The Jaccard index, for example, has no natural relationship to some underlying hypothesis of how species might be distributed among sites. Consequently, an observed value of J between sites 1 and 2 has no objective meaning for us. Much less can we associate a probability level with some observed J; there is not even an expected value from which we can detect a deviation. Typically in biogeography one uses one of these arbitrarily chosen indices to compute pairwise similarities among all of a group of sites, then ordinates the similarities to say that certain pairs of sites are biogeographically 'distant' while other pairs are 'near.' For example, Tobler, Mielke, and Detwyler (1970) compared a map of New Zealand and surrounding islands with interisland distances inversely proportional to botanical similarity to an ordinary geographic map.

Despite the arbitrary nature of classical similarity indices and their concomitant inappropriateness for direct tests of biological hypotheses, numerous biological conclusions have been extrapolated from studies using these statistics. Terborgh (1973) calculates the Sorensen index for bird species data between pairs of islands in three groups of West Indies: four large islands, six intermediate-sized islands, and nine small islands. From regressions of values of this index on inter-island distance, he concluded that for small and intermediate islands, about 93% of the avifaunal composition on any island is determined by the size and location of the island, while the corresponding figure for large islands is 80%. The remainder he apportions among competition between bird species and habitat differences between islands. Power (1975) used Preston's similarity index, (calculated from the resemblance equation) to examine pairwise similarity among the Galapagos islands for birds and plants. After regressing bird similarities and plant similarities on several physical factors, he concluded that bird similarity between islands is 'explained' by plant similarity, while plant similarity is primarily 'explained' by inter-island distance, with area and elevation also having 'significant' effects. However, setting this problem up as a regression necessitates a subjective judgment that certain variables are independent and others dependent, while the similarity index itself is an *ad hoc* statistic with no associated tests of significance. As a final example, Heatwole and Levins (1972) use the square of the Euclidean distance, another *ad hoc* similarity index, to measure similarity between mangrove island insect faunas at different times and conclude that these communities undergo a deterministic progression towards an equilibrium trophic structure. Not only was there no attempt to see whether this similarity index is sample-size dependent (it is; Simberloff 1976a), but even were it not, there are no significance tests to determine when two communities are, in fact, significantly different by this index.

For most sets of data, the various similarity indices in the literature correlate quite well with one another (e.g., Goodall, 1966; Sepkoski and Rex, 1974), suggesting that biological conclusions drawn from ordination of pairwise values among sites might be quite robust. Yet all three of the above examples rest on the differences' being real, and this is not demonstrable with these indices. Further, the different indices weight the same phenomena differently, and what weight one desires to give a specific event is, of course, a subjective judgment. Often there is no attempt at explicit justification of some particular convention. For example, Baroni-Urbani and Buser (1976) believe that the widespread use in Q-mode biogeographic analyses of indices in which the quantity A (defined above) is not used has been formally justified only once, by Green (1971). One may even have a matrix of the form of M analyzed in the Q-mode using one index and in the R-mode using another (e.g., Sepkoski and Rex, 1974).

With the above considerations in mind, we set out to find an objective, probabilistic similarity index which furthermore would allow us to test a simple null hypothesis concerning the distribution of species on islands (or other locations): namely, that observed pairwise floral or faunal similarities among islands do not differ from what one would expect if species were randomly distributed. In the similarity literature we found just two other attempts to produce probabilistic indices. First, Goodall (1964, 1966) proposed an index which, when applied in the Q-mode to matrices such as M, orders pairwise similarities by whether the species shared between a particular pair of sites are shared among many sites or few. If a pair of sites shares many species which are very rarely shared among sites, that pair of sites is very similar. This seems intuitively pleasing, not only for biogeographic analysis but for numerical taxonomy: after all, species sharing rare attributes and sites sharing rare species *do* seem to us to be likely to be related in some sense. However, the precise weight to give to a given degree of rarity is, we feel, a subjective matter. Goodall deals with this problem in what is perhaps the most objective manner possible. For a given pair of sites and one species he calculates the probability that a randomly drawn pair of sites would be as 'similar' for that species; then he multiplies these probabilities over all species to compute the final similarity index (or rather, its complement) for these two sites. This procedure certainly allows pairwise similarities to be ranked with respect to probability in some sense, but one can conceive of other desirable ways of combining the probabilities over all species. For example, a pair of sites which shares a species found nowhere else would, by this algorithm, be the most similar pair (or one of the most similar). Do we wish one shared species to have this much weight? Could not the probabilities be added? And in any event, the precise values of the probabilities which Goodall generates have no objective significance beyond allowing the ranking. They are derived only from the sample of sites at hand, and there is no significance test, for example, for the difference between two such similarity values or for whether one would have expected to find at least one similarity value as small as the smallest observed one, given certain biological assumptions.

A second attempt to derive a probabilistic similarity index for Q-mode analysis in either taxonomy or biogeography was that of Baroni-Urbani and Buser (1976). Again for subjective reasons of taste, they proposed a new similarity index based on the parameters listed above:

$$S_{**} = (\sqrt{N_c A} + N_c)/(\sqrt{N_c A} + N_1 + N_2 - N_c).$$

However, a statistical test is proposed by producing, for a given number of attributes or species, a large random sample of all possible OTU's (or, in biogeography, all possible subsets of species), then plotting the distribution of S_{**} calculated for all pairs of OTU's (or sites) among this sample. We feel that this statistical test is inappropriate for the sorts of questions we wish to ask, for two reasons. First, in nature there is no *a priori* reason to believe that every possible subset of species is equiprobable on a set of sites such as islands; surely the sizes of the subsets are not drawn from a uniform distribution if only because of the well-known species-area effect (Connor and McCoy, 1980). Second, one of the very hypotheses which begs testing in the ecological literature is that, for a given sized subset of species, certain subsets are much less likely to occur than others because of interactions among their component species, particularly diffuse competition (e.g., MacArthur and Wilson, 1967; MacArthur, 1972; Diamond, 1975). We return to this point below.

2. A PROBABILISTIC INDEX FOR Q-MODE ANALYSIS

If one assumed that all species in some species pool of size N_t where equally likely to colonize all islands in some archipelago, the expected number of species held in common between two islands with m and $n \geq m$ species, respectively, would be:

$$E[N_c(m,n)] = \frac{mn}{N_t} . \quad \text{(Simberloff, 1978).} \tag{1}$$

The variance is given by Connor and Simberloff (1978):

$$\mathrm{Var}[N_c(m,n)] = \left[\sum_{i=0}^{m} \binom{N_t - n}{i} \binom{n}{m-i} \left(m - i - \frac{mn}{N_t}\right)^2 \right] \Big/ \binom{N_t}{m} \tag{2}$$

Examining vascular plants on 29 of the Galapagos Islands, Johnson (1974) found that 369 of the 406 island pairs shared more species than expected (for 345 of these pairs, $P < .05$), while 37 pairs shared fewer species (none significantly). Updated species lists (Connor and Simberloff, 1978) do not greatly change these results. For arboreal arthropods on nine small Florida mangrove islands, Simberloff (1976b) found all 36 island pairs shared more species than expected with $P << .01$.

Of course this null hypothesis is rejected for these archipelagoes, but one would have expected this result because elementary biological considerations dictate that all species in a pool are not equally adept colonists. One can revise the null hypothesis in many ways to make it more biologically realistic. For example (Simberloff, 1978), it would be an easy matter to restrict the species pool for each island to just those species found on islands within some restricted size range; one such attempt (Johnson, 1974) produced a better fit of observed to expected number of species shared between islands for the Galapagos flora. Our tactics were different, however, and were motivated by our belief that any such restriction in size range of islands which a species can colonize is probabilistic rather than absolute for most archipelagoes and species sets. It may be that certain species are more likely to be found on islands of a given size, but anomalies are often found and usually the smaller the island, the less likely is any species to be found on it. Further, if one specifies precisely enough all of the limitations on which species may be found on which islands, one could well be left with a null hypothesis so tailored to a given data set that all variation for that set would be explained in the statement of the hypothesis and the hypothesis would permit few or no predictions for other sets!

Consequently, we wished our new null hypothesis to be that the distribution of island-pair similarities for any archipelago is approximately what one would expect if the species in the pool had colonized the islands randomly, subject only to the condition that certain species are better colonists than others. This is the original null hypothesis but with weighted instead of equal colonization probabilities. There are two aspects of species' biologies that could cause them to have different colonization probabilities: dispersal and persistence. Ideally one would wish independent assessments of the dispersal and persistence abilities for all N_t species. However, comprehensive data for an entire species pool on either ability are rare, so we have instead used the row sums, $M_i = \sum_j m_{ij}$, from the original matrix as weights for the colonization probabilities. Although this procedure clearly has an element of circularity in that the product of dispersal and persistence abilities is inferred from the matrix of species occurrences, and these products are then used to test an hypothesis about the causes of the matrix, we are not tautological here since the row sums do not uniquely specify the distribution of resulting similarities. This was easily demonstrated by repeated simulation using random matrices constructed with the appropriate distribution of row sums.

We have not yet been able to find an analytic expression for $E[N_c(m,n)]$ or its variance when the species pool is weighted, but both quantities are easily produced by repeated random draw simulation to fill the matrix with the column (island) totals specified and then examination to see how many species are shared between each pair of islands. The expected number of shared species, $E[N_c(m,n)]$, is a monotonic increasing function of both its arguments, so for our similarity index we used the deviation of observed from expected value divided by the standard deviation:

$$S = \{N_c(m,n) - E[N_c(m,n)]\}/\sqrt{Var[N_c(m,n)]}.$$

For the Galapagos plants (Connor and Simberloff, 1978), 338 of the 406 island pairs still share more species than expected (206 significantly), while 68 pairs were more dissimilar than expected

(6 significantly); generally a large fraction of the species shared between islands is simply that which one would have expected for a random arrangement with fixed column (island) totals. A similar result is found for the nine small mangrove islands discussed above, and this method of similarity analysis was also applied to successive insect communities on individual repeatedly censused mangrove islands to demonstrate that there is no apparent species compositional equilibrium toward which mangrove island communities tend (Simberloff, 1978). Finally, this similarity index was applied to the same west Indian bird data on which Terborgh (1973) used the Sørensen index, and a rather different conclusion was reached (Simberloff, 1978): whereas Terborgh felt that most species composition could be construed as completely determined by island size and isolation, with little accounted for by chance, the above analysis suggested that large islands are about as similar to one another as would be expected were the above hypothesis of random colonization with weighted probabilities correct. Small and medium-sized islands are about 2/3 as similar as would have been predicted by this hypothesis. For the West Indian birds as for the other examples, observed number of species shared between islands more often exceeded expected than fell below it, and many of the individual deviations were statistically significant. This implies that the weighted random colonization hypothesis is not correct, and that other forces must be important determinants of inter-island biotic similarity. But these endeavors should be viewed as baselines, in the sense that they demonstrate that a large fraction of colonization can be interpreted as random, and indicate how much remains to be explained in other, less parsimonious ways. Independently of this work, Grassle and Smith (1976) have suggested, for community samples in which not only species' presences are known, but also the numbers of individuals of each species, that the expected number of species shared between equal-sized samples from two communities may serve as a similarity index and have provided an estimator for this index. They have not, however, interpreted this index probabilistically as a test of an underlying hypothesis about the forces determining biotic similarity between sites.

3. R-MODE ANALYSIS OF A SPECIES-BY-SITE, PRESENCE-ABSENCE MATRIX

Formal analysis of species similarity based on shared sites is far less common than analysis of site similarity based on shared species. For example, Sepkoski and Rex (1974) analyze a presence-absence matrix by column (site) and row (species), and explicitly state that though the resulting clustering of sites has biological significance, the clustering of species is simply an arbitrary but convenient way of arranging them for visual presentation and subsequent site analysis. Fager and McGowan (1963) used an *ad hoc* similarity index, $S = [(N_1 + N_2 - N_c)/\sqrt{N_1 N_2})] - (1/2\sqrt{N_2})$, to cluster species into 'recurrent groups,' a procedure since followed occasionally by others (references in Cheetham and Hazel, 1969). Similarly Holloway and Jardine (1968) and Holloway (1970, 1973) clustered species into faunal elements using the arbitrary similarity index $S = 1-(n/N)$, where n is the number of sites in which two given taxa are found together and N is the number of sites in which either or both taxa are found. Sneath and Sokal (1973) list a few similar efforts. Finally, Williams and Lambert (1961) suggest simultaneous Q- and R-mode analysis, using an *ad hoc* similarity index, to delineate joint clusters of both sites and species, which they call 'noda.'

Our R-mode analysis (Connor and Simberloff, 1979) was motivated by a series of 'assembly rules' for bird species communities published by Diamond (1975) and based on his studies of the Bismarck Archipelago avifauna. Among these rules were two which stated that in a given archipelago, (a) certain pairs of species are never found together on an island, and (b) certain groups of *related* species are never found together on an island. Diamond attributed this state of affairs to the workings of interspecific competition, but we observed that, because of the enormous numbers of possible pairs, trios, etc. of species in any archipelago, and the much smaller number of islands, one would *expect* to find certain pairs of species, or groups of related species, not coexisting in any archipelago even if the species were distributed randomly among the islands. The critical question is, "Is the number of such non-coexisting groups significantly different from what chance alone would have predicted?"

Since the Bismarck bird data have never been published, we tested this null hypothesis for rules (a) and (b) on birds of the New Herbrides (Diamond and Marshall, 1976) and West Indies (Bond, 1971) and bats of the West Indies (Baker and Genoways, 1978). For each of these biotas, we had an observed presence-absence matrix of form M, for which we could compute number of non-coexisting pairs of species to test rule (a) and number of non-coexisting, confamilial pairs and (for the New Hebrides birds and West Indies bats) trios to test rule (b). In order to determine expected values for our null hypothesis of random distribution, we repeatedly produced random matrices for each biota, subject only to three conditions:

1) Island totals (column sums) are maintained.

2) Species totals (row sums) are maintained.

3) Each species is placed only on islands which has species numbers in the range of islands on which that species is actually found.

Condition 3 is an acceptance of the notion that species are found only on islands in certain size range; *cf.* 'incidence functions' (Diamond, 1975). For each random matrix, we found numbers of non-coexisting pairs and non-coexisting, confamilial pairs and trios, and expectations were estimated by the means over all matrices. Results (Table 1) indicate that in all instances, a very large fraction of the observed non-coexisting groups of species would have been expected under a random distribution hypothesis, even though for some data sets this hypothesis is insufficient to explain the amount of exclusion observed. However, since the lists all contain recent cases of allopatric speciation, the observed amounts of potentially competitive exclusion are exaggerated. The evolution of specific differences between allopatric, formerly conspecific populations generates an observed exclusive pair independent of interspecific competition. In any event, just as for Q-mode analysis, we suggest that this sort of R-mode analysis be performed to produce a baseline of 'expected' exclusion for randomly distributed species before less parsimonious explanations are invoked.

A more complete version of the above form of R-mode analysis would consist of determining not only observed and expected numbers of non-coexisting pairs, trios, etc. of species, but observed numbers of pairs, trios, etc. sharing only 1, 2, 3,··· islands in the archipelago. Such calculations were done for species pairs in all

TABLE 1: Observed and expected numbers of species pairs and trios which are mutually exclusive for selected taxa and archipelagoes. Parenthetic values represent one standard deviation.

Taxon	Group size	Con-familial	Total Groups	Non-coesisting Observed	Non-coesisting Expected
New Hebrides birds	Pair	No	1,540	63	63.2 (2.9)
	Pair	Yes	99	1	0.9 (0.3)
	Trio	Yes	304	7	6.4 (1.9)
West Indies birds	Pair	No	22,155	12,757	12,448.1 (79.2)
	Pair	Yes	1,029	621	437.0 (18.3)
West Indies bats	Pair	No	1,711	996	941.7 (11.6)
	Pair	Yes	499	325	208.6 (5.5)

three of the above faunas with the resulting graphs (Connor and Simberloff, 1979) showing a remarkable match over the entire domain from total exclusion (no islands shared) through total coexistence (all islands in the archipelago shared).

4. AN EXEMPLARY MATRIX ANALYZED IN BOTH MODES

A presence-absence matrix of form M was constructed for the Galapagos land birds from the data of Harris (1973). For the 15 tabulated islands there are 105 pairs, and when a Q-mode analysis is performed with a null hypothesis of random colonization by equiprobable colonists, 83.8% of the pairwise similarities between islands exceeded expected, 31.4% significantly (Connor and Simberloff, 1978). That so few of the deviations are significant is a result of the assumption of equiprobability, since some of the *Geospiza* species (Darwin's finches) are single island endemics which could not possibly contribute to observed number of species shared between any two islands. They *do* contribute to the expected number, however. When the null hypothesis is changed to one of weighted colonization probabilities, all 105 of the pairwise island similarities exceeded expected; 69.5% were significant deviations. Figure 1 plots the similarities as a function of numbers of species m and n on the pair of islands. Despite the universal excess of observed shared species over expected, it is clear (Figure 2) that a random hypothesis predicts a large fraction of the observed similarities in species composition between islands.

An R-mode analysis of the same data for pairs and trios of species among the 23 birds was performed as described above, with expected values and standard deviations calculated from ten random matrices generated subject to the constraints listed above: island totals, species totals, and incidence ranges. The results are plotted for species pairs in Figure 3, for trios in Figure 4. The number of pairs and trios of birds sharing 0, 1, 2,···, 15 islands is almost always near the expected number, all but once within two standard deviations. In particular the 0-classes, consisting of exclusive pairs and trios, are approximately what one would have predicted for randomly distributed birds. We note as above that the presence of single-island endemics allopatrically speciated from a common ancestor increases the number of observed examples of exclusion, so that even the slight excess of observed over expected in both 0-classes need not be attributed to interspecific competition.

In sum, for this fauna and archipelago, these methods of similarity analyses have allowed us to test interesting hypotheses about the nature of colonization and have suggested that a large

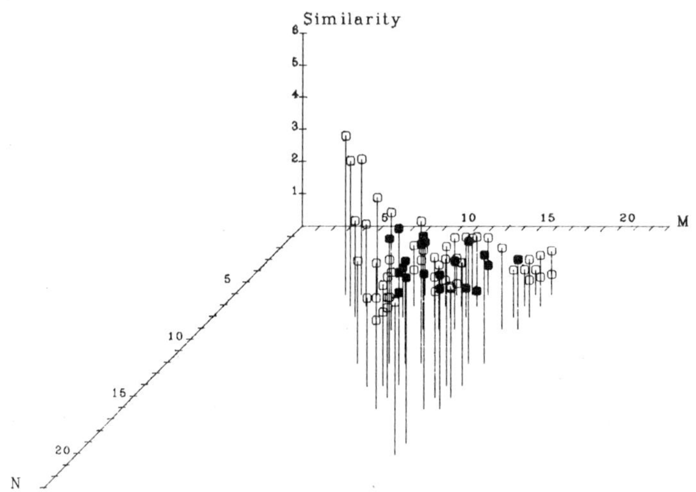

FIG. 1: Distribution of pairwise similarities of Galapagos Island avifaunas. Numbers of species on islands in any pair are M *and* N*, respectively. Darkened symbols represent multiple occurrences of the same value.*

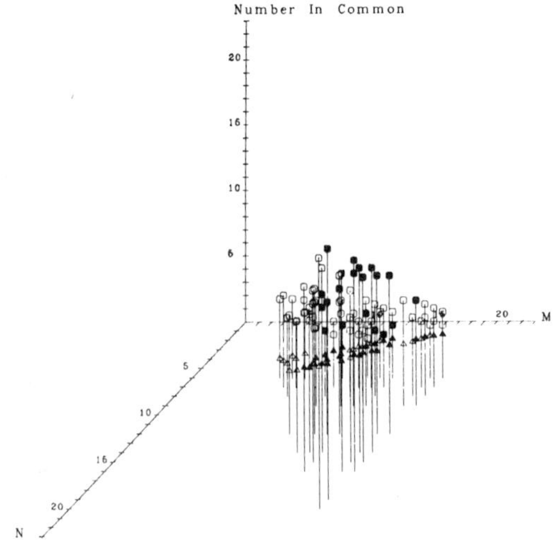

FIG 2: Observed (circles) and expected (triangles) numbers of shared bird species among pairs of Galapagos Islands. Numbers of species on islands in any pair are M *and* N*, respectively. Darkened symbols represent multiple occurrences of same value.*

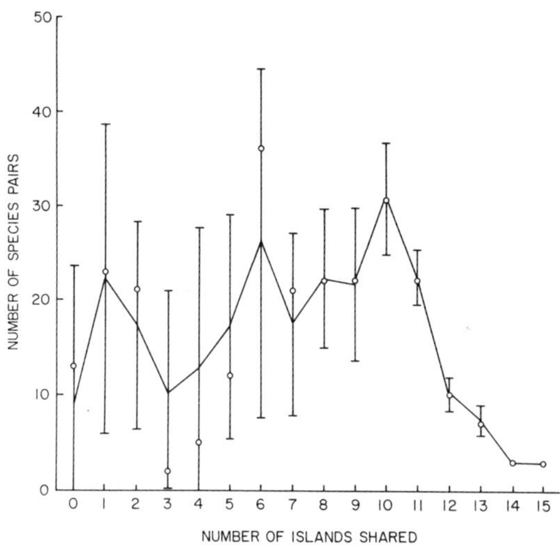

FIG. 3: Numbers of pairs of Galapagos Island land birds sharing 0,1,2,···,15 *islands. Observed values are circles, while the line connects expected values. Vertical lines represent two standard deviations about the expected values.*

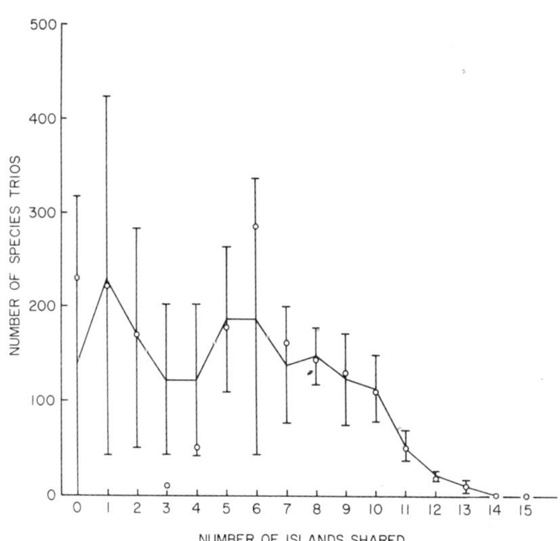

FIG 4: Numbers of trios of Galapagos Island land birds sharing 0,1,2,···,15 *islands. Same conventions as for Figure 3.*

fraction of observed distributional similarity is possibly due to chance. The analyses have also indicated the extent to which observed distributional patterns demand biological explanations.

REFERENCES

Baker, R. J. and Genoways, H. H. (1978). Zoogeography of Antillean bats. *Proceedings, Academy of Natural Sciences, Philadelphia,* in press.

Baroni-Urbani, C. and Buser, M. W. (1976). Similarity of binary data. *Systematic Zoology,* 25, 251-259.

Bond, J. (1971). *Birds of the West Indies,* 3rd ed. Collins, London.

Cheetham, A. H. and Hazel, J. E. (1969). Binary (presence-absence) similarity coefficients. *Journal of Paleontology,* 43, 1130-1136.

Connor, E. F. and McCoy, E. D. (1979). The statistics and biology of the species-area relationships. *American Naturalist,* 113, 791-833.

Connor, E. F. and Simberloff, D. S. (1978). Species number and compositional similarity of the Galapagos flora and avifauna. *Ecological Monographs,* 48, 219-248.

Connor, E. F. and Simberloff, D. S. (1979). The assembly of species communities: Chance or competition? *Ecology,* in press.

Diamond, J. M. (1975). Assembly of species communities. In *Ecology and Evolution of Communities,* M. L. Cody and J. M. Diamond, eds. Belknap Press of Harvard University Press, Cambridge, Massachusetts.

Diamond, J. M. and Marshall, A. G. (1976). Origin of the New Hebridean avifauna. *Emu,* 76, 187-200.

Dice, L. R. (1945). Measures of the amount of ecologic association between species. *Ecology,* 26, 297-302.

Fager, E. W. and McGowan, J. A. (1963). Zooplankton species groups in the North Pacific. *Science,* 140, 453-460.

Goodall, D. W. (1964). A probabilistic similarity index. *Nature,* 208, 1098.

Goodall, D. W. (1966). A new similarity index based on probability. *Biometrics*, 22, 882-907.

Grassle, J. F. and Smith, W. (1976). A similarity measure sensitive to the contribution of rare species and its use in investigation of variation in marine benthic communities. *Oecologia*, 25, 13-22.

Green, R. H. (1971). A multivariate statistical approach to the Hutchinsonian niche: Bivalve molluscs of central Canada. *Ecology*, 52, 543-556.

Hagmeier, E. M. and Stults, C. D. (1964). A numerical analysis of the distributional patterns of North American mammala. *Systematic Zoology*, 13, 125-155.

Harris, M. P. (1973). The Galapagos avifauna. *Condor*, 75, 265-278.

Heatwole, H. and Levins, R. (1972). Trophic structure stability and faunal change during recolonization. *Ecology*, 53, 531-534.

Holloway, J. D. (1970). The biogeographical analysis of a transect sample of the moth fauna of Mt. Kinabalu, Sabah, using numerical methods. *Biological Journal of the Linnean Society*, 2, 259-286.

Holloway, J. D. (1973). The affinities within four butterfly groups (Lepidoptera:Rhopalocera) in relation to general patterns of butterfly distribution in the Indo-Australian area. *Transactions of the Royal Entomological Society of London*, 125, 125-176.

Holloway, J. D. and Jardine, N. (1968). Two approaches to zoogeography: A study based on the distribution of butterflies, birds and bats in the Indo-Australian area. *Proceedings of the Linnean Society of London*, 179, 153-188.

Jaccard, P. (1908) Nouvelles récherches sur la distribution Florale. *Bulletin de la Société Vaudoise de la science naturelle*, 44, 223-276.

Johnson, M. P. (1974) Species number, endemism, and equilibrium in the Galapagos biota. Unpublished address at A.A.A.S. Galapagos Symposium, San Francisco.

MacArthur, R. H. (1972). *Geographical Ecology*. Harper & Row, New York.

MacArthur, R. H. and Wilson, E. O. (1967). *The Theory of Island Biogeography*. Princeton University Press, Princeton, New Jersey.

Power, D. M. (1975). Similarity among avifaunas of the Galapagos Islands. *Ecology*, 56, 616-626.

Preston, F. W. (1962). The canonical distribution of commonness and rarity: Part II. *Ecology*, 43, 410-432.

Sepkoski, J. J. and Rex, M. A. (1974). Distribution of freshwater mussels: Coastal rivers as biogeographic islands. *Systematic Zoology*, 23, 165-188.

Simberloff, D. S. (1976a). Trophic structure determination and equilibrium in an arthropod community. *Ecology*, 57, 395-398.

Simberloff, D. S. (1976b). Experimental zoogeography of islands: Effects of island size. *Ecology*, 57, 629-648.

Simberloff, D. S. (1978). Using island biogeographic distributions to determine if colonization is stochastic. *American Naturalist*, 112, 713-726.

Simpson, G. G. (1960). Notes on the measurement of faunal resemblance. *American Journal of Science*, 258a, 300-311.

Sneath, P. H. A. and Sokal, R. R. (1973). *Numerical Taxonomy*. Freeman, San Francisco.

Sørensen, T. (1948). A method of establishing groups of equal amplitude in plant sociology based on similarity of species content and its application to analyses of the vegetation on Danish commons. *Biologiske Skrifter*, 5, 1-34.

Terborgh, J. (1973). Chance, habitat, and dispersal in the distribution of birds in the West Indies. *Evolution*, 27, 338-349.

Tobler, W. R., Mielke, H. W., and Detwyler, T. R. (1970). Geobotanical distance between New Zealand and neighboring islands. *Bioscience*, 20, 537-542.

Williams, W. T. and Lambert, J. M. (1961). Nodal analysis of associated populations. *Nature*, 191, 202.

[*Received June* 1978. *Revised May* 1979]

G. P. Patil and M. Rosenzweig, (eds.),
Contemporary Quantitative Ecology and Related Ecometrics, pp. 139-155.

APPLICATIONS OF EIGENVECTOR ANALYSIS IN THE RESOLUTION OF SPECTRAL PATTERN IN SPATIAL AND TEMPORAL ECOLOGICAL SEQUENCES

DANIEL GOODMAN

Scripps Institution of Oceanography
LaJolla, California 92093 USA

SUMMARY. Two programs of analysis are described for resolving spatial or temporal repetitive pattern in loose species assemblages. The first, spectral analysis of orthogonal time domain components, identifies species aggregates on the basis of their associations within samples, and then analyzes the spatial pattern of these assemblages. This method forces an initial clustering, and to this extent it will possibly yield more explicit species groups, but the groups may turn out not to be representative of the dominant spatial pattern. The second program, frequency domain orthogonal components, chooses a frequency band and then analyzes the contribution of each species to presumed sinusoidal fluctuations at this frequency. This method forces a fit to a sine wave, so it is not suited to testing for periodic behavior, but if the periodicity is present, the method has the advantage of resolving each species' importance and position in the wave structure.

Use and interpretation of the methods are illustrated by application to a vegetation transect with a well-developed, two phase, mosaic pattern. In this example, both methods function successfully, and they seem complementary in elucidating somewhat different aspects of the vegetation pattern.

KEY WORDS. eigenmode, empirical orthogonal function, mosaic, pattern analysis, principal components, spatial pattern, spectral analysis.

1. INTRODUCTION

Given sufficient data, spectral analysis is the method of choice for detecting repetitive patterns in an ordered sequence of observations. Roughly, this reveals an apportionment of the variance in the record of a single variable, or the covariance in the records of two variables, over a range of frequencies. The frequencies corresponding to peaks in the distribution then indicate the dominant scales of organization in the single records, or of interaction in the pair of records. The techniques are thoroughly described in Jenkins and Watts (1968); ecological applications have been reviewed recently by Platt and Denman (1975).

In complex ecological communities, the natural applications of conventional spectral analysis to the question of community pattern involve gross, system-level biological variables, such as chlorophyll fluorescence or biomass, and their relation to environmental variables such as moisture or temperature. While power spectra successfully test for periodicity in the abundance of single species populations, the transition to analysis of community pattern from spectral treatment of species abundance records is an awkward one. Yet species populations surely are the fundamental units of which communities are composed, functionally as well as evolutionarily.

The difficulty is that the records of single species abundance, considered individually, or pairwise, will often prove too poor in information to reveal a critical pattern of interaction. In fact, this may be more a reflection of the nature of the community structure than a simple problem of sample adequacy. That is, the essential fabric of community organization may consist of very loose constellations of species, appearing together or replacing one another, in time or space. There may, perhaps, be forbidden combinations (Diamond, 1975), but a large fraction of species associations, which are real nonetheless, probably are not obligate. Thus, our task is to identify and quantify appropriate aggregate indices corresponding to these loose species groups, and it is the temporal or spatial behavior of these measures which will be submitted to frequency domain analysis.

In this paper we will consider two ways of employing eigenvector analysis to resolve the species constellations.

2. SPECTRAL ANALYSIS OF TIME DOMAIN PRINCIPAL COMPONENTS

Principal components are computed from the eigenvectors of the matrix of covariances between variables. As this matrix is symmetric, the eigenvectors are orthogonal, and so they define a linear transformation of the original variables into new variables, the components, which are uncorrelated in the data set in question. The eigenvalue associated with each eigenvector is proportional to the fraction of the variance of the original data set accounted for by that component. Usually, a few of the components, perhaps two or three, account for the preponderance of the variance, and so these constitute an economical description of the data. The absence of correlation between the components, within the actual data set used to establish the covariances between variables, is a great convenience in computing further correlations or regressions between individual components and additional variables that were not included in the covariance matrix.

Ideally, the principal component analysis will resolve physically real 'components.' Imagine, for example, a system of n source communities encompassing a total of m species, with no species appearing in more than one community. Let the species proportions within each source community be constant, so each community can be characterized by an m-vector of relative species abundances. Now, imagine that some mixing process throws together assemblages drawn at random from the source communities, with the amount drawn from any particular source community being statistically independent of that drawn from any other, and with the amount drawn from any one source community having a Gaussian distribution, over time. If principal component analysis were performed on censuses of species abundances in a sufficient number of these mixed communities, the resultant eigenvectors of the covariance matrix would, in fact, reproduce the n m-vectors of relative species abundances in the source communities.

In practice, any number of departures from these ideal conditions can confound the physical interpretability of the eigenvectors. If the species composition of the individuals drawn from a source community is not constant, but instead varies systematically with the total number of individuals drawn from that community, the trajectory of the variation contributed by that source community to the variation in the mixed communities will be curved. Axes defined by principal components are necessarily linear. In this situation, the first principal component will probably be a first order approximation to the actually nonlinear relationship representing the behavior of the dominant source community. But the residual owing to the nonlinearity will then be taken up by one or more of the succeeding principal components, which thus cannot be counted upon to represent, in a recognizable way, the compositions of the other source communities.

Another type of interference arises if, in the mixing process, the amounts drawn from the various source communities are not independent, but instead involve some pattern of correlation. Strong correlation can cause several source communities to appear collapsed in a single principal component, the eigenvector of which will be a weighted average of their respective compositions. The residual owing to the extent to which the correlation is imperfect will give rise to contributions in one or more succeeding components describing, in part, the differences between the correlated source communities.

A similar failure of isomorphism between the eigenvectors and the species composition vectors can occur if there is some overlap in the species represented in the respective source communities. Then the vectors of relative abundances cannot be orthogonal, but the vectors defining the principal components must be orthogonal. The discrepancy in outcome will depend on the degree of overlap. If some source communities share some of their more abundant species, these communities will appear collapsed into a single component with an eigenvector that is intermediate to their actual compositions. If the source communities have in common only species that are rare, or that at least are rare in all but one, their vectors of relative abundance will not be too far from orthogonal, and the principal components will still provide a close approximation to the compositions of the original source communities.

Finally, if the distributions of amounts drawn from the source communities are perversely non-Gaussian, the principal axis of variation in the mixed communities will represent polar extremes of the composition of the mixtures, but this will not correspond to the composition of any one of the source communities. This is likely to occur when the composition of the mixed communities sorts out into fairly distinct clusters (as would take place if the amounts drawn from each source community were strongly bimodal), in which case the principal component analysis will resolve the modes of variation between clusters, rather than between source communities. The effect is analogous to the way in which an outlier can cause a regression line to deviate drastically from the slope of a tight linear relation manifested near the centroid of the observations. The between-cluster pattern revealed by the principal components is certainly worth knowing, if that is the way the mixtures are structured, but the physical interpretation of the components is now quite different from that in the hypothetical case first described.

In general, the thoughtful practitioner should be able to deduce the physical significance of the first few principal components in an analysis of species abundance data. The higher order components almost certainly will not correspond to anything identifiable as a distinct constituent in the pattern of real data: they are just the result of a forced sort of residual mop-up operation. The decision as to how many of the components are informative is a matter of judgement, depending at least as much on familiarity with the ecological system as on statistical expertise.

Now, the covariances which comprised the input for the principal components analysis did not include temporal or positional information. That is, the covariances rest entirely on the relationships between species within discrete samples. Thus it is of interest to examine the spatial or temporal behavior of the derived components. This may be accomplished through spectral analysis of the sequences of values computed by transforming the original data sequence according to the appropriate eigenvectors.

Power spectra will indicate whether there is any periodic tendency in the components, and if so, at what scales. Cross spectra will indicate whether there is any lagged interaction between components. It may seem strange to speak of interaction between principal components, since these are necessarily uncorrelated. That zero correlation, however, like the covariances in the input matrix, is without regard for position or time. Thus we are assured that the cross correlation between components will be zero at zero lag; but at any scale larger than the span of one sampling interval, the cross correlation may be non-zero. Consider, for example, two components which are perfectly periodic, and with the same wavelength, but with one leading the other by one quarter of a period. As regards simultaneous observations of the two, they will display zero correlation, so they could arise as principal components; yet, at a lag of one quarter their period, their correlation is one, as would be revealed in a peak in their coherence at this frequency.

The computation of spectra, in principal, could be accomplished by fitting sine and cosine functions, the wavelengths of which form an harmonic progression, to segments of the sequence of observations. The average of the squares of the coefficients of the fitted amplitudes at a given wavelength, when suitably normalized, would be the power spectral estimate at that wavelength. Unfortunately, this does not imply that we may, without further reflection, infer from a spectral peak that the input sequence really is well described (noise aside) by an oscillatory function with the corresponding wavelength. Several features of the input sequence can give rise to misleading spectral peaks.

The sampling procedure is itself periodic. If the sampled input contains periodic variation at a wavelength roughly comparable to the sampling interval, or as much as about one order of magnitude shorter, the sequence of observations will give the impression of periodic variation with a spuriously long wavelength, in a manner analogous to the low frequency beats that result when two higher frequency oscillations are summed. This type of spurious low frequency signal is indicated if low frequency peaks in the spectrum prove sensitive to changes in the sampling interval.

A second class of misleading spectral peaks occurs if regular oscillations in the input sequence are not precisely sinusoidal. Then, there will, as expected, be a peak at the frequency corresponding to the frequency of the input oscillations, but the residuals resulting from the poor fit of the fundamental sine wave to the input wave will themselves be periodic. These will give rise to spectral peaks at harmonics of the fundamental frequency.

Finally, in short data records (which are, by force of circumstance, the rule rather than the exception in ecological studies where many species are enumerated), it is possible that an actual oscillation in the input sequence will not give rise to a well-defined spectral peak. Mathematically, the number of lags examined cannot exceed the number of observations. For statistical stability of the resultant spectrum, the maximum lag should be a good deal less than this--certainly no more than one quarter the length of the record of observations. This, then, severely limits the resolution at the low frequency end of the spectrum. In effect, a genuine low frequency periodicity may prove indistinguishable from a simple trend in the data, though to some extent it is possible to remove trends prior to calculating the spectrum.

As usual, the theoretical antidote to the possible interpretational difficulties posed by this program of spectral analysis on principal components is to apply more wisdom than anyone has to more data than anyone can get. Still, a biologist with a sound understanding of the behavior of a community should succeed in identifying the major features revealed by the analysis. These techniques can be of use in explicitly quantifying patterns that the biologist, in a rough way, already knows are present; and the analysis may suggest a specific search for some aspects of pattern that were not recognized in advance, but which are indeed found when an attempt is made to reconcile the results of the analysis with features of the original data.

To my knowledge, spectral analysis of principal components has not been attempted hitherto with ecological abundance data. The method has been employed with physical data, for example in meteorology (Trenberth, 1975).

3. PRINCIPAL COMPONENTS IN THE FREQUENCY DOMAIN

When using principal component analysis to identify species complexes, we rely on the pattern of covariances at zero lag to generate mathematical groupings of organisms, in the hope that these groupings form the basis of a pattern at non-zero lags. This hope may be frustrated if the essential pattern of the community is manifested only at scales larger than the span of one of our observations.

Consider, for example, an extreme situation in which all the spatial pattern in a community is due to a suite of species which are distributed in clumps considerably larger than the quadrats used to tally them, and where interference is so strong that no two of these species can occupy the same clump, though their distribution does not affect the spatially random arrangement of the other species. Imagine that the competitive relationships also determine which pairs of species can form adjacent clumps or neighboring clumps once removed. Then there will be a distinct pattern at this intermediate spatial scale. But the only hint of species grouping when the quadrats are analyzed without regard for position will be the co-occurrences recorded in that minority of observations where a quadrat fell on the border between two clumps; and interaction at a distance of more than one clump will be missed altogether.

So, we cannot always count on time domain principal components to identify correctly the natural species groupings in a spatially or temporally structured community. Instead, that information is to be found in the matrix of cross covariances computed at the appropriate lag.

A second drawback to the program of analysis outlined in Section 2 is the possibility of strong coherence between two components at the frequency of interest. Such interaction may correctly reflect the pattern of the community, but it abnegates one of the initial benefits of principal component analysis--namely, the orthogonality of the resultant components. That absence of intercorrelation facilitates the interpretation of multiple regression with other factors, and in a sense it is a guarantee of minimal redundancy between components. Thus, if possible, we should like a technique that will resolve components which show zero coherence, the analog of zero correlation, at the frequency of interest.

A technique that might suggest itself is eigenvector analysis of the matrix of cross covariances at a specific lag. The connection seems natural in that eigenvector analysis of a matrix of autocovariances at a sequence of lags, for one variable, is known to be mathematically equivalent to spectral analysis of that variable, in the limit as the number of lags becomes large. The analogy breaks down, however, in the multivariate example, for the matrix of cross covariances generally is not symmetric, so we would lose the orthogonality property which we desire for the eigenvectors.

In order to circumvent these difficulties, Wallace and Dickinson (1972) proposed eigenvector analysis of the cross spectrum matrix at a selected frequency. The cross spectrum matrix is Hermitian, and so its eigenvectors are orthogonal, though complex. Since the cross spectrum matrix is specific for frequency, the 'components' derived from it are likewise specific.

The analysis in essence filters the original sequence of observations, and deals with the pattern at only one frequency. Accordingly, if we wish to employ the resultant eigenvectors to transform the original variables, the eigenvectors should be applied to a filtered version of the original multivariate sequence from which variation at inappropriate frequencies has been removed. Multiplying the filtered sequence of observations by one of the complex eigenvectors will give rise to a complex series which, though awkward, preserves the mathematical properties we require.

In order to obtain a sequence of real-valued 'components,' Wallace and Dickenson employed a preliminary transformation of the filtered input sequence according to the formula

$$y_j(k) = x_j(k) - \frac{i}{f}\frac{dx_j(k)}{ds} ,$$

where $y_j(k)$ is the complex value of the j*th* transformed variable at observation (time or position) k , $x_j(k)$ is the value of the j*th* original variable in the filtered input sequence, f is the frequency of interest, and s is the positional basis (time or distance) of the original sequence. The derivative assumes an underlying continuous function which our observations sample at discrete intervals. In practice, a difference filter may be substituted for the derivate

$$\frac{dx_j(k)}{ds} \cong \frac{x_j(k+1) - x_j(k-1)}{2\Delta s}$$

where Δs is the sampling interval. The derivative is not a radical transformation of a periodic signal that is more-or-less sinusoidal, for $d(\sin x)/dx = \cos x$.

The multivariate complex-valued sequence formed by the sequence of vectors $\{\underline{y}(k)\}$ has the property that transformation by forming dot products with m of the complex eigenvectors of the cross spectrum matrix will give rise to a new, real valued, multivariate sequence, the m variables of which have no coherence at frequency f . The *jth* eigenvalue of the matrix is proportional to the amount of variance in the filtered original sequence that is accounted for by the *jth* new variable. Thus this program of analysis generates new variables which, in the frequency domain, have properties identical to those of principal components in the time domain.

The complex eigenvectors define periodic aggregates, and each complex element indicates the intensity and phase of the corresponding original variable's contribution to that aggregate. In an ecological application, where the original sequence is, for example, a record of numbers of individuals of various species along a spatial transect, species which maintain a consistent phase relationship at the frequency of interest will appear with high intensity values in the same complex eigenvector. If the wavelength chosen corresponds to the period of a multi-species patch pattern, species which appear together in the same patches will have similar phase values in their corresponding elements of the eigenvector; and pairs of species which alternate in appearance, never occupying a common patch, will be indicated by eigenvector elements that are out of phase. Species which regularly lead, or lag, the others by some fraction of a wavelength (as in a seasonal succession in a many-year time sequence) will also be indicated according to their phases in the eigenvector. If there is another aggregate of species which amongst themselves display a similar sort of patterning, but which show no consistent phase relationship with the former species groupings - sometimes overlapping one sort of patch, sometimes another - these will appear with large intensity values in a separate eigenvector which would similarly reflect their internal patch structure in the corresponding phase values.

The technique of frequency domain principal components has been applied to meterological data by Wallace (1972). Considerations bearing on the interpretability of the wave-structure components are the same as those discussed in connection with principal components and spectral analysis in Section 2. An added element of judgement enters here in the choice of frequency band.

4. COMPUTATION

The eigenanalysis of the Hermitian matrix may be undertaken directly with algorithms written for complex arithmetic. Alternatively, the problem may be transformed into one requiring eigenanalysis of a real symmetric matrix, which may be solved with widely available algorithms. Writing the m-order Hermitian matrix, $\underset{\sim}{H}$, as the sum of a matrix of real parts and a matrix of imaginary parts,

$$\underset{\sim}{H} = \underset{\sim}{A} + i\underset{\sim}{B} \quad ,$$

the real symmetric matrix of order $2m$, formed as

$$\underset{\sim}{M} = \begin{pmatrix} \underset{\sim}{A} & -\underset{\sim}{B} \\ \underset{\sim}{B} & \underset{\sim}{A} \end{pmatrix} ,$$

has m distinct eigenvalues, and these are identical to the eigenvalues of the Hermitian matrix. The associated real eigenvectors may be used to construct the complex eigenvectors of the Hermitian matrix according to the relationship

$$u_j = v_j + iv_{j+m} \quad ,$$

where u_j is the *jth* (complex) element of an eigenvector of $\underset{\sim}{H}$, and v_j is the *jth* (real) element of the corresponding eigenvector of $\underset{\sim}{M}$.

In the example presented in the next section, the alternative approach was employed in eigenanalysis of the Hermitian matrix. The eigenanalysis of the 2m-order real symmetric matrix was accomplished with the Givens-Householder algorithm, as coded by Cooley and Lohnes (1971). The eigenanalysis of the time domain covariance matrix was accomplished by the diagonalization method of Jacobi, as coded in the IBM SSP manual. The spectral analyses were carried out by the lag correlation method, using subroutine FTFREQ of the IMSL package.

5. APPLICATIONS TO AN ECOLOGICAL EXAMPLE: A TRANSECT THROUGH TWO-PHASE VEGETATION

To illustrate the methods of analysis presented in Sections 2 and 3, we apply them to a set of vegetation data generously provided by R. H. Whittaker. The sequence of observations consisted of estimates of coverage percent for all vascular plant species encountered in 107 contiguous, 1 m^2 quadrats in a transect through a mesquite grassland site in the Chaparral Wildlife Management Area, Texas. The very diverse vegetation is organized in patches of shrub scattered in open grassland. The transect intersected eight distinct shrub clumps, at an average interval of about 13 m.

Whittaker, Gilbert, and Connell (1978) have analyzed this transect with a number of ordination techniques, achieving their best results with the method of reciprocal averaging. By this latter means, all species were assigned a score from zero to 100--

low values indicating strong affinity for the shrub clumps, high values indicating strong affinity for grassy openings.

The transect included 71 species, most of which were found in more than 10% of the quadrats. To reduce the problems caused by excessive zeros in a correlation analysis, we restrict the ensuing, preliminary analysis to the eleven most frequent species--those that presented at least 1% coverage in more than 30% of the quadrats. The identities of these species, and their reciprocal averaging scores (taken from Whittaker *et al.*) are listed in Table 1. For the analysis, the percent coverage data were all augmented by one and then log transformed. This reduced skew, but the distributions remained markedly different from normal.

Time domain principal component analysis was performed using the correlation matrix to standardize the influence of each species. The first principal component clearly represents the axis of variation between the shrub and grassland vegetation types, as witness the strong negative correlation between the eigenvector elements and the corresponding species' reciprocal averaging score. This component indicates that *Dichaetophora, Chloris, Plantago, Aristida,* and *Evax* are positively correlated amongst themselves, as are *Parietaria, Monarda, Prosopis,* and *Cirsium,* but the two groups are negatively correlated. That is, the method resolves the two main vegetation types, and shows that they tend to exclude one another within quadrats.

The second component indicates in the residual a positive association among *Plantago, Oxalis,* and *Parietaria* with these negatively correlated with *Eragrostis.* Plotting the observations in the two dimensional space defined by components I and II showed a fairly ellipsoidal cloud of points, rather than the horseshoe shape that often plagues ordinations. Nine observations formed a cluster at the positive extreme of component II in this plot. *Plantago* and *Oxalis* were present in all nine of these; *Parietaria* was present in six, and *Eragrostis* was absent in all but one. There was no particular clustering at the negative extreme along component II. Examination of the ten most extreme observations at this pole showed *Eragrostis* abundant, *Parietaria* and *Plantago* absent, and *Oxalis* absent in six and sparse in four. The quadrats contributing both high and low extremes on component II were clustered along the transect. Both extremes occurred in quadrats that were predominantly transitional in their reciprocal averaging scores, with a slight grassland association.

No attempt was made to interpret components III and IV.

Spectra were computed on the basis of fifteen lags. Since the records of the original variables showed some obvious

TABLE 1: The eleven most frequent species encountered in the 107 m *transect through mesquite grassland. The* RA *score, from Whittaker, et al.* (1978) *is an index to the species' vegetation-phase affinity. The species are listed in order of increasing association with grassy openings.*

Species	Fraction of quadrats absent	Mean % coverage	RA score
Parietaria obtusa	.626	7.5	26.5
Prosopis glandulosa	.682	15.2	35.7
Cirsium texanum	.589	2.7	49.1
Monarda punctata	.664	1.0	54.9
Oxalis dilleni	.467	3.4	61.7
Eragrostis lugens	.346	17.8	66.7
Evax multicaulis	.456	1.5	69.2
Plantago sp.	.533	1.8	70.0
Aristida roemeriana	.682	3.1	72.3
Dichaetophora campes	.355	3.0	74.0
Chloris cucullata	.673	1.8	83.3

TABLE 2: Eigenvectors defining the first four time domain principal components of mesquite grassland transect data.

Species				
Dichaetophora campes	.45	-.18	.01	.28
Chloris cucullata	.38	-.06	-.27	.36
Plantago sp.	.31	.41	.35	-.11
Aristida roemeriana	.30	.21	-.22	.16
Evax multicaulis	.27	.01	.29	.38
Oxalis dilleni	.07	.41	.52	-.07
Eragrostis lugens	-.16	-.57	.25	.06
Circium texanum	-.26	.16	.25	.20
Prosopis glandulosa	-.29	.26	-.19	.62
Monarda punctata	-.30	-.15	.37	.42
Parietaria obtusa	-.34	.37	-.32	.05
Variance %	28	17	13	8

nonstationarity (e.g., almost all of the *Monarda* were encountered in the first half of the transect), the input sequences of the components were detrended. The spectra of all four components displayed distinct peaks at a wavelength of 15 m. For components II, III, and IV, this was clearly the dominant peak. For component I, there was slightly higher variance in the zero frequency band, attributable to periodicity at wavelengths too long to be resolved with this number of lags (greater than 30m).

Not counting coherence in the zero frequency band (probably due to correlation in the large scale trends across the transect), two of the cross spectra showed squared coherence greater than 0.5: one was a peak at 15 m in the cross spectrum of components I and II, and the second was at 6 m in the cross spectrum of components I and IV. Components I and II were exactly out of phase; components I and IV were in phase.

The peaks at 15 m capture the basic patch scale of the two phase vegetation, agreeing well (within limits of resolution) with the estimate of Whittaker *et al.* that the dominant scale of repetitive pattern in the community was 13-14 m. The coherence between components I and II at 15 m accords with our first thought that component II was picking out some sort of a transition association; but the revealed phase relationship forces upon us a different interpretation. That the two components are exactly out of phase implies that high values for component II occur as a distinct class of deviant quadrats *within* shrub patches, and low values occur as another sort of deviant quadrats *within* grassy openings.

Thus the spectral analysis shows that both time domain components I and II are features of the same mosaic pattern at a scale of about 15 m, despite their independence at the scale of individual quadrats.

No attempt was made to interpret the interaction between components I and IV.

All but three of the eleven species, themselves, had power spectra dominated by peaks at 15 m. The exceptions were *Aristida*, which, though it had a distinct peak at 15 m, had more variance at 6 m, *Monarda*, which had the peak shifted to the next lower frequency band (30 m), and *Parietaria*, which had the peak shifted to the next higher frequency band (10 m). *Evax* had a large secondary peak at 6 m, and *Plantago* had a large secondary peak at 5 m.

The cross spectra were computed, also at 15 lags, using *Prosopis* and *Chloris* as base series. There was strong low frequency coherence between *Chloris* and *Dichaetophora* at 15 m (in phase), between *Prosopis* and *Monarda* at 30 m (0.2 cycle phase difference), and between *Prosopis* and *Parietaria* in a broad low frequency band (in phase). At higher frequencies there was strong coherence between *Chloris* and *Oxalis* (out of phase), *Chloris* and *Prosopis* (0.2 cycles phase differences), and *Prosopis* and *Cirsium* (in phase). The low frequency coherences were consistent with the strongest patch affinities, as may be seen by comparison with reciprocal averaging scores in Table 1, or loadings on principal component I in Table 2, but the predominance of poorly defined cross spectra was dissappointing.

A matrix of all cross spectra at the 15 m wavelength was formed from calculation with 15 lags. The contribution of each species to the variance at this wavelength was standardized by dividing element a_{ij} of the matrix by $\sqrt{a_{ii}a_{jj}}$, thus normalizing the power of each variable to unity. The first three frequency domain components computed from this matrix accounted for 43%, 22%, and 17% of the variance, respectively.

The eigenvector defining the first component is plotted in amplitude phase form in Figure 1. This component was not dominated by a small subset of variables--all species except *Evax*, *Prosopis*, and *Monarda* made substantial, and roughly equivalent contributions to its variance. Neglecting these latter three, we can get some sense of the spatial pattern at the 15 m wavelength by examining the phase relationships in Figure 1. *Oxalis*, *Cirsium*, and *Parietaria* have similar phase angles, and these must correspond to the shrub patches. Their phase is opposite that of the pair, *Dichaetophora* and *Chloris*, which, evidently, comprise the spatially coherent core of the grassy opening assemblage.

Eragrostis, though not well centered, is aligned with the grassy association, and *Plantago* is aligned with the shrubs. The inclusion of *Oxalis* and *Plantago* with the shrub group might seem at odds with their loadings on the first time domain principal component, but their high loadings on the second time domain component, and the coherence between the two components revealed in their cross spectrum, indicated that *Oxalis* and *Plantago* should indeed be found in the shrub patches, albeit in atypical shrub quadrats. Similarly, the alignment of *Eragrostis* with the grass association is in qualitative agreement with the interpretation developed from spectral analysis of the time domain components.

The transitional character of *Aristida* is specific to the frequency domain analysis. The phase alignments of *Monarda* and *Prosopis* with the shrub group, and *Evax* with the grass assemblage, are in accord with the time domain analysis, despite their low amplitude-loadings in Figure 1.

Owing to the large amount of variance accounted for by frequency domain component I, and the small difference between the eigenvalues associated with frequency domain components II and III (indicative of the indistinctness of these axes), no attempt was made to interpret the higher order frequency domain eigenvectors.

Comparing the two programs of analysis, in this application, suggests that the method of Section 3 resolved more detail of the spatial pattern, whereas the method of Section 2 resolved more distinct groups. The two methods agreed in the main features of the pattern revealed. The groups formed by the first and second

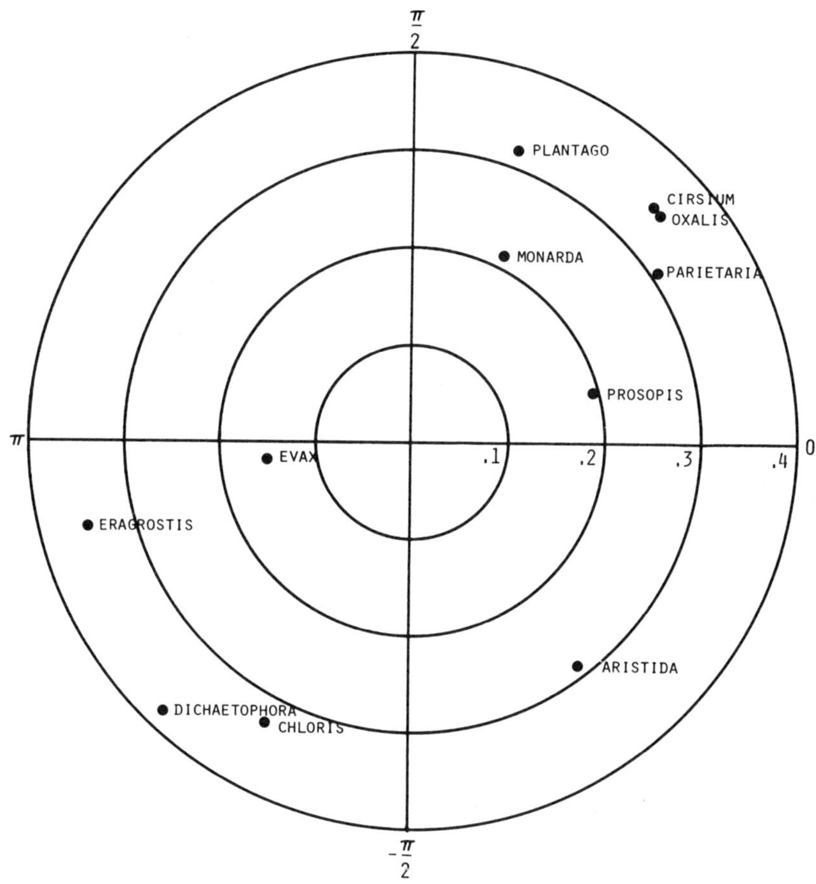

FIG. 1: First frequency domain principal component of the mesquite grassland transect. The eigenvector is plotted in polar form, with distance from the origin denoting the amplitude, and angle denoting phase.

time domain components did prove to have spatial integrity; and their compositions, once the coherence pattern was taken into account, coincided with the two main clusterings in the wave structure resolved by the first frequency domain component at the dominant frequency. The frequency domain component analysis identified species characteristic of transitions in the vegetation mosaic; the time domain analysis seemed to pick up only the two extremes of the mosaic.

The success of these techniques when applied to a distinctly patterned vegetation encourages us to attempt the same program with more complex and less obvious vegetation structures.

ACKNOWLEDGEMENTS

I am indebted to J. M. Wallace for first bringing his frequency domain techniques to my attention. R. H. Whittaker's collaboration was invaluable both in providing the transect data and in interpreting the patterns revealed therein, J. T. Enright and an anonymous reviewer made helpful comments on the original draft.

REFERENCES

Cooley, W. W. and Lohnes, P. R. (1971). *Multivariate Data Analysis*. Wiley, New York.

Diamond, J. M. (1975). Assembly of species communities. In *Ecology and Evolution of Communities*, M. L. Cody and J. M. Diamond, eds. Belknap, Cambridge, Massachusetts.

Jenkins, G. M. and Watts, D. G. (1968). *Spectral Analysis and its Applications*. Holden Day, San Francisco.

Platt, T. and Denman, K. L. (1975). Spectral analysis in ecology. *Annual Review of Ecology and Systematics*, 7, 189-210.

Trenberth, K. E. (1975). A quasi-biennial standing wave in the Southern Hemisphere and interrelations with sea surface temperature. *Quarterly Journal of the Royal Meteorological Society*, 101, 55-74.

Wallace, J. M. (1972). Empirical orthogonal representation of time series in the frequency domain. Part II: Application to the study of tropical wave disturbances. *Journal of Applied Meteorology*, 11, 883-900.

Wallace, J. M. and Dickenson, R. E. (1972). Empirical orthogonal representation of time series in the frequency domain. Part I: Theoretical considerations, *Journal of Applied Meteorology*, 11, 887-892.

Whittaker, R. H., Gilbert, L. E., and Connell, J. H. (1978). Intracommunity pattern in a mesquite grassland, Texas. *Journal of Ecology*, in press.

[*Received July* 1978. *Revised March* 1979]

G. P. Patil and M. Rosenzweig, (eds.),
Contemporary Quantitative Ecology and Related Ecometrics, pp. 157-165.

ANALYSIS OF TWO-PHASE PATTERNS

R. H. WHITTAKER

Ecology and Systematics
Cornell University
Ithaca, New York 14853 USA

Z. NAVEH

Faculty of Agricultural Engineering
Israel Institute of Technology
Haifa 32000, Israel

SUMMARY. Ordination of small quadrats (1-m^2) by reciprocal analysis was effective for pattern analysis of three communities with shrub patches in grassland matrices. The technique extracted the major axes of community differentiation from shrub cover to openings, expressed species (niche) relationships to these axes, and provided measurements of pattern diversity, intensity, and boundedness. Along the axes species centers were dispersed and species turnovers were of the order of 2-4 half-changes.

KEY WORDS. mallee, mesquite grassland, ordination, patches, pattern analysis, *Pistacia*, reciprocal averaging, two-phase vegetation.

Traditional pattern analysis deals with species one or two at a time for measurements of contagion and species association (Greig-Smith, 1964; Kershaw, 1973; Pielou, 1974). These techniques are appropriate for cases in which there is no strong coordination among the species in their patterns of occurrence. There is need also for pattern techniques that are integrative in the sense of dealing with the whole flora at once to reveal major axes of differentiation in the community, axes to which most or all species relate. One approach is the use of ordination of small quadrats by their whole species compositions, instead of species-by-species measurements. Reciprocal averaging or correspondence analysis (Hill, 1973b, 1974) seems to be a best technique for revealing, objectively and with minimum distortion, a first, major direction of differentiation in a set of community samples (Gauch *et al.*, 1977). We have used reciprocal averaging to ordinate sets of 100 contiguous one-meter square quadrats forming strip transects in communities with well-defined two-phase differentiation, and we describe results for communities on three continents--a mallee in New South Wales, Australia (Whittaker *et al.*, 1979b); a mesquite grassland in Texas, USA. (Whittaker *et al.*, 1979a); and a *Pistacia lentiscus* woodland at Bosmat Tivon here in Israel (Naveh and Whittaker, unpublished data).

Reciprocal averaging simultaneously arranges and ranks species and quadrats along the major axis of community differentiation from shaded centers of woody clumps to the full light of openings. On the first two axes the ordinations are triangular as illustrated in Figure 1, for samples in the Australian mallee. The triangular form is consequent on the fact the grassy open quadrats, on the right, share more small plant species of frequent occurrence than do the quadrats in the woody clumps. Results may be shown also in the form of a reciprocal averaging trace (Figure 2). Ordination scores for the meter-square quadrats are the vertical axis, and the sequence of 100 quadrats on the horizontal. The trace is a graphic depiction of the pattern as recorded in the field, with quadrats primarily in the open above the dashed line and those in shrub clumps below it. The characteristics of the patches, including their sizes and relative intensities are shown, together with such special features as a single shrub in the open, partly modifying its undergrowth toward that of the shrub clumps (roman numeral VIII, quadrat 76), and a large shrub patch with a break in its canopy below which the flora varies toward that of the openings (roman numeral X, quadrats 100 and 101).

Reciprocal averaging, spectral analysis, and traditional measurements were used to study species association. The species are not organized into two or three distinct groups corresponding to the openings and the shrub patches, or these and the borders. They are, instead, dispersed along the axis of community differentiation as represented by reciprocal averaging. Figure 3 shows species ordinations for the mesquite grassland with lines representing two measurements of association--the Cole (1949) index and percentage similarity of distribution (Whittaker and Fairbanks, 1958). These and other measures of association give patterns that are generally similar, but differ in detail, from one measure to another. These measures, like the reciprocal averaging, show no strongly defined, coherent groups of species. There are instead many relatively weak associations among species that are widely distributed along the axis and centered near the middle of the axis,

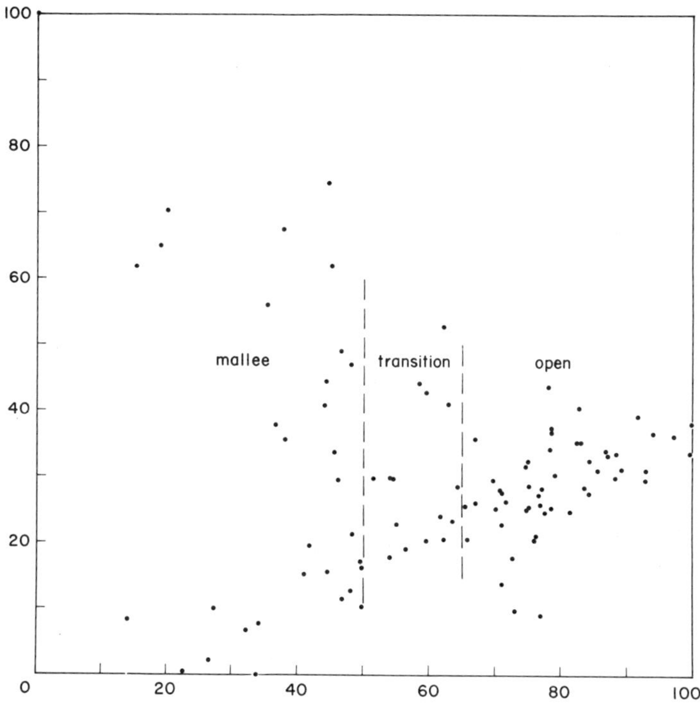

FIG. 1: Reciprocal averaging ordination of 100 *quadrats (each one square meter) of a strip transect in a mallee community, New South Wales (Whittaker et al.,* 1979b). *First axis = abscissa and second = ordinate.*

and fewer, weaker associations among species occurring primarily in the openings or in the shrub clumps. A number of measures of species contagion and scale were used (Krishna Iyer, 1948; Jones, 1955-1956; Greig-Smith, 1964; Hill, 1973a; Platt and Denmann, 1975; Goodman, 1979). As would be expected, most of the species show positive contagion, but the suggested scales of species contagion showed little consistency with one another or with the scale of the community pattern. The pattern is striking on the community level as recognized by reciprocal averaging, and yet stochastic and variable on the level of individual species.

The reciprocal averaging permits a number of other measurements and characterizations of the pattern. A first property is the relative boundedness vs. continuity of the two kinds of patches with each other. The RA scores are assessments of relative compositional separation of samples along the major axis of community

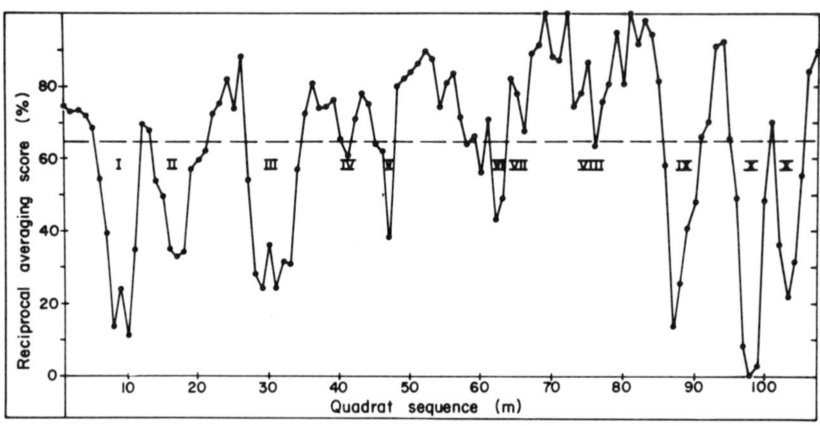

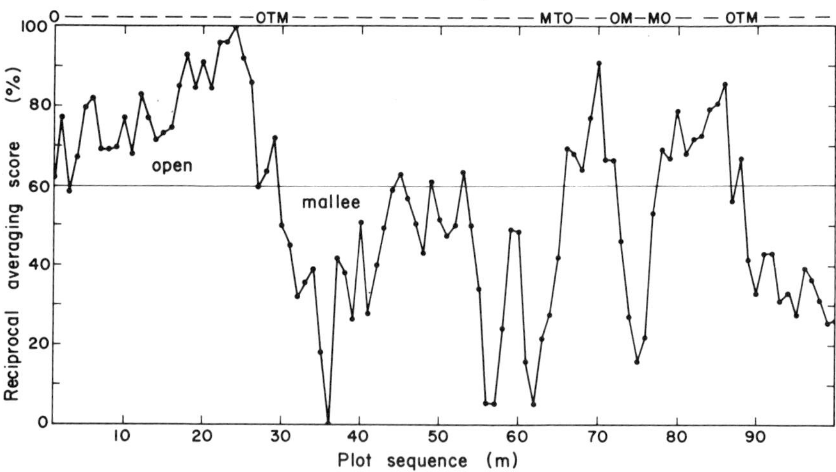

FIG. 2: Reciprocal averaging traces for strip transects in two-phase vegetation: a mesquite grassland in Texas, above, and a mallee community in New South Wales, below (Whittaker et al., 1979a,b)

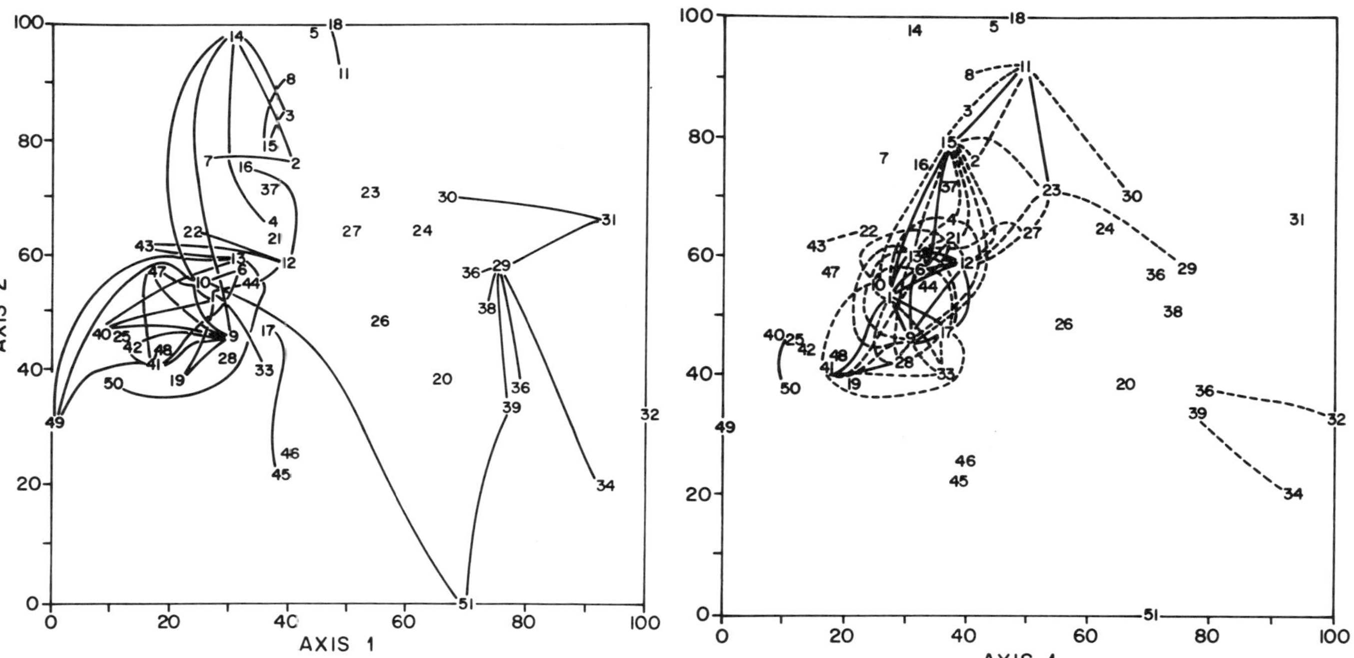

FIG. 3: Reciprocal averaging ordinations of species in a Texas mesquite grassland (Whittaker et al., 1979a). Connecting lines indicate high values for species association: Left -- Cole (1949) indices over 60%; right -- percentage similarity of distribution over 60% (solid) or 50% (broken).

differentiation. These scores can then be used as expressions of rate of change from one quadrat to the next within the patches and across their borders. In the mallee the mean difference of RA scores of successive quadrats was 6.86 in the openings and 11.86 in the shrub patches vs. 14.66 in the transitions. Corresponding values for the mesquite grassland are 4.97 and 11.28 vs. 24.66. As already observed, the quadrats within the shrub clumps differ from one another more than those within the openings. However, the ratio of the rate of change in the transitions to the mean rate within the two phases gives an expression of relative boundedness of the pattern; these ratios are 1.57 for the mallee and 3.04 for the mesquite grassland. The phases are more sharply bounded in the mesquite grassland than in the mallee with its larger, taller, more diffuse clumps. For the *Pistacia lentiscus* woodland the three values are 4.35 in the openings, 5.58 within the shrubs, and 49.31 for transitions; the corresponding ratio is 9.9. This woodland has shrubs, or small trees, that are not erect but convex, almost hemispheric from the ground upward; their surfaces are trimmed by goat browsing, and over the whole hemispheric surface the branches, foliage, and shade are relatively dense. The flora changes abruptly across the edges of the shrubs; this is the most sharply bordered two-phase vegetation known to us in the world.

Other uses of the RA scores can be suggested. Correlations of quadrats at increasing distances can be used to seek the scale of the pattern. For the mesquite grassland peak similarities of quadrats by minimum difference in RA scores, as well as by the traditional measures of percentage similarity and coefficient of community, are at 13-14 m. This agrees with the observed spacing of 8 shrub clumps (excluding separate small shrubs in the openings) in the 107-m strip. A measure of period intensity subtracts this minimum difference in RA scores--21.9% at 14 m--from the background difference for quadrats compared (25-28%, mean 27%). This difference, which is only 5%, indicates that the periodicity of this vegetation (with its patches of irregular size variously intersected by the strip transect) is weak, despite the strong floristic contrast of the two phases.

Intensity of contrast can be measured by the difference between the mean RA scores for quadrats within the openings and within the shrub patches; these values are 43% for the mallee, 51% for the mesquite grassland, and 87% for the *Pistacia* woodland. This woodland again seems one of the most strongly developed two-phase patterns in the world. These values are affected by decisions on which quadrats are transitional. Perhaps more effective is an assessment of the length of the first axis of community differentiation as measured in half-changes of species turnover along it. This, which we define as pattern diversity (Whittaker, 1972; cf. Pielou, 1966), can be estimated from the first eigenvalue of the RA by the relationship $HC = \sqrt{12EV/(1-EV)}/1.349$ (M. O. Hill, personal communication), and checked by a graphical estimation of half-change number (Whittaker, 1960). These values are 1.79 HC for an eigenvalue of 0.328 in the mesquite grassland, 2.3 HC for an eigenvalue of 0.444 in the mallee. Thus the mallee has stronger pattern differentiation, despite weaker periodicity and patch boundedness, than the mesquite grassland. The *Pistacia* woodland gives much the strongest pattern diversity of the three communities--about 4 HC.

We suggest some conclusions. *First,* the approach of using floristic ordinations of quadrats by reciprocal averaging as a basis of a community-level or synthetic pattern analysis is effective. For our cases the approach is much more effective than an effort to build from species-by-species measurements to characterization of community pattern. The approach seems most effective for patterns that are both well-defined and floristically rich; trial applications to ordinary forest undergrowth have not seemed rewarding. *Second,* results of interest in relation to theory of community organization are suggested. It is striking that the degrees of species turnover along this internal pattern axes - 2 to 4 half-changes - are of the same order as those along topographic moisture gradients connecting different kinds of communities in mountains (Whittaker, 1960; Whittaker and Niering, 1965). Along these internal axes species are not organized into discrete groups but are dispersed and 'individualistic' in their distributions. We presumably deal with effects of evolution--the tendency of species to disperse themselves, with consequent reduction of competition, along a resource gradient--in this case a gradient of microhabitat differentiation within the community. *Third,* we note the wide occurrence of two-phase patterning in vegetation consequent on undulant microtopography, on frost effects or bog succession, on local disturbance by tree-falls or animals, or on effects of woody plants dispersed in an open matrix. The last effects, of which we describe three cases, are widespread in semiarid climates. The community characteristics we call 'patterns' and 'mosaics' are not a single phenomenon but a congeries of different yet variously overlapping kinds of relationships of populations to environment and one another (Whittaker and Levin, 1977). The approach we describe is adapted to some kinds of pattern and not to others. For two-phase vegetation, however, it offers: effective characterization of a given pattern, clarification of species relationships that may be the subject of autecological research, measurements by which patterns that share no species may be compared, and perhaps some insight into the organization of plant communities in which pattern is a major axis of niche differentiation.

ACKNOWLEDGEMENTS

Research supported by the National Science Foundation USA, and the US-Israel Binational Science Foundation.

REFERENCES

Cole, L. C. (1949). The measurement of interspecific association. *Ecology*, 30, 411-424.

Gauch, H. G., Jr., Whittaker, R. H., and Wentworth, T. R. (1977). A comparative study of reciprocal averaging and other ordination techniques. *Journal of Ecology*, 65, 157-174.

Greig-Smith, P. (1964). *Quantitative Plant Ecology*, 2nd ed. Butterworths, London.

Goodman, D. (1979). Applications of eigenvector analysis in the resolution of spectral pattern in spatial and temporal ecological sequences. In *Contemporary Quantitative Ecology and Related Ecometrics*, G. P. Patil and M. Rosenzweig, eds. Satellite Program in Statistical Ecology, International Cooperative Publishing House, Fairland, Maryland.

Hill, M. O. (1973a). The intensity of spatial pattern in plant communities. *Journal of Ecology*, 61, 225-235.

Hill, M. O. (1973b). Reciprocal averaging: an eigenvector method of ordination. *Journal of Ecology*, 61, 237-249.

Hill, M. O. (1974). Correspondence analysis: a neglected multivariate method. *Journal of the Royal Statistical Society, Series C*, 23, 340-354.

Jones, E. W. (1955-6). Ecological studies on the rain forest of southern Nigeria. IV. The plateau forest of the Okomu Forest Reserve. *Journal of Ecology*, 43, 564-594; 44, 83-117.

Kershaw, K. A. (1973). *Quantitative and Dynamic Plant Ecology*, 2nd ed. Elsevier, New York.

Krishna Iyer, P. V. (1948). The theory of probability distribution of points on a line. *Journal of Indian Society of Agricultural Statistics*, 1, 173-195.

Pielou, E. C. (1966). Species-diversity and pattern-diversity in the study of ecological succession. *Journal of Theoretical Biology*, 10, 370-383.

Pielou, E. C. (1974). *Population and Community Ecology: Principles and Methods.* Gordon & Breach, New York.

Platt, T. and Denman, K. L. (1975). Spectral analysis in ecology. *Annual Review of Ecology and Systematics*, 6, 189-210.

Whittaker, R. H. (1960). Vegetation of the Siskiyou Mountains, Oregon and California. *Ecological Monographs*, 30, 279-338.

Whittaker, R. H. (1972). Evolution and measurement of species diversity. *Taxon*, 21, 213-251.

Whittaker, R. H. and Fairbanks, C. W. (1958). A study of plankton copepod communities in the Columbia Basin, southeastern Washington. *Ecology*, 39, 46-65.

Whittaker, R. H. and Levin, S. A. (1977). The role of mosaic phenomena in natural communities. *Theoretical Population Biology*, 12, 117-139.

Whittaker, R. H., and Niering, W. A. (1965). Vegetation of the Santa Catalina Mountains, Arizona. (II). A gradient analysis of the south slope. *Ecology*, 46, 429-452.

Whittaker, R. H., Gilbert, L. E. and Connell, J. H. (1979a). Analysis of two-phase pattern in a mesquite grassland, Texas. *Journal of Ecology* (in press).

Whittaker, R. H., Niering, W. A. and Crisp, M. D. (1979). Structure, pattern, and diversity of a mallee community in New South Wales. *Vegetatio* (in press).

[*Received July* 1978. *Revised March* 1979]

G. P. Patil and M. Rosenzweig, (eds.),
Contemporary Quantitative Ecology and Related Ecometrics, pp. 167-196.

ECOLOGICAL PARAMETERS INFERRED FROM SPATIAL CORRELOGRAMS

ROBERT R. SOKAL

Department of Ecology and Evolution
State University of New York
Stony Brook, New York 11794 USA

SUMMARY. Spatial autocorrelation is the dependence of the values of a variable on values of the same variable at geographically adjoining locations. An account of measures of spatial autocorrelation and the computation of correlograms is followed by a discussion of the structural inferences that can be made about ecological response surfaces from analyses of their correlograms. The nature of the patterns and estimates of patch sizes are reflected in correlograms. These points are illustrated by means of surfaces containing artificial patterns and patches varying in size and shape. The ecological implications of spatial autocorrelation can be deduced from sign, magnitude and order of autocorrelation coefficients, from analyses of statistical heterogeneity of data as contrasted with spatial patterns, and from the analysis of more than one variable measured for the same population. These considerations are focused on those variables likely to be of special interest in ecology -- abundance patterns, marker variables such as gene frequencies or morphometric data that estimate underlying ecological parameters, and species associations in studies of the distribution of various species making up a community. The methodology is illustrated with an analysis of human population changes in the counties of the Republic of Ireland during the last 100 years.

KEY WORDS. spatial autocorrelation, spatial variation patterns, spatial correlograms, patch size, population changes in Ireland.

1. INTRODUCTION

Spatial autocorrelation is the dependence of the values of a variable on values of the same variable at geographically adjoining locations. Although techniques for estimating spatial autocorrelation date back to the work of Moran (1950) and Geary (1954), and although its biological implications were noted quite early (Whittle, 1954; Matern, 1960), the recent interest in this subject dates to the important contribution by Cliff and Ord (1973), whose monograph on the subject furnished measures of estimation of the dependence and tests of its significance. Biologists were quick to recognize the importance to their own field of this method developed by statisticians and geographers. The range of application of spatial autocorrelation to biology is considerable. Biological materials manifestly obey the first law of geography, as stated by Tobler (1970); "Everything is related to everything, else, but near things are more related than distant things." Jumars, Thistle, and Jones (1977) introduced the method to modern ecology by an analysis of spatial autocorrelation in abundance data of marine benthos. A follow-up study by Jumars (1978) included vertical and horizontal structure of a benthic community. Sokal and Oden (1978a), working independently, presented the technique to population biologists, and extended it to include an analysis of spatial correlograms. In a companion paper (Sokal and Oden, 1978b), they analyzed and interpreted a series of data sets from population genetics, population ecology, and community ecology. Other applications of this technique can be found in Sokal (1978). These methods are currently being tried out by numerous workers, in a variety of applications.

In their papers, Sokal and Oden (1978a,b) emphasized population genetic inferences that could be made from spatial autocorrelation studies. In the present paper, I shall stress ecological parameters that can be inferred by autocorrelation analysis. Among the parameters that lend themselves to such analysis are vagility of organisms, the patch size, grain and gradients of ecological factors, and patterns of differential mortality (selection).

2. MEASURES OF SPATIAL AUTOCORRELATION

The computation of spatial autocorrelation requires a set of localities represented as points in the plane, and one or more variables mapped onto these points with one value per variable for each point. The autocorrelation for one variable is computed over those pairs of points that are connected, i.e., considered neighbors in some sense. Rules for connecting localities are given by various authors. Tobler (1975) lists eight different techniques. One of these is the so-called Gabriel graph or Gabriel network, developed by Gabriel and Sokal (1969), whose properties have recently been examined by Matula and Sokal (1979). Such connections between localities can be indicated graphically by a line (edge) joining the pair of localities concerned, whereas localities not considered neighbors will not be so connected. The resulting graph can be represented in matrix form by a so-called connectivity or adjacency matrix $\underset{\sim}{W}$. The simplest such matrix is binary, indicating absence of connections between pairs of localities by zeros and the presence of connections by ones. In many instances, the adjacency matrix becomes a weight matrix, with the ones in the matrix replaced by weights w_{ij} which are some function of the geographic or ecological distance between i and j, the pair of localities indicated.

Spatial autocorrelation for interval (and ranked) data, such as population densities, physical variables, or gene frequencies, is estimated by either of two coefficients. Moran's coefficient is computed as

$$I = n\Sigma_{ij}w_{ij}z_iz_j \;/\; W\sum_{i=1}^{n} z_i^2 \;,$$

where n is the number of localities in the study; Σ_{ij} indicates summation over all i from 1 to n and over all j from 1 to n, $i \neq j$; w_{ij} is the weight given to an edge between localities i and j (w_{ij} need not equal w_{ji}); $z_i = Y_i - \bar{Y}$, where Y_i is the value of variable Y for locality i, and $\bar{Y}$ is the mean of Y for all localities; and $W = \Sigma_{ij}\, w_{ij}$, the sum of the matrix of weights (except for the diagonal entries, if any). Although for large sample size this coefficient ranges from -1 to +1, the expected value (no autocorrelation) approaching zero, significant spatial patterns of $|I|$ typically range only to 2/3.

The formula for Geary's coefficient is

$$c = (n-1)\,\Sigma_{ij}\, w_{ij}(Y_i-Y_j)^2/2W \sum_{i=1}^{n} z_i^2 \;.$$

All terms in this formula have already been explained. Geary's c will range from zero for perfect positive autocorrelation to an unbounded positive value for negative autocorrelation, the expected value in the absence of autocorrelation being one. The coefficient is frequently multiplied by minus one, to reflect its range and make it correspond in direction with I (thus a change to negative autocorrelation will make the coefficient more negative). Such a transformed c, as employed in this paper, ranges from 0 to −1 to <−1 for perfect positive, no, or negative autocorrelation, respectively.

Moran's coefficient is related to the product moment correlation coefficient, and is strongly affected by covariant departures from the mean, that is, if extreme values are spatial neighbors, they will appreciably contribute to the magnitude of the autocorrelation coefficient I. Even great similarity among spatial neighbors showing values near the mean will not contribute much to this coefficient. By contrast, Geary's coefficient c measures the similarity of neighboring values as a distance function, regardless of the size of the deviation of these values from the mean. High values of I not accompanied by high values of c would imply a deterministic patterning of extreme values, whereas the opposite relation would imply that extreme values were chance phenomena not spatially related, whereas more central observations were spatially aggregated.

Another way of interpreting the relations between the two coefficients has been proposed by Neal L. Oden (in litt.). Let us set

$$T = (n-1) \sum_{i=1}^{n} (w_{i\cdot} + w_{\cdot i})\, z_i^2 \,/\, 2W \sum_{i=1}^{n} z_i^2 \ .$$

The summation in the numerator of this quantity is the product of two terms. The first expresses a measure of the importance (weights) of any given locality ($w_{i\cdot} + w_{\cdot i}$, the sum of the row weights plus that of the column weights, will be $2w_{i\cdot}$ in the case of a symmetrical weight matrix; with a binary connectivity matrix this quantity is a measure of the number of connections entering any one locality). The second term is the square of the deviation $Y_i - \bar{Y}$. Thus T will be large when important (highly connected) localities also have large deviations from the mean of the variable studied. It can be shown that

$$c = T - (n-1)I/n \ .$$

Defining $d = 1-c$ and $J = I + (n-1)^{-1}$ will make the two transformed coefficients change in the same direction with changes in the autocorrelation coefficients, and expected values of each coefficient are zero. For large n

$$d - J \doteq 1 - T \ .$$

It can be shown (under normality and without spatial autocorrelation) that the expected value of T is $(n-1)/n$. Thus the difference between d and J, and by extension between c and I, is due to differences in T. When there is positive autocorrelation (d and J both > 0), $T > 1$ will result in $d < J$, while $T < 1$ will result in $d > J$. With negative autocorrelation (d and J both < 0). $T > 1$ will result in $|d| > |J|$, whereas $T < 1$ results in $|d| < |J|$. Thus we can make inferences about the magnitude of T from the sign and magnitude of the differences between c and I, and in turn can ascribe in T to the size of the deviations at important, highly connected localities.

The analysis of nominal data, such as individuals representing different species or different genotypes, is carried out by calculating join counts, a join being a synonym for an edge connecting two localities considered neighbors. Joins can connect localities with like or unlike values of the nominal variable. Thus, if there are two color morphs, black and white, a count of the number of edges connecting like individuals (BB and WW), and those connecting unlike morphs (BW), can be compared against expected counts on the assumption of spatial independence of the observations. Departures from expectation can be tested for both the interval and the nominal case by dividing the deviations by their appropriate standard errors which are given by Cliff and Ord (1973) and are summarized for the nominal case and for Moran's I by Sokal and Oden (1978a).

The analysis can be extended to include not only those pairs of localities directly connected by an edge, but also those that can be connected indirectly by passing through other points in the study. Autocorrelation can thus be computed for all pairs of points in a study, so long as the set of points and edges represents a connected graph. If the weight matrix is binary, then edge length between a pair of directly connected localities is 1 and the minimum distance between any pair of localities i and j is the least number of edges one needs to traverse to move from locality i to locality j. Alternatively, one can assign real values as edge lengths, usually the distance between the localities or some transformation of this distance. One can then compute the minimum distance between all pairs of localities measured along the connecting graph.

A frequency distribution of interlocality distances is set up for a given study, and the average autocorrelation for each distance class is computed. A graph of spatial autocorrelation coefficients against distance classes is known as a spatial correlogram. Correlograms summarize the patterns of geographic variation exhibited by the response surface for any given variable, and are thus simple analogs to spectral analysis of response surfaces. Correlograms describe the underlying spatial relationships of a surface rather than its appearance, and for this reason they are probably closer guides to some of the processes that have generated the surface than are the surfaces themselves.

3. STRUCTURAL INFERENCES FROM CORRELOGRAMS

A high positive or negative autocorrelation for any given distance class is relatively easy to interpret. Thus, when low order autocorrelations (those between near neighbors) are high and positive, it implies that neighbors are similar to each other. If such autocorrelations are high and negative, neighboring localities are strongly dissimilar. However, relations become more complex when the correlogram is considered as a whole. The shape of the correlogram will depend on the autocorrelation coefficients for the separate distance classes and the coefficients in turn will be a function of the proportion of locality pairs a given distance apart that are similar in the variable being studied (i.e., they are in the same 'patch,' as regards the variable being observed), versus the proportion of locality pairs that are dissimilar, (i.e., that fall into different patches). The nature of the sampling system and the connection matrix imposed upon the localities will determine the distribution of distance classes, while the topography of the response surface of the variable over the localities will determine the relative proportions within and among patches.

Considering the response surface as a continuum, Sokal and Oden (1978a) examined a series of artificially generated patterns superimposed on a Gabriel-connected graph of 53 localities. The patterns had been generated earlier by Royaltey, Astrachan, and Sokal (1975) in a study developing tests for the significance of geographic variation patterns. The patterns were generated by assigning ranks from 1 through 53 to the vertices of this graph in such a manner as to achieve the desired pattern. The results of applying autocorrelation analysis to the five artificial patterns are shown in Figure 1 as correlograms based on a binary adjacency matrix. The cline, a gradual transition of high values in the north to low values in the south, yielded a marked decline in the correlogram from significant positive autocorrelation up to distance class 3, to significant negative autocorrelation for distance classes 5 through 9. Thus it is difficult to travel to a place more than 3 edge lengths away and remain in a region where values of the variable are more or less similar to those at the point of origin. The depression, which was a circular cline, showed a correlogram quite similar to that of the regular cline, with the lowest negative autocorrelation at distance class 7, the approximate radius of the depression. In the double depression, a pattern of two largely overlapping circular clines, the autocorrelations at higher distances are not so strongly negative, since dissimilar

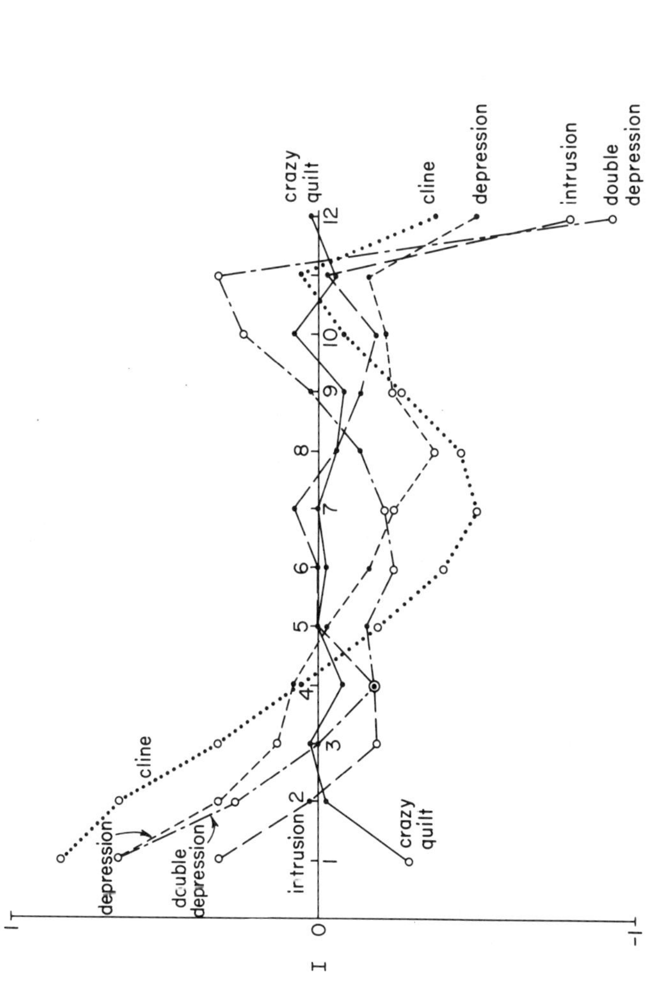

FIG. 1: Correlograms for five artificially generated variables intended to simulate distinct patterns of geographic variation. The variables had been mapped onto fifty-three localities in eastern North America connected by a Gabriel network and shown in Sokal and Oden (1978b, *Figure* 7). *Abscissa: distance classes in edge lengths (the correlograms differ in this respect from another analysis of the same data, where distances are expressed in kilometers published in Sokal and Oden,* 1978b, *Figure* 8). *Ordinate: Moran's autocorrelation coefficient* I. *Significant autocorrelation coefficients* ($p \leq 0.05$) *are indicated by open circles.*

localities that are far apart (mostly center versus periphery) are to some extent cancelled out by similar pairs in diametrically opposite regions of the circle.

The intrusion is an area of high rank in the southwest of the map, with a sharp abutment (step cline) bordering on low ranked localities and consequently shows positive autocorrelation up to distance class 2 because of the homogeneity within the compact southwestern region. Significant negative autocorrelation is shown at distance class 4, because travel over that distance tends to lead from one area to another differing greatly in the value of that variable. The crazy-quilt pattern was arranged in such a way that a traveller across the graph would encounter alternating high and low ranks. Consequently, its correlogram yielded a significant negative autocorrelation at distance class 1. At greater distances, the number of high and low ranks encountered would cancel out because the connecting graph was not a regular figure and the crazy-quilt pattern was not perfect. For this reason the autocorrelation coefficients for all distance classes ≥ 2 are low and not significant.

To obtain a clearer understanding of the relation of surface topography to correlogram shape, I investigated the properties of patches of various sizes, and configurations. A series of experiments was undertaken to determine the effect of patch size, patch shape, and nature of pattern on the resulting correlogram. For this study we employed a rectangular 12 × 12 grid consisting of 144 localities which are connected by queen's moves, thus making each point in the center of a 3 × 3 subset of the grid connected to 8 surrounding points. The edges of the lattice were connected to each other to form a torus in order to avoid edge effects. The surface grid of the torus was first divided into regular square patches of four sizes: 36 patches of dimension 2 × 2 (containing 4 localities), 16 of dimension 3 × 3, 9 of dimension 4 × 4, and 4 of dimension 6 × 6. For the study of the 2 × 2 patches the numbers 1 through 36 were assigned to the 36 patches. The manner in which these numbers were assigned determined the pattern of the variable on the surface. The patterns for patches of dimension 2 × 2 are shown in Figure 2. A helical cline radiating out from a high center is illustrated in Figure 2a, a ridge cline with a ridge of high values in rows 3 and 4 of the grid and low values in rows 1 and 6 in Figure 2b, and a random pattern generated by permuting the numbers 1 through 36 over the 36 patches in Figure 2c. Helical clines were also produced for the 3 × 3, 4 × 4, and 6 × 6 patches and random patterns for the 3 × 3 and 4 × 4 patches.

The results are shown in Table 1 which shows the numerical values for correlograms for all distance classes. Note that distance classes only go to 6 edge lengths. Since the lattice is a torus no pair of localities can be more than 6 edge lengths apart.

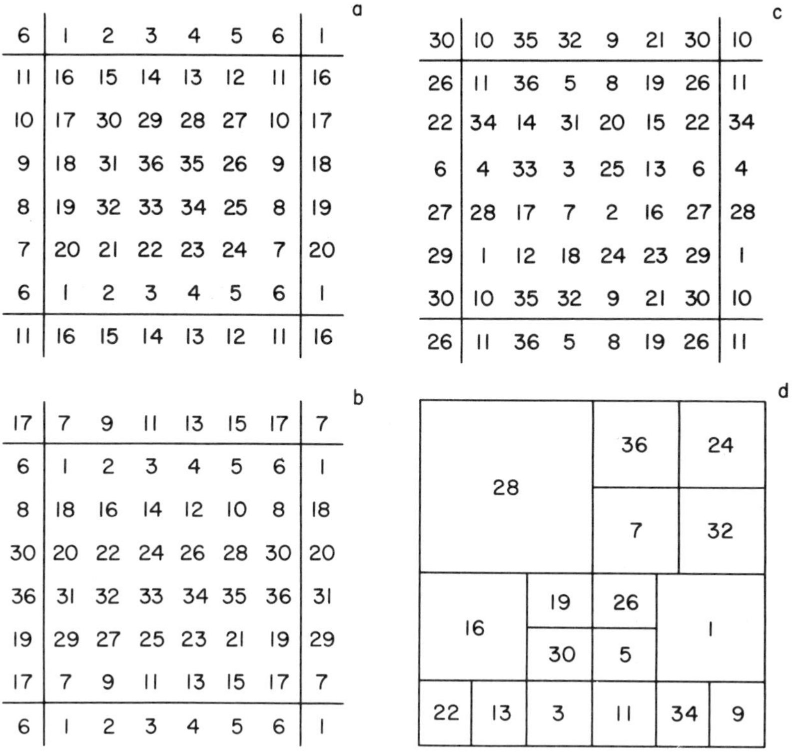

a

6	1	2	3	4	5	6	1
11	16	15	14	13	12	11	16
10	17	30	29	28	27	10	17
9	18	31	36	35	26	9	18
8	19	32	33	34	25	8	19
7	20	21	22	23	24	7	20
6	1	2	3	4	5	6	1
11	16	15	14	13	12	11	16

c

30	10	35	32	9	21	30	10
26	11	36	5	8	19	26	11
22	34	14	31	20	15	22	34
6	4	33	3	25	13	6	4
27	28	17	7	2	16	27	28
29	1	12	18	24	23	29	1
30	10	35	32	9	21	30	10
26	11	36	5	8	19	26	11

b

17	7	9	11	13	15	17	7
6	1	2	3	4	5	6	1
8	18	16	14	12	10	8	18
30	20	22	24	26	28	30	20
36	31	32	33	34	35	36	31
19	29	27	25	23	21	19	29
17	7	9	11	13	15	17	7
6	1	2	3	4	5	6	1

d

28 36 24 7 32 16 19 26 1 30 5 22 13 3 11 34 9

FIG. 2: Patterns imposed on a 12 × 12 *grid surface on a torus. The first three patterns are mapped onto patches of size* 2 × 2. a. *Helical cline;* b. *Ridge cline;* c. *Random pattern;* d. *Variable patch sizes as described in text with random pattern superimposed.*

Because of the large number of connections in the matrix even autocorrelation coefficients as low as .05 are significant in Table 1. The results are quite clear. The autocorrelation coefficients for the helical and ridge cline patterns for patches of dimension 2 × 2 show a regular clinal decrease in magnitude, the negative autocorrelations commencing with distance class 4. By contrast, the random pattern for size 2 × 2 shows significant positive autocorrelation only for distance class 1 with weak but significant negative autocorrelation for distance classes 2 and 3, and no other correlations at greater distances.

TABLE 1: Moran's coefficient of spatial autocorrelation at different distances and for different patch sizes for patches on a 12 × 12 *lattice mapped onto a torus.* (L, M, H *indicate values significant at* .02<P≤.05, .01<P≤.02, *and* P≤.01, *respectively. For description of variable size patches, see text.)*

Patch size	Distance class 1	2	3	4	5	6
2 × 2						
Prop. internal edges	.375	0	0	0	0	0
I (ridge cline)	.77H	.56H	.26H	-.06H	-.26H	-.44H
I (helical cline)	.71H	.45H	.17H	-.10H	-.21H	-.29H
I (random)	.29H	-.09H	-.05L	-.00	.00	-.03
3 × 3						
Prop. internal edges	.556	.222	0	0	0	0
I (helical cline)	.67H	.38H	.11H	-.04	-.18H	-.29H
I (random)	.57H	.22H	-.04	-.08H	-.09H	-.09H
4 × 4						
Prop. internal edges	.656	.375	.156	0	0	0
I (helical cline)	.67H	.38H	.11H	-.12H	-.17H	-.19H
I (random)	.71H	.42H	.14H	-.12H	-.19H	-.23H
6 × 6						
Prop. internal edges	.764	.556	.375	.222	.097	0
I (helical cline)	.75H	.50H	.25H	.00	-.25H	-.48H
1 × 4						
Prop. internal edges	.188	.062	.021	0	0	0
I (random)	.20H	-.07L	-.02	-.03	-.02	-.05
1 × 6						
Prop. internal edges	.208	.083	.042	.021	.008	0
I (random)	.31H	-.04	.02	-.10H	-.03	.05M
1 × 12						
Prop. internal edges	.250	.125	.083	.062	.050	.087
I (random)	.15H	.13H	-.16H	-.26H	.23H	-.06L
2 × 4						
Prop. internal edges	.500	.125	.042	0	0	0
I (random)	.40H	-.05	-.06L	-.05L	-.02	.01
2 × 6						
Prop. internal edges	.542	.167	.083	.042	.017	0
I (random)	.47H	.00	-.08H	-.11H	-.02	.06H
2 × 12						
Prop. internal edges	.625	.250	.167	.125	.100	.174
I (random)	.61H	.23H	.04L	-.11H	-.12H	-.10H
Variable size patches						
Prop. internal edges	.580	.278	.128	.056	.024	0
I (ave. rank based on patch size)	.59H	.28H	.04L	-.10L	-.14H	-.11H
I (ranks unrelated to patch size, ave. of 5 runs)	.55H	.20H	.01	-.08H	-.09H	-.11H

The patterns for the other size patches were generated by similarly arranging numerical values on the grid. The same range of numerical values was maintained in order to keep constant the mean and variance of the artificial variables being autocorrelated. The results obtained for larger patch sizes agree with those already noted. The clines decrease regularly becoming negative (0 in case of the 6 × 6 patch) by distance class 4. It may seem surprising at first that the results for clines of patch sizes 3 × 3 and 4 × 4 should be virtually identical. The actual value of the autocorrelation coefficient is a function of the similarity of the patches at distances under consideration and of the percentage of within-patch connections of a given length. As can be seen from Table 1 the proportion of within-patch connections of a given length increases as patch size increases. This is reflected in the increase in positive autocorrelation with patch size of the coefficients for distance classes 1 and 2 in the random pattern. It would be true of higher distance classes as well but would require larger patch sizes imbedded within a larger lattice for demonstration. In the clinal patterns there is no similar increase because as the patch size increased the variable was modified to maintain mean and variance and thus adjacent patches became less similar.

In fact an epistomological problem appears in connection with inferences concerning patch size by autocorrelation techniques. Although the notion of a patch as a clearly delimited area containing a population, a resource, or some other ecological factor, is well established in ecological thinking, some reflection makes it apparent that patches as such have no independent existence but are defined by the variable that is being mapped. Thus in a study of soil moisture patches in a meadow, we can map these patches only to the degree that there are differences in soil moisture and sharp boundaries delimiting areas of higher or lower moisture. Obvious ecological patches such as hummocks of vegetation in an area simply are sharply delimited boundaries of population density of the organism or organisms defining the hummock. Thus in clinal situations the boundaries between patches formally laid down in a simulation study become fuzzy and it is not surprising therefore that when the variable is clinal there is no marked drop in autocorrelation at the patch diameter; whereas in random patterns, patch sizes are clearly indicated by the fall from positive to zero or negative autocorrelation.

Next the patch shape was modified by producing rectangular patches of dimensions 1 × 4, 1 × 6, 1 × 12, 2 × 4, 2 × 6, and 2 × 12, with random patterns being generated on these patches. The results differed from those on square patches but were consistent with the earlier interpretations on the relations between correlograms and within- and among-patch connections. The strips one unit wide have appreciable positive autocorrelation only at distance class 1, with some suggestion of appreciable positive autocorrelation for the 1 × 12 patch in distance class 2. This obviously relates to the proportion of 'trips' one or two edges long that can be taken within these patches. Thus irregularly shaped patches reflect the number of trips that can be taken by populations within the patch without leaving it, relative to the number of trips of the same distance that would result in leaving the patch. The 1 × 12 patch has some unusual structure because of the nature of the lattice onto which the patches had been mapped. Since the lattice is a 12 × 12 torus, strips twelve units long border upon themselves, and thus what would appear to be an inter-patch connection in the direction of the long axis of the patch is in reality an intro-patch link, since it connects two points in the same patch. The positive autocorrelation at distance class 5 in the 1 × 12 patches is due to the particular realization of the randomization procedure. By chance parallel strips 5 units apart were similar in rank scores assigned to them.

Considering the 2 × 4 patches, they seem to follow the general rules enunciated earlier and clearly show positive autocorrelation at distance class 1, with good evidence of positive autocorrelation at distance class 2 for the 2 × 12 patch. Not knowing that these patches were rectangular an observer, judging from the correlogram alone, might conclude that the 2 × 4 and 2 × 6 patches were of dimension 2, whereas the 2 × 12 patch was of dimension 3. The patch size indicated by the resulting correlogram is clearly closer to the shorter width of the rectangle than to the longer one.

A test of a surface with patches of differing sizes was also undertaken. A size distribution in the shape of a reversed J resulted from dividing up the torus into one patch of dimension 6 × 6, two of dimension 4 × 4, four of dimension 3 × 3, and ten of dimension 2 × 2. The average dimension of these patches is 2.7 unweighted, or 3.7 when patch size is weighted by proportion of area covered. The results for a random pattern are shown in Table 1. The correlograms are quite similar, regardless of whether the ranks were assigned as averages based on patch size or were assigned sequentially regardless of patch size. Five replicates of the latter approach yielded quite comparable results. Strong positive autocorrelation is shown through patch size 2, but, while significant, the autocorrelation is only slight at patch size 3. This is undoubtedly due to only 12.8% of the paths three edges long remaining within the same patch. From these results, and knowing nothing about the underlying distribution of patch sizes, one would conclude that the patch dimension in the surface containing variably-sized patches was 3, an estimate that is quite close to average patch size.

4. ECOLOGICAL IMPLICATIONS OF SPATIAL AUTOCORRELATION IN POPULATIONS

Although many variables of ecological interest can be profitably analyzed by spatial autocorrelation techniques, the majority of the ecological applications will employ three classes of variables. The first comprises autocorrelation studies of abundance patterns in animals and plants with the variable under study being population density or counts at sampling stations. Second are studies of 'marker variables' such as gene frequencies or morphometric data to estimate underlying ecological parameters that are hard to measure (e.g. vagility, mortality). Third are species associations in studies of the distribution of the various species making up a community. The last class of variable will necessarily use autocorrelation techniques for nominal data.

Of these variables, population density will undoubtedly be the most important in ecological work. If we interpret spatial autocorrelation as predictability of variates a given distance apart, we may summarize interpretations of various types of autocorrelations in population densities as in Table 2.

TABLE 2: Ecological interpretations of different spatial autocorrelations of population density. (PD = *patch diameter,* LD = *interlocality distance.)*

	Autocorrelation	
Sign	Low order	High order
Positive	1. Dispersal 2. Large favorable patches PD > LD Gaps > PD 3. Gradient	Symmetrical surface or patch arrangement
Negative	Heterogeneous small patches PD < LD	Gradient

Low order (short distance) positive autocorrelation can be produced by three distinct ecological phenomena. The first of these is dispersal such as settlement in adjoining areas by populations that emigrate from an overpopulated area (i.e., migration pressure is a function of population density); or such as radiation by a population growth process, as in a colonial organism. A second model for producing low order positive autocorrelations is by large homogeneous patches, where large is defined as a diameter greater than inter-locality distance, and separated by gaps that are greater than patch diameter. Such patches could be of nutrient substances that are nonrandomly distributed in the environment, or they may be host plants or other environmental factors with a patchy distribution. A third cause of low order positive autocorrelation in population density may be by gradients of ecological factors affecting population density. Examples could result from diffusion of a metabolite, or through the existence of temperature or salinity gradients that affect population density.

Low order negative autocorrelations imply heterogeneous small patches, differing in amount of resources, with interlocality distance greater than patch diameter. An extremely patchy environment with a great deal of heterogeneity among patches could give rise to such a situation. Organisms that are aggregated into clumps, with empty or near-empty interstices, will exhibit low order negative autocorrelation.

High order (long distance) positive autocorrelation should be casued by symmetrical surfaces or symmetrical ordering of patches leading to similarity between far-away population densities. Such symmetries might be observed on both sides of a mountain at corresponding elevations.

High order negative autocorrelation will take place in a model where patches have been ordered in a gradient manner or where there is a gradient in the density due to a corresponding gradient in ecological resources.

Frequently estimates of population densities will be available for the same area over two or more time periods. In such cases a single surface of ΔN can be analyzed rather than the two separate density surfaces. The implications of low and high order autocorrelations for ΔN surfaces do not correspond fully to those for population size discussed above. Low order positive autocorrelation for ΔN implies an aggregation (in patches or along a gradient) of those ecological conditions that brought about the population change. Low order negative autocorrelation in ΔN would imply compensatory population growth - that is a population nearby an increasing one would be decreasing and vice versa. Such a phenomenon could be brought about with nomadic populations that shifted their areas of high settlement and density over time, coupled with an exhaustion of local resources. High order positive autocorrelation in ΔN will again imply a symmetrical surface or patch arrangement of those ecological conditions that brought about the change in population density. Finally, high order negative correlation implies a gradient of those ecological conditions that brought about the population change.

The interpretation of a given autocorrelation situation can be aided by an examination of the dispersion of the sample of counts or densities. Sokal and Oden (1978b) and Jumars *et al.* (1977) both have pointed out that the heterogeneity or dispersion of a sample can be considered independent of its spatial pattern. Thus two samples of counts may have the same variance (and means), yet may represent drastically differing spatial patterns. It is therefore profitable to examine in any given study of population counts or densities whether the samples obtained are homogeneous, i.e. could have been sampled at random from a population with a common mean density. A Poisson distribution would be a common but not the only plausible model. If the realized Poisson variates were distributed at random over the study area we could consider the population counts to be homogeneous and the spatial pattern (autocorrelations) to be nonsignificant. By contrast a contagious distribution would show more samples with high and more with very low counts. These heterogeneous results could again be associated with a random (nonsignificant) spatial pattern or with a nonrandom (significant) spatial pattern. Four possible outcomes are summarized in Table 3.

Heterogeneous counts (observed counts are clumped when tested against a Poisson distribution) coupled with a significant spatial pattern as shown by autocorrelation analysis is likely to be the typical case in studying populations density over an area. This combination may reflect 1. asynchrony of population growth over the entire area studied with patchy synchrony among local population samples; 2. incomplete mixtures of populations from different sources (and different densities); 3. autocorrelation of resources determining population size.

Heterogeneous counts combined with nonsignificant patterns may occur when resource quality is heterogeneous but arranged in a spatially random manner. Population densities track resource quality hence are heterogeneous but show no spatial pattern. A second model leading to this combination is the result of random settlement by mixtures of populations from different sources (and different densities) with little subsequent dispersal.

Homogeneous counts with significant spatial patterns are an unlikely result in nature. It would be as though the entire area (grid) had been settled on the basis of a random (e.g., Poisson) process, but that the choice of a quadrant in which to settle was a function of the magnitude of the population density of adjoining quadrants.

Homogeneous counts coupled with nonsignificant patterns would imply that population density is a truly random phenomenon caused by spatially random population growth processes or by a random patterning of environmental resources.

The interpretation of population changes ΔN will be similar. Heterogeneous population changes ΔN coupled with a significant spatial pattern may reflect 1. asynchrony of population changes among local population samples; 2. differential emigration or immigration with spatial autocorrelation (perhaps by bands of

TABLE 3: Ecological interpretations of different combinations of patterns and heterogeneities of population counts.

	Population counts	
Pattern	Heterogeneous	Homogeneous
Significant	The typical case: 1. Asynchrony of population growth among local population samples 2. Incomplete mixtures of populations from different sources (densities) 3. Autocorrelation of resources	Unlikely - random allocation of population size, coupled with spatially nonrandom settlement
Not significant	1. Random arrangement of resources. Population density tracks resource quality which is heterogeneous 2. Results of random settlement by population mixtures, with little subsequent dispersal.	1. Spatially random population growth processes 2. Random patterning of environmental resource

TABLE 4: Ecological interpretation of similarities and differences in variation patterns and spatial correlograms of pairs of variables (one being population size, the other a phenetic character or gene frequency.

	Variation patterns	
Correlograms	Similar	Different
Similar	1. Common response to patterned ecological factor 2. Density dependent selection	Unlikely: Similar autocorrelation of different environmental factors causing population growth and selection.
Different	Impossible	Differential patterns for selection factors and resources affecting population size.

organisms that leave or settle in one area); 3. heterogeneous and autocorrelated resources that determine population change.

Heterogeneous values of ΔN with a nonsignificant pattern will occur when there is a random patterning of heterogeneous resources that determine population change. A second model would be random settlement in (or emigration from) very small areas in which the patch diameter is less than interlocality distance.

Homogeneous values of ΔN with a significant spatial pattern are an unlikely outcome in nature. The model implies random change coupled with spatially nonrandom occurrence. By nature of the definition of the random process, the nonrandom spatial occurrence would be likely to follow the random generation of the change, and this would imply that such an event should occur rarely in most population models. Randomly generated numbers of emigrants that settle near each other might be one plausible interpretation.

Homogeneous values of ΔN lacking significant patterns imply random change, randomly patterned, a model in which population changes are quite localized, random, and do not affect events at neighboring localities.

Among the ecological variables about which inferences can be made by autocorrelation analysis, studies of population density and counts will hold a preeminent place. Numerous sampling studies, typically on more or less regular grids, abound in the literature and there is some hope that analyses of their patterns may lead to inferences about the causes of their population structure. For those data in which changes over time have been observed, prospects are considerably brighter that the underlying processes may be revealed. Studies involving ΔN not only can be analyzed by the routine spatial autocorrelation techniques but can be subjected to special null hypotheses which deal with specific population processes that seem reasonable from a knowledge of the material. For example, to test the assumption that there has been migration in a certain compass direction, the connectivity matrix can be designed with this hypothesis in mind and autocorrelation can then be tested against such a connectivity matrix.

Analyses involving more than one variable studied in the same population in the same area hold considerable promise. Inferences about several population processes can be made by combining studies on density with those of morphometric and/or population genetic variables. Comparisons, of course, can also be made for similar populations found in different areas where presumably differences in the correlograms should reflect differences in population structure.

When a concomitant variable such as a gene frequency or a morphometric variable is studied together with population density for a given population in one area, there are four conceivable outcomes of the analysis. The surfaces described by the two variables may be similar (highly correlated) or different (uncorrelated). We shall call these situations similar and different patterns, respectively. The correlograms resulting from these surfaces may also be similar (parallel) or different (markedly nonparallel).

Similar surfaces should yield similar correlograms. Finding such a situation in population density and a genetic or phenetic character of the population implies a common response to a patterned ecological factor or possibly density-dependent selection. The combination 'similar variation patterns with different correlograms' should be impossible since identical surfaces yield a single correlogram and similar surfaces are expected to yield similar correlograms. That variation patterns of population density and the concomitant variable should be different yet produce similar correlograms is possible but unlikely. It would imply that the generating process for the two surfaces was related. While Sokal and Oden (1978b) were able to postulate a plausible explanation for a similar situation in population genetics (migration necessarily affecting more than one allele), no suitable model comes to mind in the present situation. One would have to assume similar autocorrelation of different environmental factors, one causing population growth, the other selection. The combination 'different variation patterns accompanied by different correlograms,' is the outcome most likely to be observed. This phenomenon should be due to differential patterns for selection factors and resources affecting population size. These relationships are summarized in Table 4.

Another type of study of two variables might be of abundances of two different species which are either competitors or in a prey-predator or host-parasite relationship. Depending on the nature of the coactions, correlograms describing abundances of these species may or may not be similar. Given the nature of the species coactions, it might well be of interest to lag the abundances of one of the species by various time periods if such data are available. Thus in a prey-predator model the lagged abundance of the predator might have a correlogram similar to one of the prey a given time period earlier.

Correlogram analyses of genetic and phenetic variables, and of population densities as well, should be able to lead to estimations of patch sizes if the environment is indeed patchy and the variable studied responds to this patchiness. Sharp declines in the correlograms may indicate distances equivalent to patch diameters.

Estimates of the vagility of organisms by autocorrelation analysis will require markers for individuals, such as genetic markers, and cannot be based on abundance data alone, because the changes in density brought about by the dispersing individuals will be insufficient to reflect this vagility pattern.

The various combinations in which species occur in nature can be tested through nominal spatial autocorrelation analysis. It will be of less interest to discover autocorrelations at great distances, but departures from homotypic and heterotypic expectations between species pairs at close distances should reveal information on allelopathy in plants and animals and about the preferences that species have for each other and for certain spatially patterned environmental factors.

5. AN EXAMPLE

As an example of a design common in ecology, I have analyzed a data set from human demography. The model could easily be translated in terms of animal or plant populations studies. The data (Anonymous, 1963, 1976) are population changes expressed as percentages in the counties of the Republic of Ireland for census periods from 1871-81 to 1966-71. The connectivity matrix was constructed by connecting all contiguous counties, i.e., those that have a common boundary. The actual kilometric distances between the centers of pairs of contiguous counties were computed and a minimum path distance matrix was obtained containing distances of all pairs of localities measured along the connectivity paths. The spatial autocorrelations of the percentage change in population were computed and the resulting correlograms examined. Because there is much overlap in the correlograms making it difficult to absorb the information they contain, they are summarized in tabular form in Table 5. Six representative maps of these changes are shown in Figure 3 and the correlograms should be interpreted in conjunction with these maps.

In most counties and for most of this time there was a net population loss due largely to emigration. But it can be seen that during the period under study considerable changes in the variable took place. During 1871-81 (Figure 3a) the greatest loss occurred in most of Leinster Province with the exception of County Dublin which showed a net gain. A central tier of counties running north-south shows intermediate losses, while the smallest losses and the only other gain are shown by counties largely in the western and southern provinces. Note that these data have virtually no significant or even appreciable autocorrelation. Although there is some suggestion that contiguous counties are similar to each other, the pattern is broken up sufficiently so that the positive

TABLE 5: Spatial autocorrelation coefficients at different distances for percentage population changes in the counties of the Republic of Ireland over census periods from 1871 *to* 1971. *Distances, in kilometers, are upper class limits not class marks;* L, M, H, *as in Table* 1.

	Geary's coefficient c(-1)						Moran's coefficient I					
Census	70	130	190	250	310	370	70	130	190	250	310	370
1871-1881	-.64	-.93	-.92	-1.39L	-1.10	-.54	.19	-.02	-.10	-.14	-.11	-.02
1881-1891	-.77	-1.32	-.87	-1.04	-.57	-.95	.09	-.11	-.10	.04	-.01	-.23
1891-1901	-.84	-1.20	-.84	-1.21	-.56	-.90	.04	-.02	-.06	-.14	.04	-.20
1901-1911	-.77	-1.22	-1.00	-1.02	-.74	-.50	.31H	-.06	-.19	-.07	-.07	-.00
1911-1926	-.90	-.96	-.91	-1.35	-.77	-2.47L	.02	.15M	-.12	-.21	-.04	-.99L
1926-1936	-.57	-1.11	-.83	-1.43	-.83	-1.33	.20L	.05	-.07	-.26M	-.13	-.42
1936-1946	-.62L	-1.03	-1.05	-1.31	-.82	-1.02	.55H	-.02	-.22L	-.31M	-.02	-.26
1946-1951	-.47M	-1.06	-.96	-1.39	-.94	-.65	.39H	.06	-.18	-.29M	-.10	-.07
1951-1956	-.56M	-.82	-.98	-1.43H	-1.30	-2.16L	.33H	-.16M	-.12	-.38H	-.29	-.83
1956-1961	-.59L	-1.02	-.96	-1.27	-1.12	-1.72	.28M	.05	-.08	-.27L	-.27	-.61
1961-1966	-.43H	-.99	-.93	-1.48M	-1.09	-1.40	.41H	.06	-.11	-.34H	-.25	-.45
1966-1971	-.55M	-.92	-.98	-1.41L	-1.26	-1.43	.45H	.09	-.13	-.37H	-.31	-.46

TABLE 6: Spatial autocorrelation coefficients testing a hypothesis about migration patterns in the Republic of Ireland over census periods from 1871 *to* 1971. (L, M, H *as in Table* 1.)

Census	c(-1)	I	Census	c(-1)	I
1871-1881	-1.05	.01	1936-1946	-1.52H	-.40H
1881-1891	-1.46L	-.20H	1946-1951	-1.69H	-.50H
1891-1901	-1.39	-.15L	1951-1956	-1.31H	-.33H
1901-1911	-1.64H	-.42H	1956-1961	-1.25L	-.29H
1911-1926	-1.40M	-.21H	1961-1966	-1.58H	-.51H
1926-1936	-1.63H	-.42H	1966-1971	-1.77H	-.68H

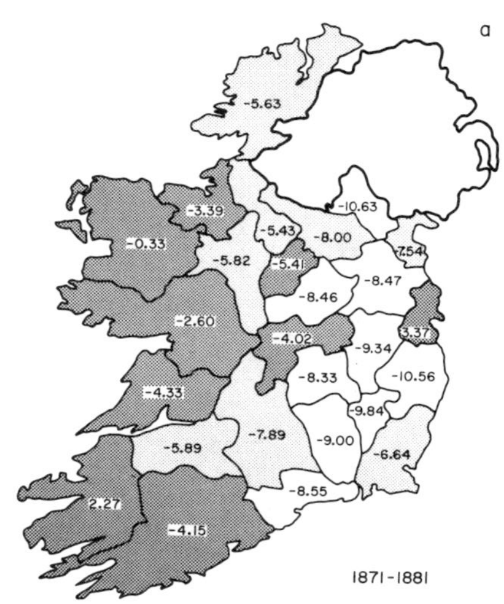

FIG. 3: Maps depicting percentage population changes in the counties of the Republic of Ireland during selected census periods. The changes have been grouped into terciles with the highest third shaded darkest, the middle third shaded lighter and the lowest third represented as white. Figures in each country describe the percentage change during the given interval. The maps represent the following census periods:
a. 1871-81; b. 1891-1901; c. 1911-26; d. 1936-46; e. 1951-56; f. 1966-71; g. *this final map locates names of counties and provinces.*

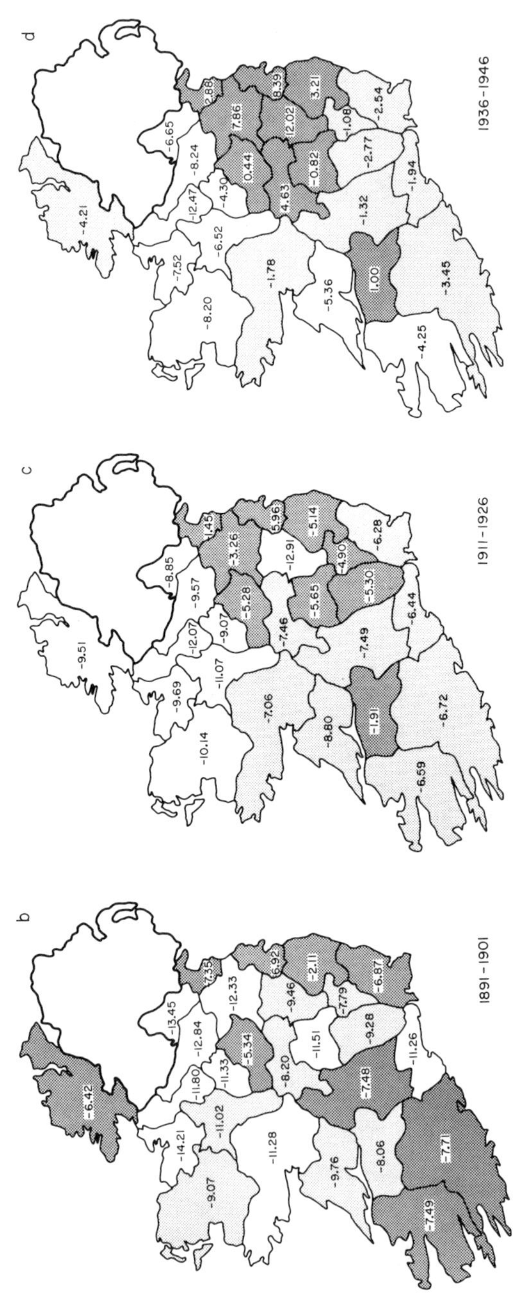
b
1891-1901
7.35
6.92
-2.11
-6.87
-13.45
-12.33
-9.46
-7.79
-12.84
-5.34
-11.51
-9.28
-6.42
-11.80
-11.33
-8.20
-11.26
-7.48
-14.21
-11.02
-11.28
-9.76
-8.06
-7.71
-9.07
-7.49
c
1911-1926
1.45
-3.26
5.96
-5.14
-6.28
-12.91
-4.90
-8.85
-9.57
-5.28
-5.65
-5.30
-9.51
-12.07
-9.07
-7.46
-7.49
-6.44
-11.07
-9.69
-7.06
-1.91
-6.72
-10.14
-8.80
-6.59
d
1936-1946
2.88
7.86
8.39
12.02
3.21
-6.65
-8.24
0.44
-0.82
-1.08
-2.54
-2.77
-4.21
-12.47
-4.30
4.63
-1.32
-1.94
-6.52
-7.52
-1.78
1.00
-3.45
-8.20
-5.36
-4.25

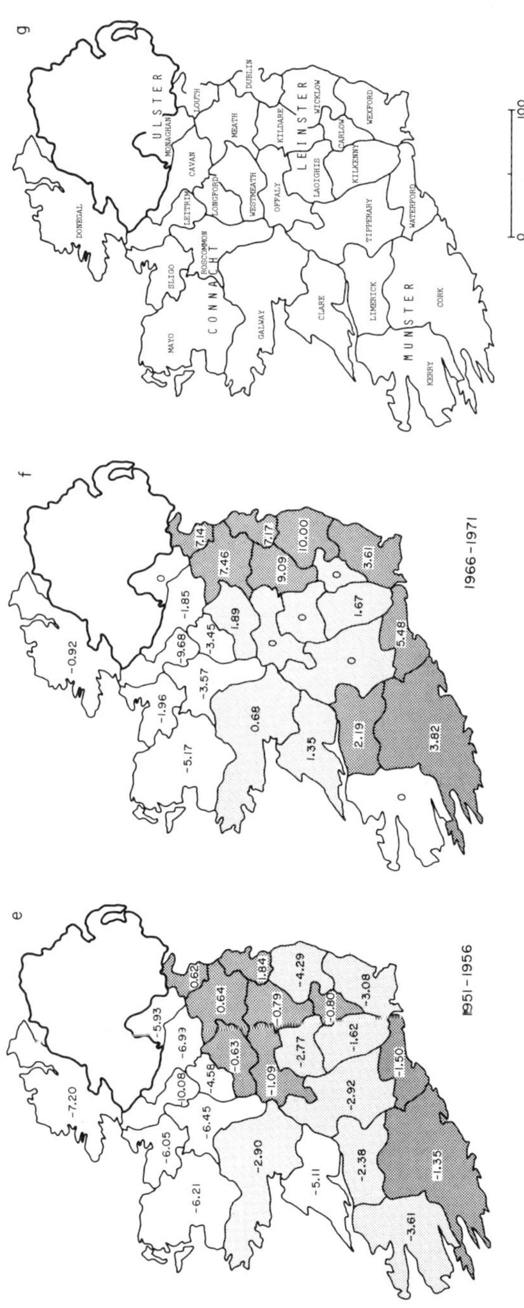
g
ULSTER
DONEGAL
MONAGHAN
LOUTH
DUBLIN
CAVAN
MEATH
LEITRIM
LONGFORD
WESTMEATH
KILDARE
WICKLOW
LEINSTER
OFFALY
LAOIGHIS
CARLOW
KILKENNY
WEXFORD
SLIGO
ROSCOMMON
CONNACHT
MAYO
GALWAY
CLARE
TIPPERARY
WATERFORD
LIMERICK
MUNSTER
CORK
KERRY
0
km
100
f
1966-1971
-0.92
0
-1.85
-9.68
-3.45
7.14
7.46
7.17
9.09
10.00
3.61
1.89
1.67
5.48
-3.57
-1.96
-5.17
0.68
1.35
2.19
3.82
e
1951-1956
-7.20
-5.93
-6.93
-10.08
-4.58
0.62
0.64
-0.63
1.84
-0.79
-1.09
-4.29
-0.80
-3.08
-2.77
-1.62
-1.50
-2.92
-6.45
-6.05
-6.21
-2.90
-5.11
-2.38
-1.35
-3.61

spatial autocorrelation at distances up to 70 km is not significant. The significant Geary coefficient at distances up to 250 km which is not matched by a significant Moran coefficient deserves some comment. Apparently many of the values contributing to comparisons in the 250 kilometer class are between counties in the "corners" of the study area. A situation in which Geary's coefficient is high but Moran's is not, will usually involve similarities between central values. I have examined the map in detail and found that only half of the extreme values (high gains or great losses in population) involve the 250 km comparisons. The difference between c and I, and consequently between d and J, is not very large in this case. It is not surprising therefore that a pattern of extreme values at highly connected counties cannot be found on visual inspection of the map. The earlier pattern no longer held during the periods 1881-91 and 1891-1901 (Figure 3b) when large losses appeared in the western counties contrasted with small losses in the eastern counties. However, the changes are very much a patchwork and no clear pattern emerges.

From 1901-11 on, a new pattern emerges (Figure 3c) which becomes progressively intensified by 1936-46 (Figure 3d) and during the rest of the period studied (Figure 3e). The counties with the greatest population losses are in the west and in Republican Ulster, whereas the lowest losses and some gains are shown in the northern part of Leinster Province and along the southeast coast in Counties Cork and Waterford. This trend has resulted in sharp gradients from the relatively most positive values in eastern Ireland to strong negative values in the west and northwest and to intermediate values in the south and center of the island. These data consequently show positive autocorrelation at least at distance class 70 km and negative autocorrelation at distances of 250 km. In the last of these maps (1966-71) consolidation of the areas of similar population change has taken place so firmly that two of the tercile classes in Figure 3f are contiguous and the third is represented by two clumps.

The positive correlation for short distances can be interpreted as aggregation of the ecological resources (economic conditions leading to emigration or immigration); the negative autocorrelation at distance class 250 km relates to the peculiar geography of Ireland. Travelling 250 km from anywhere in the Republic will land one in a county where socioeconomic conditions are likely to be markedly different.

The population changes were analyzed statistically by carrying out a G-test of independence for an $R \times C$ model (Sokal and Rohlf, 1969), testing the thirteen censuses against the twenty-six counties. The value of $G = 1348.7422$ indicates very significant ($P \ll 0.001$) heterogeneity among counties and years. However, when successive pairs of census years are analyzed in this manner, none of the tests yield significant values of G. It would appear that the changes occurring in any one census interval are not sufficiently differentiated among counties and the accumulation of several census intervals is required before a differential pattern of response among the counties can be substantiated statistically. Despite the nonsignificance of successive changes between census times, I would prefer to consider the population changes to be heterogeneous among the counties in view of the obvious and highly significant trend over the entire period studied. Therefore, when considering the combination of homogeneity of changes together with significance of pattern, it would appear that during the earlier census intervals of the study heterogeneous population changes were coupled with nonsignificant patterns indicating a random patterning of the conditions that determined the population changes; whereas starting with the 1901-11 interval, there seems to be autocorrelation of ecological and socioeconomic conditions that determine the population changes that took place. Further research would require an investigation into the agricultural and industrial economic conditions in various counties that either lost or attracted populations and would undoubtedly demonstrate the regional grouping of such conditions.

These data also permit a test of a particular ecological hypothesis. Census records for the counties are sufficiently detailed so that it is obvious that the net losses in population are not an excess of deaths over births but are the result of emigration. However, no records are available of immigration into counties and it is impossible to evaluate from these data how many emigrated from one Irish county to another and how many left the Island entirely. It is known from historical records that large numbers of Irish left predominantly for North America in the 19th and early 20th Century, and later in large numbers for Britain. Nevertheless, the development of the urban centers on the east coast, notably Dublin, Cork, and Waterford, and to a lesser degree Galway and Limerick in the west attracted migrants from the impoverished rural counties of the west and northwest. The general urbanization pattern familiar for much of western Europe resulted in spillover effects to counties surrounding Dublin.

To test the hypothesis that the increase in the favored counties is a result of emigration from the unfavorable counties I constructed a connectivity matrix testing this hypothesis. By considering Counties Louth, Dublin, Meath, Kildare, Wicklow, Wexford, Waterford, and Cork as 'sinks,' Counties Galway and Limerick 'semi-sinks,' and all other counties as 'sources,' a network was constructed by first connecting every source county to every sink county. The idea behind this scheme was that once a migrant decided to move to the attractive sink counties it did not matter very much which was the closest sink, but that any of them were equally attractive regardless of their distance from the source. The semi-sink counties are connected only to those source counties immediately contiguous to them. The idea here was that these less attractive centers would attract only migrants from nearby counties. Semi-sink counties are connected to sink counties only if they are contiguous.

The autocorrelations for population changes over the census period are given in Table 6. The results are clear cut. As predicted there are significant negative autocorrelations between sources and sinks. This trend was absent in 1871-81, commenced in the next decade, but did not become clearly established until 1901-11. A general increase in the tendency is observed throughout the period studied. Note that the trend could not be shown for the earliest of the census periods which clearly has a very different demographic pattern. The trend could be demonstrated considerably earlier than significant patterns (or correlograms) could be demonstrated (see Table 5). Some question may be raised about the validity of the significance tests in this example. Were the data not used to suggest the hypothesis? I believe this is true only in the most general sense. The specific hypothesis emerged from considerations of Irish and European historical demography, extrinsic to the present data base. Nevertheless, the specific significance values should be viewed with caution. The trend in magnitude of the coefficients is quite clear, however.

Thus, spatial autocorrelation analysis seems useful indeed as a hypothesis testing tool. Clearly the method does not prove that migrants from the less favored counties move to the favored ones. It is conceivable that emigration from the source counties out of the Island took place coupled with immigration to the sink counties from outside the Island. Such a situation, too, would yield negative autocorrelation of ΔN . But given our knowledge of the ecology of a given situation (in our case a knowledge of historic patterns of Irish demography) such an alternative explanation can be ruled out or made very unlikely.

6. DISCUSSION AND CONCLUSIONS

Another common instance of the application of spatial autocorrelation techniques to ecology is to so-called marker variables such as gene frequencies and morphometric variables. Since the investigation of such cases as well as the spatial analysis of species association patterns have already been described in a previous publication (Sokal and Oden, 1978b), there is no need to repeat their discussion here.

A number of problems remain in the evaluation of spatial autocorrelations for ecological and evolutionary hypotheses. Earlier statements about the identity of correlograms from surfaces generated by the same stochastic processes are strictly true only in the case of stochastic stationarity, that is, homogeneity of the surfaces with respect to means, variances and autocovariances. Although stationarity may be a reasonable assumption for some restricted cases in ecology, we cannot in general assume that the phenomena observed by us have had sufficient time for stationarity to develop. Furthermore, apparent patterns may well be the results of trends imposed by some external constraint. Thus there may be a gradient of a natural resource to which population density responds by a gradient of abundance modulated by stochastic fluctuations in population density. I am currently assuming, but cannot prove, that separate distributions of variables subject to the same stochastic processes for equal amounts of time will result in surfaces with similar spectral properties that will in turn yield similar correlograms. Thus, I assume with Sokal and Oden (1978b), that nonstationary surfaces which represent similar processes will yield similar correlograms. Some surfaces may indeed be stationary after an initial trend has been removed. In such a case, residuals can be tested for their autocorrelations and correlograms as described by Cliff and Ord (1973).

An important aspect of the reasoning advanced above is that parallel correlograms from different surfaces may indicate identical stochastic processes. However, how are we to determine whether two correlograms are parallel or not? The identity of two autocorrelation coefficients from different surfaces can be tested by conventional means using their standard errors as defined by Cliff and Ord (1973). However, autocorrelation coefficients from a single correlogram are not independent since the structure at greater distances is in part a function of structure at lower distances. My approach so far has been to cluster correlograms by the k-means method (Hartigan, 1975). This approach has yielded classes of grossly differing correlograms which have so far sufficed for purposes of analysis. Nevertheless, the development of more appropriate methods of testing the parallelism of correlograms is urgently needed and I have initiated a study of this problem by computer simulation.

Whenever there are many edges in a connectivity matrix, as in the regular grid employed in the earlier simulation experiments to determine effects of patch size on correlogram shape, even very low autocorrelation coefficients are statistically significant. This is not true for more sparsely connected sets of localities, as in the typical haphazardly located sites in an irregularly shaped geographical area. Thus the magnitude of autocorrelation coefficients needs to be considered as well as their significance.

When using spatial autocorrelation to test various hypotheses in population biology, there is no a priori limit to the number of alternative hypotheses thay may be tested by separate connectivity networks. However, there are obvious statistical objections to testing all possible such hypotheses. One must, therefore, be reasonably certain about an initial hypothesis before testing it and one should not overwork one's data in this regard.

For interval data one will almost always employ either Moran's or Geary's coefficient. Results for the two coefficients will differ as already described. We have found patterns characterized by Moran's coefficient to be more interpretable. Possibly this is related to the emphasis most observers would place on autocorrelated extreme values when evaluating a surface. The preference for I over c can also be defended on grounds of the asymptotic relative efficiency of tests based on the two statistics (Cliff and Ord, 1973, p. 137).

ACKNOWLEDGMENTS

Contribution number 295 from the Program in Ecology and Evolution at the State University of New York at Stony Brook. This study was supported by grant number DEB77-04824 from the National Science Foundation. The invaluable technical assistance of Jacqueline Bird is greatly appreciated. The computations were carried out with programs prepared by Neal L. Oden and modified by David W. Mallis. I have benefited from the constructive suggestions of J. K. Ord, Neal L. Oden, Peter J. Diggle, and S. A. L. M. Kooijman. The illustrations were prepared by Joyce Roe and the manuscript was typed by Christie Helquist and Barbara McKay.

REFERENCES

Anonymous. (1963). *Census of Population of Ireland 1961*. Stationery Office, Dublin.

Anonymous. (1976). Ireland. In *The New Encyclopaedia Britannica. Macropaedia, Vol. 9*. Encyclopaedia Britannica Inc., Chicago. 880-889.

Cliff, A. D. and J. K. Ord. (1973). *Spatial Autocorrelation*. Pion, London.

Gabriel, K. R. and Sokal, R. R. (1969). A new statistical approach to geographic variation analysis. *Systematic Zoology*, 18, 259-270.

Geary, R. D. (1954). The contiguity ratio and statistical mapping. *The Incorporated Statistician*, 5, 115-145.

Hartigan, J. A. (1975). *Clustering Algorithms*. Wiley, New York.

Jumars, P. A., Thistle, D., and Jones, M. L. (1977). Detecting two-dimensional spatial structure in biological data. *Oecologia*, 28, 109-123.

Jumars, P. A. (1978). Spatial autocorrelation with RUM (Remote Underwater Manipulator): vertical and horizontal structure of a bathal benthic community. *Deep-Sea Research*, 25, 589-604.

Matern, P. (1960). Spatial variations; stochastic models and their application to some problems in forest surveys and other sampling investigations. *Matter Meddelanden fran Statens Skogsforskingsinstitut*, 49, 1-144.

Matula, D. W. and Sokal, R. R. (1979). Properties of Gabriel graphs relevant to geographical cluster analysis and the clustering of points in the plane. Submitted to *Geographical Analysis*.

Moran, P. A. P. (1950). Notes on continuous stochastic phenomena. *Biometrika*, 37, 17-23.

Royaltey, H. H., Astrachan, E., and Sokal, R. R. (1975). Tests for patterns in geographic variation. *Geographical Analysis*, 7, 369-395.

Sokal, R. R. (1978). Population differentiation: Something new or more of the same? In *Ecology and Genetics: The Interface*, P. Brussard and O. Solbrig, eds. Springer Verlag, New York, 1-25.

Sokal, R. R. and Oden, N. L. (1978a). Spatial autocorrelation in biology. 1. Methodology. *Biological Journal of the Linnean Society*, 10, 199-228.

Sokal, R. R. and Oden, N. L. (1978b). Spatial autocorrelation in biology. 2. Some biological applications of evolutionary and ecological interest. *Biological Journal of the Linnean Society*, 10, 229-249.

Sokal, R. R. and Rohlf, F. J. (1969). *Biometry*. Freeman, San Francisco.

Tobler, W. R. (1975). Linear operators applied to areal data. In *Display and Analysis of Spatial Data*, J. C. Davis and M. J. McCullogh, eds. Wiley, London.

Whittle, P. (1954). On stationary processes in the plane. *Biometrika*, 41, 434-449.

[*Received December* 1978. *Revised January* 1979]

G. P. Patil and M. Rosenzweig, (eds.),
Contemporary Quantitative Ecology and Related Ecometrics, pp. 197-210.

CURRENT APPROACHES TO THE NON-LINEARITY PROBLEM IN VEGETATION ANALYSIS

M. P. AUSTIN

Division of Land Use Research, CSIRO
P. O. Box 1666
Canberra City, A.C.T. 2601 Australia

SUMMARY. The problem of the shape of species response curves and their influence on ordination results is discussed. The idea of bimodal response curves is reviewed and experimental evidence for their existence presented and discussed. Results of multi-species competition experiments along environmental gradients are discussed in relation to problems of standardization and data-smoothing. An example of an alternative vegetation analysis using a regression approach is presented. The need for better statistical methods and experimental designs is emphasized, together with the development of better conceptual models of species response to environmental gradients before the methods of vegetation analysis can be improved.

KEY WORDS. ordination, species response curves, multivariate analysis, regression, environmental gradients, competition.

1. INTRODUCTION

In the last twenty years, there has been a vast expansion in the numerical analysis of vegetation data, that is, analysis of field observations on variables concerning floristic composition or physiognomic structure. A large literature now exists on ordination and classification methods. Recently, numerous attempts have been made to compare these methods and examine their underlying assumptions. These studies using artificial data sets demonstrate, I believe (Austin, 1976a,b), inadequacies in current approaches which, if not overcome, may retard the development of vegetation analysis.

The best prospects for advance are likely to be through much greater synthesis of approaches used by different branches of ecology. In this paper, I should like to present some examples of problems which indicate the need for close cooperation between the descriptive (observational) ecologist using exploratory methods of data analysis, the theoretical ecologist with his ideas on niche-packing along environmental gradients, the experimentalist with his statistical designs for competition experiments, and the systems analyst with his models of environmental processes.

2. MULTIVARIATE ANALYSIS AND SPECIES RESPONSE CURVES

Most community ecologists with interests in multivariate methods are aware of the limitations of current methods. Linear models and associated statistical assumptions of multivariate normality are inappropriate for analysis of vegetation responses to environmental gradients (Whittaker, 1973). Principal components analysis is the method on which this has been most extensively demonstrated using artificial data assuming an ecological model of species behaviour based on a bell-shaped response curve for each species in relation to the environmental gradient (Noy-Meir and Austin, 1971; Austin and Noy-Meir, 1971; Gauch and Whittaker, 1972). The inappropriate linear model can produce spurious components and curvilinear distortions of the data.

Current discussions center upon the assumption of a bell-shaped or gaussian response curve. This has become almost universal as an ecological assumption on which to base research amongst both vegetation analysts (Whittaker, 1973) and niche theoreticians (May, 1973). "Everyone believes in it. For the field ecologists fancy that it is a theoretical principle and the theoretical ecologists that it is a field observation" (a paraphrase of a remark of Lippman (in Austin, 1976a)). A recent review (Austin. 1976a) of published ordination results suggests that a wide variety of asymmetric response curves have been obtained (Table 1). Bimodal curves are as common as symmetric unimodal (gaussian) curves. The possible existence of bimodal curves poses very difficult problems.

My suggestion (Austin, 1976a) that mathematical models which explicitly allow for bimodal curves may be required in developing better ordination methods has drawn criticism from one biometrician: "What he [Austin] fails to observe is that allowing such complexity can, in the end, lead to total indeterminacy" (Hill, 1977). Hill suggests that any symmetric curve can be converted to a bimodal curve with a suitable

TABLE 1: Summary of published species response curves, classified visually from both ordination and gradient analysis studies. (*From Tables* 1 *and* 2 *of Austin,* 1976a)

	Great Smoky Mountains[a]	Siskiyou Mountains[a]	Others	
Symmetric	8	14	Linear	7
Skewed	6	16	Gaussian	3
Shouldered	10	8	Symmetric	8
Plateau	2	1	Skewed	15
Bimodal	9	12	Bimodal	16

[a]Whittaker, personal communication.

transformation and any bimodal curve probably reflects poor specificity of the 'true' ecological gradient. Searching for a 'holy grail' of the ideal environmental gradient with its pure symmetric response curves for all species requires some biological justification. On the other hand, a search for a universal ordination method providing all curves for all men by means of a multitude of disposable coefficients is equally uninviting. Such methodological debate is sterile, and progress in vegetation analysis now requires recourse to ecological theory or evidence from suitable experiments. These, in turn, generate futher mathematical problems.

3. EXPERIMENTAL STUDIES ON SPECIES RESPONSE CURVES

Theoretical or experimental work on the structure of plant communities is mainly limited to very general, highly idealized, niche studies (May 1973) or to the 'simple' two-species competition experiment in a uniform environment (Harper, 1977). European plant ecologists, however, have had a well-developed conceptual framework for some years regarding species response curves (see Mueller-Dombois and Ellenberg, 1974). They clearly distinguish between the physiological response of a species when in monoculture and its ecological response when exposed to multi-species competition (Figure 1). Three types of optima for the ecological response curve are recognized. They may be

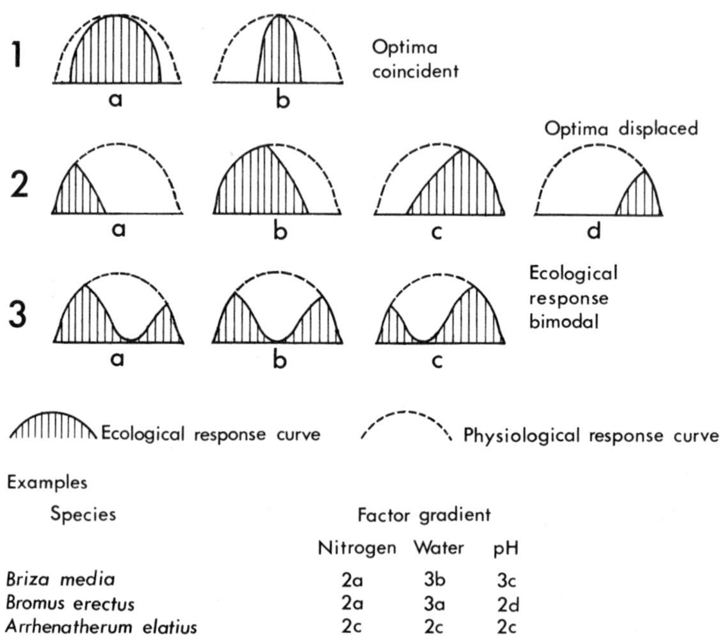

Examples

Species	Factor gradient		
	Nitrogen	Water	pH
Briza media	2a	3b	3c
Bromus erectus	2a	3a	2d
Arrhenatherum elatius	2c	2c	2c

FIG. 1: Schematic representation of types of ecological response in relation to a given factor gradient (modified from Mueller-Dombois and Ellenberg, 1974).

coincident with the physiological optimum, displaced towards an extreme, or display a variety of bimodal forms. The physiological response curves are defined as a form of bell-shaped curve. In this conceptual framework, competition is recognized as determining the shape of the response curve. Theoretical concepts for suggesting a possible shape for the physiological response curve appear to be lacking however what evidence there is suggests they are generally skewed and non-gaussian. This theoretical, though descriptive, framework is based on personal observation of the field behaviour of species and not on statistical analysis or detailed experimental studies, and hence is susceptible to bias. Its validity is open to objection.

Experimental evidence for bimodal curves does, however, exist. Ellenberg (1953) provides one example of a two-species competition experiment along a pH gradient which demonstrates the occurrence of a bimodal response under competition (Figure 2). These results support the conceptual framework.

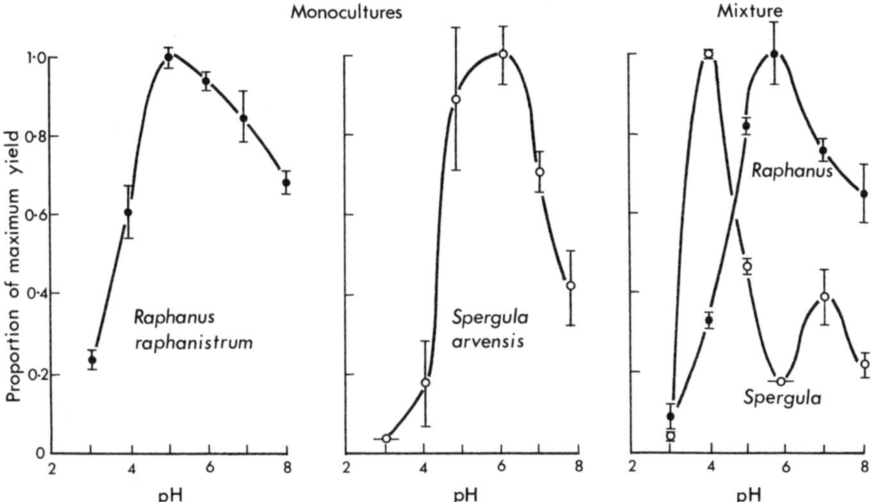

FIG. 2: Performance of two species Raphanus raphanistrum *and* Spergula arvensis *in monoculture and mixture. Range of variation for the two replicates per treatment indicated by bars (modified and redrawn from Ellenberg,* 1953*).*

Other results have been obtained from multi-species mixtures under experimental conditions (Ellenberg, 1953, 1954; Mueller-Dombois and Sims, 1966; Mueller-Dombois, 1964; Austin and Austin, unpublished manuscript). The work of Ellenberg and students has used an experimental set-up involving an environmental gradient of depth to permanent water table. My own recent work (manuscript) has involved a relative nutrient gradient produced by serial dilution of a general nutrient solution. Ellenberg (1953) used a six-species mixture and I have used both five- and ten-species mixtures. Both studies used grass species. Figure 3 provides an example of the type of results which are obtained for the nutrient gradient (see also Figure 4). The general conclusions from the studies can be summarized as: (1) the physiological optimum of all species tend to coincide at some intermediate position along the gradient; (2) the ecological optima vary between species; (3) the shapes of the ecological response curves are quite varied.

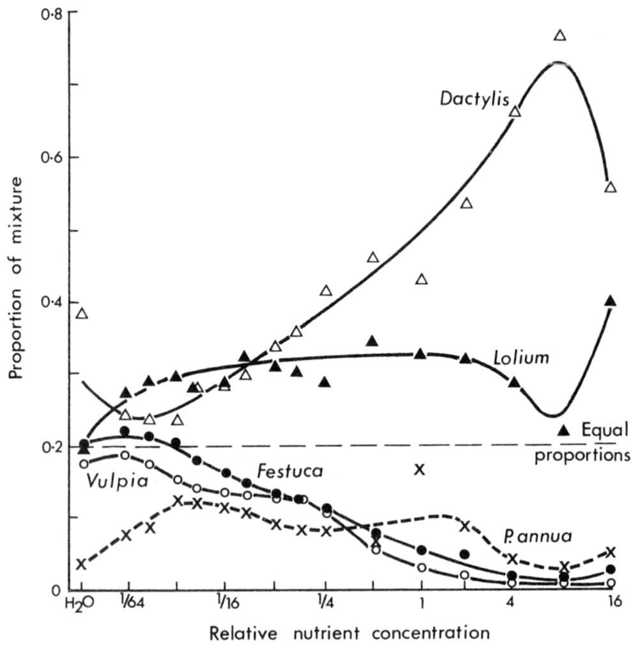

FIG. 3: Ecological response curves for five species when grown together as a mixture along a nutrient gradient (16 treatment levels). Species Dactylis glomerata, Lolium perenne, Festuca ovina, Vulpia membranacea, Poa annua *(from Austin and Austin ms).*

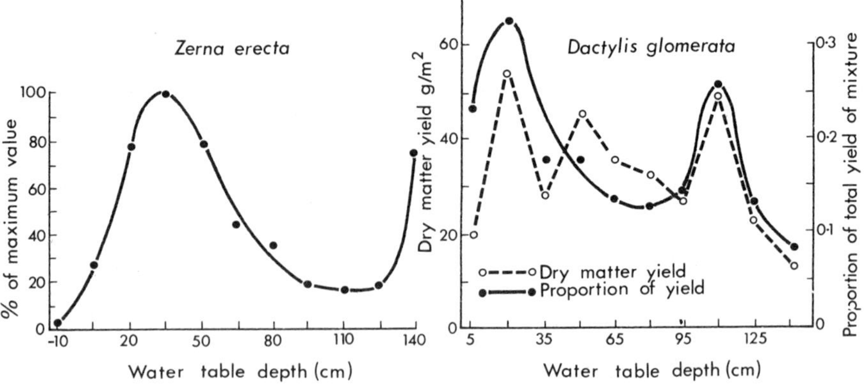

FIG. 4: Ecological response curves for two species when grown in a six-species mixture along a depth to water table gradient. Species selected to show bimodal response: (a) Zerna erecta *(redrawn from Ellenberg,* 1954*); (b)* Dactylis glomerata *showing effect of standardization (redrawn from Ellenberg,* 1953*).*

The water table gradient studies show more complex behaviour with a number of response curves showing bimodality (Figure 4). The graph of *Zerna erecta* (*Bromus erectus* in Ellenberg, 1953, 1954) shows possible bimodality but only at the most extreme position on the gradient (Figure 4a); this is, however, consistent with field observations. For the species *Dactylis glomerata* (Figure 4b), the dry matter yield shows an erratic response to the gradient, but, if expressed as a proportion of the total yield of the mixture, more consistent results are obtained. Ellenberg (1953, 1954) discusses many of the ecological complications which are associated with interpreting species performance along such complex gradients.

To the statistician or mathematician, there are many other problems. In both studies, no replication was performed. The existence of a bimodal curve in Figure 4a is dependent on one unreplicated observation, though the consistency of the other observations indicates that experimental error may be small. The standardization in Figure 4b may be justified in terms of the importance of relative performance rather than absolute yield for species survival. Support for the results is provided by the consistent improvement obtained for all species response curves in the experiment. Theoretical development of ideas concerning the relationship between total biomass per unit area (carrying capacity?) and position along an environmental (resource) gradient is needed to provide a more rigorous justification for such a standardization. The question remains, however, of how best to test the statistical significance of a bimodal curve based on two data points. Statistical analysis of constrained, non-independent values has then to be examined as a topic in vegetation analysis. These problems apply to data obtained from field observations and experiments.

One approach would be to greatly increase both replication and number of treatment levels. An alternative approach is to increase treatment levels and rely on data-smoothing techniques to produce suitable estimates of species performance. Polynomial regression techniques would require knowledge of the appropriate shape of the response curve which we do not have. A non-linear data-smoothing technique developed and suggested by Velleman (1975) and based on work of Tukey (Beaton and Tukey, 1974) was applied to shoot yields of the individual species in the nutrient gradient experiments. The procedure consisted of applying a series of smoothers in sequence to the residuals from an initial estimate of the curve. The smoothers were running medians of various length (4,3,2,3) with arbitrary conventions for endpoint values (Austin and Austin, unpublished manuscript). The use of medians was required because of unequal intervals between treatments. They provide a non-linear smoother relatively insensitive to outliers as compared to conventional least-squares methods. Figure 5a provides an example of the relatively

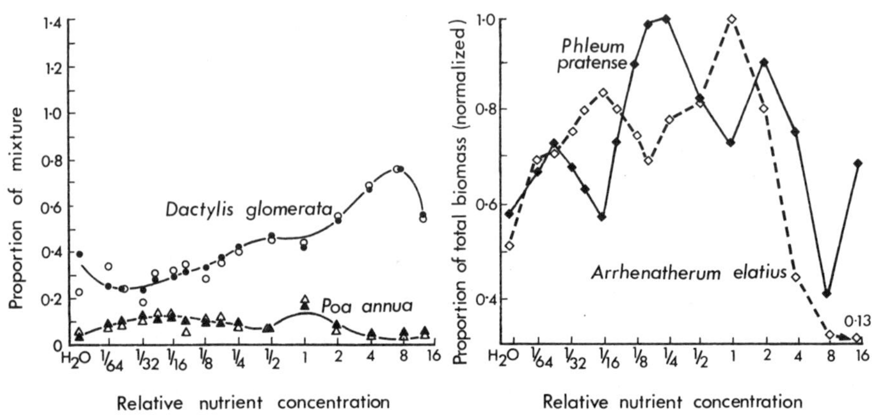

FIG. 5: Problems of data analysis of gradient observations (unpublished data, Austin and Austin). (a) Examples of influence of non-linear smoothing; ●, ▲ smoothed, o, Δ original observations; (b) interpretation problem of the significance of alternating relative performance in ten-species mixture.

conservative nature of the procedure. The relative performance of such smoothing procedure is not yet well understood (P. F. Velleman, personal communication) and obvious difficulties exist in 'cascading effects' from the endpoint conventions with short data sequences (16 points). Medians should be relatively robust against such problems (D. Jupp, personal communication). Clearly, research is needed on data-smoothing for short data sequences if analysis of vegetation response to environmental gradients is to become more rigorous.

The difficulties that may arise in future can be suggested by a speculative interpretation obtained by using such smoothing methods on data from a ten-species mixture along a nutrient gradient (Austin and Austin, unpublished manuscript). The results in Figure 5b suggest that two species alternately replace each other along the gradient even though each contributes only a small proportion of the total biomass of the mixture (maximum observed for either of the species <15%). Supporting evidence can be obtained from another ten-species mixture experiment in which *Phleum pratense* was grown in the absence of *Arrhenatherum elatius*. Without *Arrhenatherum*, *Phleum* has an ecological optimum at a relative nutrient concentration of 2x rather than (1/4)x and its response curve is relatively smooth. What, therefore, would constitute an efficient experimental design for testing such an hypothesis and what would be an appropriate statistical test?

The experimental study of species behaviour along environmental gradients suggests that bimodal and more complex curves may be common, but suitable methods of analysis of such experiments can hardly be said to have been developed as yet.

4. REGRESSION AS A METHOD OF VEGETATION ANALYSIS

If multivariate methods of vegetation analysis are based on inappropriate ecological models and experimental study of multi-species mixtures has as yet no well-defined statistical procedures for analysis, are any other approaches viable at the present time? Regression analysis of unplanned observations, particularly when coupled with testing all possible regression equations, is often condemned but often used in vegetation analysis. If regression is combined with careful definition of the appropriate domain for analysis, selection of independent variables based on awareness of relevant environmental processes and acceptance of the role of biotic influences, then it can be used as an exploratory data-analysis tool rather than as a probablistic hypothesis-testing procedure.

One example of this is a study (Austin, 1971) which examined the use of regression in detecting possible factors controlling the distribution of *Eucalyptus rossii* in a region of south-eastern Australia. It was necessary to define a domain using gradient analysis (Whittaker, 1973) within which *E. rossii* was likely to occur in order to avoid the 'zero-value' or 'naughty nought' problem in which large numbers of zero values are obtained for the dependent variable over a wide range of values of an independent variable. The degree of absence of a variable cannot be measured and the distribution is truncated at zero with corresponding distortion of the relationships.

No success with exploratory regression analysis was achieved with the simple environmental variables usually recorded on vegetation surveys. When a numerical classification of the vegetation was used to stratify the data into sets within which similar ecological processes might be operating, some success was obtained (Figure 6). This success depended on three distinct factors:

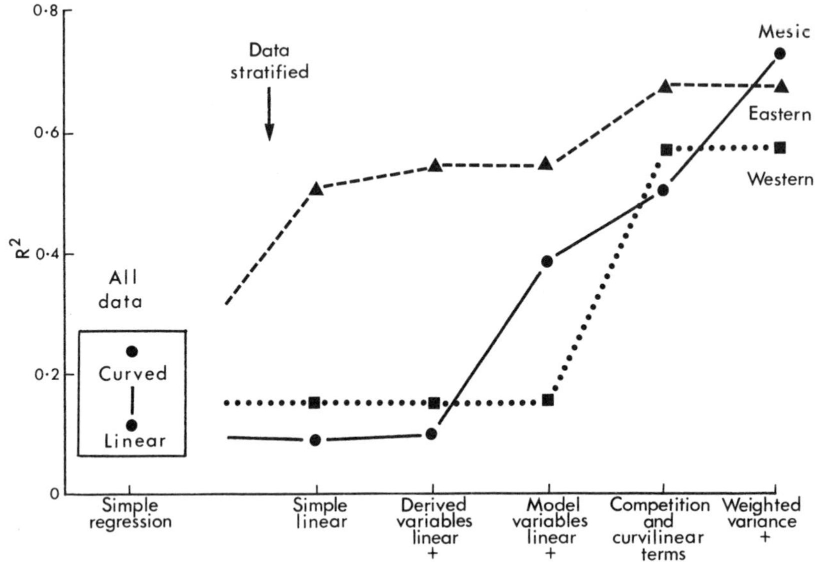

FIG. 6: Relative success of exploratory regression analysis based on increasingly relevant ecological models. Eastern - eastern dry sclerophyll forest; Mesic - mesic dry sclerophyll forest; Western - western dry sclerophyll forest (results of Austin, 1971*).*

1. The testing of curvilinear relationships using backwards elimination. If the species ecological response does approach a symmetric bell-shaped curve, then, unless both x and x^2 terms are tested simultaneously, this may not be detected.

2. Use of an environmental process model or scalar to estimate the combined effects of aspect, slope, available soil moisture, and local rainfall and evaporation on moisture stress in plants using known physical processes.

3. The inclusion of the amounts of other species as a variable providing a measure of possible competition.

Several types of regression equation were examined:

(a) Simple linear equation using environmental variables as measured on vegetation surveys.

(b) Equation based on derived variables obtained from measured variables using known physical relationships, e.g., use of aspect, slope, and latitude to calculate a radiation index.

(c) Equation of model variables in which derived variables are used to estimate the dynamic behaviour of plant moisture stress over an average year and variables based on seasonal stress included in the equation.

(d) Competition equations in which the density of other eucalypt species were included.

Three distinct hypotheses for the factors controlling *E. rossii* distribution emerged from the following exploratory results:

1. Eastern dry sclerophyll forest. Stratification produced immediately an increase in R^2 for equations containing simple environmental variables. Addition of variables based on scalars using known processes combining the effects of aspect, soil moisture, and seasonal rainfall produced little improvement. Further improvement was obtained by incorporating competitive terms for the dominant eucalypt species of this forest type.

2. Western dry sclerophyll forest. No improvement was found until moisture stress variables were incorporated into the regression equations and further improvement was obtained when curvilinear terms for moisture stress were tested. No competitive terms were found to be 'significant.'

3. Mesic dry sclerophyll forest. No success was obtained until all the components, i.e., competition, moisture stress scalars, and curvilinear terms were incorporated.

In western dry sclerophyll forest, *E. rossii* is the most abundant eucalypt (dominant?) species; other species have little influence on it, hence the lack of a competition variable in the equation. The environment is the most extreme of those studied, hence the importance of the stress variables. The significance of curvilinear terms suggests that the optimum of *E. rossii's* ecological response curve falls in this range of the moisture stress factor. The eastern dry sclerophyll forest, though less extreme, is also characteristic of a different seasonal rainfall climate and has different dominant eucalypt species. Under the changed conditions, *E. rossii's* performance becomes sensitive to simple changes in environmental conditions and to the competition of other species better able to utilize the environment. The mesic dry sclerophyll forest is similar climatically to the western dry sclerophyll forest but less extreme. No simple process has a dominant influence, and all must be considered before the influence of any one can be detected.

The analysis has a number of obvious inadequacies, but suggests that more attention should be paid to regression methods. One possible improvement would be to use a simultaneous regression approach with the performance of the various eucalypt species as the endogenous variables and the moisture stress variables as exogenous variables. Apart from one suggestion (Yarrenton, 1967), I know of no such study in plant ecology. Many of the techniques are well defined in econometrics, though the existence of asymmetric response curves may complicate application. This analysis was based on unplanned observations collected as part of a vegetation survey; clearly an appropriate sampling design needs to be developed for any future studies.

5. CONCLUSION

Ordination is a valuable technique of vegetation analysis but encounters increasing difficulties as the number of factors and range of communities in a data set increases. Not merely gaussian curvilinearities but skewed and bimodal departures from bell-shaped form set limits on the efficacy of ordination in some circumstances. Subdivision of the data set and regression analysis of the subsets will in some cases escape these difficulties and provide results not obtainable from ordination.

The methods used in vegetation analysis are essentially static and descriptive. They are exploratory in nature. If vegetation analysis is to become predictive, then statistical designs and tests will need to be developed. These are unlikely to be effective unless explicit models of vegetation behaviour can be stated. This in turn requires experimentation for which current methods appear inadequate when required to analyze multi-species communities. An optimistic view of current prospects would be to assume that all three types of analysis will see major improvements in the development of new conceptual frameworks, mathematical methods and statistical designs for testing hypotheses. Whatever happens, I hope ecologists and mathematicians will manage to avoid making errors of the third kind - 'solving the wrong problem.' This type of error has been prevalent in statistical ecology, particularly in ordination methodology.

REFERENCES

Austin, M. P. (1971). Role of regression analysis in ecology. *Proceedings of the Ecological Society of Australia*, 6, 63-75.

Austin, M. P. (1976a). On non-linear species response models in ordination. *Vegetatio*, 33, 33-41.

Austin, M. P. (1976b). Performance of four ordination techniques assuming three different non-linear species response models. *Vegetatio*, 33, 43-49.

Austin, M. P. and Noy-Meir, I. (1971). The problem of non-linearity in ordination: experiments with two gradient models. *Journal of Ecology*, 59, 762-773.

Beaton, A. E. and Tukey, J. W. (1974). The fitting of power series, meaning polynomials illustrated on band-spectroscopic data. *Technometrics*, 16, 147-185.

Ellenberg, H. (1953). Physiologisches und okologisches Verhalten deselben Pflanzenarten. *Berichte der Deutschen Botanischen Gesellschaft*, 65, 351-362.

Ellenberg, H. (1954). Uber einige Fortschritte der kausalen Vegetationskunde. *Vegetatio*, 5/6, 199-211.

Gauch, H. G. and Whittaker, R. H. (1972). Comparison of ordination techniques. *Ecology*, 53, 868-875.

Harper, J. L. (1977). *Population Biology of Plants*. Academic Press, New York.

Hill, M. (1977). Use of simple discriminant functions to classify quantitative phytosociological data. In *First International Symposium on Data Analysis and Information, Versailles*. 7-9 September 1977. E. Diday, L. Lebart, J. P. Pageis, and R. Tomassone, eds. Institut de Recherche d'Informatique et d'Automatique, Domaine de Voluceau, Rocquencourt, B. P. 105, 78150 le Chesnay, France.

May, R. (1973). *Stability and Complexity in Model Ecosystems*. Princeton University Press, Princeton, New Jersey.

Mueller-Dombois, D. (1964). Effect of depth to water table on height growth of tree seedlings in a greenhouse. *Forest Science*, 10, 306-316.

Mueller-Dombois, D. and Sims, H. P. (1966). Response of three grasses to two soils and water-table depth gradient. *Ecology*, 47, 644-648.

Mueller-Dombois, D. and Ellenberg, H. (1974). *Aims and Methods of Vegetation Ecology*. Wiley, New York.

Noy-Meir, I and Austin, M. P. (1971). Principal component ordination and simulated vegetation data. *Ecology*, 51, 551-552.

Velleman, P. F. (1975). Robust non-linear data-smoothing. Technical Report No. 89, Series 2, Department of Statistics, Princeton University, Princeton, New Jersey.

Whittaker, R. H. (1973). Ordination and classification of communities. *Handbook of Vegetation Science*, 5, Dr. W. Junk, b.v., The Hague.

Yarrenton, G. A. (1967). Organismal and individualistic concepts and the choice of methods of vegetation. *Vegetatio*, 15, 113-116.

[*Received July* 1978. *Revised January* 1979]

G. P. Patil and M. Rosenzweig, (eds.),
Contemporary Quantitative Ecology and Related Ecometrics, pp. 211-213.

A PROBABILISTIC COMPUTATION FOR THE RESEARCH OF 'OPTIMAL CUTS' IN VEGETATION STUDIES

M. GODRON

Centre national de la recherche scientifique
Centre d etudes phytosociologiques
et ecologiques Louis Emberger
Route de Mende-B.P. 5051
34033 Montpelier Cedex, France

1. METHOD

In order to study the horizontal structure of vegetation, observation of a simple linear sample often suffices. For this reason, the presence of species in equal sized segments of the line is the sampling method which may be very efficient if there is an ecological gradient. The statistical efficiency of the sampling method may in fact be computed (Thionet, 1953), but we should rather consider directly the results of the observed set of segments. The results are given in the form of a matrix where presences and absences may be represented by 1 and •, respectively. An example taken from *Pinus sylvestris* forests of Sologne region, central France, illustrates such a matrix.

Pinus silvestris	1 1 1 1 1 1 1 1 1 1 1 1 1 1 1 1 1 1 1 1 1 1 1 1 1 1 1 1 1 1 1 1
Pleurozium schreberi	1 1 . 1 . 1 1 1 1 1 1 1 . 1 . . 1 1 1 1 . . 1 . . . 1 . 1 1 1 1
Pseudoscleropodium purum	. 1 1 . . 1 1 . 1 . . . . 1 . . . .
Hypnum cupressiforme	. 1 1 1 1 1 1 1 1 1 1 1 1 1 1 1 1 1 . 1 1 1 1 1 1 1 1 1 1 1 1 1
Dicranum scoparium	. 1 1 1 1 1 1 1 1 1 . 1 1 1 1 1 1 1 1 1 1 1 . 1 1 1 1 1 . 1 . 1
Erica cinerea	. . . . 1 . 1 . . 1 . . . . . 1 1 . . . .
Cladonia silvatica	. . . . 1 . . 1 . . . 1 . 1 . . . .
Cladonia furcata	. . . . 1 . . . .
Polytrichum juniperinum	. . . . 1 . . . .

This matrix is the basis of four types of analysis: structure within each row, comparison of rows, structure within each column, comparison of columns. A method based on the comparison of adjacent columns, considering Hamming's distances, has been discussed in a previous paper (Gautier and Godron, 1976). It seems possible to improve it now by the computation of the probability of the 'ruptures' observed at each point. A 'rupture' is a local heterogeneity corresponding to one presence and one absence in the two adjacent segments situated on each side of a point.

The following symbols will be used for a species E:

E - the number of presences of the species,
S - the number of segments of the line,
S-E - the number of absences of the species E,
I - the abscissa of the point called I,
N - the number of species.

We may consider that all the S segments are placed in a Bernoulli's urn; then the chance to find a presence of E to the left of the point I is E/S, and the one to find an absence to the right of the point I is (S-E)/(S-1). The probability to have, by this way, a rupture for the species E is then E(S-E)/S(S-1). Symmetrically, the probability of a presence of E on the right of the point I and an absence of E on the left of this point has the same value. Finally, the probability that the number of ruptures equals 1 is 2E(S-E)/S(S-1).

If we consider another species called F , present in F segments, the chance to find a rupture at the same point for F is equal to 2F(S-F)/S(S-1). For the two species, the chance to find two ruptures is 4EF(S-E)(S-F)/S(S-1)S(S-1), and the same approach may be used for all the N species.

Our null hypothesis is that the vegetation is homogeneous around the observed point. Then, the probability of the set of ruptures observed around this point is the product of the probabilities of all the ruptures observed at this point. For example, at the first point, there are three ruptures; their probabilities are 0.31, 0.12, and 0.27. The probability to have these three ruptures is:

$$P_1 = 0.31 \times 0.12 \times 0.27 = 0.01.$$

At the second point, there is only one rupture, whose probability is $P_2 = 0.46$. So, one can say that the vegetation is more homogeneous around the second point than around the first one.

For a complete discussion, it would be possible to compute the probabilities of all the types of distributions of the presences and absences at any point, and to build tests. This would not be too useful though, because we are interested only in the information given by the observed ruptures at each point, and we know that this information is equal to $\log 1/P$, if P is the probability of the ruptures observed at this point.

2. APPLICATION

The method has been applied to 39 'relevés' taken in Pine forests of Sologne, realized with 32 segments whose length was 2m. Among these relevés, twelve gave no point where P, the probability of the observed ruptures, was less than 1%; seven relevés contained one point where P was less than 1%; five had two points where P was less than 1%, etc. The most heterogeneous relevé had fifteen points where P was less than 1%.

On the whole, a probability less than 1/100,000 was observed only at three points. It is noticeable that these three points were all inside blocks of points where P was less than 1%. More generally, it seems possible to tell that the points where there is a great number of improbable ruptures are not isolated, but, on the contrary, are in a heterogeneous zone.

3. DISCUSSION

The main use of this method is that it helps to find the heterogeneous zones where a boundary line would have to be placed if a map had to be drawn. It seems possible to improve the approach by taking into account the number of species observed in the relevés. The easiest method would be to compare the information given by the observed ruptures with the information that would be obtained if all the species gave a rupture at the observed point.

Another improvement would be to do the computations after grouping the segments two by two, three by three, etc., in order to observe if the vegetation appears fine-grained or coarse-grained.

REFERENCES

Gauthier, B. and Godron, M. (1976). La recherche de limites ou de coupures optimales; application à un relevé phytosociologique.

Thionet, P. (1953). *La Théorie des Sondages*. Ed. INSEE, Paris.

[*Received April* 1979]

G. P. Patil and M. Rosenzweig, (eds.),
Contemporary Quantitative Ecology and Related Ecometrics, pp. 215-228.

THE $\overset{*}{M}$-M METHOD FOR ANALYZING THE DISTRIBUTION PATTERNS OF SINGLE- AND MIXED-SPECIES POPULATIONS

SYUN'ITI IWAO

Entomological Laboratory
College of Agriculture
Kyoto University
Kitashirakawa, Kyoto 606, Japan

SUMMARY: A comprehensive system for analyzing the distribution pattern of single-species populations as well as spatial association between two species has been developed using the relationships of intra- and interspecies mean crowdings to mean density. The basic methods hitherto derived are briefly reviewed and some applications of these methods to analyze ecological processes are mentioned.

KEY WORDS. spatial distribution pattern, $\overset{*}{m}$-m method, interspecies mean crowding, spatial overlap, spatial correlation, niche overlap.

1. INTRODUCTION

Currently an increasing attention has been paid to the spatial aspects of ecological processes such as reproduction, dispersal, intra- and interspecies competition, and predator-prey interaction. Then, it is important to develop appropriate statistical methods of analyzing the spatial pattern of single-species populations as well as the spatial association between two species. We have developed a series of analytical methods relevant to these purposes using the relationships of intra- and interspecies mean crowdings to mean density (Iwao, 1968, 1972, 1977, 1977a; Iwao and Kuno, 1971).

In this paper, fundamental relationships are briefly reviewed and a few examples are given to show how the methods can be applied to ecological problems as mentioned above.

2. BASIC RELATIONSHIPS IN SINGLE-SPECIES POPULATIONS

Lloyd's (1967) parameter 'mean crowding' is defined as "the mean number of other individuals per individual per quadrat" and expressed by

$$\overset{*}{m} = \frac{\sum_{j=1}^{Q} x_j(x_j-1)}{\sum_{j=1}^{Q} x_j} \qquad (1)$$

where Q is the total number of quadrats in the area, and x_j is the number of individuals in the j*th* quadrat.

Lloyd derived this parameter as a measure of the degree of crowding experienced by the average individual in a population, and hence the mean crowding in the original sense can be applied properly only to freely moving, nongregarious animals that are living in continuous and uniform habitats. As pointed out by Iwao (1968) and Pielou (1969), however, the concept of mean crowding as defined by equation (1) is generally valid for any kind of organisms and for any size of quadrat unit, without special implication on the effect of crowding. Thus the mean crowding gives an absolute measure of aggregation.

The ratio of $\overset{*}{m}$ to mean (m) is called patchness by Lloyd and it serves as a relative measure of aggregation which is essentially identical with Morisita's (1959) I_δ and Kuno's (1968) C_A indices. The ratio $\overset{*}{m}/m$ equals unity in random (Poisson) distribution, and is larger and smaller than unity in aggregated and uniform distributions respectively. This ratio is not affected by mean density in the sense that its value does not change if individuals are randomly removed with a given probability from a population (Pielou, 1969; Iwao and Kuno, 1971). This does not necessarily mean, however, that any change in $\overset{*}{m}/m$ with m is due to some biological cause such as operation of density-related mortality. Actually, $\overset{*}{m}/m$ will change with m for purely statistical reasons if the distribution is colonial.

Iwao (1968) found that the change of $\overset{*}{m}$ with m can be fitted to a single linear regression in a variety of both theoretical and biological distributions. Namely, the relation is shown by

$$\overset{*}{m} = \alpha + \beta m, \qquad (2)$$

where the intercept α indicates the basic component of the distribution; $\alpha = 0$ when the single individual is the component, $\alpha > 0$ when several individuals are distributed in groups or colonies, and $-1 \leq \alpha < 0$ when there exists some repulsive interaction between individuals. The slope β indicates the distribution pattern of basic components over the quadrats; $\beta = 1$ in random (Poisson) distribution, $\beta > 1$ in aggregated distributions, and $0 \leq \beta < 1$ in uniform distributions. Thus α and β show the different aspects of aggregation. The relation can be applied to a set of frequency distributions obtained by using the same quadrat size and is called the series $\overset{*}{m}$-m relation.

When the $\overset{*}{m}$-m relation is examined by successive changes of quadrat size in a single population, we get the unit-size $\overset{*}{m}$-m relation, which is not necessarily identical with the series relation even in basic distribution models. The unit-size $\overset{*}{m}$-m relation is curvilinear if there are clumps. The ρ index proposed by Iwao (1972) is useful in detecting the approximate area occupied by the average clump (or colony), intraclump distribution of individuals, and the distribution pattern of clumps themselves. It is given by

$$\rho_i = \frac{\overset{*}{m}_i - \overset{*}{m}_{i-1}}{m_i - m_{i-1}}, \tag{3}$$

where $i = 1,2,3,\cdots$ stands for the ascending order of quadrat size and $\rho_1 = \overset{*}{m}_1/m_1$ for the smallest quadrat size. The behavior of ρ in typical distribution patterns is shown in Figures 1 and 2.

For detection of clump areas, ρ-index is much better than Greig-Smith's (1952) method of mean squares, the latter being efficient only when the clumps are distributed uniformly (Iwao, 1972).

3. ASSOCIATION BETWEEN TWO SPECIES

Interspecies mean crowding is defined as "the average number of individuals of the other species per individual of the subject species per quadrat" (Lloyd, 1967). The mean crowding on sp. X by sp. Y and that on sp. Y by sp. X are given by

$$\overset{*}{m}_{XY} = \sum_{j=1}^{Q} \chi_{Xj}\chi_{Yj} / \sum_{j=1}^{Q} \chi_{Xj}$$

and

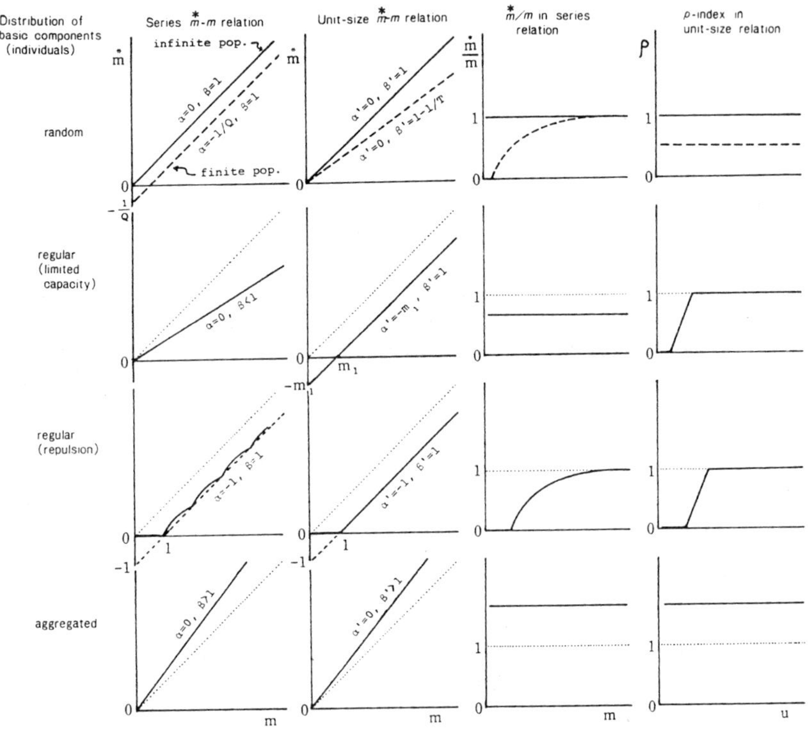

FIG. 1: Scheme showing the series and the unit-size $\overset{*}{m}$–m *relationships in typical distribution patterns comprising single individual as the basic component. If a compact colony, instead of single individual, is the basic component, the regression line in both series and unit-size relations move upwards corresponding to the value of* $\overset{*}{m}_c$*, the mean crowding for colony-size distribution.*

Q : *total no. of quadrats;* T : *total no. of individuals in a population;* u : *quadrat size. (After Iwao,* 1977a*).*

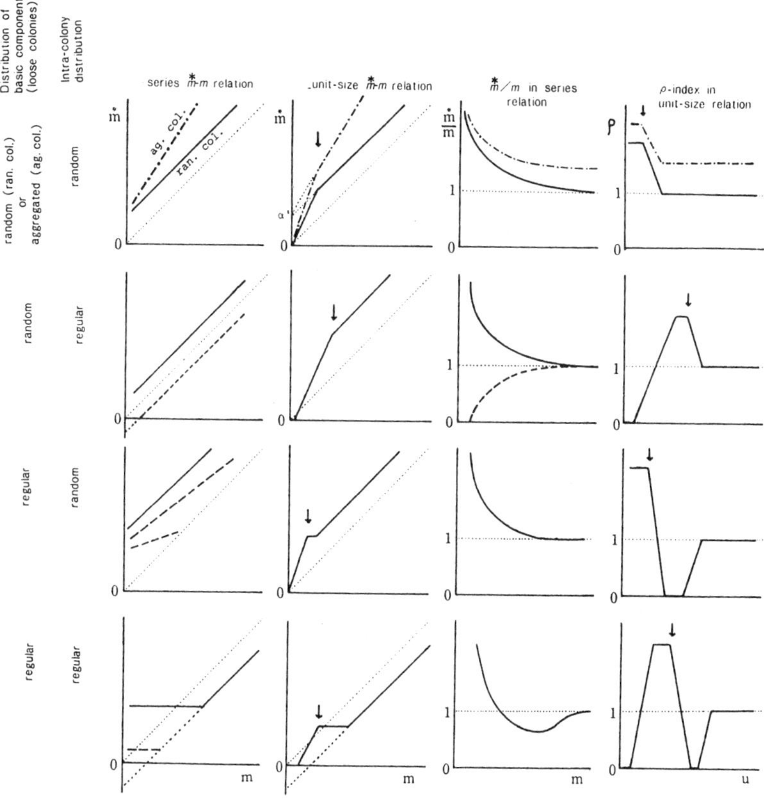

FIG. 2: Scheme showing the series and the unit-size $\overset{*}{m}$-m *relationships in typical distribution patterns comprising loose colonies. In the series relation, the intercept* α *of the regression becomes large as quadrat size increases unless quadrat size is sufficiently large relative to the mean colony area. In the unit-size relation, the regression line would bend around the quadrat size corresponding to the mean colony area (indicated by arrow) and the slope* β *for the larger quadrat sizes indicates the distribution pattern of colonies. The intercept* α *extrapolated from this part of the regression would be more or less smaller than the actual colony-size mean crowding,* $\overset{*}{m}_c$*. Characteristics of colony distribution and intra-colony distribution of individuals as well as colony area (area occupied by average colony) can be detected by* ρ*-index plotted against quadrat size. Broken lines in the series relation show the relations expected when the quadrat size is smaller than the colony area (After Iwao,* 1977a*).*

$$\overset{*}{m}_{YX} = \sum_{j=1}^{Q} \chi_{Xj}\chi_{Yj} / \sum_{j=1}^{Q} \chi_{Yj} \tag{4}$$

respectively, where χ_{Xj} and χ_{Yj} are the numbers of individuals of sp. X and sp. Y in the *jth* quadrat.

If there is no spatial overlapping between spp. X and Y, obviously

$$\overset{*}{m}_{XY} = \overset{*}{m}_{YX} = 0.$$

And, if the distributions of both species are completely overlapped (i.e., the ratio χ_{Xj}/χ_{Yj} remains constant in every quadrat), the relations $\overset{*}{m}_{XY} = \overset{*}{m}_{Y} + 1$ and $\overset{*}{m}_{YX} = \overset{*}{m}_{X} + 1$ will be expected.

Then, we have the following index to show the degree of overlapping:

$$\gamma = \sqrt{\frac{\overset{*}{m}_{XY}\ \overset{*}{m}_{YX}}{(\overset{*}{m}_X+1)(\overset{*}{m}_Y+1)}} = \frac{\sum_j \chi_{Xj}\ \chi_{Yj}}{\sqrt{\Sigma\chi_{Xj}^2\ \Sigma\chi_{Yj}^2}} . \tag{5}$$

Its value ranges from the maximum of 1.0 in the case of complete overlapping to the minimum of 0 when there is complete exclusion (Iwao, 1977). This index is essentially identical with Pianka's (1973) measure of niche overlap which is derived from Levin's (1968) index of alpha.

When two species are distributed independently of each other, the value of γ changes with mean densities m_X and m_Y. Since in such a case the relations $\overset{*}{m}_{XY} \approx m_Y$ and $\overset{*}{m}_{YX} \approx m_X$ are expected, γ for independent distribution is given by

$$\gamma_{(ind)} = \sqrt{\frac{m_X\ m_Y}{(\overset{*}{m}_X+1)(\overset{*}{m}_Y+1)}} = 1 \Bigg/ \sqrt{\frac{(\overset{*}{m}_X+1)}{m_X}\ \frac{(\overset{*}{m}_Y+1)}{m_Y}} . \tag{6}$$

As a measure of the degree of spatial correlation between two species or the degree of overlapping relative to the independent occurrence, Iwao (1977) proposed ω index which is given by

$$\omega_{(+)} = \frac{\gamma-\gamma_{(ind)}}{1-\gamma_{(ind)}} = \sqrt{\frac{\overset{*}{m}_{XY}\,\overset{*}{m}_{YX}-m_X m_Y}{(\overset{*}{m}_X+1)(\overset{*}{m}_Y+1)-m_X m_Y}}, \quad \gamma \geq \gamma_{(ind)},$$

$$\omega_{(-)} = \frac{\gamma-\gamma_{(ind)}}{\gamma_{(ind)}} = \sqrt{\frac{\overset{*}{m}_{XY}}{m_X}\frac{\overset{*}{m}_{YX}}{m_Y}} - 1, \quad \gamma \leq \gamma_{(ind)}. \qquad (7)$$

The value of ω changes from the maximum +1.0 for complete overlapping, through 0 for independent occurrence, to the minimum -1.0 for complete exclusion.

When $\overset{*}{m}_{XY}$ (or $\overset{*}{m}_{YX}$) is plotted against m_Y (or m_X) on a graph, the point would fall around the line passing through the origin with the slope of 1 if two species are distributed independently of each other. Conveniently, the line also shows the random distribution of conspecific individuals when $\overset{*}{m}_X$ (or $\overset{*}{m}_Y$) is plotted against m_X (or m_Y) as mentioned before. Thus, the $\overset{*}{m}$-on-m graph can show both interspecific relationship and the distribution pattern of each composite species.

4. SOME APPLICATIONS

4.1 Spatio-temporal Pattern of Egg-laying by Females of the Potato Lady Beetle, Henosepilachna vigintioctomaculata, *in a Potato Field.* Post-hibernated female adults of this univoltine insect lay eggs in masses on potato plants in May-June. Distributions of adults and egg masses deposited were counted at weekly intervals in a small potato field. Since the duration of egg stage in the field was longer than 1 week, we obtained the complete record of the cumulative number of egg masses deposited on each plant by respective census dates. Double count was avoided by tagging each egg mass discovered. The $\overset{*}{m}$-m relation obtained for the cumulative numbers of egg masses on successive sampling days indicates the spatio-temporal pattern of egg-mass accumulation in the field. As shown in Figure 3, it is fitted well to a linear regression with $\hat{\alpha} = 0.33$ and $\hat{\beta} = 1.16$, indicating that the females tend to lay egg-masses in small groups (average being 1.3 or so) nearly at random over the field.

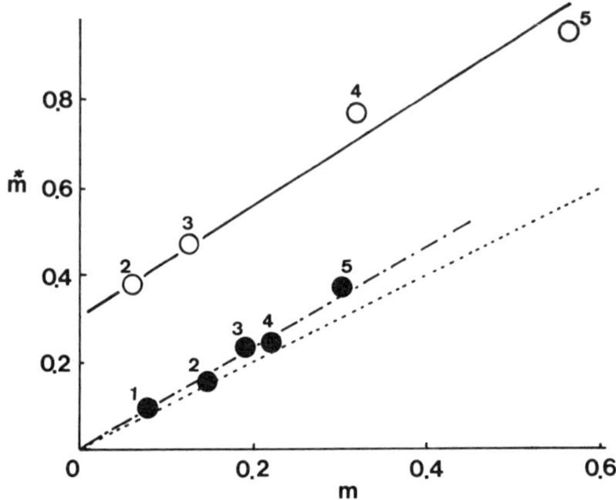

FIG. 3: The series $\overset{*}{m}$–m *relations for the distributions of egg masses (open circles) and adult females (solid circles) of the potato lady beetle,* Henosepilachna vigintioctomaculata *in a potato field. Numericals in the figure indicate the order of weekly censuses.*

To confirm this, the pattern of utilization of potato plants by females are examined by calculating $\overset{*}{m}$ and m for the cumulative totals of female adults discovered on respective plants. If the females tend to prefer some particular plants, the distribution pattern would become more aggregative at later census dates. As is apparent from Figure 3, the $\overset{*}{m}$-on-m regression for females are almost parallel with that for egg masses ($\hat{\beta} = 1.14$) but its intercept α becomes almost zero ($\hat{\alpha} = 0.006$). Observation of marked adults indicated the small mobility of females; the average distance moved by females was only 1.6 m in a week and none of marked adults moved to other fields in the surrounding area.

Then it is assumed that individual females would move within a limited range while they were laying egg masses, though every plant in the field was utilized with nearly equal probability as the whole. If so, the distribution of adult females should be overlapped with that of egg masses in a relatively small spatial unit. This is tested using the unit-size relationships of intra- and interspecies mean crowdings to mean densities of adults and egg masses. The result is shown in Figure 4. Intraspecies relations for adults ($\overset{*}{m}_X - m_X$) and for egg masses ($\overset{*}{m}_Y - m_Y$) indicate the absence of any appreciable clump structure in any

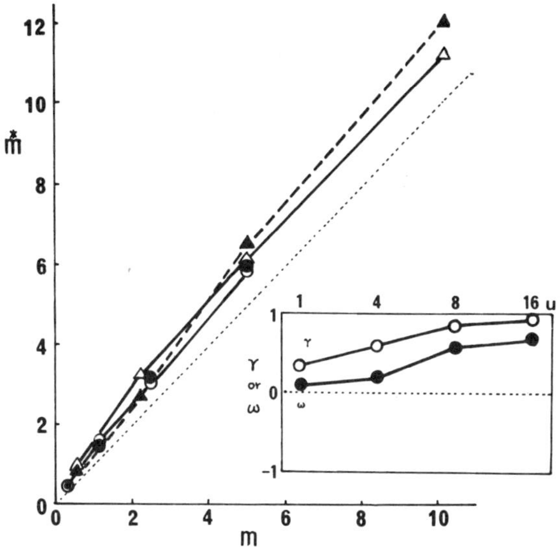

FIG. 4: The unit-size $\overset{*}{m}$-m *relations for egg masses and female adults at 5th census and the spatial association between them. Open circles:* $\overset{*}{m}_X-m_X$ *(female adults); open triangles:* $\overset{*}{m}_Y-m_Y$ *(egg masses); solid circles:* $\overset{*}{m}_{YX}-m_X$*; solid triangles:* $\overset{*}{m}_{XY}-m_Y$*. Inset figure shows the changes in* γ *and* ω *with quadrat size.*

unit size ranging from 1 plant to 16 plants. Interspecies relation, $\overset{*}{m}_{XY}-m_Y$ or $\overset{*}{m}_{YX}-m_X$, indicates the positive association between adults and egg masses. From change in ω value with unit size, it is clear that association becomes quite high at 8-16 plant units, which is roughly in agreement with the average range of movement by females in 1-2 weeks.

4.2 Oviposition by a Syrphid Fly, Episyrphus balteatus, *in Relation to Aphid Density on Cabbage Plants.* It is known that many species of predators tend to concentrate around high-density patches of their prey (e.g., Rogers and Hubbard, 1974). In aphidophagous syrphids, Chandler (1968) found that several species including *Episyrphus balteatus* lay eggs near the aphid colonies. Ito and Iwao (1977) conducted a simple experiment in a field cage to determine the relation of oviposition by *E. balteatus* to population density of green peach aphid, *Myzus persicae*, on 9 cabbage plants. For each plant, intra- and interspecies mean crowdings for aphids (X) and syrphids (Y) are calculated and plotted (Figure 5). In this experiment, each cabbage plant had only 9 leaves, and hence the expected relation for random distribution of individuals per leaf is given by the binomial, i.e., $\overset{*}{m} = -0.11 + m$. The $\overset{*}{m}_Y - m_Y$ relation for syrphid eggs is fitted to a regression with $\hat{\alpha} = -0.06$ and $\hat{\beta} = 1.57$. This means that this syrphid lays eggs singly on leaves with a definite tendency of aggregation. On the other hand, the $\overset{*}{m}_X - m_X$ relation for aphids is fitted well to the regression with $\hat{\alpha} = 26.97$ and $\hat{\beta} = 1.11$ as shown in Figure 5, indicating that colonies comprising 20-30 aphids each are distributed nearly at random on the leaves.

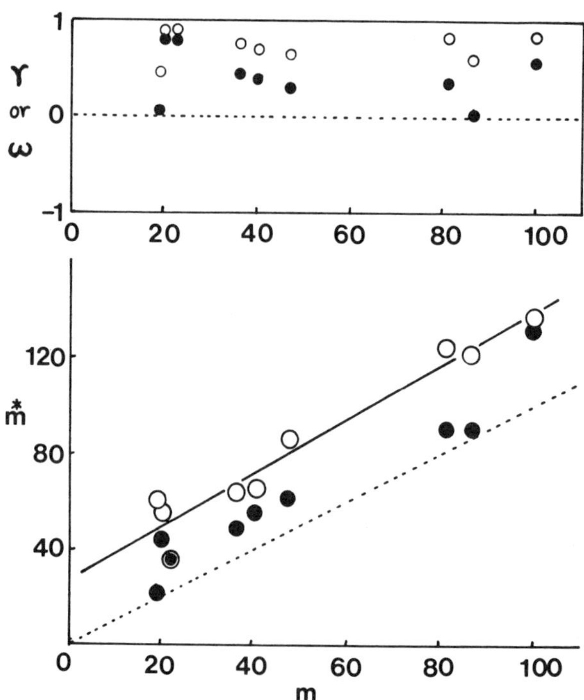

FIG 5: The series $\overset{*}{m}$-m *relation in the number of aphids per leaf and the interspecies relation between aphids and* Episyrphus *eggs. Open circles and solid circles in the lower graph indicate the relation for aphids* ($\overset{*}{m}_X - m_X$) *and that of* Episyrphus *eggs to aphids* ($\overset{*}{m}_{YX} - m_X$) *respectively. Open and solid circles in the upper graph show* γ *and* ω *indices respectively. (Modified from Ito and Iwao, 1977).*

The interspecies relation between aphids and syrphid eggs is shown also in Figure 5 by plotting $\overset{*}{m}_{YX}$ on m_X. Seven out of total 9 observations are well above the random line and hence there seems to exist a positive association between aphids and syrphid eggs. This is more obvious in the upper graph in Figure 5 showing changes of γ and ω with aphid density. The value of ω index varies from 0 to 0.8 with an average of 0.41, but there is no definite relation between ω and aphid density. The value of ω calculated for the distribution per plant is 0.87, which is higher than the values for the distribution on leaves within a plant. Thus syrphid flies seem to choose the plants with high aphid infestation, and they further prefer to lay eggs on leaves infested with high density of aphids within the plant.

4.3 Distributions of Nests of Two Ant Species in Cocoa Farms. Majer (1976) carried out field investigations on the distributions of ants in cocoa farms in Ghana. From his Figure 3 showing the distributions of ant nests in 10-m^2 quadrats, intra- and inter-species mean crowdings are calculated for the nests of two dominant species, *Oecophylla longinoda* (X) and *Macromischoides aculeatus* (Y). As seen in Figure 6, nests of both species are distributed aggregatively among 10-m^2 quadrats, but interspecies relation seems to be negative in many cases. When ω values are plotted against total nest densities (m_{X+Y}), it is apparent that the relation is rather neutral at low nest densities except for one point at the lowest density, but it becomes negative at higher densities of nests. The negative value of ω at the lowest-density point may be incidental, because only 3 quadrats each are occupied by each ant species. Thus, the two species of ants would be competitive as their nest densities increase.

In this instance, data are not suitable for the analysis of the unit-size relation. If such an analysis were possible, we could distinguish competitive exclusion from differential habitat requirements between species (Iwao, 1977). Namely, mutual exclusion would be more obvious in small spatial units if competitive interaction were operating.

5. CONCLUDING REMARKS

In the foregoing sections, only a few aspects of the $\overset{*}{m}$-m statistics and their application are mentioned. Other aspects, including density-related and contagious mortality processes, are given in earlier papers (Iwao, 1970; Iwao and Kuno, 1971). Also,

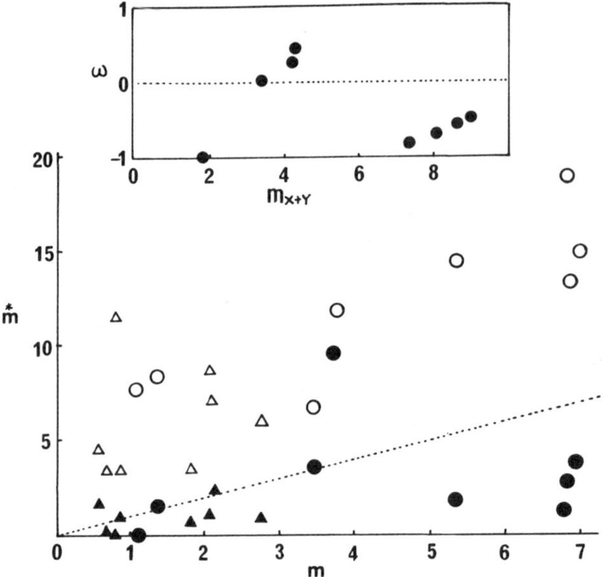

FIG 6: The intraspecies and interspecies $\overset{*}{m}$-m *relations for the distributions of nests of ants,* Oecopylla longinoda (X) *and* Macromischoides aculeatus (Y) *in a cocoa farm (Data from Majer, 1976). Open circles:* $\overset{*}{m}_X - m_X$; *open triangles:* $\overset{*}{m}_Y - m_Y$; *solid circles:* $\overset{*}{m}_{YX} - m_X$; *soild triangles:* $\overset{*}{m}_{XY} - m_Y$.

applications to various sampling problems are recently reviewed by Iwao (1977a).

Recently, Taylor *et al.* (1978) have criticized our method of the series $\overset{*}{m}$-m relation given by equation (2), mainly on the ground that our model fits less well than Taylor's to empirical distributions they tested. They also pointed out that $\overset{*}{m}$-m relation often shows a curvilinear tendency. Although I do not intend to discuss these problems here, it should be said that the degree of fitness to empirical data is not the essential point; the usefulness of any statistical method lies in how much biologically meaningful information one can get from its application. Also, we never argued that the $\overset{*}{m}$-on-m regression is always linear, but pointed out that a tendency for curvature may indicate some change in biological mechanism determining distribution pattern (e.g., Iwao and Kuno, 1971). Thus, the curvilinear

$\overset{*}{m}$-m relation for wireworm distributions given by Taylor *et al.* may indicate the operation of density-dependent process at high densities, as in the example of *Callosobruchus* sp. given by Iwao and Kuno (1971). This provides a working hypothesis for further, biological investigation. Taylor's model seems not to be based on sound theoretical background, as already pointed out (Iwao, 1977; Iwao and Kuno, 1971, etc.). The so-called Δ-model proposed by Taylor and Taylor (1977) seems unrealistic and is based on unjustified assumptions.

REFERENCES

Chandler, A. E. F. (1968). The relationship between aphid infestation and oviposition by aphidophagous Syrphidae (Diptera). *Annals of Applied Biology*, 61, 425-434.

Greig-Smith, P. (1952). The use of random and contiguous quadrats in the study of the structure of plant communities. *Annals of Botany, New Series*, 16, 293-316.

Itô, K. and Iwao, S. (1977). Oviposition behavior of a Syrphid, *Episyrphus balteatus*, in relation to aphid density on the plant. *Japanese Journal of Applied Entomology and Zoology*. 21, 130-134. (In Japanese with English summary).

Iwao, S. (1968). A new regression method for analyzing the aggregation pattern in animal populations. *Researches on Population Ecology*, 10, 1-20.

Iwao, S. (1970). Analysis of contagiousness in the action of mortality factors on the western tent caterpillar population by using $\overset{*}{m}$-m relationship. *Researches on Population Ecology*, 12, 100-110.

Iwao, S. (1972). Application of the $\overset{*}{m}$-m method to the analysis of spatial patterns by changing the quadrat size. *Researches on Population Ecology*, 14, 97-128.

Iwao, S. (1977). Analysis of spatial association between two species based on the interspecies mean crowding. *Researches on Population Ecology*, 18, 243-260.

Iwao, S. (1977a). The $\overset{*}{m}$-m statistics as a comprehensive method for analyzing spatial patterns of biological populations and its application to sampling problems. *JIBP Synthesis*, 17, 21-46. Tokyo University Press, Tokyo.

Iwao, S. and Kuno, E. (1971). An approach to the analysis of aggregation pattern in biological populations. In *Statistical Ecology, Vol. 1*, G. P. Patil, E. C. Pielou, and W. E. Waters, eds. The Pennsylvania State University Press, University Park. 461-513.

Kuno, E. (1968). Studies on the population dynamics of rice leafhoppers in a paddy field. *Bulletin of the Kyushu Agricultural Experiment Station*, 14, 131-246. (In Japanese with English summary).

Levin, R. (1968). *Evolution in Changing Environments*. Princeton University Press.

Lloyd, M. (1967). Mean crowding. *Journal of Animal Ecology*, 36, 1-30.

Majer, J. D. (1976). The maintenance of the ant mosaic in Ghana cocoa farms. *Journal of Applied Ecology*, 13, 123-144.

Morisita, M. (1959). Measuring of the dispersion of individuals and analysis of the distributional patterns. *Memoirs of the Faculty of Science, Kyushu University, Series E (Biology)*, 2, 215-235.

Pianka, E. R. (1973). The structure of lizard communities. *Annual Review of Ecology and Systematics*, 4, 53-74.

Pielou, E. C. (1969). *An Introduction to Mathematical Ecology*. Wiley-Interscience, New York.

Taylor, L. R. and Taylor, R. A. J. (1977). Aggregation, migration, and population mechanics. *Nature*, 265, 415-421.

Taylor, L. R., Woiwod, I, P., and Perry, J. N. (1978). The density-dependence of spatial behaviour and the rarity of randomness. *Journal of Animal Ecology*, 47, 383-406.

[*Received May* 1979]

SECTION III

MODELING

AND

ECOSYSTEMS MODELING

G. P. Patil and M. Rosenzweig, (eds.),
Contemporary Quantitative Ecology and Related Ecometrics, pp. 231-250.

ON A PARADIGM FOR MATHEMATICAL MODELING

D. L. SOLOMON

New York State College of
Agriculture and Life Sciences
Cornell University
Ithaca, New York 14853 USA

SUMMARY. The role of mathematical modeling in science is explored with emphasis on its creative aspect - the modeler's art. A paradigm is described which emphasizes the cyclic nature of the process and whose components are abstraction, prediction, interpretation, testing and revision. Abstraction and interpretation allow movement between the real-world and the symbolic; prediction occurs in the mathematical realm and is governed by its laws; testing occurs in the real-world and provides the basis for revision.

The abstraction step requires inductive insight and some recent techniques for its teaching are described. The prediction step is deductive and the choices between stochastic and deterministic, between analytic and simulative, and between discrete and continuous formulations are discussed. Finally, an analysis of the weight of model predictions is given.

KEY WORDS. mathematical modeling, abstraction, teaching, modeling, induction.

1. INTRODUCTION

"Models are, for the most part, caricatures of reality, but if they are good, then, like good caricatures, they protray, though perhaps in distorted manner, some of the features of the real world."

Thus Mark Kac (1969) elegantly summarizes the role of mathematical models in science. It is my purpose here, to explore that role, but since such exploration must follow a personal course, shaped by the author's experience, I shall no doubt leave important issues untouched. For example, I shall avoid a discussion of the role of formal axiom systems in mathematical modeling. Such a discussion would be too technical for our purposes and is well treated elsewhere, e.g., Maki and Thompson (1973), or more thoroughly in Wilder (1965), and the not well enough known work of Boldrini (1972). On the other hand, I shall emphasize the creative aspect of the modeler's art. Much of what I say here will have been said before, but perhaps not all in one place.

1.1 Some Definitions. To give any definition of a model is to invite exception, but a reasonable attempt is represented by Friedenberg's (1968) universal definition of a model as a "simplified representation of a real physical system." Here, we shall limit our discussion of models to *mathematical* models, for after all, the double helix of molecular genetics is too a model as are wind tunnel prototypes.

Bender (1978) somewhat circularly defines a *mathematical* model as "an abstract, simplified, mathematical construct related to a part of reality and created for a particular purpose," while Kac (1969) simply takes mathematical models to be "models which can be described symbolically and discussed deductively."

1.2 The Value of Modeling. Why build mathematical models rather than study Nature first-hand? Again we turn to Kac (1969) who observes that the primary purpose of modeling is to "polarize thinking and to pose sharp questions." As a statistician, I would complete the statement by observing that the sharp questions are then put to Nature directly using carefully designed experiments.

In mathematical modeling we must be precise. That we are compelled to simplify our view of a physical system, to strip away all but the most essential features of reality in order to be able to formulate a model precisely, can expose the fertile ground required for the emergence of deeper insights.

In formulating a model mathematically, we provide a language with which we can manipulate (subject to the laws of mathematics and logic) relationships amongst the primitives of our subject matter, to establish, unambiguously, implications of the theory on which the model is based, and so to test that theory. The method forces us to identify and label assumption, supposition, and idealization. Modeling can also be used to synthesize and organize existing knowledge, to integrate independent findings and thus assess their compatability.

Finally, modeling provides an economy of thought and a common language for scientists in diverse disciplines. Lurking behind every biological model is an economic model - a change of nouns effects the conversion.

1.3 Two Motivations. There are many ways in which one can classify mathematical models: simple versus complex, descriptive versus conceptual, analytic versus simulative, or by technical criteria, e.g., deterministic versus stochastic, and so on. It is also important to classify a modeling effort according to the motivation which prompts it, i.e., as either motivated by a desire to understand the real world or motivated by a desire to provide a basis for prediction and (perhaps) control. An example of the latter motivation is a wish to predict the effect on the dynamics of some biological population of a proposed environmental perturbation. This can sometimes be done (and is often attempted, e.g., with linear regression models) with little biological understanding. Another example is embodied in the choice between determining the temperature in a room with a thermometer (prediction and control) or by measuring the motion of the air molecules (understanding). Gold (1977) refers to the two types as *explanatory* and *correlative*. The correlative model is required only to reflect an observed relation among variables, while the explanatory model must, in addition, reflect a causal mechanism underlying the relation.

A similar scheme, described by Lucas (1964) classifies models as *rational* or *empirical*. Rational (or heuristic) models are *Gedanken* models, derived from theory and conjecture about the real system, making careful use of known characteristics of the system, and couched in meaningful terms. Empirical models to the contrary are exemplified by 'curve-fitting' and pay little attention to underlying mechanisms. Of course, most models while ideally rational, have empirical components, reflecting lack of theory, and the modeling process aids us in identifying these gaps. We are cautioned to introduce as much rationality as possible into our models, insofar as such models are likely to prove even better predictors than their empirical counterparts when extrapolating outside the data regions on which the empirical models are based.

2. STRATEGIES FOR MODELING

The quality of a modeling effort can be measured in several ways, but the final test is how able a model is to make correct predictions. Nevertheless, numerous attributes of models can be identified and an appropriate modeling strategy should be determined by the specific objectives of the investigation.

Maxims for modeling have been proposed by numerous authors, including Noble (1967), Hammersley (1973a,b, 1974), Lin and Segel (1974), Aris (1976), Haberman (1977), and the recent monograph by Aris (1978) which has a bibliography of 261 items. Levins (1966) argues, in the context of population biology, that models should be constructed to maximize *generality*, *realism*, and *precision*. Noting that these are competing objectives, he gives examples in each of which the modeling strategy is to sacrifice one quality to the other two.

Generality, sometimes referred to as robustness, allows a model to make predictions over broad regions of time or parameter values. Robust models are relatively insensitive to minor changes in assumptions. Indeed, the reality which a model attempts to mimic may itself be quite sensitive to certain types of perturbation, and we should not fault a model of such a system for sharing this property. Thus a dynamical model of a commercial fishery might become valuless for prediction in the face of a discontinuous change in the environment brought about by a changed political condition, for example international fishing regulations.

The most visible property of a model is the extent to which it is *realistic* - to which it does not make simplifying assumptions. Equivalently, this is a measure of the model's complexity, and it is the level of resolution required by the subject matter that must determine an appropriate trade-off between tractability and reality. As we will observe later, modeling is a cyclic process, commonly beginning with simple models and adding complexity (by e.g., dropping assumptions) as understanding is gained and new questions raised. It is crucial that the modeler be able to move easily between the model and the real world, in order to assess sensitivity and to revise.

In many modeling circumstances, it is *precision* which can be sacrificed since often only qualitative predictions are required - Will the proposed perturbation make the population rise or fall? - Then too, the quality of data on which parameter estimates are based may not justify an attempt at precise prediction.

Modeling strategy also includes choice of mathematical tools - stochastic or deterministic, differential equations versus difference equations, etc. But we postpone that discussion temporarily.

3. A PARADIGM

Although there are numerous variants in the literature, the components of the mathematical modeling process are perhaps most clearly depicted by Roberts (1976) as follows:

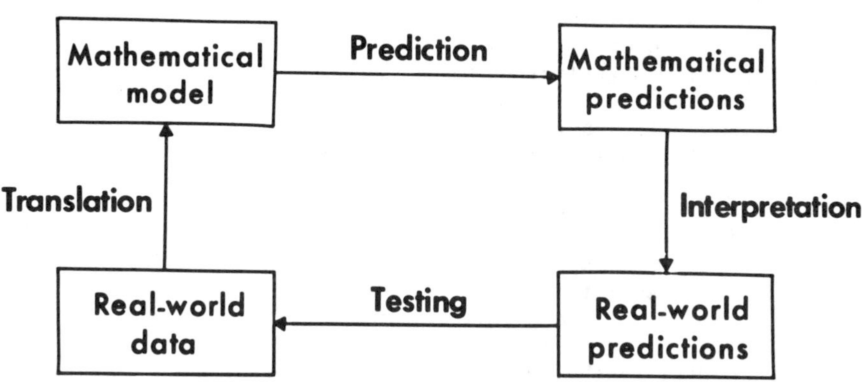

The figure emphasizes the cyclic nature of the process, and our choice of a starting point for its discussion is arbitrary.

We shall, in fact, begin with the real-world system to be modeled. By a procedure which Roberts calls translation and which others call abstraction, the modeler represents the real-world symbolically. This is the *inductive* step and the one which cannot be taught except perhaps by apprenticeship. It is, in the brain model recently popularized by Carl Sagan (1977), a right hemisphere activity. It is an art and at best we might provide conditions under which is will flourish.

The predictive step on the other hand is *deductive*, and it is here that we can bring all the power of existing mathematics to bear to arrive at the logical implications of the mathematical model. It is the left hemisphere step. These implications are then translated back to the language of reality by interpretation of the symbols. This provides the real-world predictions and is the entry point for the statistician who designs real-world experiments with which to test the predictions. The adequacy of the model is thus assessed, our understanding of reality is modified, and the cycle begins anew.

We shall, in what follows, look a bit more closely at each step in the process, but first we offer a very simple example to fix in mind some of the terms which have been introduced.

Real-World: The phenomenon to be modelled is the growth of a bacterial colony.

Translation: To formulate a mathematical model we identify:

Supposition (the primary hypothesis being entertained): The rate of growth of the colony is proportional to the number of members currently present.

Idealization (known not to be valid but offering simplification): Colony size is a continuous variable.

Assumptions (validity unknown - might be relaxed in later iterations):

a. The environment is homogeneous.
b. All individuals have the same propensity for birth (cell division).
c. Individual life histories are independent.

Choice of Tools:

a. Deterministic rather than stochastic
b. Analytic rather than simulative
c. Continuous time rather than discrete

Choice of Symbols:

$N(t)$ is the number of organisms present at time $t \geq 0$,
N_0 is the number of organisms inoculated at time $t = 0$,
r is the growth rate per individual.

Mathematical Model: A mathematical description of the supposition above is

$$\frac{dN(t)}{dt} = rN(t) \quad ; \quad r > 0 \quad , \quad t \geq 0 \quad \text{or}$$

$$\frac{dN(t)}{dt} \bigg/ N(t) = r \quad ;$$ i.e., the growth rate per individual is constant, independent of time and current colony size.

Mathematical Prediction: The solution to the differential equation is

$$N(t) = N_0 e^{rt} \quad ; \quad t \geq 0 \quad ,$$

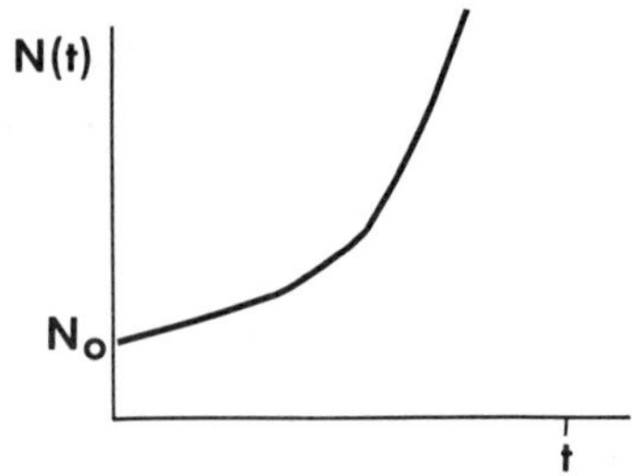

Real-World Prediction: The population will grow exponentially at rate r.

Real-World Data: Biological experiments suggest that the prediction is realistic early in the growth process, but then growth is damped by competition for limited resources.

Revision: Add realism by acknowledging density dependence.

Translation: Assumptions, idealization and tools remain unchanged.
Supposition: The growth rate per individual is a decreasing function of population size.
Symbols: K is a population size above which the environment can no longer support growth - the 'carrying capacity' of the environment.

Mathematical Model: The simplest model meeting the supposition has the growth rate per individual decreasing *linearly* with population size.

$$\frac{dN(t)}{dt} \Big/ N(t) = r\,[1 - \frac{1}{K}N(t)] \quad ; \quad r > 0,\ K > 0,\ t \geq 0.$$

Mathematical Prediction:

$$N(t) = \frac{N_0 K}{N_0 + (K-N_0)e^{-rt}} \quad ; \quad t \geq 0 \ .$$

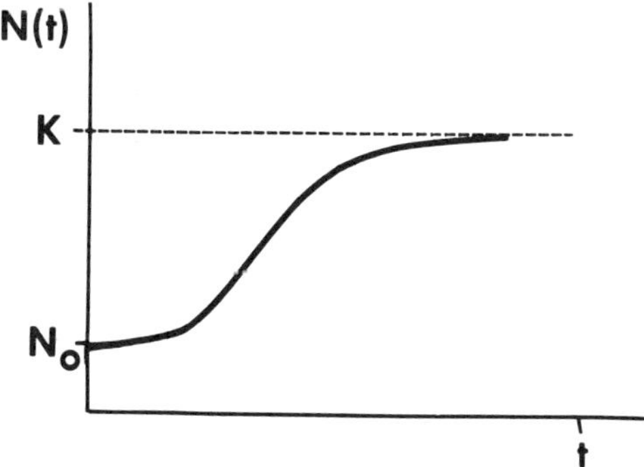

Real-World Predictions: The population increases monotonically from N_0 at time $t = 0$; in the early stages growing exponentially at rate r. As resources become limiting, the rate of growth decreases, and the population size asymptotically approaches the carrying capacity K.

4. ABSTRACTION

The translation or abstraction step which takes us from the real world to the mathematical model has two components - the inductive insight and the choice of mathematical formalism. The inductive aspect is the creative one, the one exceptionally difficult to teach, the one even difficult to describe. The choice of mathematics, on the other hand, I will soon argue, is largely preconditioned.

4.1 Induction. Leon Eisenberg (Maugh, 1974) has observed that the teaching of creativity is complicated by the fact that insight is largely preverbal and thus any verbal description of the conditions leading to the birth of an original idea may well be faulty. If true, this mitigates against the teaching of modeling even by apprenticeship. Paul Halmos (1968) in a beautiful lecture titled "Mathematics as a Creative Art," described as follows the labor pains of insight as experienced by a mathematician.

> The mathematician at work makes vague guesses, visualizes broad generalizations, and jumps to unwarranted conclusions. He arranges and rearranges his ideas, and he becomes convinced of their truth long before he can write down a logical proof. The conviction is not likely to come early - it usually comes after many attempts, many failures, many discouragements, many false starts. It often happens that months of work result in the proof that the method of attack they were based on cannot possibly work, and the process of guessing, visualizing, and conclusion-jumping begins again. A reformulation is needed - and - and this too may surprise you - more experimental work is needed. To be sure, by "experimental work" I do not mean test tubes and cyclotrons. I mean thought-experiments. When a mathematician wants to prove a theorem about an infinite-dimensional Hilbert space, he examines its finite-dimensional analogue, he looks in detail at the 2- and 3-dimensional cases, he often tries out a particular numerical case, and he hopes that he will gain thereby an insight that pure definition-juggling has not yielded. The deductive stage, writing the result down, and writing down its rigorous proof are relatively trivial once the real insight arrives; it is more like the draftsman's work, not the architect's.

Induction and creativity have, of course, long been subjects for philosophers and psychologists, and we do not pretend to those titles here. However, there have been recent attempts to teach the mathematical modeler's art in college curricula, and I should like to describe some of the techniques proposed.

Frauenthal and Saaty (1976) offer a collection of apparently difficult and confusing problems. The problems share the property that a crucial insight suggests a mathematization which renders the problem trivial, usually soluble without pencil and paper. A sample illustrating the role of symmetry and exemplifying analytic versus synthetic thinking follows:

> *The Cup of Coffee and the Cup of Milk.* Imagine you are given a gup of coffee and a cup of milk, with equal amounts of liquid in the two cups. A spoonful of milk is transferred from the milk cup to the coffee cup, stirred, and then a spoonful of the mixture is returned to the milk cup so at the end the amount of liquid in the two cups is still the same. Is there more milk in the coffee cup or more coffee in the milk cup or what?
>
> Most people say there is more milk in the coffee cup, a few say the reverse and still fewer say they are equal. The feeling about this problem is that the first transfer of milk to the coffee cup so dilutes the spoonful in coffee that the next transfer of the mixture cannot take back much of it, hence leaving more milk in the coffee cup than coffee in the milk cup. Of course, not being able to take back much of it should make it possible to take a lot more coffee in the spoonful. But people don't think of it that way.

milk cup

coffee cup

(Perhaps mathematicians will be more vulnerable than others, but if we are not careful at this point, there is the temptation to, e.g., "let x be the amount of milk ..." and be off doing algebra of ratios and proportions - but...)

Insight: Notice that after both transfers have been made, the amount of coffee missing from the coffee cup must be the same as the amount of milk in the coffee cup. This in turn is the amount of milk missing from the milk cup and thus the amount of coffee in the milk cup.

Solution: It is therefore obvious that there are equal amounts of coffee in the milk and milk in the coffee.

One can verify this with algebra but with more effort. However, the algebra assumes homogeneity of the mixture in the coffee cup after stirring. This assumption is artificial but unfortunately is needed to carry out the algebraic argument that the second spoonful has the same ratio of coffee to milk as there is in the entire coffee cup.

The object of the approach is to "... illustrate to the student an incisive way of thinking which can be carried over into more difficult problems ... to improve intuition by alerting the student to principles which operate in a domain which is apparently finer than that encountered in daily discourse and in common thought."

Morris (1967) suggests a procedure for modeling which includes the following recommendations to novice modelers:

a. *Factor the problem into simpler ones.*

b. *Establish a clear* (but perhaps tentative) *statement of the deductive objectives.* The final objective may prove to have been unforseen.

c. *Seek analogies.* This too is an intuitive step, but worth focusing on formally.

d. *Consider a specific numerical instance.* This may uncover hidden assumptions, suggest generalization and at least provide initial notation.

e. *Establish symbols.* This requires that we clearly identify the objects to be mathematized. It often involves idealizations that the experienced modeler performs by second nature but should be called to the attention of the novice. I use the word idealization as in 'ideal gas.' Similarly, in studying spatial pattern, to identify plants as points is an idealization.

f. *Write down the obvious.* That is, write in terms of the symbols the obvious aspects of the numerical example in hopes of suggesting generalization.

g. *If a tractable model is obtained, enrich it. Otherwise, simplify.* We shall return to this point in our discussion of model revision.

Pollock (1976) has adapted Morris' principles to create, in coursework fashion, a modeling studio. He argues that if modeling is an art, then the principles for teaching the traditional arts of music, painting, and sculpture should apply. Although I will not detail those principles here, an example is the *encouragement of excess.* This might be achieved by extremes of color or size in the visual arts - to test the limits of the artist and the medium. The implication for mathematical modeling is that such daring invites fresh perspectives and innovative approaches. Another important ingredient of Pollock's modeling studio is criticism by other modelers and by oneself.

I have not discussed the commonly used 'case study' approach to teaching modeling. I would argue that such a method cannot succeed in that it gives 'the answer.' In fact, I find it dangerous in that it suggests that there *is* an answer and that it is unique - the model model.

4.2 Choice of Tools for Deduction. The second part of the abstraction step of the modeling paradigm is the choice of a mathematical formulation - the choice of tools for the deductive step. An appropriate choice is of course not unique, and in fact is largely conditioned by the modeler's mathematical training. I saw an excellent illustration of this when Pollock, in lecturing to a diverse collection of faculty and students from applied mathematics, statistics, and operations research, proposed a phenomenon to be modeled and invited the audience's insights. The mathematicians immediately saw the situation as natural for description by a system of differential equations, while the statisticians envisioned a stochastic birth and death process, and to the operations researchers, it was obvious that the appropriate tools were those of queuing theory.

4.3 Stochastic versus Deterministic Models. As promised at the outset, this discussion is following a personal course, and so in considering choice of tools I should like to briefly consider the choice between stochastic and deterministic model formulations. Variability in the biological world is well known to all of us. Maynard Smith (1974) in the context of mathematical ecology observes that deterministic models fail to mirror reality in assuming infinite population sizes and in ignoring random fluctuations in the enviornment. May (1974) expands, observing that "birth rates, carrying capacities, competition coefficients, and other parameters which characterize natural biological systems all, to a greater or lesser degree, exhibit random fluctuations." Ehrlich and Birch (1967) boldly assert "models must be stochastic, not deterministic."

If stochastic models are admittedly more realistic, what then the justification for deterministic approaches? A primary one is tractability. By itself, of course, mathematical convenience has no place in model building, but deterministic models have often proven adequate for mimicing biological systems and providing biological insights. Deterministic models are often taken as first approximations to stochastic models, and study of deterministic models can also provide answers to questions about the behavior of stochastic ones. For example, May (1971a,b) describes an m-species population model with random environment showing that the conditions for the existence of an equilibrium probability distribution are identical to those for a stable equilibrium in the deterministic case. He furthermore concludes that if complex natural ecosystems are stable, then the interactions in them are essentially non-random.

Voltaire has said "Chance is a word void of sense; nothing can exist without a cause." In contrast Charles Dickens has observed: "Accidents will occur in the best regulated families." It is beyond our scope to address seriously the fundamental but irresolvable issue of determinism versus indeterminism inasmuch as it is an issue which we hold is largely irrelevant to applications. That is, even in a deterministic universe, in which (I quote Bartlett (1960) after Schrödinger (1944)) "... a multiplicity of detailed causes is operating to produce the observed broad class of events, it is often an economy of thought in the sense of Mach to ignore these and appeal merely to the operations of chance and the laws of averages." Thus, for example, in coin tossing or gamete pairing, even if we could argue that sufficient knowledge of the physical forces surrounding the event would completely determine the outcome, nevertheless from our level of resolution, we perceive the process as random.

In his treatment of the role of stochastic elements in biological models, Lucas (1964) expands on and formalizes these notions by proposing that the universe, which clearly possesses deterministic features, might also possess truly random ones, and if so, these would be characterized by inherent unexplainability and thus be outside the realm of science. However, because of our current state of ignorance, or because of our failure to take certain knowledge into account, some of the deterministic features appear to us as random. To these, Lucas ascribes the name pseudo-random and defines the role of science as that of diminishing the amount of pseudo-randomness in the world, while true randomness, if it indeed exists, sets the bounds on explainability and predictability.

Indeed, some (including Kac) would argue that the concept of chance is not operationally defined inasmuch as, e.g., when presented with a sequence of numbers we can in no way ascertain if the process which generated it was random. That we cannot dismiss these matters as *merely* metamathematics is emphasized by noting the existence of phenomena which we perceive as regular but which can be produced by random processes and phenomena which we perceive as random but are produced by strictly deterministic means.

Examples of the former are provided by certain animal population cycles which tend to show peaks of about 3 years and by sunspot and other natural cycles of length about 10 years. It is not difficult to show (e.g., Cole, 1954) that such cycle lengths are to be expected in discrete time stationary random processes. Here the mean distance between peaks in a long sequence of numbers is approximately 3 and between dominant peaks (a peak higher than its two neighboring peaks) about 10.

Examples of apparently random, deterministic phenomena are found in a recent series of publications (e.g., Oster, 1974; Marsden and McCracken, 1976; May, 1976; May and Oster, 1976; and Guckenheimer *et al.*, 1977) in which simple deterministic models are presented which exhibit behavior described as chaotic; that is, behavior in which the time path of the system could be taken for a sample path of a stochastic process. Perhaps the most simple example of this type is the difference equation $x_{n+1} = \lambda x_n(1-x_n)$, which for $0 \leq \lambda \leq 4$ maps the unit interval into itself. The behavior of the sequence $\{x_n\}$ depends dramatically on the value of λ and on the starting value x_0. For example, with $\lambda = 4$, $x = 3/4$ is a fixed point $(x_{n+1}=x_n)$ but is unstable so that, e.g., putting $x_0 = .7499999$ generates a most erratic sequence. At $\lambda = 3$, there are starting points which produce sequences of period 2 $(x_{n+2}=x_n)$, for $\lambda = 1 + \sqrt{6} \doteq 3{\cdot}45$ sequences of period 4 appear, and so on. But of special interest here, for values of $\lambda > 3{\cdot}57$, there are (infinitely many) initial conditions which produce sequences which are non-periodic, i.e. which, even though always contained in [0,1], never repeat! (for example, see figure below.) How could a trajectory of this completely determined process ever be distinguished from the sample path of a stochastic one? The implications of these observations are not yet well studied, but some useful observations have been made by Aris (1978).

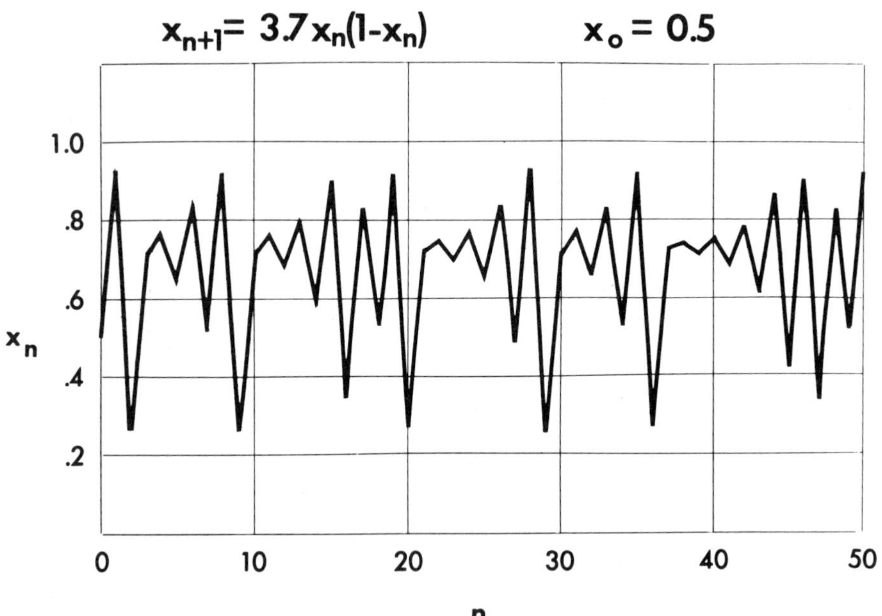

4.4 Simulation Models. When the mathematical problems associated with a model are too difficult to handle or there is a lack of fundamental theory, the modeler may turn to simulation. The role of the 'pilot plant' for simulation in physical model building is played by the computer in mathematical model building. We shall not pursue the subject at length here, but would observe that simulation models tend to share the weaknesses of all empirical models. In ecosystem simulation models for example, there are often huge numbers of parameters, and many different sets of their values, though theoretically contradictory, may adequately simulate observed systems.

Advantages of simulation modeling are that they allow us to deal with much more complex systems than we could by analytic means, and that they can usually be designed to allow easy variation of parameter values, thus enabling assessment of sensitivity of the modeled system to its parameters.

4.5 Discrete or Continuous? As a final note on the choice of mathematical tools, I would call attention to the choice between discrete and continuous formulations for a model. The distinction arises in many settings. In studying spatial pattern in plant communities for example, we decide between number of stems and percent cover. In stochastic process models, we choose between discrete and continuous states as well as between discrete and continuous time. In deterministic models, the choice of discrete or continuous time becomes one of difference equations or differential equations.

Difference equations are usually viewed as convenient tools for approximating solutions to differential equations, which are themselves of primary interest. However, for many biological situations, it is the discrete formulation which is in fact the more realistic, the differential equations model being the approximation. A difficulty which arises when using differential equations representations of discrete phenomena is that a given differential equation corresponds to many difference equations, all with the same limiting differential equation, but themselves perhaps having widely different solutions. Some examples of the implications of this fact are given in the expository works of Van der Vaart (1973) and of Frauenthal (1976).

5. MATHEMATICAL AND REAL-WORLD PREDICTIONS

We now apply our mathematical tools to the mathematical model to make mathematical predictions. The innovative part here is deciding on the subject matter questions to ask in the mathematical language. These predictions are then translated into the subject discipline to produce real-world predictions. Note that we do not here claim to real-world *explanation* - only predictions. As we shall emphasize later, different models may make identical predictions but provide different explanations. As Bender (1978) says, "The mechanism is irrelevant when dealing with predictions, but the nature of the mechanism is the heart of an explanation."

6. MODEL TESTING

The final step in the modeling cycle is to compare model prediction with reality. This typically requires experimental design and statistical inference, which topics we shall not pursue here. We only note in this regard that a decision must be reached as to how consistent with reality the model need be. This, of course, is determined by the reason for which the modeling effort was mounted. It should be emphasized that statistics is no substitute for the experienced scientist's intuition and wisdom of subject.

The question of whether a *model* is 'right' lies outside of mathematics, and in some sense no model can be right. A mathematical *prediction* is correct only insofar as its mathematical counterpart is logically deduced from the axioms of the model. A subject matter *conclusion* derived from a model with gross assumptions is weak. A conclusion derived from a general model and insensitive to its assumptions or one corroborated by several models is robust.

There are numerous illustrations of the fact that several sets of underlying assumptions about a real phenomenon can lead to the same mathematical model, and thus the same predictions. It follows that even if a model is consistent with observation, it cannot be concluded that the assumed mechanisms on which the model is based are realistic.

One of my favorite such illustrations provides several sets of assumptions - mutually contradictory - for the spatial distribution of cabbage butterfly eggs on individual cabbage plants. One model proposes that adult female butterflies visit individual plants 'at random,' laying clusters of eggs. The environment is homogeneous in that all plants are equally attractive to the females. It is also supposed that the number of eggs in a cluster follows the same probability distribution for each cluster, and that the number of eggs in a given cluster is neither dependent on the number in any other nor on the number of clusters.

Contrary to the assumptions in the first model, a second supposes that eggs are not laid in clusters, but are spatially distributed at random on the cabbage plants. Furthermore, unlike in the first model, individual heads are not equally attractive to the female butterflies, differing in size, condition and location (e.g., border versus interior of the plot or orientation with respect to the sun), and so the mean number of eggs varies from head to head.

Connecting the biological assumptions to mathematical ones leads us to conclude *for both models* that the number of eggs per plant should follow a negative binomial distribution, a prediction well confirmed by experiment (Harcourt, 1961; Kobayashi, 1965). Thus additional information would be required to decide between the two proposed mechanisms (neither of which need, of course, be correct). We emphasize that the biological assumptions on which the two models are based are not only different but are in fact *contradictory*. In the absence of additional information we might, nevertheless, make predictions about, say, the mean number of eggs per plant and thus about future butterfly population sizes. Such predictions could have implications for control decisions with important economic consequences even with the mechanism not fully understood.

For the curious, we remark that for this particular population, experiments have been performed in a net house to observe the detailed behavior of the female cabbage butterfly (Kobayashi, 1966). Peripheral plants and those nearer the light source were favored, but under uniform light conditions, the butterflies visited interior plants at random (Poisson). The independence assumption made in the first model was also tested and found tenable. Thus the first model proved the more appropriate description of the biological mechanism. A third set of assumptions about the cabbage butterfly system also leading to the negative binomial distribution, together with the mathematical details for all three models, may be found in Solomon (1976). An additional collection of three temporal (as opposed to spatial) negative binomial models also appears there.

7. MODEL REVISION

We have now come full circle and the path of science directs us to revise hypothesis, modify the model, and begin anew. We recall that the last step in Morris' (1967) modeling scheme enjoins us to enrich a tractable model, simplify an intractable one. Some means of simplification are 'making variables into constants; eliminating variables; using linear relations; adding stronger assumptions and restrictions; suppressing randomness.' To enrich, we perform the opposite modifications.

We close with the injunction that we must all be alert for new tools for the modeling studio and wary of stagnation in our modeling. New mathematics (the finite element method) or new ways to use existing mathematics (catastrophe theory) should continuously enrich our repertoire. (See Williams (1977) for a somewhat unorthodox expansion of these comments.)

REFERENCES

Aris, R. (1976). How to get the most out of an equation without really trying. *Chemical Engineering Education*, 10, 114.

Aris, R. (1978). *Mathematical Modelling Techniques*. Vol. 24 in the series "Research Notes in Mathematics." Pitman, London.

Bartlett, M. S. (1960). *Stochastic Population Models.* Methuen and Co., London.

Bender, E. A. (1978). *An Introduction to Mathematical Modeling.* Wiley, New York.

Boldrini, M. (1972). R. Kendall, trans. *Scientific Truth and Statistical Method.* Hafner, New York.

Cole, L. C. (1954). Some features of random cycles. *Journal of Wildlife Management,* 18, 2.

Ehrlich, P. R. and Brich, L. C. (1967). The 'Balance of Nature' and 'Population Control.' *American Naturalist,* 101, 97-107.

Frauenthal, J. C. (1976). Difference and differential equation population growth models. *Modules in Applied Mathematics.* Mathematical Association of America.

Frauenthal, J. C. and Saaty, T. L. (1976). Foresight-Insight-Hindsight. *Modules in Applied Mathematics.* Mathematical Association of America.

Friedenberg, R. M. (1968). *Unexplored Model Systems in Modern Biology.* Vol. 1 in the series "Pioneering Concepts in Modern Science." Hafner, New York.

Gold, H. (1977). *Mathematical Modeling of Biological Systems.* Wiley, New York.

Guckenheimer, J., Oster, G., and Ipaktchi, A. (1977). The dynamics of density dependent population models. *Journal of Mathematical Biology,* 4, 101.

Haberman, R. (1977). *Mathematical Models.* Prentice-Hall, Englewood Cliffs, New Jersey.

Halmos, P. R. (1968). Mathematics as a creative art. *American Scientist,* 56, 375-389.

Hammersley, J. M. (1973a). Maxims for manipulators. *Bulletin of the Institutional Management Association,* 9, 276; 10, 368.

Hammersley, J. M. (1973b). How is research done? *Bulletin of the Institutional Management Association,* 9, 214.

Hammersley, J. M. (1974). Poking about for vital juices of mathematical research. *Bulletin of the Institutional Management Association,* 10, 235.

Harcourt, D. G. (1961). Spatial pattern of the imported cabbageworm, *Pieris rapae* (L.). (Lepidoptera: Pieridae), on cultivated cruciferae. *Canadian Entomologist*, 43, 945-952.

Kac, M. (1969). Some mathematical models in science. *Science*, 166, 695-699.

Kobayashi, S. (1965). Influence of parental density on the distribution pattern of eggs in the common cabbage butterfly, *Pieris rapae crucivora*. *Researches on Population Ecology*, 7, 109-117.

Kobayashi, S. (1966). Process generating the distribution pattern of eggs of the common cabbage butterfly, *Pieris rapae crucivora*. *Researches on Population Ecology*, 8, 51-61.

Lin, C. C. and Segal, L. A. (1974). *Mathematics Applied to Deterministic Problems in the Natural Sciences*. Macmillian, New York.

Levins, R. (1966). The strategy of model building in population biology. *American Scientist*, 54, 421-431.

Lucas, H. L. (1964). Stochastic elements in biological models; their sources and significance. In *Stochastic Models in Medicine and Biology*, John Gurland, ed. University of Wisconsin Press, Madison.

Maki, D. P. and Thompson, M. (1973). *Mathematical Models and Applications*. Prentice-Hall, Englewood Cliffs, New Jersey.

Marsden, J. E. and McCracken, M. (1976). The Hopf bifurcation and its applications. *Applied Mathematical Sciences*, Springer-Verlag, Heidelberg.

Maugh, T. H., II (1974). Creativity: Can it be dissected? Can it be taught? *Science*, 21 June 1974, 1273.

May, R. M. (1971a). Stability in model ecosystems. *Proceedings of the Ecological Society of Australia*, 6, 18 56.

May, R. M. (1971b). Stability in multi-species community models. *Mathematical Biosciences*, 12, 59-79.

May, R. M. (1974). *Stability and Complexity in Model Ecosystems*. Princeton University Press, Princeton.

May, R. M. (1976). Simple mathematical models with very complicated dynamics. *Nature*, 261, 459.

May, R. M. and Oster, G. F. (1976). Bifurcations and dynamic complexity in simple ecological models. *American Naturalist,* 110, 573.

Maynard Smith, J. (1974). *Models in Ecology.* Cambridge University Press, Cambridge.

Morris, W. T. (1967). On the art of modeling. *Management Science,* 13, B-707-717.

Noble, B. (1967). *Applications of Undergraduate Mathematics in Engineering.* Macmillan, New York.

Oster, G. (1974). Stochastic behavior of deterministic models. In *Ecosystem Analysis and Prediction,* Proceedings of a SIAM-SIMS Conference held at Alta, Utah.

Pollock, S. M. (1976). Mathematical modeling: Applying the principles of the art studio. *Engineering Education,* November, 1976, 167-171.

Roberts, F. S. (1976). *Discrete Mathematical Models.* Prentice-Hall, Englewood Cliffs, New Jersey.

Sagan, Carl (1977). *The Dragons of Eden.* Random House, New York.

Schrödinger, E. (1944). The statistical law in Nature. *Nature,* 153, 704.

Solomon, D. L. (1976). The spatial distribution of cabbage butterfly eggs. *Modules in Applied Mathematics.* Mathematical Association of America.

Van der Vaart, H. R. (1973). A comparative investigation of certain difference equations and related differential equations: Implications for model-building. *Bulletin of Mathematical Biology,* 35, 195-211.

Wilder, R. L. (1965). *Introduction to the Foundations of Mathematics,* 2nd ed., Wiley, New York.

Williams, M. B. (1977). Needs for the future: Radically different types of mathematical models. In *Mathematical Models in Biological Discovery,* D. L. Solomon and C. Walter, eds. Volume 13 in the series *Lecture Notes in Biomathematics.* Springer-Verlag, Berlin.

[*Received July* 1977. *Revised March* 1979]

G. P. Patil and M. Rosenzweig, (eds.),
Contemporary Quantitative Ecology and Related Ecometrics, pp. 251-269.

ECOLOGICAL MODELS AND THE STOCHASTIC-DETERMINISTIC QUESTION

ROBERT W. POOLE

Section of Population Biology and Genetics
Division of Biology and Medicine
Brown University
Providence, Rhode Island 02912 USA

SUMMARY. The question of whether deterministic or stochastic models are more appropriate in ecology is discussed within the context of a classification and characterization of ecological modeling. The philosophies and usages of ecological models are divided into four categories: 1) Lotka-Volterra heuristic models, 2) systems models and simulation, 3) forecasting, and 4) optimization and control. The paper begins with a short discussion of prediction as a stochastic process and tries to demonstrate that the 'average behavior' of a stochastic model is not easily comparable to the solution of an analogous deterministic model. Following this introduction, each of the four categories is discussed in terms of the philosophy and purpose of the models and is related to the pragmatic field problems involved with each. For each category the relative advantages and disadvantages of stochastic and deterministic models are explored. Two principal, if rather obvious, conclusions are reached. First, no single category of modeling can be identified as best, nor is there a clear cut advantage to either deterministic or stochastic models. Each model category is characterized by unique properties and practical limitations making it appropriate for one specific form of ecological prediction. The choice of a modeling philosophy depends on the purpose of the model and the pragmatic problems of field research. Secondly, the first three categories of prediction seem to constitute a natural progression in the analysis, representation, and forecasting of ecological phenomena rather than mutually exclusive methods of attacking the same problem.

KEY WORDS: Modeling, deterministic, stochastic, predictions, hypothesis generation, simulation, forecasting.

1. INTRODUCTION

Periodically the question of whether stochastic or deterministic models are more appropriate in ecology raises its head in public. The stochastic-deterministic 'controversy' appears to be only part of a larger question: what is a model, what good is it, and what practical considerations limit its use. The philosophy of modeling cannot be discussed without recourse to generalities and statements of the obvious. In fact, the general theme of this paper is perhaps the most self evident of all: the choice of a modeling philosophy depends in part on the intended use of the model. An ecological model has no intrinsic reality or value beyond its ability to fulfill the task it was created for. Therefore, the nature of the job determines the philosophical basis of the model.

All uses of ecological models fall into the broad category of prediction. Prediction is a nebulous and overworked word I prefer to divide into four different categories: 1) Lotka-Volterra hypothesis generation, 2) systems simulation, 3) forecasting, and 4) optimization and control. These four categories are defined later in the paper. All four have been called prediction, although they are different problems calling for different approaches. Other classifications are possible and the dividing lines between these four are not always clear-cut.

Idealistically, all ecological prediction problems are stochastic. The only determinism in the real world is that nothing is certain. However, a deterministic outlook may be either pragmatically or scientifically more appropriate in some modeling situations. The purpose of this paper is to present one view of ecological prediction and to relate different types of prediction to the deterministic-stochastic dilemma. Ecological prediction is first briefly outlined as a problem in stochastic theory and some general comments are made about comparing deterministic solutions with the mean behavior of an analogous stochastic model. Secondly, each of the four prediction categories is considered in terms of intended use, philosophy, and the practical problems limiting it.

2. ECOLOGICAL PREDICTION AS A STOCHASTIC PROCESS

The general prediction problem can be summarized as one aspect of stochastic processes. An ecosystem is composed of a multitude of fluctuating variables such as the abundances of the species in the community, nutrient levels, the biomass of the primary producers, and so forth. Let the ecosystem 'state' variables actually studied be $Y_1(t), Y_2(t), \cdots, Y_n(t)$. There are other measured environmental variables such as temperature, humidity, immigration, incident sunlight, and so forth independent of the ecosystem dynamics so called exogenous and input variables labeled $X_1(t)$, $X_2(t), \cdots$, $X_m(t)$. Finally, there is a third class of variables, $N_i(t)$, which might be called noise, representing the combined effects of unstudied variables and purely chance events. Traditionally, time t is the present and t+L represents L time units in the future. For notational simplicity assume each of the variables is continuous between zero and infinity. If both discrete and continuous components are assumed, the conclusions are unaffected but the notation becomes a bit messy.

Let F[] stand for the probability distribution of one or more variables and f[] the corresponding density. Denote a general random variable by a capital letter and an actual value by the corresponding lower case letter. If we know nothing about the past and present values assumed by the exogenous, input, or state variables of the system, then at time t+L the unconditional joint probability distribution of the n state variables of the system is

$$F[Y_1(t+L) \leq y_1(t+L), Y_2(t+L) \leq y_2(t+L), \cdots, Y_n(t+L) \leq y_n(t+L)] \quad .$$

The unconditional density of one variable $f[Y_i(t+L)]$ might look like curve a in Figure 1. Usually, however, some of the values taken by the system, input, and exogenous variables are known. Normally the state variables are given up to time t and the exogenous and input variables may or may not be known from t to t+L as well. The specified values of Y_1, $Y_2, \cdots, Y_n$; X_1, $\cdots, X_m$ will be simply indicated by the notation $y_1(\xi), \cdots, y_n(\xi)$; $x_1(\xi), \cdots, x_m(\xi)$. The prediction problem in simplest terms is find the conditional probability density (or distribution) of the state variables $Y_1, \cdots, Y_m$ at time t+L given $y_1(\xi), \cdots$, $Y_n(\xi)$; $x_1, \cdots, x_m(\xi)$, i.e.,

$$f[Y_1(t+L),\ Y_2(t+L), \cdots,\ Y_n(t+L) \mid y_1(\xi),\ y_2(\zeta), \cdots, y_n(\xi);$$
$$x_1(\xi), \cdots, x_m(\xi)] \quad .$$

Specifying part of the past history of the ecosystem has the effect, of course, of assigning higher probabilities to some outcomes and lower probabilities to others because the past determines the future to some degree. The overall effect is to decrease the spread of the probability density. For example, the marginal conditional density for the variable Y_i at time t+L might be curve b in Figure 1.

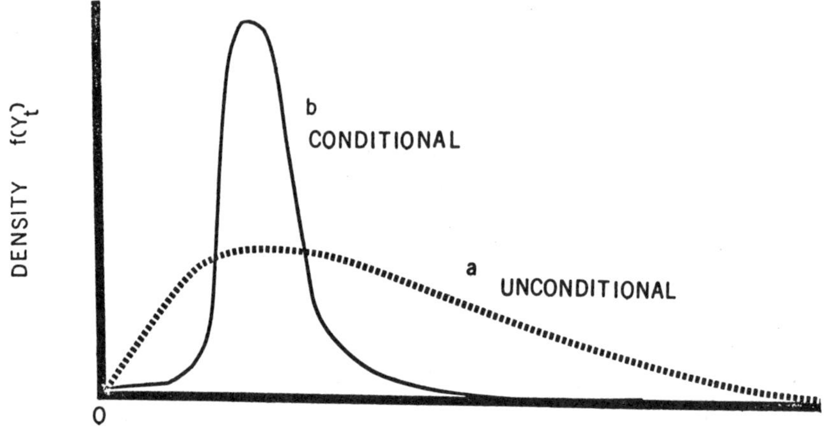

FIG. 1: A diagrammatic comparison of the unconditional probability density of an ecological variable at time t+L *with the conditional density that might occur if some of the values of the state input, and exogenous variables affecting it were known.*

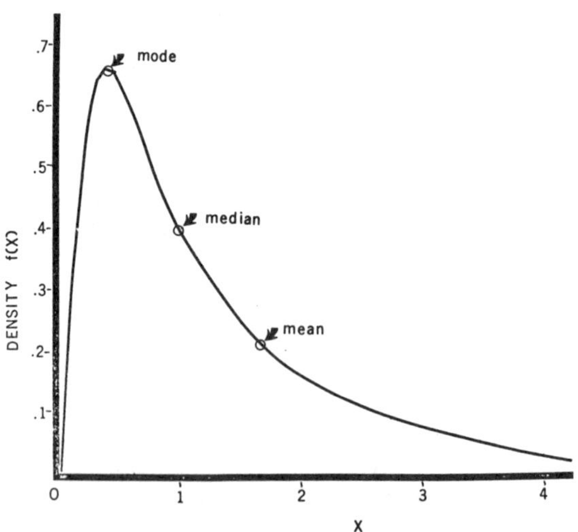

FIG. 2: A lognormal probability density with a median and variance of one to illustrate the differences in the mode, median, and mean of a non-normal distribution. The mode is 0.3678, *the median is* 1.0000, *and the mean is* 1.6487.

A stochastic model is a probabilistic biological or statistical representation of the relationships between the measured variables $Y_1,\cdots,Y_n;\ X_1,\cdots,X_m$ generating the conditional probability distribution. However, stochastic models are usually analytically difficult or hopeless if the models are meaningful biological representations. In many contexts the 'average' future behavior of the ecosystem variables might be enough. The trajectory of a deterministic model is sometimes claimed to be equivalent to the 'average behavior' of a stochastic model derived from the same biological assumptions. A deterministic model may be sufficient or even superior to a stochastic model in many modeling problems. However, the 'average behavior' claim, I feel, is not necessarily a good justification of deterministic models for two reasons:

1. It usually isn't true.

2. The intuitive meaning of average is ambiguous.

Point two is discussed first followed by point one.

There are at least three intuitive and equally reasonable interpretations of the 'average behavior' of a stochastic process at time t; the mode, the median, and the mean of $f[Y(t)]$. Suppose Figure 2 is the conditional density of population abundance at time t. The mode of a density is the most likely value. The median is the value of $Y(t)$ such that 50 percent of the outcomes fall to the left of the point and 50 percent to the right. Finally, the mean is the expectation of the density. All three parameters are candidates for intuitive 'average behavior.' Although the mean is the true average of a random variable, the mode or median are usually far more satisfying intuitively.

The most commonly used definition of 'average,' primarily for analytical reasons, is the expectation or mean function of the random variable $Y(t)$ from t to $t+L$. Is the trajectory of a deterministic model equivalent to the mean function of a stochastic model built on identical assumptions? The answer is almost always no except for linear models with orthogonal error terms. This contention will be demonstrated with a general example and then illustrated by a simple model. Consider then n multispecies homogeneous model

$$\begin{aligned} \frac{dY_1}{dt} &= f_1(Y_1,Y_2,\cdots,Y_n) \\ \vdots \quad & \quad \vdots \\ \frac{dY_n}{dt} &= f_n(Y_1,Y_2,\cdots,Y_n) \end{aligned} \tag{1}$$

and the related system of stochastic differential equations

$$\begin{aligned}\frac{dY_1}{dt} &= f_1(Y_1, Y_2, \cdots, Y_n) + g_1(Y_1, Y_2, \cdots, Y_n)w_1(t)\\ &\vdots \qquad\qquad\qquad\qquad \vdots\\ \frac{dY_n}{dt} &= f_n(Y_1, Y_2, \cdots, Y_n) + g_n(Y_1, Y_2, \cdots, Y_n)w_n(t) \quad ,\end{aligned} \tag{2}$$

where $w_1(t), w_2(t), \cdots, w_n(t)$ are zero mean, unit variance Gaussian white noise inputs and the $g_i(Y_1, \cdots, Y_n)$ are functions relating the variance of the stochastic elements to the present abundance levels of the populations. The solution and interpretation of stochastic differential equations is beside the point (see May, 1973, for example), but a partial differential equation can be directly written down for the joint conditional probability density. On the other hand, this equation is almost always unsolvable analytically.

However, a system of differential equations can be derived for the mean functions, i.e., $E[Y_1(t)], \cdots, E[Y_n(t)]$, of each variable [see Sage and Melsa (1971) for a derivation]. For species i,

$$\begin{aligned}\frac{dE[Y_i(t)]}{dt} &= f_i\{E[Y_1(t)], E[Y_2(t)], \cdots, E[Y_n(t)]\}\\ &+ \frac{1}{2}\sum_{j=1}^{n}\sum_{k=1}^{n} \frac{\partial^2 f_i\{E[Y_1(t)], \cdots, E[Y_n(t)]\}}{\partial E[Y_j(t)]\ \partial E[Y_k(t)]} \operatorname{Cov}[Y_j(t), Y_k(t)]\\ &+ \text{terms involving all higher order central moments.}\end{aligned} \tag{3}$$

This system of equations involves not only the mean functions at time t, but also the variance of each species, the covariance between variables, and an infinite number of terms in the third, fourth, and all higher order central moments. If the mean function $E[Y_i(t)]$ of the stochastic model were equivalent to the trajectory $Y_i(t)$ of the corresponding deterministic model, only the first term on the right would appear. Consequently, if we equated $E[Y_i(t)]$ with $Y_i(t)$ in the deterministic model we would be wrong. Similar results have been found for nonlinear birth-death processes (see Bharucha-Reid, 1960).

By way of example consider the logistic model

$$\frac{dY}{dt} = Y(r - bY) \quad .$$

Plugging into equation (3) for a corresponding stochastic model and letting $\mu(t) = E[Y(t)]$,

$$\frac{d\mu(t)}{dt} = [r - b\mu(t)]\mu(t) + b\mathrm{Var}[Y(t)] \quad .$$

The mean function of the stochastic model is not identical to the deterministic model's trajectory. Also suppose the solution of the stochastic model approaches a stationary distribution with time, i.e. $\mu(t) \to \mu$ and $\mathrm{Var}[Y(t)] \to V$ as $t \to \infty$. The admissible stationary value of the mean is

$$\mu = \frac{K + (K^2-4V)^{1/2}}{2}$$

where $K = r/b$. Clearly the stationary mean decreases relative to the deterministic equilibrium point K as the stationary variance increases. Candidly, the difference is not great in this example because the nonlinearity of the deterministic model is not large. For example, if $K = 100$ and $V = 900$, then $\mu = 90$ while the deterministic equilibrium point is $Y^* = 100$.

Equation (3) suggests two general rules of thumb for anticipating how closely a deterministic trajectory will match the mean function of a corresponding stochastic model:

1. The larger the variances and covariances within and between variables, the greater the discrepancy, i.e., the greater the variability of the variance and the more closely the dynamics of the system are intertwined, the worse the comparison.

2. The more nonlinear the model, the worse the match due to the increasing importance of higher order partial derivatives and central moments in equation (3) as the degree of nonlinearity increases.

3. LOTKA-VOLTERRA HYPOTHESIS GENERATION

Lotka-Volterra models are so typical of this category of prediction that I am using them as a sobriquet for reductionist hypothesis generation. The purpose of a reductionist approach is best defined by describing the problems justifying the model. The natural world is a very complex place. There may be thousands of interacting species in an ecosystem of all age classes and sizes spread out in a spatially heterogeneous environment along with hundreds of nutrients and a multitude of physical environmental variables. Each variable fluctuates in abundance, level, or concentration both in time and in space, sometimes in apparently capricious ways. To preserve our sanity, we conclude that the species composition of the community in part and at least some of the fluctuations we see in species abundances, nutrient levels, and energy flow can be traced to interactions between the physical and biological variables of the ecosystem. If communities are not random collections of species there are at least three fundamental ecological questions to resolve:

1. How do interactions between species and the physical environment determine the behavior of the ecosystem and its individual components?

2. How do interactions between species and the physical environment affect the composition of the community?

3. How do interactions between species and the physical environment affect the evolutionary history of the community?

There are two ways to attack these three problems: 1) head on with a systems analysis approach, or 2) by reducing the ecosystem to a set of idealized components and studying each simplified part separately. The Lotka-Volterra school of ecological modeling is the reductionist approach.

Reductionist models are based on highly simplified and general assumptions of how the individuals in an isolated component should interact. Detailed representations encompassing specific biological characteristics of particular species and including interactions with other species and the rest of the ecosystem are deliberately avoided. The purpose of reductionist models is twofold. First, by isolating a single component of an ecosystem, such as a predator-prey interaction, general properties of the interaction may be found that can later be expanded upon and used as basic blocks in more specific and detailed studies of specific ecosystems. Secondly, the analysis of highly simplified and idealized models of species interactions may suggest general hypotheses about population fluctuations, the composition of species groups, or evolutionary trends. The hypothesized behavior can never be observed in exactly the form postulated in real populations or species because specific detail has been deliberately avoided. However, if enough similar empirical populations are studied, the hypothesized behavior may appear as a general characteristic amid a myriad of detail. If somehow this general trend can be separated from the detail, and it is not obvious that it can, then one might be willing to call the hypothesized behavior a 'general principle' if it occurred in some given proportion of empirical situations, say 50 percent. If nothing else, reductionist models allow us to forget for the moment the apparent impossibility of ecological problems, to define a starting point in an overwhelming maze of interactions, and to define sharply how one goes about studying anything as complex as an ecosystem.

A critical underpinning of reductionist modeling is whether or not the predictions of the models are testable against empirical data (see Peters, 1976; Ferguson, 1976; Stebbins, 1977; Caplan, 1977; Castrodeza, 1977 for a recent debate). I don't wish to become entangled in the logical arguments involved but offer these practical considerations instead. From a pragmatic point of view the question of testability is composed of two problems: Are the predictions of the model physically testable, and do the empirical data have any real relationship to the hypotheses it purports to test? It is appallingly easy to generate predictions that are interesting but impossible to test. For example, a valid prediction might be that the number of tree species in Rhode Island will decline by 30 percent in the next two thousand years. The prediction may be interesting, but is pragmatically untestable. The second aspect of testability appears more important to me because of the nature of reductionist arguments. Any deductive or deductive-inductive model starts by postulating particular patterns of interactions between species or species and physical environment. The specified interactions determine the behavior of the model and the hypotheses generated. The hypotheses are tested against empirically derived data. However, no real species population or ecosystem component exactly matches the pattern of interactions postulated by the model. Therefore, the hypotheses are untestable in the strict sense of the word and the empirical data are not a valid test of the model's predictions. The problem is vaguely analogous to the statistical testing of a simple null hypothesis versus a composite alternative. If the null hypothesis is 'the effect of factor a is exactly c,' then the null hypothesis will always be falsified if enough data are collected because nothing in the universe is exactly equal to the infinitely small point on the line c. What most people really mean is 'for all practical purposes the effect of a is not enough different from c to be important.' This definition of significance is not the same as statistical significance but is pragmatically more appropriate. Consequently, in testing ecological hypotheses, the logical formalism of testability should probably be replaced with a more subjective concept defining the phrase 'close enough.' More specifically is the behavior of the empirical data close enough to the postulated behavior for all practical purposes.

The key idea in testing the hypotheses of reductionist models, therefore, is not whether the predictions of the model are right or wrong. By definition they are wrong. The main point is how 'wrong' they are and why they are wrong. If this line of argument is accepted, the predictions of the Lotka-Volterra class of models are heuristic devices. The hypotheses of the model define the problem by initially restricting the problem to one sharply focused question. The observed discrepancies between the predictions and empirically derived data are explained by matching the biology of the species against the assumptions of the model. Hopefully this procedure leads to new insights and suggests new patterns of interactions to serve as the basis of a new model and a new round of heuristic prediction. The analysis moves from the general to the specific and from purely heuristic models to more specific systems type models as more and more detail and complexity is added. Meanwhile, if the simple models appear to produce properties present as basic components in most populations or ecosystems components of the type studied, then this general trend might be termed a 'general principle.'

The uses of Lotka-Volterra heuristic models are limited by some practical considerations:

1. The empirical data needed to compare with the model's predictions must be physically gatherable.

2. Alternative models based on different assumptions must produce observably different behaviors. For example, if two predator-prey models based on different patterns of interactions both produced the same stable limit cycle behavior, then a comparison with empirical data cannot distinguish between the two alternative hypotheses.

3. Different types of hypothesized behavior generated by alternative models must be distinguishable after the specific details and stochastic factors present in every population or ecosystem component are added. Very often they cannot (see Poole, 1977).

4. A standard definition of 'good enough' should be agreed upon. What may look like excellent agreement to some may not seem particularly good to others, particularly if one has preconceptions of what is 'true.'

If the primary purpose of reductionist prediction is heuristic, is a deterministic or a stochastic basis more appropriate? There is, of course, no right or wrong answer. The situation always dictates the choice. However, I believe that deterministic models are more applicable in the majority of cases. The analytical difficulty of stochastic models is usually given as a major reason for preferring deterministic models. There is, I think, a more important point in favor of deterministic models. Heuristic predictions are created to compare and analyze empirical data. Consequently, the more complex the predictions are, the more difficult they are to use. For example, a deterministic model might produce the prediction 'the limiting number of coexisting cicada species is 2.76.' Field observations yield the figure 5 species. We have a firm basis for comparison and can ask why the prediction differs from the observed value. Perhaps the species populations are not in equilibrium or spatial heterogeneity exists. A stochastic model, on the other hand, might produce the prediction 'the probability of three species is 0.40, the probability of four species is 0.30, and the probability of five species is 0.20.' The stochastic prediction is not a firm standard for comparison. The occurrence of five species may be fortuitous, there is a postulated 20 percent chance of five co-existing species, or the assumptions of the model may be badly in error. The only way to distinguish between these two possibilities is to replicate independently the field observations enough times to estimate the probability distribution of co-occurring species. Unfortunately, replication is generally impossible except in those rare field situations which can be experimentally repeated. One might argue that stochastic prediction is more realistic. However, realism is not necessarily a quality demanded of heuristic models. Heuristic predictions are created to be disproven. The more difficult they are to disprove, the less valuable they are. If an empirical set of data can be replicated, however, repetition may provide a rough, although subjective, estimate of the importance of stochastic events.

A stochastic outlook in heuristic modeling can at times be absolutely essential, of course. Two examples are the many important insights on the general shapes of probability distributions of population abundance and the probability of extinction produced by birth-death models (see Bharucha-Reid, 1960) and more recently the theoretical limits established on the limiting similarity of competing species as a function of environmental variability (see May, 1973, for example). Nevertheless, the very nature of heuristic modeling is more in keeping with a deterministic philosophy than a stochastic one.

4. SYSTEMS MODELING AND SIMULATION

At first glance systems models are merely elaborate versions of their Lotka-Volterra cousins. However, their purpose and philosophical basis are remarkably different. The systems model rests on two fundamental tenets:

1. An ecological variable such as population abundance or calcium concentration is directly or indirectly influenced by other components of the ecosystem. Consequently, the behavior of a variable depends on both the internal dynamics of the component to which it belongs and the interconnections between this component and the remainder of the ecosystem. In short, the whole is not equal to the sum of the isolated parts.

2. Realistic predictions about the potential behavior of a population or nutrient must be based on their specific dynamical properties.

The purpose of heuristic, reductionist models is to uncover and analyze the consequences of simple, non-specific patterns of interrelationships within an isolated part of an ecosystem. A system model, however, assumes the interactions are correctly specified by the model. The assumed interrelationships are usually based on previous reductionist modeling efforts. Heuristic modeling seeks general patterns amid masses of specific detail. The systems model is more concerned with the detail. If general characteristics appear in comparing several systems models of major eco-systems, so much the better. Generalization, however, is not the main purpose of systems modeling.

The primary use of a systems model is the generation of 'what if' or 'scenario' predictions, usually by simulation. The adjectives 'what if' and 'scenario' are justified by the use of simulation to analyze systems models. A model is constructed based on our best knowledge of the internal dynamics within isolated ecosystem components and the components are interconnected. Although systems models differ greatly in size and format, the model is usually based on a group of interconnected 'state' or 'endogenous' variables such as the abundances of the age classes of some species in the community, nutrient concentrations, or energy levels. Various inputs and outputs such as nutrients and energy entering and leaving the ecosystem or immigrating and emigrating individuals are usually also part of the model. In addition to the state variables, inputs and outputs, there are exogenous variables such as temperature, humidity, wind velocity, incident sunlight, and economic factors. Their effects are incorporated in the model by specifying some of the parameters as functions of the exogenous variables. Given the model the constant parameters are estimated or guessed. The model is now ready for prediction. Analytical methods play a minor role in prediction and numerical simulation is the dominant force because of the size and complexity of the models.

Simulation begins by specifying initial values for the system variables. In a deterministic model the initial conditions and the values of the exogenous and input variables from time t to $t+L$ exactly specify the behavior of the system up to time $t+L$. The 'what if' label specifies the types of predictions made by a systems model simulation. Suppose we had constructed a systems model of a pond. Some of the important questions which might be asked of the model are:

1. Suppose the values of the exogenous variables are specified from time t to $t+L$. What will be the behavior of

phytoplankton biomass or zooplankton phosphate content if temperature is raised three degrees centigrade?

2. Suppose light intensity decreases because of the runoff of silt into the pond. How will the species composition of the ecosystem be affected?

3. Suppose the biological interactions between two zooplankton species is altered. How will the change affect the ecosystem as a whole?

4. Suppose one or more of the system variables can be manipulated. How should be manipulate nitrate in runoff, for example, to maximize the biomass of a commercially valuable fish species without unduly upsetting the remainder of the ecosystem?

5. How sensitive is one component of the ecosystem to fluctuations or sudden changes in the birth and death rates of some species in another component?

All of these questions fall into the category of 'what if' or 'scenario' prediction. We ask 'what would be the behavior of the ecosystem given the initial values of the state variables if the exogenous and input variables and the parameters take specified values from t to $t+L$.' The last possible use of systems models, 'what will' prediction or forecasting, is discussed in the next section.

The practical criteria limiting the usefulness of systems models in 'what if' prediction are:

1. The biological and physical dynamics between and within the ecosystem components specified in the model must accurately match the ecosystem the model is attempting to represent. 'Realism' is far more important in a systems model than in heuristic modeling.

2. The model parameters should be reasonable estimates. Consequently, the parameters must be estimable from the type of data it is feasible to gather in the field. Arbitrary specification of parameters should be avoided because a large model with a multitude of parameters, no matter how absurd it may be, can always be made to fit any set of data by fiddling with its parameters.

3. The exogenous variables are exactly specified from the present to the end of the simulation and should, therefore, be biologically reasonable.

4. If the model has a stochastic element, accurate estimates of the variances and atuocorrelations of the error components must be known.

The strengths and weaknesses of deterministic versus stochastic systems models seem to me to be about equally divided. Analytical considerations are usually fairly immaterial because both deterministic and stochastic versions of the same system must be simulated. Two arguments can be advanced in favor of deterministic models. First, there is something in the human spirit that rebels against probabilistic prediction. A deterministic model produces a unique point prediction from time t to $t+L$. If a large number of ecosystem variables are predicted simultaneously, the deterministic trajectories are much easier to comprehend than the welter of marginal and joint probability densities produced by a stochastic model. The essential patterns emerging from the complex predictions of a stochastic model are extremely difficult to grasp intuitively. Also if the variances of the 'state' variables from t to $t+L$ are small relative to their means, the entire distributions of the random variables $Y_1(t), Y_2(t),\cdots, Y_n(t)$ may not add appreciably to the accuracy or value of the predictions and may only tend to cloud the essential issues. The second argument in favor of deterministic models is monetary. Numerical simulation of a stochastic model is no more difficult than for a deterministic analogue. However, good estimates of the probability distributions of the random variables $Y_1(t),\cdots, Y_n(t)$ for all t between t and $t+L$ may require several hundred replications of the same simulation. The deterministic model, by definition, requires but one. If the model is large, and it usually is, the cost of the simulations can be enormous and even, perhaps, beyond the capacity of the computer. On the other hand, a single stochastic simulation can give a fair idea of what effect the addition of stochastic factors has on the deterministic solution.

There are also strong arguments in favor of stochastic models. The first is greater realism. If the variances of the state variables $Y_1(t),\cdots, Y_n(t)$ between t and $t+L$ are large, and they often are in ecology, the deterministic trajectories may be a snare and a delusion. Specifying a single point prediction does not provide any indication of the range or probabilities of possible outcomes and might lead to disaster if managerial decisions must be made on the basis of the deterministic predictions. Also a deterministic trajectory cannot be equated with the mean behavior of an analogous stochastic model as noted in the introduction, nor is it clear what average behavior even is unless the model is linear and the error terms are normally and independently distributed. The cost of repeated stochastic simulations may prove less important some day as the speed and storage capacity of computers increases.

5. FORECASTING

The characteristics of forecasting distinguishing it from hypothesis generation and simulation are defined by forecasting's purpose and the practical problems limiting it. The forecasting problem is easily described. A population or nutrient is of ecological and/or economic importance. The population or nutrient and possibly other environmental variables affecting it have been studied and recorded up to the present time t. The problem is 'predict the abundance or concentration of the variable at some future time t+L.' Superficially, this statement of the problem seems little different than the simulation use of systems models. However, forecasting and simulation are quite distinct because of the practical problems involved. Simulation asks what can happen if the parameters and input and exogenous variables were to take specified values from t to t+L. Forecasting, on the other hand, asks 'what will happen' once the parameters and exogenous variables are no longer under arbitrary control.

Forecasting is the category of prediction most beset by practical problems some of which are:

1. The information used in forecasting is almost always limited to the present and past values of the predicted variable and the recorded variables believed to affect it. In rare cases one or more variables interacting with the predicted variable can be controlled (e.g. fishing effort in fisheries models). This type of variable, however, is relatively rare in ecology. Of course interacting variables can themselves be forecast into the future based on their present and past values and the forecasts used in predicting the ecological variables of primary interest. However, the errors made in forecasting these variables add to the overall variance of the ultimate prediction. In short, exogenous variables of the ecosystem are not within our control. The only solid information available for forecasting is the present and past.

2. Most projects requiring forecasting are limited by money and manpower and only a few important variables are normally recorded over a long enough period of time to make accurate forecasts. Forecasts, therefore, must usually be based on a limited number of variables and these variables must be amenable to accurate and inexpensive monitoring. The effects of unrecorded variables and chance events lends an inevitable stochastic element to forecasting.

3. A population or ecosystem component as defined in the field is not equivalent to the theoretical non-spatial nature of most representational models. A forecasting method must conform with a population or component defined by an arbitrarily designated geo-

graphical area. Although some populations are defined on biological grounds, many others are delimited by economic, political, or practical reasons.

4. The parameters of a forecasting model must be easily and accurately estimable from field data. It is distressingly easy to create models with parameters that cannot be accurately estimated.

5. Forecasting methods should be as general and as intuitively simple as possible within the limits of accurate forecasting. The more complex and specialized a method is, the less likely that it will be adopted by those agencies needing it.

Forecasting is the one category of prediction, I feel, falling strictly in the domain of stochastic modeling. The uncertainty of the future is due to the inherent stochastic nature of the environment and to our inability to study all of the variables and interactions affecting a population or ecosystem component. A particularly strong stochastic component is added by the difficulty of accurately forecasting the future behavior of important exogenous variables such as weather and economic pressure.

Because of the inherent variability of ecological variables, the best any forecasting method can do is to calculate the probabilities of possible outcomes. Consequently, forecasting is essentially the original prediction problem outlined in the introduction: calculate the conditional probability distribution of the predicted variable at time $t+L$ given any available information on the past and present histories of the recorded variables. The more information available, presumably, the narrower the spread of the conditional distribution. But no matter how much data we gather, there is an irreducible minimum variance to the future since there is always some uncertainty in any prediction. The task of forecasting, therefore, is to extract the maximum amount of information about the future from a minimum amount of data.

A forecasting method can be either a representational model or can be based on the exploitation of the statistical properties of the recorded variables. It is my personal and highly biased opinion that the use of representational models for forecasting is nearly impossible because of the pragmatic constraints. If one accepts the proposition that forecasting is equivalent to the computation of the conditional distribution of the predicted variable at time $t+L$, then the analytical and pragmatic problems confronting the use of representational models for forecasting are truly formidable.

6. Optimization and Control

Although the problem is well defined, optimization and control problems can usually be divided into the three previously defined categories. In optimization and control, one or more ecosystem variables are manipulated directly or indirectly, usually subject to some constraints, to achieve some purported optimal property. The classic example is manipulation of fishing effort to maximize sustainable fish catch. A typical control problem might be to minimize the abundance of an insect pest species subject to cost constraints on the control agents employed. The methodology of optimization and control in ecology has recently been reviewed by Clark (1976) and is too varied and complex to go into in any detail. I might note in passing, however, that optimality is an ambiguous term since optimality has different interpretations economically, politically, and environmentally. An economically optimal fishing policy is not necessarily optimal ecologically.

The models used in optimization and control can fall into any of the three previous categories of prediction. For example, a possible sequence of steps in developing an optimization program might be:

1. A preliminary analysis of the problem using simplified models of an isolated part of the ecosystem containing the variable to be controlled and the controlling variable. What is optimality and how can the controlling variable be manipulated to achieve our concept of optimality? Are there general principles involved in optimization problems and what broad problems affect implementation of the control or optimization program?

2. Increase the 'realism' of the model by adding details on the unique biological or economic factors involved and by specifying the interrelationships between the controlled variable, the remainder of the ecosystem, the controlling variable, and the economy, and the environment. Does the control policy adopted in step one still work within the context of the entire ecosystem? What effect does the optimization program have on the remainder of the ecosystem? What effect does the control or optimization procedure have on the environment or economy? If ill effects are observed, can the procedure be modified to reduce or eliminate these effects?

3. Implement the program. Suppose the first two steps have outlined the interrelationships between variables, established an optimal trajectory for the controlled variable, and discovered any possible negative effects of the program. Once the control or optimization technique is put into operation, it will, of course, be impossible to achieve perfect control because of the effects of stochastic events and unstudied variables. What is the

probability distribution of actual values of the controlled variables at t+L given the control scheme? How does one minimize the variance of the possible values of the controlled variables about the optimal value?

The first step in the sequence outlined above obviously corresponds to the heuristic predictions of the Lotka-Volterra models. Step two is one aspect of systems modeling and step three is a modification of the forecasting problem.

Since optimization and control may correspond to either reductionist modeling, systems analysis, or forecasting, the practical limitations on each also hold. The comments made on the respective suitability of stochastic and deterministic models in each case also apply to a limited extent. There is, however, one further justification for deterministic models in optimization and control: the analytical and numerical problems are horrendous. Although stochastic control models are theoretically possible (e.g. see Kushner, 1971), I suspect that deterministic models will play a dominant role in optimization and control in the foreseeable future.

7. DISCUSSION AND CONCLUSIONS

I find it distressing whenever I see one of the first three categories of modeling touted as ecology's one true religion. Lotka-Volterra, systems, and forecasting models are not mutually exclusive alternatives. Indeed I feel that the three form a natural progression from the initial study to the final comprehension and prediction of ecological phenomena. Each category is uniquely suited to its own type of ecological prediction and is best defined by and subject to the practical considerations delimiting the problem. Therefore, the choice of a modeling basis and ultimately the deterministic-stochastic question itself rests not so much on the philosophies behind each type of model, but on whether or not the model is capable of performing the task it was created to do.

REFERENCES

Bharucha-Reid, A. T. (1960). *Elements of the Theory of Markov Processes and their Application.* McGraw-Hill, New York.

Caplan, A. L. (1977). Tautology, circularity, and biological theory. *American Naturalist,* 111, 390-393.

Castrodeza, C. (1977). Tautologies, beliefs, and empirical knowledge. *American Naturalist,* 111, 393-394.

Clark, C. W. (1976). *Mathematical Bioeconomics.* Wiley Interscience, New York.

Ferguson, A. (1976). Can evolutionary theory predict? *American Naturalist,* 110, 1101-1104.

Kushner, H. J. (1971). *Introduction to Stochastic Control.* Holt, Rinehart, and Winston, New York.

May, R. M. (1973). *Stability and Complexity in Model Ecosystems.* Princeton University Press. Princeton, New Jersey.

Peters, R. H. (1976). Tautology in evolution and ecology. *American Naturalist,* 110, 1-12.

Poole, R. W. (1977). Periodic pseudoperiodic, and chaotic population fluctuations. *Ecology,* 58, 210-213.

Sage, A. P., and Melsa, J. L. (1971). *Estimation Theory with Applications to Communications and Control.* McGraw-Hill, New York.

Stebbins, G. L. (1977). In defense of evolution: Tautology or theory. *American Naturalist,* 111, 386-390.

[*Received June* 1977. *Revised April* 1979]

G. P. Patil and M. Rosenzweig, (eds.),
Contemporary Quantitative Ecology and Related Ecometrics, pp. 271-277.

ON THE ROLE OF TIME AND INTERACTION IN ECOSYSTEM MODELLING

ROBERT ROSEN

Department of Physiology and Biophysics
Dalhousie University
Halifax, Nova Scotia, Canada B3H 4H7

SUMMARY. Time appears in dynamical modelling as a gratuitous parameter, in the sense that the rates at which trajectories are traversed are not fixed with respect to clock time by the dynamical equations. We consider therefore the following problem: to what extent does an autonomous dynamics fix an intrinsic time scale, and what is the relation of such a scale to clock time? The starting point for the analysis is a pattern of interactions between the variables involved, which is supposed to depend on state but not on time. We show that an arbitrary interaction pattern does not, in general, yield the kinds of closed-form rate equations of the kind customarily considered in ecosystem modeling, but we obtain conditions necessary and sufficient for this to be so. We show that such a pattern of interactions gives rise to an intrinsic rate at which trajectories are traversed, which is not in general uniform with respect to clock time. The problem of temporal calibration of dynamical models is briefly discussed.

KEY WORDS. ecosystem modelling, time, stability.

1. INTRODUCTION

The point of departure for ecosystem modelling, and dynamical modelling in general, is a system of rate equations of the form

$$dx_i/dt = f_i(x_1, \cdots, x_n) \tag{1}$$

defined on some appropriate manifold in Euclidean n-dimensional space. It seems to have first been pointed out by Higgins (1967) that, in this context, it is meaningful to introduce relations of activation and inhibition between the state variables x_i in (1). Specifically, we may say that x_j *activates* x_i is a state $(x_1^o, \cdots, x_n^o)$ if the derivative

$$\frac{\partial}{\partial x_j}\left(\frac{dx_i}{dt}\right)$$

evaluated in that state is positive, and that x_j *inhibits* x_i if this derivative is negative when evaluated in that state. Intuitively, if x_j activates x_i, then an increase in x_j increases the *rate* at which x_i is produced; conversely, a decrease in x_j decreases the rate at which x_i is produced.

Thus if we are given a dynamical system (1), we may define the n^2 functions

$$u_{ij}(x_1, \cdots, x_n) = \frac{\partial}{\partial x_j}\left(\frac{dx_i}{dt}\right) \tag{2}$$

which specify the relations of activation and inhibition between the state variables in every state. A great deal of information about the qualitative behavior of the system (1) can be obtained from the functions (2) (Higgins, 1967; Rosen, 1970); for many purposes it suffices to know just the signs of the functions u_{ij}, and not their specific numerical values.

We propose to deal with the following inverse question; to what extent does information about the functions u_{ij} determine the properties of an associated system of rate equations of the form (1)? This kind of question is particularly important in ecosystem modelling, since very often we are in a position to observe patterns of activation and inhibition, but not directly to write down rate equations (see for instance Levins, 1976). It is thus of great interest to determine how far we may go in passing backwards from the functions (2) to the dynamical system (1). This will be undertaken in Section 2. We will then see how we may define transit time along trajectories in an intrinsic fashion; this will be the substance of Section 3.

2. ACTIVATION-INHIBITION PATTERNS AND RATE EQUATIONS

Let us suppose that we are given some initial state $(x_1^o, \cdots, x_n^o)$ of (1). By the existence and uniqueness theorems for systems of first-order ordinary differential equations (cf. Rosen, 1971), there will be a unique trajectory passing through this initial state. If this state moves to a nearby state $(x_1^o + dx_1, \cdots, x_n^o + dx_n)$, then for each $i = 1, \cdots, n$ we have

$$df_i = f_i(x_1^o + dx_1, \cdots, x_n^o + dx_n) - f_i(x_1^o, \cdots, x_n^o),$$

or

$$df_i = \sum_{j=1}^{n} \frac{\partial f_i}{\partial x_j} dx_j, \tag{3}$$

up to higher-order terms. On the other hand, from (1), we know that the motion of the system along its trajectory is such that $dx_j = f_j(x_1^o, \cdots, x_n^o)dt$ and that therefore we can write

$$df_i/dt = d^2x_i/dt^2 = \sum_{j=1}^{n} f_j \frac{\partial f_i}{\partial x_j} . \tag{4}$$

This result shows that the functions u_{ij} defined above determine the rates at which the derivatives dx_i/dt are changing; i.e., they determine the accelerations of the state variables, and hence the curvatures of trajectories.

For our purposes, the main interest reposes in (3) above. This relation asserts that the differential change in the velocity of the *ith* state variable is the sum of activations and inhibitions seen by that variable in a given state, each weighted by the differential change in the corresponding state variable. If we think of the differential changes dx_j in the state variables as incoming signals to x_i at some instant, each flowing along a channel whose properties are determined by the corresponding function u_{ij}, then the right-hand side of (3) can be regarded as the net activation seen by x_i at that instant. The response to this net excitation is a change in rate; i.e., an acceleration or deceleration of x_i.

Now let us conversely suppose that a family of functions u_{ij} are given, and that these specify an activation-inhibition pattern holding among the variables $(x_1, \cdots, x_n)$. If we wish (2) to be satisfied, we can write

$$dx_i/dt = u_{ij}dx_j + \phi_{ij}(x_1, \cdots, x_{j-1}, x_{j+1}, \cdots, x_n),$$

where ϕ_{ij} is an arbitrary function. If we do this for each j, we find that

$$dx_i/dt = \frac{1}{n} \int \sum_{j=1}^{n} u_{ij}dx_j + \Phi(x_1, \cdots, x_n). \tag{5}$$

By differentiating (5) partially with respect to each x_j and comparing with (2), we see that the arbitrary function $\phi_{ij}(x_1, \cdots, x_n)$ must in fact be

$$\Phi(x_1, \cdots, x_n) = \frac{n-1}{n} \int \sum_{j=1}^{n} u_{ij}dx_j + C_i,$$

where C_i is a constant.

We thus see that the choice of activation-inhibition functions u_{ij} determines a corresponding set of rate equations up to n arbitrary constants. However, we see that our ability to write (5) in a closed form analogous to (1) depends on our ability to integrate the differential forms

$$\sum_{j=1}^{n} u_{ij}dx_j. \tag{6}$$

As noted above, these differential forms represent the set activation seen by the state variable x_i in a state, for $i = 1, \cdots, n$. We then must require that these net activations are *perfect differentials*. That is, there must exist functions $f_i(x_1, \cdots, x_n)$ such that

$$df_i = \sum_{j=1}^{n} u_{ij}dx_j, \quad i = 1, \cdots, n.$$

Otherwise, there is no globally valid system of rate equations (1) arising from the presumed activation-inhibition pattern given by the functions u_{ij}.

There are some well-known conditions on a system of function u_{ij} such that the differential form $\sum_{j=1}^{n} u_{ij}dx_j$ is perfect. For instance, if the manifold in which we are working is simply connected, the differential form will be exact if, for each i, the Jacobian matrix

$$\left(\frac{\partial u_{ij}}{\partial x_k}\right)$$

is symmetric. The symmetry of these Jacobian matrices is highly reminiscent of the phenomenon of *reciprocity* as postulated by Onsager for irreversible thermodynamic processes, but we shall not explore this relation here.

Let us briefly note some further ramifications of the above situation.

1. We can show immediately that an activation-inhibition pattern (i.e., a pattern of interactions) is more sensitive to arbitrary perturbations than is a system of rate equations. For if we perturb (1) to obtain new equations of the form

$$dx_i/dt = f_i(x_1,\cdots,x_n) + \varepsilon\ \phi_i(x_1,\cdots,x_n),$$

then we obtain a corresponding perturbation of the activation-inhibition pattern in which the exactness of the differential forms is retained. On the other hand, if we are initially given an activation-inhibition pattern u_{ij} and perturb these arbitrarily to obtain a new pattern of the form

$$u_{ij}' = u_{ij} + \varepsilon\ \psi_{ij}(x_1,\cdots,x_n),$$

then we do not even in general obtain a system of rate equations for the perturbed system. This is another form of a result we obtained previously (Rosen, 1975) by a rather different argument, and shows how careful one must be in using dynamical arguments based on structural stability.

2. Activation-inhibition patterns of the type considered above are highly reminiscent of other types of communication networks found in biology; for instance in the modelling of neural, chemical, or genetic networks. Indeed, the above considerations were framed initially in order to be able to convert dynamical interactions into a form pertaining to information flow. This fact allows us to develop a general network theory of such dynamical interactions, which allows us to treat all of these diverse situations in a unified fashion.

3. THE ROLE OF TIME

In general, time appears in a system of rate equations like (1) simply as an arbitrary parameter. For instance, we can multiply each of the rate equations by the same non-vanishing function $a(x_1,\cdots,x_n)$ without changing the trajectories or the stability properties of the system. Such a function merely modifies the rates at which the trajectories are traversed, relative to some fixed time-scale. Thus, relative to that fixed time-scale, we have no way of choosing this function within the context of the rate equations alone. To gratuitously choose $a(x_1,\cdots,x_n)=1$ is an arbitrary assumption for which there is no theoretical basis.

Moreover, let us observe that in dynamical modelling, if we write down a rate equation of the form $dx/dt = f(x)$, we tacitly suppose that, roughly speaking, dx and dt are *known*, and that $f(x)$ is their ratio.

The picture is quite different when we consider activation-inhibition patterns. Here the functions $u_{ij}(x_1,\cdots,x_n)$ are supposed given. We then construct the differential forms

$$\sum_{j=1}^{n} u_{ij}dx_j,$$

where initially the differentials dx_j are *arbitrary*. If however each of these differential forms if perfect, then there are n function $f_i(x_1,\cdots,x_n)$ such that

$$df_i = \sum_{j=1}^{n} u_{ij}dx_j, \quad i=1,\cdots,n. \tag{7}$$

Moreover, the differentials dx_i are no longer arbitrary; they must satisfy the n relations (7). In other words, the specification of an activation-inhibition pattern determines (indeed, generates) the differentials dx_i in each state.

Now all of the above arguments are independent of time. However, a notion of time can be introduced naturally by recognizing that we want to identify the functions f_i, whose differentials df_i are given in (7), with the velocities dx_i/dt. It should be carefully noted that here, in contradistinction to the usual situation, it is the df_i and the dx_i which are known, and it is the time differential dt which is to be defined.

That is, we wish to specify a parameter t whose differential dt simultaneously satisfies the n local relations

$$dx_i = f_i(x_1, \cdots, x_n)dt.$$

Thus, the parameter t is no longer arbitrary; its character is, so to speak, created locally by the character of the dynamics; i.e., by the local pattern of activation and inhibition.

Thus, the rate equations (1) which arise from an activation-inhibition pattern are correct *only* if we use the local time-scale defined by the dynamics itself. If we use any other time-scale, the dynamical equations will contain a factor $a(x_1, \cdots, x_n)$ which expresses the relation between the local intrinsic time created by the dynamics, and the fixed time-scale we are using. This implies, for instance, a profound distinction between measured dynamical parameters (rate constants) and the actual intrinsic rates at which the dynamics is proceeding.

One immediate corollary of the above argument is the following. If our fixed time-scale is very fast compared to the intrinsic time generated by a dynamics, the built-in conversion factor $a(x_1, \cdots, x_n)$ will be very close to zero. Thus we will be driven to suppose that our system is at a steady state, when in fact it can be arbitrarily far from being so. It is obvious that this kind of temporal artifact can have the most profound consequences for dynamical modelling in ecological and evolutionary situations. This kind of artifact can be avoided by taking an activation-inhibition pattern as primary, and referring all dynamical arguments to the intrinsic time-scale generated thereby. It is clear that this would involve a substantial reinterpretation of data pertaining to rates, which presently are always referred to a fixed time-scale which is in general unrelated to the actual dynamics.

REFERENCES

Higgins, J. (1967). The theory of oscillating reactions. *Industrial and Engineering Chemistry*, 59, 18-62.

Levins, R. (1977). Qualitative analysis of complex systems. In *Mathematics in the Life Sciences*, D. E. Matthews, ed. Springer-Verlag, New York. 153-199.

Rosen, R. (1970). *Dynamical System Theory in Biology*. Wiley, New York.

Rosen, R. (1975). On interactions between dynamical systems. *Mathematical Biosciences*, 27, 299-308.

G. P. Patil and M. Rosenzweig, (eds.),
Contemporary Quantitative Ecology and Related Ecometrics, pp. 279-293.

ON THE PARAMETER STRUCTURE OF A LARGE-SCALE ECOLOGICAL MODEL

EFRAIM HALFON

National Water Research Institute
Canada Centre for Inland Waters
Burlington, Ontario, Canada, L7R 4A6

SUMMARY. A model of a lake ecosystem (Lake Ontario) is analyzed to find the level of aggregation in the model itself and its morphic relationship with a minimal realization. This study has been carried out by an analysis of the parameter structure of the model. Statistical methods such as correlations and principal components analysis have been used for the purpose. Results indicate that the model is highly aggregated and is close to a minimal realization. Redundant features are related to the zooplankton and soluble phosphorus compartments, and to sedimentation rates. Also many parameter combinations give a good fit to a given data set, therefore, no individual model can be considered best, indeed, the number of good models complicates the choice of the model to be used for prediction.

KEY WORDS. parameter, optimization, cake model, aggregation problem, minimal realization, ecology.

1. INTRODUCTION

The parameter set is one important feature of a mathematical model. Parameters describe fluxes among state variables and, depending on the degree of model aggregation, may contain information on state variables not explicitly described; for example, in some lake models, bacteria and micro-organism influences are buried in a parameter. Simulation practitioners know that free parameters are very valuable. A free parameter may take an arbitrary or quasi-arbitrary value. This parameter may completely modify the behavior of a model and therefore is very useful in fitting a data set, and in some instances, e.g., astronomy, in providing that a theory is valid.

In ecology, as in other sciences, models represent physical systems and parameters must represent physical events. However, a parameter cannot be an abstract entity, but must be related to the system itself. Most publications of mathematical models include a table which presents parameter values, but their variance or our incomplete knowledge is usually not included. The effects of this variability can be studied numerically by sensitivity analysis.

When models are used for prediction, the parameter variability is usually not taken into account and therefore the model is assumed to be deterministic. Recently, ecologists have analysized the effects of uncertainty in parameters, initial states, inputs, and model structure, on model prediction, by means of numerical techniques such as Monte Carlo simulations (O'Neill *et al.*, 1979) or Kalman filters (Beck and Halfon, in preparation). These methods allow for the definition of confidence boundaries around the nominal trajectory. In these applications the covariance of the parameters among themselves should also be accounted for. Some parameter values depend on other parameter values; for example, in a single-population model, equilibrium may be maintained with different combinations of birth and death rates. When large-scale models are developed, the parameters are assumed to be independent to one another, i.e., modification of one parameter does not imply that any other one should be changed to keep the model realistic.

In most ecological models, free parameters are generally considered to be eliminated by the physical constraints of the problem, however, is this the case? All models are idealizations of reality. As such, all models are homomorphs of the real system, they are lumped models. Zeigler (1976, 1979) studied the effects of lumping models. If the models are homomorphic representations, then the state variables and parameters contain information on the dynamic behavior of the real system. Since much information on the real system is included in a model, we can expect that each parameter would contain as much information as possible; thus parameters should be independent of each other. In fact, correlation among parameters would imply that even if much information on the system is lumped together, some parameters represent the same process. This correlation represents some overdetermination in the model and points towards the presence of free (redundant) parameters which help in obtaining a good fit to the data. Computationally, this overdetermination may be costly since a larger number of equations than are necessary need to be solved to receive the same answer. A smaller or more lumped model would be as adequate. As indicated earlier, this problem of model aggregation has received the attention of theoreticians (e.g., Zeigler, 1976) and now of the systems ecologists (Wiegert, 1975; Halfon and Reggiani, 1978; Halfon *et al.*, 1979; O'Neill and Rust, 1979; Cale and Odell, 1979; Zeigler, 1979).

In this paper a large-scale model of Lake Ontario is analyzed to obtain information on the variance-covariance of its parameters and on the degree of aggregation of the model, actual and possible.

2. LAKE ONTARIO MODEL

The model was developed in 1977 and 1978 at the Canada Center for Inland Waters in Burlington, Ontario. Complete documentation of the model is included in an Unpublished Report (Simons *et al.*, 1978). Table 1 describes the parameters of the model and Figure 1 shows a graphical description of the model structure. The model has nine compartments. Phosphorus and carbon cycle through the model but are only weakly coupled as recent limnological research suggests (Lean, 1973, and personal communication). Since carbon and phosphorus are considered to be

TABLE 1: Description of parameters of Lake Ontario model (Simons et al., 1978).

Parameter No.	Meaning
1	Michaelis-Menten constant of phosphorus uptake by phytoplankton.
2	Michaelis-Menten constant of phosphorus release to water by phytoplankton.
3	Zooplankton feeding rate during cold months.
4	Michaelis-Menten constant of zooplankton feeding.
5	Zooplankton excretion rate.
6	Zooplankton death rate.
7	Phytoplankton excretion rate.
8	Mineralization rate of detrital organic matter.
9	Transformation rate of soluble organic phosphorus into soluble reactive phosphorus.
10	Zooplankton feeding rate during warm months.
11	Phytoplankton sedimentation rate in second layer (20-40m).
12	Phytoplankton sedimentation rate in third layer (40m-bottom).
13	Detritus sedimentation rate in first layer (0-20m).
14	Detritus sedimentation rate in second layer (20-40m).
15	Detritus sedimentation rate in third layer (40m-bottom).

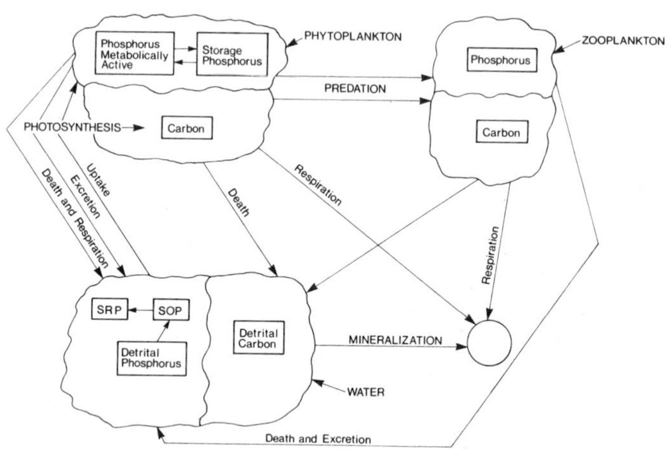

FIG. 1: A schematical diagram of a Lake Ontario model (Simons et al., 1978). There are nine components to the model signified as boxes. Phosphorus and carbon cycle through the model with only carbon being driven by light. Phosphorus uptake rates depend on the metabolic status of the phytoplankton cells. The two nutrient (P,C) cycles are only weakly coupled. Phosphorus limitation within phytoplankton cells limits net algal carbon growth by increasing respiration. Parameter values for the same processes, e.g., predation, death rate, etc., are the same for carbon and phosphorus. Flux rates, however, are different since the nutrient pools are different.

almost independent of each other, the ratio between carbon and phosphorus may change in time. Phyotplankton and zooplankton are represented by three and two compartments, respectively. These compartments are either in phosphorus or carbon units and interact with one another to determine the lake biomass. Lake Ontario is divided into three layers, 0-20 meters, 20-40 meters, and 40 meters to bottom. Plankton in the middle layer is still influenced by the phenomena taking place in the epilimnion, e.g., light, high temperatures, and low nutrients in the summer, but the stable hypolimnion supplies nutrients and cold water. Horizontally the lake is considered fully mixed. This assumption is strictly true in the winter but can be considered approximately true also in summer since nearshore waters are a minimal volume of the whole lake. A hydrodynamical model (Simons *et al.*, 1978) is used for computation of exchanges among layers.

3. MODEL PARAMETER ESTIMATION

When an objective function of several variables is minimized, the computed minimum is usually local rather than global. Indeed, the more variables to be reconciled, the more difficult it is to find a global minimum since the hyper-dimensional surface is not convex. Several numerical methods of optimization and function minimization exist. The purpose is to find the minimum (or a maximum) of a function of several variables $(\underset{\sim}{p})$

$$y = f(\underset{\sim}{p})$$

by modifying $\underset{\sim}{p}$ until a minimum is found. The parameters can be modified according to some information on the function hyper-dimensional topography: gradients $\partial f/\partial \underset{\sim}{p}$ provide this information. Some methods use a gradient computed analytically, others, a gradient computed numerically. Some methods do not use gradient information at all, but rather search for the minimum in a random mode. Two or more of these methods can be also combined. In this exercise a random search method was used. This technique allows a complete search over the field of interest, and therefore the likelihood of finding a global minimum is increased. However, the procedure is time-consuming, because the search for the optimum is random and therefore computationally expensive.

3.1 Random Search. Each variable is allowed a range of variation between zero and 400. Zero is chosen as the lowest positive number for a parameter with a physical meaning since negative numbers do not respect this constraint. Four hundred is an arbitrary large number (most parameter values are between zero and 20). The controlled random search procedure for global optimization (Price, 1977) is used to develop a topography of the hyperdimensional surface. This procedure combines the random search and mode-seeking routines into a single, continuous process which makes implicit the choice of clusters for further investigation. The algorithm is programmed to keep the best 250 parameter combinations which minimize the objective function. Convexity of the surface is estimated by comparing the objective function value at the 'global' minimum and at all the other local minima. If the value of the objective function is similar, then a flat surface is expected; under this binding the 'global' minimum is not much different or better than any local minima. The program set 250 random initial conditions in the parameter space. Minima were searched for in a random mode and the algorithm performed 11,450 function evaluations which lead to the identification of 250 local minima, one of which might be the global minimum, given the thoroughness of the search. A function evaluation consisted of a comparison of a one-year model run with data (297 points:

33 points at each of the nine points in time) observed in Lake Ontario during the International Field Year on the Great Lakes (IFYGL) by a joint Canada-USA research program in 1972-1973.

3.2 Objective Function. The objective function to be minimized is Theil's inequality coefficient (Theil, 1970) which is recommended for the validation of simulation models by Kheir and Holmes (1978). This index is a weighted least squares, in which the weights are the sums of squares of the predicted (P) and observed (A) points:

$$U = \frac{\sqrt{(\frac{1}{n}\Sigma\ (P_i - A_i)^2)}}{\sqrt{(\frac{1}{n}\Sigma\ P_i^2)} + \sqrt{(\frac{1}{n}\Sigma\ A_i^2)}}\ ; \qquad 0 \leq U \leq 1,$$

when n is the number of sampling points, in this case $n = 9$. The index is valid for one state variable. In the case when many state variables are present the multiple Theil's inequality coefficient is used

$$U = N/D; \qquad 0 \leq U \leq 0.5$$

where $$N = \Sigma_\omega\{\sqrt{(\frac{1}{n}\Sigma(P_i - A_i)^2)}\}$$

and $$D = 2 \cdot \Sigma_\omega\{\ \sqrt{(\frac{1}{n}\ \Sigma P_i^2)} + \sqrt{(\frac{1}{n}\ \Sigma A_i^2)}\}$$

where ω represents the set of state variables considered in the model. This index takes into consideration differences among means, differences among standard deviations, and imperfect covariation among the predicted and observed data. It, therefore, contains some information about statistical properties of the model and of the observations. In ecological models state variables range over several orders of magnitude in absolute value. The weights in Theil's index make each state variable equally important so that the goodness of fit is good for both large and small state variables. More details, including a geometric interpretation of the coefficients, are given in Theil (1970) and Kheir and Holmes (1978).

3.3 Results. Random search over the parameter space yielded the results that the range of the multiple Theil's inequality coefficient is between 0.1233 and 0.1166. Since the possible range of U is between 0 and 0.5, all of the best 250 different combinations of parameters (parameterizations) stored are included in a range of .0067 or 1.34% of the whole spectrum. Thus, the best combination of parameters can not be uniquely identified. This result is of concern since apparently many models exist which fit the data equally well (or equally bad). Please note the assumption that a model with a different set of parameter values is a different model.[1] The output of the 250 model parameterizations are different from one another. The parameterization vectors are therefore (output) distinguishable. According to Cobelli *et al.*, (1970) "a fixed structure system is said to *almost everywhere globally identifiable* if it is globally identifiable in every parameterization vector of the space P, except at most in a subspace of zero measure . . ." Therefore, according to their Definition 4, "a compartmental system is said to be *uniquely structurally identifiable* if it is almost everywhere globally identifiable." Should the search for the 'global' minimum have been performed with a, e.g., gradient method, a unique parameterization would have been found. My approach however shows that even if the model structure is appropriate and identifiable, a number of parameterizations exist which are output distinguishable, none can be arbitrarily preferred over the others. The 250 models may simulate 250 different futures for the ecosystem studied. Not very promising if we are interested in prediction. Indeed, we could use the parameter combination identified at the global minimum 0.1166. The range of values in the objective function, however, is so small that one should consider with great caution any results of a prediction simulation. The range of each parameter is also very small in relation with the starting range of 0.-400.; the model structure constrains the parameter values so that no strange values are observed (Table 2). If the random search showed that the model was capable of giving a good fit to the data with a large range of parameter values, confidence in the model would be reduced because the values would have been outside of the range usually ascribed to them in modeling exercises.

3.4 Parameters Variance and Covariance. Since the objective function had such a small range for successful model candidates, all models can be assumed to be equally good descriptions of the data. Therefore, a matrix of variances-covariances can be computed. For eacy display and to make comparison easier this matrix

---

[1]Two hundred and fifty parameterizations need not lead to 250 distinct model realizations. Indeed, if a direct parameter tradeoff exists (e.g., a new parameter may be found that is a combination of model parameters) an infinity of parameterizations can be determined leading to a single model prediction. Principal components analysis is used later in the paper to find this tradeoff with generally negative results. Also note that the large number of distinct realizations exist because the data are noisy.

TABLE 2: Range of parameter values which give the best fit to IFYGL *Lake Ontario data. These are the best* 250 *combinations. Range has been estimated by random search.*

Parameter No. (for meaning see Table 1)	Range	Units
1	0.3579- 4.8682	$\mu gP\ell^{-1}$
2	9.8662- 20.9897	$\mu gP\ell^{-1}$
3	0.1529- 0.7756	day^{-1}
4	111.7980-310.2340	$\mu gC\ell^{-1}$
5	0.0126- 0.0598	day^{-1}
6	0.0111- 0.1437	$\ell\ \mu gP^{-1}\ day^{-1}$
7	0.0020- 0.0174	day^{-1}
8	0.0033- 0.0061	day^{-1}
9	0.0047- 0.0624	day^{-1}
10	0.3124- 3.9375	day^{-1}
11	0.0311- 0.4472	$m\ day^{-1}$
12	0.0021- 0.6513	$m\ day^{-1}$
13	0.0787- 0.3022	$m\ day^{-1}$
14	0.2765- 0.7156	$m\ day^{-1}$
15	0.0389- 0.7656	$m\ day^{-1}$

has been transformed into one of correlations (Table 3). Since the matrix is symmetrical, only the lower triangular part is shown. The significance of the correlation may be tested only on a single prechosen pair. In this instance, with 233 degrees of freedom, the null hypothesis that two parameters are independent of each other can be rejected if the absolute value of the correlation coefficient is greater than .1678 at the 0.01 level, and .1280 at the 0.05 level. When we start making comparisons among all parameters, the significance level becomes hopelessly incorrect. If the significance of all correlation is of interest, a simultaneous inference test, such as a Bonferroni approach (Games, 1977) or Scheffe's (1953) method, should be used. For a large number of parameters, the numbers of all possible contrasts θ, defined as $\theta = \sum_i^k c_i \bar{x}_i$, where $\sum_i^k c_i = 0$, and $\bar{x}_i$ is the mean value of parameter p_i, and k is the number of

TABLE 3: Correlation matrix among parameters. Two hundred and fifty combinations which give a good fit to lake Ontario data are used to produce this matrix. Meaning of parameter number is presented in Table 1.

Parameter Number	1	2	3	4	5	6	7	8	9	10	11	12	13	14	15
1	1.000														
2	0.494	1.000													
3	0.085	0.186	1.000												
4	0.161	0.093	0.268	1.000											
5	-0.162	-0.010	0.138	-0.107	1.000										
6	-0.004	-0.214	0.038	-0.216	-0.322	1.000									
7	-0.158	0.879	-0.152	0.044	-0.162	-0.145	1.000								
8	-0.275	0.072	-0.068	0.084	-0.263	0.023	0.509	1.000							
9	0.072	0.110	-0.103	0.194	-0.126	-0.040	0.276	0.062	1.000						
10	0.045	0.038	0.340	0.003	-0.093	0.350	-0.018	0.075	-0.080	1.000					
11	0.053	0.076	0.195	0.234	0.037	0.069	-0.098	0.006	0.046	-0.196	1.000				
12	-0.021	0.006	-0.064	-0.114	0.203	-0.073	-0.018	-0.286	0.152	0.041	-0.164	1.000			
13	-0.305	-0.202	-0.118	-0.150	0.400	-0.056	-0.204	-0.143	0.009	-0.019	0.127	0.163	1.000		
14	-0.252	-0.154	-0.052	-0.067	0.060	0.026	0.064	0.092	0.023	0.254	-0.082	0.255	0.161	1.000	
15	-0.019	-0.103	0.035	0.093	-0.307	0.064	0.310	0.093	0.153	0.000	-0.104	0.106	-0.333	-0.965	1.000

Parameter Number

parameters, is very large (on the order of hundreds)[1].

One alternative way of solving the question, if we can reduce the number of correlated parameters by changing the model structure, is principal component analysis. This method is used to analyze the interdependence of parameters. The advantage of having fewer parameters is that, if revised, the new model configuration would be more aggregated. For some large-scale models reduction may be necessary and desirable.

4. PRINCIPAL COMPONENTS ANALYSIS

This statistical technique is used to pick apart the dependence structure of the data when the responses are symmetric in nature or when no *a priori* patterns of causality are apparent (Morrison, 1967). A principal component of the observations x, in this case the parameter values obtained during the random search, is the linear compound:

$$y_k = a_{1k}x_1 + \cdots + a_{pk}x_p$$

where the coefficients a_{ik} are the elements of the characteristic vector associated with one of the characteristic roots ℓ_i for the sample covariance matrix (S) of the responses. The a_{ik} are unique up to multiplication by a scale factor, and if they are scaled so that $a_k'a_k = 1$, the characteristic root ℓ_i is interpretable as the sample variance of y_k (modified from Morrison, 1967). The 250 values of each parameter are here considered as random variables with mean $\bar{x}_i$ and variance s^2, therefore, in this analysis each combination of 15 model parameters is treated as a response to a particular minimization experiment. The advantage of working with the artificial variate y is that if most of the variation in a system of 15 responses (parameters) could be accounted for by a simple weighted average of the response (parameter) values, then the variation could be expressed along a smaller continuum (i.e., smaller than 15) rather than in fifteen-dimensional space. The coefficients of the

---

[1]Note that the Bonferroni and Scheffe methods have been developed for the problem of multiple comparisons of means. The extension of the method to the multiple comparisons of correlations has been suggested by an anonymous referee and I have not been able to confirm it or deny it by reading the appropriate statistical literature.

responses also indicate the relative importance in the new derived component.

The sample covariance of the *ith* component is ℓ_i and the total system variance is thus

$$\ell_1 + \cdots + \ell_p = \text{trace } S.$$

The importance of the *ith* component in a more parsimonious description of the system is described by

$$\frac{\ell_i}{\text{tr } S}$$

The algebraic sign and magnitude of a_{ij} indicate the direction and importance of the contribution of the *ith* response to the *jth* component.

4.1 Results. Table 4 shows the results of the principal components analysis of the Lake Ontario water quality model. The percentage of total variance explained by each component is usually small and quite uniform. These results indicate that the model is highly aggregated and agrees with the modeling objectives which called for a model as reduced as possible but which would still simulate some interesting lake features. If only a few principal components were able to describe the total variation, a much more reduced model, e.g., with four or five parameters would have been sufficient.

To extract information from Table 4, I analyzed the components which are able to explain the least amount of variance. These components represent parameters that are superfluous to the model. In fact, components which explain relatively large percentages would indicate relations among parameters important for the model. Furthermore, since the model structure is complex, these components indicate that model behavior is regulated by most parameters without showing in what ways.

Component 15 which explains only 1.1% of total variance is a linear combination of several parameters. The highest correlation with component 15 is observed with parameters 3, 6, and 10. These parameters (see Table 1) are associated with the zooplankton feeding, growth and death rates.

Component 14 is correlated with parameters describing flows related to phosphorus in water, for example, mineralization of organic matter and zooplankton excretion rate.

TABLE 4: Component correlations of the Lake Ontario model (Simons et al, 1978). The component coefficients a_{ij} *are the elements of the characteristic vector of the correlation matrix (Table 3) among parameters. Each element is associated with a characteristic root of that matrix.*

	Component														
Parameter No.	1	2	3	4	5	6	7	8	9	10	11	12	13	14	15
1	0.22	-0.39	-0.22	0.40	0.00	0.21	-0.12	-0.01	-0.05	-0.07	-0.11	0.07	0.54	0.45	-0.1+
2	0.23	-0.32	0.10	0.28	-0.14	0.44	0.11	0.34	0.22	0.07	0.15	0.29	-0.26	-0.31	0.32
3	0.07	-0.38	0.08	-0.18	0.26	-0.28	0.50	0.38	-0.16	0.25	0.09	0.03	0.05	-0.03	-0.+3
4	0.25	-0.21	0.21	-0.08	0.40	0.05	0.26	-0.62	0.00	0.08	0.22	-0.16	0.11	-0.03	0.35
5	-0.43	-0.17	0.29	0.01	-0.14	0.01	0.21	-0.01	0.30	0.34	-0.48	-0.09	-0.05	0.37	0.24
6	0.02	0.18	-0.53	-0.08	0.39	-0.03	-0.07	0.38	0.10	0.32	0.04	-0.22	0.11	0.06	0.4+
7	0.34	0.30	0.37	-0.02	-0.17	0.02	-0.04	0.15	0.25	0.14	-0.19	-0.30	0.93	-0.33	-0.03
8	0.31	0.29	0.21	-0.36	-0.11	0.28	0.14	0.18	0.05	-0.04	0.30	0.02	-0.11	0.63	-0.03
9	0.17	0.10	0.35	0.29	0.39	0.08	-0.42	0.04	-0.29	0.42	-0.13	-0.03	-0.32	0.09	-0.15
10	-0.03	0.33	-0.32	0.17	0.17	0.40	0.33	-0.28	0.37	0.18	-0.10	0.07	-0.11	-0.09	-0.4+
11	0.02	-0.26	0.08	-0.35	0.43	0.12	-0.33	0.11	0.46	-0.42	-0.23	-0.0+	-0.49	-0.02	-0.1+
12	-0.21	0.08	0.19	0.54	0.13	-0.24	0.07	0.14	0.30	-0.26	0.43	-0.39	-0.07	0.1+	-0.06
13	-0.47	0.04	0.20	-0.13	0.12	0.14	-0.26	-0.02	0.14	0.26	0.44	0.43	0.39	-0.05	-0.04
14	-0.18	0.31	0.21	0.14	0.37	0.19	0.33	0.20	-0.34	-0.40	-0.27	0.2+	0.19	-0.03	0.09
15	0.33	0.18	0.01	0.15	0.08	-0.56	0.02	-0.07	0.35	0.01	-0.11	0.59	-0.03	0.12	0.12
Characteristic root	2.325	2.148	1.659	1.487	1.253	1.291	0.970	0.874	0.703	0.571	0.520	0.446	0.363	0.293	0.266
Percentage of total variable	15.5	14.3	11.1	9.9	8.4	8.0	6.5	5.8	4.7	3.8	3.5	3.0	2.5	2.0	1.1
Cumulative variance	15.5	29.8	40.9	50.8	59.1	67.2	73.6	79.4	84.1	87.9	91.4	94.4	96.9	98.9	100.0

Component 13 is correlated with parameters similar to those of 14 but with more correlation with phytoplankton than with zooplankton.

Component 12 relates to sedimentation rates. The sedimentation rates of detritus in all three layers contribute similarly to the component. The sedimentation rates of phytoplankton are negatively correlated and in one instance (parameter 11) the correlation is not significant.

5. DISCUSSION

The relation between the level of model aggregation and the parameter structure of an ecosystem model is analyzed. A unique set of parameter values which would optimally fit the data cannot be determined. The objective function which describes the goodness of fit is a very flat hyper-dimensional surface which implies that many combinations of parameters are equally adequate. Since the parameters are correlated among themselves, model parameters should be identified as ranges rather than scalars. Both our inadequate knowledge of the ecosystem under study and the conceptualization process during modeling bring us to this conclusion. We can also question the goodness of fit to data as the only criterion for validation. An alternative view has been discussed by Halfon and Reggiani (1978).

Correlation among parameters raised the question of free parameters, i.e., parameters that are not really necessary for the model. A multivariate statistical method, principal components analysis, was used to test the hypothesis that if a component could be found which was a linear combination of parameters, then perhaps the model could have been further reduced to a minimal realization. Results showed that the Lake Ontario model considered in this study was already very aggregated. The analysis did not indicate new lumped parameters which were linear combinations of several others. However only little information was included in four parameters associated with the zooplankton compartment.

Other possibilities are related to the compartments describing phenomena in water: detrital carbon, detrital phosphorus, soluble organic phosphorus, and soluble reactive phosphorus. These compartments are necessary for an understanding of the phenomena taking place in the lake that they may be excessive if only long-term prediction is sought. The final point concerns sedimentation rates of detritus; even if their ranges in the three layers overlap only slightly (Table 2) these three parameters (13, 14, and 15) may be considered into one by an appropriate modification of the equations describing the sedimentation processes. All of these modifications would then eliminate the implicit free parameters which allow us to obtain a good fit to the data.

How much is to be gained by making the model more lumped, that is, more abstract? A more lumped model would be more general and the assumptions would stand out more clearly. However, we would lose some of the detail that is included in a less aggregated model. The analysis performed in this paper provides an impression for the level of abstraction of a model after it has been developed and validated, and therefore, can then lead to research towards a more aggregated or less aggregated model as deemed necessary to simulate natural phenomena.

ACKNOWLEDGEMENTS

Ideas on the analysis of some model properties originated during discussions I had at the International Institute of Applied Systems Analysis, Laxenburg, Austria, with the resident and visiting scientists. Particularly, I would like to thank Dr. Dominic M. DiToro for his clues and the National Water Research Institute for granting me leave. Thanks are also due to Dr. Sylvia Esterby for suggestions on the problem of multiple comparisons and parameter correlations, to Dr. Robert V. O'Neill and Andrew S. Fraser for pointing out specific topics, and to three anonymous referees for constructive comments.

REFERENCES

Cale, W. G. and Odell, P. L., (1979). Concerning aggregation in ecosystem modeling. In *Theoretical Systems Ecology: advances and case studies*, E. Halfon ed. Academic Press, New York and London. 55-77.

Cobelli, C., Lepschy , A. and Romanin-Jacur, (1979). Identifiability of compartmental systems and related structural properties. *Mathematical Biosciences*, 4, 1-18.

Games, P. A. (1977). An improved table for simultaneous control on g contrasts. *Journal of the American Statistical Association*, 172, 531-534.

Halfon, E. and Reggiani, M. G. (1978). Adequacy of ecosystem models. *Ecological Modelling*, 4, 41-50.

Halfon, E., Unbehauen, H., and Schmid, C. (1979). Model order estimation and system identification theory and application to the modeling of ^{32}P kinetics within the trophogenic zone of a small lake. *Ecological Modelling*, 6, 1-22.

Kheir, N. A. and Holmes, W. M. (1978). On validating simulation models of missile systems. *Simulation*, 30, 117-128.

Lean, D. R. S. (1973). Phosphorus dynamics in lake waters. *Science*, 179, 678-680.

Morrison, D. F. (1967). *Multivariate Statistical Methods*. McGraw-Hill, New York and Toronto.

O'Neill, R. V., Gardner, R. H., and Mankin, J. B. (1979). Propagation of parameter error in a nonlinear model. *Ecological Modelling* (in press).

O'Neill, R. V. and Rust, B. (1979). Aggregation error in ecological models. *Ecological Modelling* (in press).

Price, W. L. (1977). A controlled random search procedure for global optimization. *The Computer Journal*, 20, 367-370.

Scheffe, H. R. (1953). A method of judging all contrasts in an analysis of variance. *Biometrika*, 40, 86-104.

Simons, T. J., Boyce, F. M., Fraser, A. S., Halfon, E., Hyde, E., Lam, D. C. L., Schertzer, W. M., El-Shaarawi, A. H., Willson, K., Warry, D. (1978). Assessment of water quality simulation capability for Lake Ontario. Canada Centre for Inland Waters, Unpublished Report.

Theil, H. (1970). *Economic Forecasts and Policy*. North-Holland Publishing Company, Amsterdam and London.

Wiegert, R. G. (1975). Simulation modeling of the algal-fly components of a thermal ecosystem: effects of spatial heterogeneity, time delays and model condensation. In *Systems Analysis and Simulation in Ecology, Vol. 3*, B. C. Patten, ed. Academic Press, New York. 157-181.

Zeigler, B. P. (1976). *Theory of Modeling and Simulation*. Wiley-Interscience, New York and Toronto.

Zeigler, B. P. (1979). Multilevel multiformalism modeling: an ecosystem example. In *Theoretical Systems Ecology: advances and case studies*, E. Halfon, ed. Academic Press, New York. 17-54.

[*Received October* 1978. *Revised June* 1979]

G. P. Patil and M. Rosenzweig, (eds.),
Contemporary Quantitative Ecology and Related Ecometrics, pp. 295-318.

EVALUATION OF ECOLOGICAL SIMULATION MODELS

GORDON SWARTZMAN

Center for Quantitative Science
University of Washington
Seattle, Washington 98195 USA

SUMMARY. Model evaluation was originally done by comparing model output with data traces. More recently, model evaluation methods in addition to this have been suggested as they relate to model objectives. The objectives seen as general to ecological simulation models are as follows: (1) replicate ecosystem behavior, (2) further understanding of ecosystem mechanisms, (3) organic information and data on ecological processes, (4) pinpoint areas for future research, (5) generalize the model beyond a single site, and (6) investigate the effect of manipulation and perturbations.

Methods discussed in this paper include model corroboration, sensitivity analysis, model recalibration, simulation experiments, key process identification, and equation rationale evaluation. These are defined and problems with their use and recommended techniques for their application are discussed. Examples of their use from the literature are given and are used to explain how the methods relate to various model objectives. Due to the complexity of many of the systems modeled, there is difficulty in assessing when model objectives have actually been achieved. Also the methods for evaluation often attempt to corroborate the model as well as investigate the implications of the model assuming its validity which can lead to circular reasoning.

KEY WORDS. model evaluation, validation, simulation models, ecological models, simulation experiments, sensitivity analysis.

1. INTRODUCTION

Since the inception of the use of ecological simulation models to represent the seasonal dynamics of energy and material flow in large ecosystems around the beginning of this decade, the use of such models has burgeoned (e.g., Steele, 1974; Innis, 1978; Kremer and Nixon, 1978; Steele and Frost, 1977; Park *et al.*, 1975; Coniferous Forest Biome Modeling Team, 1977; MacCormick *et al.*, 1972,; Bledsoe, 1976; Di Toro *et al.*, 1975; Andersen and Ursin. 1977). The primary means for evaluation (validation) of these models has been comparison of model output to field data, preferably data independent from that used to calibrate the model. Recently, a number of criteria for model evaluation have been suggested in addition to model predictivity. These criteria relate to model objectives, since models must stand or fall on their ability to satisfy the objectives for which they were built. This paper suggests methods for evaluation relative to the suggested criteria and presents examples from various models in the literature. Strengths and weakensses of the methods are discussed. The paper also discusses the difficulty of evaluating when model objectives have been achieved.

2. HISTORY OF MODEL EVALUATION

Until recently, there has been a wide gap between the expectations of ecological models and the criteria used to demonstrate their value. Proposals for large-scale modeling efforts promised and fully expected to develop models predictive of system behavior under ordinary conditions as well as under a wide range of natural and managed perturbations. As such, they were to prove the vanguard in man's renewed awareness of his need to responsibly manage ecosystems by studying the impact of proposed changes before the changes were implemented. They have since been demonstrated to be only one tool in this worthy but still nebulous quest for responsible impact assessment--indicating trends rather than yielding accurate predictions.

On the other hand, when models were first implemented, there seemed no way to demonstrate their credibility other than by how closely their output agreed with data traces obtained from field sampling. Rationale for model mechanisms was hidden in a morass of nearly indecipherable equations often accompanied by typographic errors, drastically reducing the usefulness of the equations. Mention was seldom made of how data sources were used or of which changes resulted in improved model behavior or proved abortive. Skellam (1972), in a paper on model philosophy, seemed to imply that the effective modelers were guarding this information as if it were proprietary. However, the information was probably not given because the reality of the situation was so far from the ideals set in prognostications and promises that it was feared honest discussion of the state of the art would be letting the cat out of the bag. Meanwhile, it was apparent that modeling had desirable side effects (e.g., education, serving as a communication link between researchers engaged in previously separate disciplines, organizing data and information on ecosystems, etc.) and to have a backlash on model credibility would not allow the scientific community sufficient exposure time to modeling for its apparent benefits to sink in.

3. EVALUATION CRITERIA RELATED TO MODEL OBJECTIVES

Discussions of means for evaluating models have appeared in recent literature that bridge the gap between the realities of ecological modeling and the results originally expected. Several articles (Innis, 1976; Woodmansee, 1978) have suggested that models should be evaluated on how well they meet their objectives. This opens the door to a wide range of alternative criteria for evaluation of which validation (i.e., how well the model fits data traces) is only one. The question arises, however, whether ecological models are similar enough in their objectives that criteria for their evaluation can be discussed as a whole rather than separately for each model. Swartzman (1978a) reviewed objectives for a number of large-scale ecological models and found them to be very similar. From this, he formulated several objectives common to most models reviewed. These include (1) to replicate system behavior in comparison with field data, (2) to further understanding of ecosystem behavior, (3) to organize information and data on processes, (4) to pinpoint areas for future research, (5) to generalize the model beyond a single site, and (6) to investigate the effect of manipulation and perturbation on system behavior.

Bledsoe (personal communication) asserts that model evaluation in an ongoing part of model development. Evaluation criteria are any means which increase the modeler's confidence that model objectives are being satisfied. Evaluation as seen from outside the modeling process consists of ways of increasing the reviewer's confidence that model objectives have been or are being satisfied. Thus, in the end, evaluation criteria relate back to model objectives--whether in internal or external review--and the methods used in assessing the model with respect to such criteria should be independent of who is doing the evaluating.

Wiegert (1975) and Caswell (1976) suggest that ecosystem models may range from strictly theoretical to entirely predictive, the theoretical aspect of the model being nonfalsifiable by comparison of output to data traces. Wiegert suggests that model evaluation comprises assessing confidence in the theoretical as well as the predictive function of the model. But how does one assess confidence in the theoretical part of a model? Relevant criteria are the generalizability of a model, the ability of a model to aid in the understanding of ecosystem behavior and to allow the comparison of alternative hypotheses and the scientific respectability of the sources of information used.

Kitchell *et al.* (1974) suggest model evaluation criteria similar to Wiegert and suggest that a successful model should aid in ecosystem understanding (as well as generate new hypotheses or use data in new ways), should have predictivity (with respect to data traces independent of those used to derive the model), and scientific respectability (it should include a significant proportion of important mechanisms and define the parameter measurement techniques). Clearly these criteria relate directly to the objectives cited by Swartzman (1978a).

Figure 1 indicates methods used on the above discussion and experience of the author, which are here suggested to apply as aids in measuring model performance relative to model objectives.

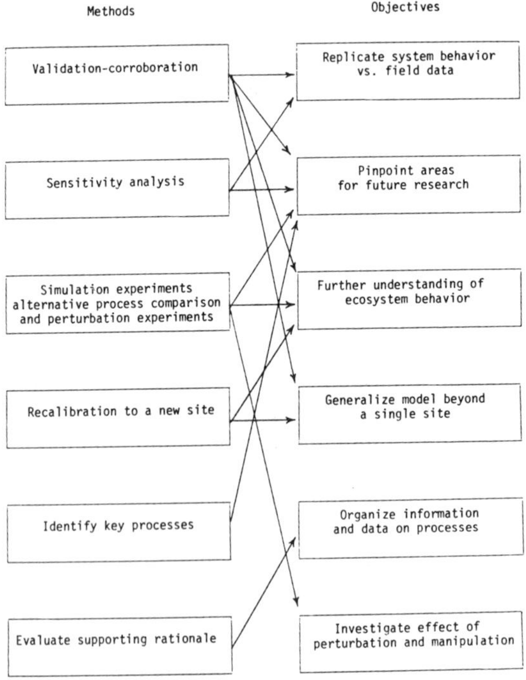

FIG. 1: Relationship of evaluation methods to simulation model objectives.

The body of the text will treat these methods in detail with examples and will attempt to clearly indicate the connection between the methods and objectives. *Validation* or *corroboration* involves a group of methods comparing model output to field data sets, preferably those independent of data used to develop the model. *Perturbation* simulation involves examining model behavior under perturbations--both abiotic and biotic, natural and man-made--and comparing results with either data or expectation. New site *recalibration* of a model involves extending the model to a new site by changing only parameter values. *Sensitivity analysis* as used here involves examining model behavior when parameter values are changed over some given range of values. The method of *key processes identification* involves establishing which process hypotheses are both least well known and most integral to model behavior. *Simulation experiments* involve considering the implications of several alternative hypotheses to model behavior to decide which hypothesis is most appropriate for use in the model or experiments designed to investigate the role a given process or organism plays in model dynamics. Most of the above methods are predominantly *simulation methods* in that they are done with model runs. In addition to these simulation methods, model evaluation must also consider such data and scientific related questions as data variability, data availability for parameter estimates, simplicity of equation forms, and quality of supporting evidence. These are encompassed in evaluation of the *supporting rationale* for model equations.

A *document* for model evaluation should contain the results from studies on the model conducted using the aforementioned methods. It is recognized that all models are subject to improvement; however, until a clearcut set of evaluation methods are utilized, model improvement will remain highly subjective and the arguments between alternative formulations will continue to be academic.

Let us now discuss each of these methods in some detail. We will follow the format of introducing the method, discussing where and how it has been applied in the past, problems associated with the method, and examples where appropriate. Finally, we will discuss how the methods relate to the various objectives found general to most simulation models (the arrows in Figure 1).

3.1 Preditivity-Model Corroboration (Validation). Model validation by comparison of model output with independent data traces is an important measure of model predictivity. Much recent literature is devoted to means for validating models. On one hand, it has been asserted in criticizing validation as a useful tool for model evaluation (e.g., Garfinkel *et al.*, 1975) that a model with enough parameters can be made to fit any data trace. On the other hand, in defense of model corroboration as a tool in model evaluation are the facts that many of the parameters in ecological models can be realistically varied only over quite limited ranges for a particular ecosystem, that there are a great many output variables in many ecological models, making simultaneous model fitting to a large number of data traces impractical. Garratt (1975), Steinhorst, and Garratt (1976), Kitchell *et al.* (1974), and others have emphasized that a data set independent of the one(s) used to develop the model is necessary for validation. Certainly use of an independent data set for model corroboration limits parameter adjustment as a means of insuring model corroboration. One argument against using validation as a sole means of model evaluation is that the model can give the right output for the wrong reasons (Andersen and Ursin, 1977).

The most common validation methods are either statistical or graphical methods. Simulation model output is looked at as deterministic and as such any statistical comparisons of model output and data traces question whether or not the model and data agree within the statistical variation of the data (Steinhorst, 1977). However, assuming that model output represents a transformation of environmental data via deterministic mechanistic equations (processes) with parameters estimated from data obtained from experiments on the relevant processes, we could speak in theory of the statistical distribution of model output. This model output distribution would be extremely difficult to obtain since two potentially estimable statistical distributions--that of the environmental data and the parameters is transformed in a complex fashion by a series of mostly nonlinear equations (and presumably known--a tenuous assumption). Argentesi and Olivi (1978) present means for investigating the statistical distribution of the model output time trace using a polynomial approximation technique assuming a deterministic environment.

Tawari (1974) and Overton (1973) looked at model validation as a goodness-of-fit criterion where the model output time traces for selected output variables are evaluated on how closely they follow the data traces, closeness being defined by the least-squares distance between the two curves. Garratt (1975) suggested several statistical approaches to validation, most of which assume both model output and data to be a series of paired points to be compared for overall closeness, irrespective of their time relationship. This assumes temporal independence of both data and model output traces--tenuous assumptions in either case. Other statistical tests suggested by Steinhorst and Garratt (1976) and Steinhorst (1979) compare model output with field data estimating the covariance matrix between field data at different times.

Rather than implementing a rigorous quantitative validation scheme, a graphical comparison of the timing and magnitude of model changes with those in the ecosystem is often used. Since our eyes and minds are geared to the recognition of pattern in graphical displays, inspection of model output versus data traces might be an acceptable means of model validation, especially when differences between model output and field data can be traced to specific processes or omissions in the model. However, when a number of different models are to be compared to a data set, graphical comparison is more difficult and a number of indices of model performance are often compared with data. Most ecological models have several variables to be compared with field data. This argues for a table of comparison of model performance with field data according to selected indices, perhaps supplemented by comparison of model output with data traces for those variables having data available.

Examples of graphical comparison of model output to data time traces is given in Figure 2 from plankton models for Lake Ontario (Scavia *et al.*, 1976) and Narragansett Bay (Kremer and Nixon, 1978). The Lake Ontario model compares model output to a zone of data variability based on five years of data looked at as five replicates of average seasonal dynamics. The Narragansett Bay model compares one year of data with model output. Here it is apparent--and observation of such comparisons from many models support the observation--that models of this sort perform better under comparison with data averaged over several years than with data for a specific year.

An example of tabular comparison of model indices is taken from Andersen and Ursin (1977) (Table 1) comparing fisheries yields over a 20-year period with model output for 11 species of commercially important fish in the North Sea.

In all these examples, some parameter adjustment was done by the modelers to make model output more closely fit data traces, as is done in all ecological model development at present. What is to be done if the data sets don't agree or only partially agree even after parameter adjustment? Regier and Rapport (1978) see systems models (compartment-flow analysis) comparison with data resulting in model revision and certainly this is a possibility, though it could be an unending cycle of model adjustment. Another possibility is to use such anomalies as an indicator of areas for further research into particular processes. This is possible only if the anomalies can be traced back to specific processes. One example of this is from Kremer and Nixon's (1978) Narragansett Bay plankton model, where they found that the prebloom winter phytoplankton biomass is insufficient to maintain the observed zooplankton stock, suggesting that alternative food sources are being used at that time. This indicates further research is needed on the seasonal change of diet of dominant zooplankton.

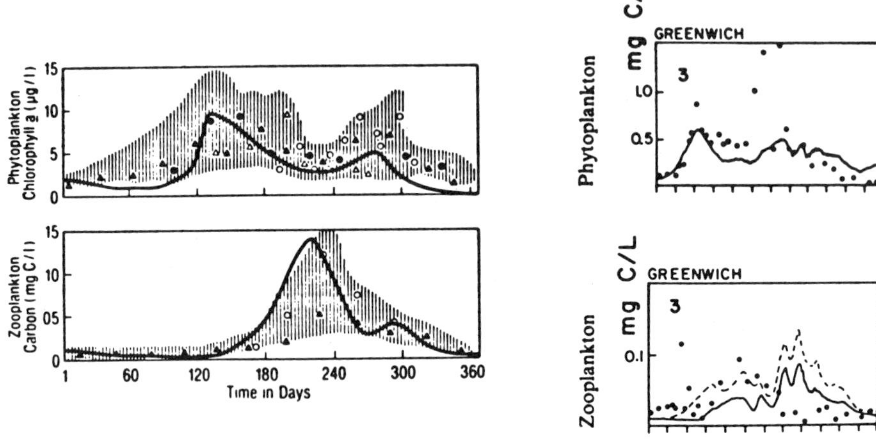

FIG 2: Data trace and model comparison from (a) Lake Ontario (Scavia et al., 1976) and (b) Narragansett Bay (Kremer and Nixon, (1978).

TABLE 1: Fishery yields on North Sea vs. model output for 11 commercially important fish species (from Anderson and Ursin, 1977).

Annual yield (1000 tons) of North Sea fisheries for 11 species (from Bull. Stat. with amendments).

	1960	1961	1962	1963	1964	1965	1966	1967	1968	1969	1970
Plaice	87	86	88	108	110	97	100	109	111	121	130
Dab*	11	11	11	11	11	11	11	11	11	11	11
Long rough dab*	1	1	1	1	1	1	1	1	1	1	1
Saithe	29	31	22	28	55	69	87	73	97	106	170
Cod	105	108	91	110	125	182	229	250	285	199	229
Haddock	67	69	53	60	202	225	272	167	140	640	675
Whiting	55	85	64	99	88	110	158	91	145	199	183
Norway pout	27	26	137	110	44	36	47	178	437	103	217
Mackerel	73	86	66	73	115	208	530	930	822	739	322
Herring	696	697	628	716	871	1169	896	696	718	547	563
Sandeels	121	178	109	162	128	131	142	189	194	113	191
11 species, total	1272	1378	1270	1478	1750	2239	2473	2695	2961	2779	2692

*Accurate information missing, but small figures anyway.

Calculated annaul yield (1000 tons) of North Sea fisheries for 11 species.

	1960	1961	1962	1963	1964	1965	1966	1967	1968	1969	1970
Plaice	94	86	100	111	116	103	102	138	101	97	118
Dab	9	9	10	11	11	12	12	12	11	11	11
Long rough dab	0	1	1	1	1	1	1	1	0	1	1
Saithe	31	29	28	25	42	52	75	73	77	90	151
Cod	108	103	89	100	130	152	184	173	218	208	220
Haddock	63	57	43	61	164	214	307	162	195	607	627
Whiting	35	51	50	77	62	82	109	59	95	130	148
Norway pout	12	21	48	53	17	5	6	44	224	581	592
Mackerel	74	75	76	76	120	206	523	917	825	689	291
Herring	618	762	778	601	970	1418	980	865	896	564	578
Sandeels	138	117	104	91	71	50	42	66	142	272	345
11 species, total	1188	1317	1332	1212	1709	2298	2346	2514	2790	3255	3089

3.2 Model Recalibration. It is the hope of ecological modelers that ecological simulation models be generalizable beyond the system for which they were built since, if models must be revamped from one ecosystem to another, the potential for using these models for resource management or in understanding ecosystem is limited indeed.

Model recalibration in its strictest sense involves transporting a model to a new site, changing the initial conditions, weather data (driving variables), and parameter values based on data from the new site if available and then comparing model output to data traces. Unfortunately, this process almost never happens. Either the recalibration serves as a first step in changing the model (O'Connor *et al.*, 1975) or the agreement between model output and data traces is poor (DiToro *et al.*, 1975; Eggers, 1975; Swartzman, 1978b). Also, often difficult data are available on different sites, which makes parameter estimation and model corroboration as conducted on one site difficult to transport to another.

Recalibration efforts claiming some success are reported by Desormeau (1978), Kitchell *et al.* (1974), Lehman *et al.* (1975), and O'Connor *et al.* (1975). Table 2 shows an example from O'Connor *et al.* (1975) showing parameter values for a model reparameterized to several lake and river sites. While comparisons to field data traces were made for all these models, parameter estimates were not made from data taken at the sites modeled and model equations were changed between models, especially those relating to the role of nutrients. O'Connor *et al.* (1975) adjusted parameter values to calibrate the model to fit field data traces. As such, their approach is not strictly model recalibration as outlined above, although it represents a good effort at generalizing a single model, with modest alteration, to a wide range of sites. Desormeau (1978) reports success with strict recalibration for a model applied to a couple of alpine lake ecosystems. As an example of how recalibration might relate to the objective of understanding ecosystems, the role different nutrients play in control of seasonal dynamics on different lakes is indicated in Table 2. The additions of phosphate on Lake Eric and silicate on Lake Ontario as controlling nutrients in the model in addition to nitrate and ammonia shows the relative importance of the various nutrients on different sites.

Figure 3 shows the model of Thomann *et al.* (1975) of Lake Ontario run on Lake Washington (Swartzman, 1978b), altering only the initial conditions and weather data. While the fit is surprisingly good for the first part of the year, it fails to replicate the observed drop in phytoplankton in mid-summer. In fact, no parameter alteration could replicate this major change in phytoplankton since it was due in some way to a storm event which is

TABLE 2: Parameter value comparison for a plankton model calibrated to a number of sites (from O'Connor et al., 1975).

Parameter	Units	Symbol	River	Estuaries		Lakes	
			San Joaquin	Delta	Potomac	Erie	Ontario
Physical							
Temperature range	°C	T	8-25	7-26	5-30	2-25	1-20
Solar radiation range	Langleys/day	I	200-300	160-720	135-530	170-180	100-630
Detention time	Days	t_o	30	30	60	45	7.9 yrs
Extinction coefficients	1/m	K_e	4.0	5.5	3.5	1.2	0.22
Phytoplankton							
Saturating light intensity	Langleys/day	I_s	300	300	300	350	350
Saturated growth rate	1 day @ 20°C	K_r	2.0	2.5	2.0	1.3	2.1
Engogenous respiration rate	1 day @ 20°C	K_1	0.10	0.10	0.10	0.08	0.10
Nitrogen Michaelis constant	μg/l	K_{mn}	25.0	25.0	25.0	25.0	25.0
Phosphorus Michaelis constant	μg/l	K_{MP}	---	---	5.0	10	2.0
Settling velocity	m/day	W	---	---	---	---	0.1
Zooplankton							
Grazing rate	ℓ/mg carbon day @ 20°C	C_g	0.13	0.18	---	0.25	1.2
Endogenous respiration rate	1 day @ 20°C	K_2	0.075	0.10	---	0.16	0.02
Assimilation efficiency	mgC/mgC	a_1	0.60	0.60	---	0.65	0.60
Grazing Michaelis constant	μg chlor/ℓ	K_{MP}	50	50	---	50	10
Stochiometric Ratios							
Carbon/chlorophyll	mg/mg	a_{ZP}	50	50	50	50	50
Nitrogen/chlorophyll	mg/mg	a_{NP}	7	7	10	7	10
Phosphorus/chlorophyll	mg/mg	A_{pp}	---	---	1	1	1

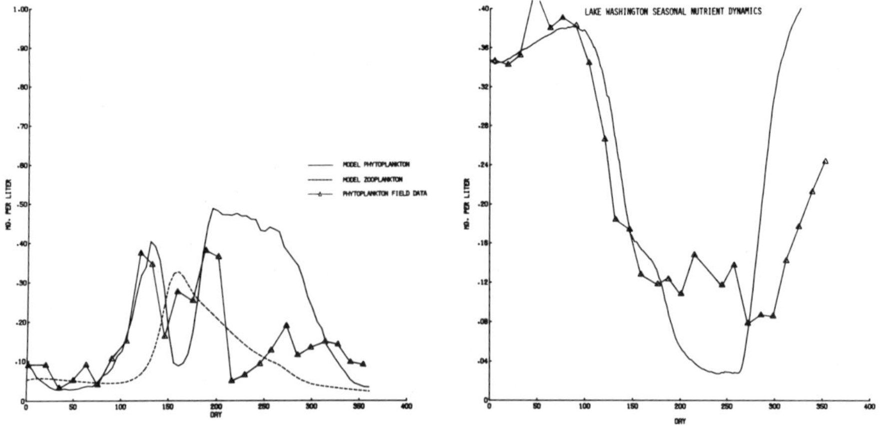

FIG 3: Model of Thomann et al. (1975) for Lake Ontario recalibrated to Lake Washington, Seattle, Washington.

not represented as a driving variable in the model. This also points to the general inability of whole ecosystem simulation models to replicate individual time traces (see also Figure 2).

General indications from the available literature are that ecological models can be successfully recalibrated in a strong sense only to a limited number of systems quite similar to the original system. Weaker recalibration involving limited equation changes and some parameter alteration to improve fits to data traces on the new sites should work over a broader range of systems (e.g., northern temperate lakes of a certain size range) but only in the sense of replicating average seasonal dynamics.

3.3 Sensitivity Analysis. Sensitivity analysis consists of methods for evaluating how sensitive model output is to changes in parameter values. In its most straightforward guise, sensitivity study involves changing parameter values, singly and in various combinations, by a constant percentage and observing changes in model behavior. This gives an idea of the relative sensitivity of model output to each of the parameters.

There are several major problems with this approach. First, there are too many parameters to change them all. Secondly, the output of the model is not a single time trace but an ensemble of time traces, making it difficult to evaluate the effect of the parameter changes. Furthermore, response to change in a parameter is usually nonlinear, having small impact within one range and greater effect in another. Finally, changing each parameter by a constant percentage neglects the fact that the parameters have different variances associated with their estimates.

Sensitivity analysis of differential equation models is treated theoretically by Tomović (1963) but his methods are deemed impractical for large simulation models with many parameters (Steinhorst *et al.*, 1978).

Another method due to Kitchell *et al.* (1977) is to conduct a conventional sensitivity analysis (plus and minus constant percentage) and classify system response as either high, moderate or low, and to compare this with the relative variance in the parameter estimate. This is used to indicate which parameters are not well known and when changed result in large changes in model output, thus indicating directions for future research. Table 3 shows Kitchell's method applied to his fish growth model (Kitchell *et al.*, 1977).

In models with large numbers of parameters, there is a problem inherent in the choice of which parameters to vary. One method for narrowing the field of parameters to investigate by

TABLE 3: Sensitivity analysis and comparison with field data variability (from Kitchell et al., 1977).

Parameter	Sensitivity	Availability of estimate	Need for further research
Consumption			
a_1	H	L[a,b]	H
b_1	H	M[a]	H
P	H	L[b,c]	H
T_o	M	H	L
T_m	L	L[a]	M
Q	M	M	M
Respiration			
A	H	L[b,c]	H
a_2	H	M[a]	H
b_2	M	M[a]	M
T_o	M	H	L
T_m	L	H	L
Q	M	H	L
S	M	M[a,b]	M
Waste losses			
a_1	L	L[a]	M
β_1	L	L[a]	M
γ_1	L	L[a]	M
α_a	M	L[a]	H
β_2	M	L[a]	H
γ_2	L	L[a]	M
T_p	L	H	L
B_o	L	H	L

sensitivity analysis is to perturb macroparameters (i.e., groups of 10 parameters) singly and in combination according to a fractional factorial analysis of variance (Shannon, 1975) and then conduct further sensitivity analysis singly on parameters inside those macroparameters having most sensitive response (Steinhorst *et al.*, 1978). Another method (key process identification aids in narrowing the choice of parameters for sensitivity analysis) is discussed later in this paper. Since parameters may not be independent either in the system or in the model in their effect on model output, changing parameters in various combinations as in a factorial design is preferable to changing them singly (Steinhorst, 1979; Swartzman, 1978b). The problem with such a model experiment is in the choice of parameters to vary this way and in the interpretation of the results especially when multiple output criteria are considered (Steinhorst *et al.*, 1978).

Attaching variances to parameters in a model may be justified by two alternative interpretations of the parameter estimate: either the parameter variation represents imprecision in the initial parameter estimate (i.e., the parameter is not known precisely but rather with some uncertainty due to measurement techniques), or it represents variation in the parameter due to actual differences in the process from place to place, individual to individual, or species to species. An estimate of the parameter value variance may be made from the data used for initial parameter estimation. However, the fact that parameters are not usually estimated from data on a study site but rather from values reported in laboratory experiments on related (or even unrelated) species leads to an unclear picture of what the variance of a parameter estimate means. One method of sensitivity analysis called "scientific sensitivity analysis" (L. J. Bledsoe, personal communication) takes the parameters' variability into account by picking values not necessarily at equal intervals but at intervals weighted by the variance. A version of this method, varying parameters not by plus and minus a constant percentage but by plus and minus a standard deviation estimated from the data used for parameter estimation was used by Swartzman (1978b) in a comparative study of plankton models. This method is a quantitative analog to the qualitative method in Table 3.

Sensitivity analysis can serve to corroborate a model if model behavior is relatively insensitive to parameter value changes within one standard deviation of their means. However, as shown by Table 3, sensitivity analysis is most likely to indicate future studies needed rather than serve to corroborate a model. Sensitivity analysis has also been used in understanding ecosystem behavior by indicating the relative sensitivity of ecosystem dynamics to process rates (Brylinski, 1972; Patten *et al.*, 1975). The problematic side of sensitivity analysis is that it is being used to guide future research and draw implications about ecosystem behavior *assuming* the model is true (i.e., that it has been corroborated while the fact that future research is suggested indicates that the model has not been corroborated in certain areas.

3.4 Evaluate Supporting Rationale. Evaluate supporting rationale involves a process-by-process review of the reasoning behind model equations, taking into account scientific information available about that process and whether it is considered in the equation. The process of model building should involve an evaluation of this sort in the choice of model equations. Such evaluations were presented in development of models by Noble (1975), Swartzman and Bentley (1977), Kremer and Nixon (1978), and Andersen and Ursin (1977).

The problem in choice of a preferable equation is one of detail and realism versus simplicity. Considerations on evaluating scientific rationale are (1) inclusion of details of scientific import or of known phenomena, (2) numbers of parameters to estimate and availability of data for these estimates, and (3) whether different constructs result in different model behavior and, if so, which are most realistic. This last point requires a simulation comparison of alternative candidates when there is ambivalence of choice after (1) and (2) are considered. As this is frequently the case, simulation comparison of alternative equations is usually called for (and rarely attempted). Ideas regarding simulation study of this sort for deciding on which of several models is most useful are given in Mankin *et al.* (1977).

Noble (1975), Kremer and Nixon (1978), Steele (1974), Steele and Frost (1977), and Swartzman (1978b) show simulation comparison of some alternative formulations as they contributed to the choice of a process equation. An example of a process comparison is given in Figure 4 where the light effect on phytoplankton photosynthesis is shown for formulations by Steele (1962) and Smith (1936). While each has the same number of parameters,

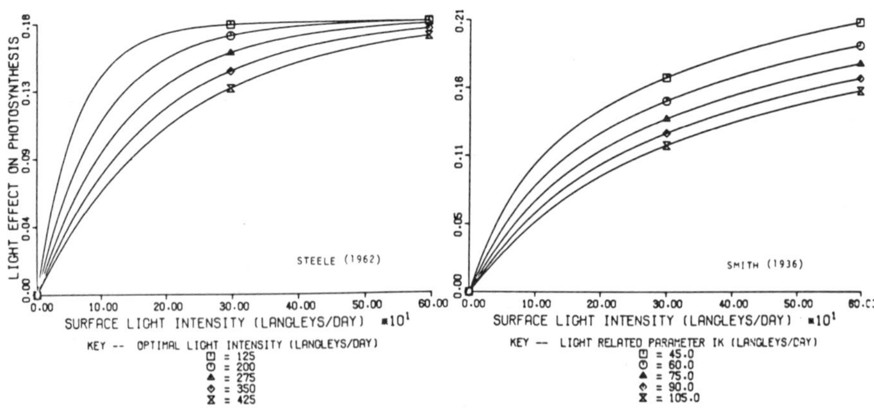

FIG. 4: Comparison between light effects on phytoplankton photosynthesis by Steele (1962) and Smith (1936).

Steele's equation includes the observed phenomenon of photoinhibition at high light intensities and, as such, appears preferable, though comparison with experimental data has indicated preference for Smith's equation in some cases (Groden, 1977). The main value in rationale evaluation, besides its role in deciding between alternative equations for the same process, is its relationship to the model objective of organizing data and information about an ecosystem.

3.5 Key Process Identification. We mentioned earlier the role that sensitivity analysis can play in indicating areas for further research. However, the choice of parameters for a sensitivity analysis is not straightforward. The author has observed that model behavior for most, if not all, models is highly sensitive to only a small fraction of the processes considered and terms these *key processes*. These processes can be uncovered simply by keeping a record of changes on parameter values and process equations during model development and then looking for which processes were changes (or had parameters changed) most often. Figure 5 shows an example from Conifer (Coniferous Forest Biome Modeling Team (1977) where the numbers of changes to the processes are depicted. It can be seen that only a small fraction of the equations were often changed.

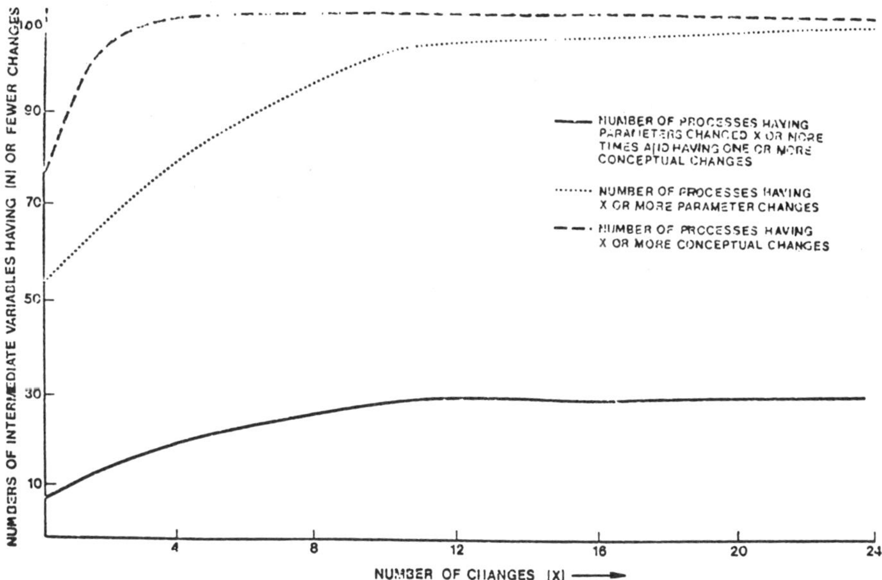

FIG. 5: Numbers of processes having varying numbers of changes in Conifer *(Coniferous Forest Biome Modeling Team,* 1977*).*

As a test of whether these key processes were indeed those most controlling model behavior, sensitivity experiments were conducted on CONIFER: one in which 20 parameters were taken only from key process and another in which parameters were chosen by one of the modelers solely on whether he thought they would be sensitive. Results of this study convinced us that key processes were indeed the most sensitive processes in the model.

Thus, key process identification can be conducted during model development and can be used to indicate future research needs by showing which of the relatively poorly understood processes have the strongest bearing on whole ecosystem dynamics.

3.6 Simulation Experiments. Simulation experiments include (1) running a model with various alternative equations substituted for a given process (see supporting rationale discussion), (2) simulating the effect of a perturbation or a management decision, and (3) designing a sensitivity experiment involving both alternative process formulations and a range of parameter values in various combinations--combining method (1) with sensitivity analysis.

As an example of an alternative equation substitution, Swartzman (1978b) looked at the effect of using Q10 (exponential) (Kremer, 1975) versus a linear (DiToro *et al.*, 1971) temperature effect on photosynthesis in a model of Lake Ontario plankton dynamics. Figure 6 shows the results of this experiment compared to five years' average data which clearly favor the Q10 formulation. (The equations were standardized to give the same photosynthesis rate at 20°C.)

Designed experiments of management alternatives or perturbations are commonly done in ecological simulation models. Examples include Andersen and Ursin (1977) (shut off fishing), Nixon and Oviatt (1973) (dam a salt marsh to eliminate tidal action), Patten *et al.* (1975) (control water level on a damned reservoir), Innis (1978) (change intensity of grazing and add fertilizer to a grassland), Sollins *et al.* (1978) (defoliation in woodlands), and O'Connor *et al.* (1975) (eutrophication of a lake).

Table 4 shows a series of perturbation experiments on a grassland performed by Innis (1978) where indices of model output are compared with normal runs of the model and with field data where available. This is one of the few examples where any corroborative data is available on the effects of perturbations on an ecosystem.

Recognizing the facts that (1) parameters in a model are related and (2) a number of interrelated processes may comprise

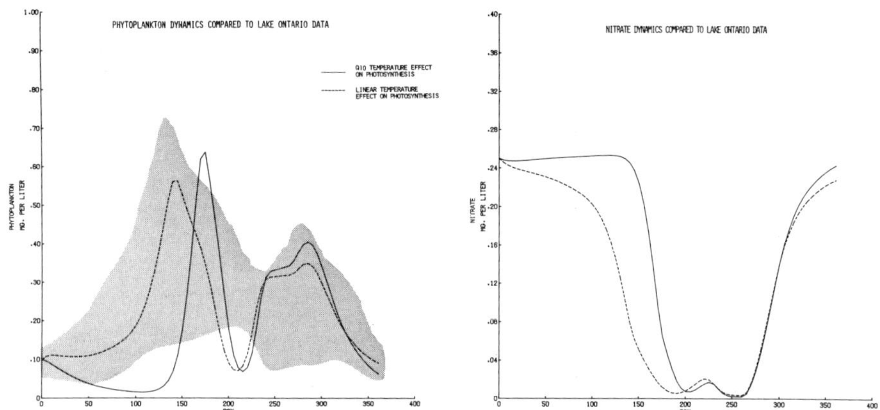

FIG. 6: Lake Ontario model comparison of **linear** *vs.* Q10 *effect of temperature on photosynthesis compared with five years of phytoplankton field data.*

TABLE 4: Perturbation experiments on grazing land model (adapted from Innis, 1978).

Exerpiment Number	Treatment Description	Water Total (cm)	Net primary production (g dry wt/m^2)	Gross primary production (g dry wt/m^2)	Live[a,b] peak (g dry wt/m^2)	Date
1	1972 weather	27.1	707.0	1086.0	170.0	256
2	Water added to keep tension > -0.8 bars	88.0	3494.0	5151.0	620.0 388.0	286
3	N fertilizer applied in June	27.1	612.0	1009.0	161.0 290.0	258
4	E and F combined--different initial conditions	85.9	3190.0	4673.0	604.0 863.0	288
5	1972 weather 0.082 cows/ha for 6 mos.	27.1	715.0	1087.0	139.0 182.0	258
6	0.2 cows/ha for 6 mos	27.1	823.0	1294.0	169.0 170.0	258
7	All consumers removed 1972 weather	27.1	711.0	1092.0	172.0	256
8	0.04 cows/ha for 365 days	27.1	689.0	1071.0	165.0	256
9	0.3 cows/ha for 140 days	27.1	552.2	901.0	119.0	188
10	Temperature raised 2°C	27.1	649.0	1060.0	163.0	258
11	Temperature lowered 2°C	27.1	706.0	1031.0	170.0	254
12	Rainfall reduced to 25% of 1972 level	6.86	49.0	134.0	63.2	174
13	1972 weather	27.1	713.0	1090.0	170.0	256

[a] Where multiple entries occur, model values are given above and field-determined means are given below.

[b] Model computes g C(grams of carbon), whereas g dry weight are quoted; g dry wt = g C · 2.5.

a field of scientific research and are observed together in the field, led to the design of simulation experiments involving groups of parameters and process equations taken in various combinations. Such an experiment can be used to investigate the combined effect of a group of related processes on model behavior and to help formulate testable hypotheses about this group of processes or part of the ecosystem.

For example, an experiment by Adams and Swartzman (unpublished manuscript) was conducted to see how various parameters controlling nutrient and light effects on phytoplankton growth influenced model behavior differently under thermal loading (increased temperature) from normal conditions. A Lake Ontario model was simulated with and without thermal loading ($\Delta T = 8°C$) with various combinations of light nutrient and temperature effects on photosynthesis. The runs were then subjected to canonical analysis and observation was made about which combinations resulted in relatively higher net production under thermal loading than under normal conditions. This led to hypotheses on which types of phytoplankton would be favored under thermal loading--a testable hypothesis. Results indicated that organisms more severely limited by phosphate but having a lower optimum light intensity for growth would be more favored under thermal loading--a testable hypothesis. Results indicated that organisms more severely limited by phosphate but having a lower optimum light intensity for growth would be more favored under thermal loading than normal conditions. This in fact has been observed in the increased occurrence of blue-green algal blooms in thermally loaded environments (most common blue-green algae being more severely phosphate limited (and nitrogen fixing) and able to grow under low light intensities better than most other algae.

Model experiments relate to a large number of model objectives. They serve to organize information about an ecosystem by showing the logical consequences of replacing one hypothesis by another, thus putting rationale evaluation into a simulation perspective. They can aid in guiding future research by helping to formulate testable hypotheses (Adams and Swartzman, unpublished manuscript; Swartzman, 1978b) and by indicating these processes that most strongly control ecosystem behavior and are least well known (as an expanded form of sensitivity analysis). An example of such a process is the effect of prey density on the zooplankton grazing rate in plankton models (Swartzman and Bentley, 1977). This process which has widely varying parameters from one model to another (Swartzman, 1978b) (and when one equation is substituted for another) results in widely different model behaviors.

Perturbations studies relate directly to objectives of investigating effects of perturbation and manipulation. An example like that in Table 4 (Innis, 1978) shows how models might be used to choose between management alternatives.

Simulation experiments as described above can contribute to an increased understanding of ecosystem behavior by helping to formulate hypotheses about the role various processes play. For example, a designed experiment conducted by Swartzman (1978b) on the zooplankton grazing process indicated that zooplankton which are voracious feeders should not graze detritus under normal conditions. This hypohtesis about an organism within the system resulted from examining which combinations of simulation runs (parameter values and equation forms) resulted in realistic behavior. Again, let us warn that using an experiment as described above both to generate theories about a system from a model and to indicate future experiments is corroborated and, as such, runs the risk of circular reasoning.

4. CONCLUSIONS

We have presented a group of model objectives for present-day simulation models and have also presented a number of methods for evaluation of these models which relate to various of the objectives. The examples given here show that many of these methods have been used and are feasible for general application and it is our conclusion that they should be used--if only as a start in addressing the question of whether simulation models are achieving their objectives.

The presentation here is only a start toward putting modeling on a more scientific footing. There are some disturbing and confusing strains--for example, the need to assume the model is a valid or corroborated model of the ecosystem in order to use some of the methods presented. However, this what if mentality can lead to some alternative means of corroborating or invalidating a model. Isn't it the way of science to assume a construct or theory true and then treat them as anxioms generating consequences which can be said to either fit or not fit observation? One dissimilarity on the surface between science and ecosystem modeling is the large number of interlocking hypotheses on which ecosystem simulation models are built.

The picture in modeling at present is like that of an archer aiming at his target. His tool (bow-simulation modeling) is well-developed and can aim the arrows true at the target (objectives) except the target is blocked by a thick brick wall (the morass of ignorance about the system). And yet a vague shimmering light penetrates the wall. The shimmering waxes and wanes and we see it to be the evaluation methods trying to clear a way through the opacity of the brick. The picture of the target is far from clear yet and it is through the mists our archer must aim for no other path is open to him. As he aims, he is distracted from the target by the shimmering colors of the mists beckoning the objectives and partially replacing the target (the temptation for the evaluation methods to become ends in themselves, losing sight of the target). At last, the arrows are loosed and they speed through the wall and beyond. And now we are left with the dilemma of not being able to see clearly if our arrows have hit the target since the very mists we aimed through are the only way we can see the target irrespective of whether we are aiming (modeling) or evaluating.

Somewhere, way off behind the target, is the real system. Even if we were able to penetrate the brick wall between modeler and target (here the dialogue is sounding like Kafka's "Before the Law"), we might find the real system blocked from our view by the target. The whole method--the scientific method as applied to the multiplicity of hypotheses comprising our model--may be giving an obscured picture of the real system.

REFERENCES

Adams, V. D. and Swartzman, G. L. (1978). Simulating the effects of increased temperature in a plankton ecosystem. *Limnology and Oceanography* (submitted).

Andersen, K. P. and Ursin, E. (1977). A multi-species extension to the Beverton-Holt theory of fishing, with accounts of phosphorous curculation and primary production. Meddelelser Fra Danmarks Fiskeri-og Havundersøgelser.

Argentisi, F. and Olivi, L. (1978). Statistical analysis of simulation models--an RSM approach. In *Summer Computer Simulation Conference, Proceedings 1978*. Newport Beach, California.

Bledsoe, L. J. (1976). *Simulation of a grassland ecosystem.* Ph.D. dissertation, Colorado State University, Fort Collins.

Brylinski, M. (1972). Steady state sensibility analysis of energy flow in a marine ecosystem. In *Systems Analysis and Simulation in Ecology, Vol. II,* B. C. Patten, ed. Academic Press, New York. 81-139.

Caswell, H. (1976). The validation problem. In *Systems Analysis and Simulation in Ecology, Vol. IV,* B. C. Patten, ed., Academic Press, New York 313-325.

Coniferous Forest Biome Modeling Group (1977). Conifer: a model of carbon and water flow through a coniferous forest. Coniferous Forest Biome. University of Washington and Oregon State University, Bulletin 8.

Desormeau, C. J. (1978). Mathematical modeling of phytoplankton kinetics with application to two Alpine lakes. CEM Report 4. Center for Ecological Modeling, Rensselaer Polytechnic Institute, Troy, New York.

DiToro, D. M., O'Connor, D. J., and Thomann, R. V. (1971). A dynamic model of the phytoplankton populations in the Sacramento-San Joaquin Delta. In *Nonequilibrium Systems in Natural Water Chemistry*. Advances in Chemistry Series 106, American Chemical Society. 131-150.

DiToro, D. M., O'Connor, D. J., Thomann, R. V., and Mancini, S. L. (1975). Phytoplankton-zooplankton-nutrient interaction model for western Lake Erie. In *Systems Analysis and Simulation in Ecology, Vol. III*, B. C. Patten, ed. Academic Press, New York 424-473.

Eggers, D. M. (1975). *Limnetic feeding behavior of juvenile sockeye salmon in Lake Washington and predator avoidance*. Ph.D. dissertation, University of Washington, Seattle.

Garfinkel, D., McLeod, J., Pring, M., and DiToro, D. (1975). Application of computer simulation to research in the life sciences. *Simulation Today*, 5, 17-20.

Garratt, M. (1975). Statistical techniques for validating computer simulation models. US/IBP Grasslands Biome Technical Report 286. Colorado State University, Fort Collins.

Innis, G. S. (1975). The role of total systems models in the grassland biome study. In *Systems Analysis and Simulation in Ecology, Vol. III*, B. C. Patten, ed. Academic Press, New York. 13-47.

Innis, G. S., ed. (1978). *Grassland Simulation Model*. Ecological Studies, Vol. 26. Springer-Verlag, New York.

Kitchell, J. F., Koonce, D. F., O'Neill, R. V., Shugart, H. H., Magnuson, J. J. and Booth, R. S. (1974). Model of fish biomass dynamics. *Transactions of the American Fisheries Society*, 4, 786-796.

Kitchell, J. F. and Stewart, D. J. (1977). Applications of bioenergetics model to yellow perch (*Perca flavescens*) and walleye (*Sitzostedion vitreum vitreum*). *Journal of the Fisheries Research Board of Canada*, 34, 1922-1935.

Kremer, J. (1975). *Analysis of a plankton-based temperate ecosystem: an ecological simulation of Narragansett Bay.* Ph.D. dissertation, University of Rhode Island.

Kremer, J. N. and Nixon, S. W. (1978). A coastal marine ecosystem: simulation and analysis. In *Ecological Studies, Analysis and Synthesis, Vo. 24,* W. D. Billings, F. Golley, O. L. Lange, and J. S. Olson, eds, Springer-Verlag, Berlin.

Lehman, J. T., Botkin, D. B. and Likens, G. E. (1975). The assumptions and rationales of a computer model of phytoplankton population dynamics. *Limnology and Oceanography,* 20, 343-364.

MacCormick, A. J. A., Loucks, O. L., Koonce, D. F., Kitchell, J. F., and Weiler, P. R. (1972). An ecosystem model for the pelagic zone of Lake Wingra. Eastern Deciduous Forest Biome Memo Report 72-122.

Mankin, J. F., O'Neill, R. V., Shugart, H. H. and Post, B. W. (1977). The importance of validation in ecosystem analysis. In *New Directions in the Analysis of Ecological Systems,* George S, Innis, ed. Simulation Council, La Jolla, California. 63-71.

Nixon, S. W. and Oviatt, C. A. (1973). Ecology of a New England salt marsh. *Ecological Monographs,* 43, 463-498.

Noble, I. (1975). *Computer simulations of sheep grazing in the arid zone.* Ph.D. dissertation, University of Adelaide, Australia.

O'Connor, D. J., DiToro, D. M. and Thomann, R. V. (1975). Phytoplankton models and eutrophication problems. In *Ecological Modeling in a Resource Management Framework,* S. Russel, ed. Resources for the Future, Inc., Washington, D. C.

Overton, W. S. (1973). Sensitivity analysis as propagation of error and model validation. Coniferous Forest Biome Internal Report No. 1. University of Washington, Seattle.

Park, R. A. *et al.* (1975). A generalized model for simulating lake ecosystems. *Simulation,* 23(2), 33-50.

Patten, B. C., Egloff, D. A., and Richardson, T. H. (1975). Total ecosystem model for a cover in Lake Texoma. In *Systems Analysis and Simulation in Ecology, Vol. III,* B. C. Patten, ed. Academic Press, New York. 206-415.

Regier, H. A. and Rapport, D. J. (1978). Ecological paradigms, once again. *Bulletin of the Ecological Society of America*, 59, 2-6.

Scavia, D., Eadie, B. J. and Robertson, A. (1976). An ecological model for Lake Ontario model formulation, calibration and preliminary evaluation. NOAA Technical Report ERL 371-GLERL 12.

Skellam, J. G. (1972). Some philosophical aspects of mathematical modeling in empirical science with special reference to ecology. In *Mathematical Models in Ecology*, J. N. R. Jeffers, ed. Blackwell, Oxford. 13-28.

Smith, E. L. (1936). Photosynthesis in relation to light and carbon dioxide. *Proceedings of the National Academic of Sciences, USA*, 22, 504-511.

Sollins, P., Goldstein, R. A., Mankin, J. B., Murphy, C. E., and Swartzman, G. (1978). Applicability and behavior of some complex ecosystem models. In *Woodland Synthesis*, Dowden, Hutchinson, and Ross, eds. (in press).

Steele, J. H. (1962). Environmental control of photosynthesis in the sea. *Limnology and Oceanography*, 7, 137-150.

Steele, J. H. (1974). *The Structure of Marine Ecosystems*. Harvard University Press, Cambridge.

Steele, J. H. and Frost, B. W. (1977). The structure of plankton communities. *Philosophical Transactions of the Royal Society of London*, 280, 485-534.

Steinhorst, R. K. and Garratt, M. (1976). Validation of deterministic system simulation models. In *August ASA Proceedings, Statistical Comp Section*.

Steinhorst, R. K., Hunt, H. W., Innis, G. S., and Haydock, K. P. (1978). Sensitivity analysis of the ELM model. In *Grassland Simulation Model*, G. S.Innis, ed. Springer-Verlag, New York. 231-255.

Swartzman, G. and Bentley, R. (1977). A comparison of plankton models with emphasis on application to assessing non-radiological nuclear plant impacts on plankton in natural ecosystems. Technical Report UW-NRC-1, Center for Quantitative Science, University of Washington, Seattle.

Swartzman, G. L. (1978a). Simulation modeling of material and energy flow through an ecosystem: methods and documentation. *Journal of Ecological Modeling* (in press).

Swartzman, G. L. (1978b). A comparison of plankton simulation models emphasizing their applicability to impact assessment. *Journal of Environmental Management* (in press).

Tawari, P. V. (1974). *A systematic procedure for validating and evaluating dynamic models*. Ph.D. dissertation, University of Washington, Seattle.

Thomann, R. V., DiToro, D. M., Winfield, R. P. and O'Connor, D. J. (1975). Mathematical modeling of phytoplankton in Lake Ontario. I. Model development and verification. Grosse Ile Laboratory, National Environmental Research Center, Grosse Ile, Michigan.

Tomović, R. (1963). *Sensitivity Analysis of Dynamic Systems*. McGraw-Hill, New York.

Wiegert, R. G.(1975). Simulation. Models of ecosystems. *Annual Review of Ecology and Systematics*, 6, 311-338.

Woodmansee, R. G. (1978). Critique and analysis of the grassland ecosystem model ELM 73. In *Grassland Simulation Model, Ecological Studies, Vol. 26*, G. S. Innis, ed. Springer-Verlag, New York.

[*Received January* 1979. *Revised April* 1979]

G. P. Patil and M. Rosenzweig, (eds.),
Contemporary Quantitative Ecology and Related Ecometrics, pp. 319-341.

MODELING COASTAL, ESTUARINE, AND MARSH ECOSYSTEMS: STATE-OF-THE-ART

RICHARD G. WIEGERT

Department of Zoology
University of Georgia
Athens, Georgia 30602 USA

SUMMARY. Simulation models of marsh-estuarine and coastal marine ecosystems can 1) suggest new hypotheses, 2) predict the results of perturbations and 3) summarize large complex data sets. The seven models discussed in this paper were chosen to illustrate the relationship between these three overlapping objectives, the type of model (theoretical versus empirical) and the amount and kind of biological information included in construction of the model.

KEY WORDS. estuary, marine, salt marsh, simulation, model, ecosystem, ecology, population, succession.

1. THE APPROACH OF THIS REVIEW

Simulation models of coastal waters, estuaries, and salt marshes are fewer and tend to be less complete than those of freshwater and terrestrial ecosystems. A recent comprehensive review of mathematical modeling in ecology, O'Neill, Ferguson, and Watts (1977), for example lists 946 papers, none of which deal with models of estuaries or marshes. The index does not contain the words *coastal, marine,* or *salt marsh*. The earliest coastal marine and estuarine models have emphasized the open water food chain comprising phytoplankton, grazers, and predators. Only recently has serious consideration been given to the detritus-benthic part of these communities. The few models devoted to the salt marsh have been more comprehensive in coverage but are site specific, often with heavy emphasis on management or exploitation of the marsh.

A few models have also been used to formulate testable hypotheses and guide the course of basic research. By way of example I hope to show a bit of the diversity of approach, scope and objectives that characterize those models of nearshore marine and salt-marsh ecosystems published or currently under development. I want to compare the form, structure, and particularly the objectives of each model, emphasizing not only the differences but, more important, their similarities. Coastal marine and estuarine ecosystems are large; they are difficult to study, and their dynamic behavior is conditioned strongly by physical influences such as tides, storms, currents, salinity changes, and variation in temperature. Ideally, they are studied by multidisciplinary teams of scientists. This is precisely the kind of cooperative study that can most benefit from concurrent modeling with feedback from the model simulations to the field/laboratory experiments and vice versa. I shall try, wherever possible in this paper, to reemphasize this point and discuss mechanisms for facilitating this interaction.

2. MODEL ORGANIZATION

A model represents an abstraction of the real world. It is a summary statement of knowledge and assumption. An empirical model summarizes the correlation between measured variable and system behavior; it does not explain and is useful only to the extent its predictions are correct. A dynamical model is a summary based not on observed correlation, but on independent measurement, observation, and intuition. The dynamic model attempts explanation. In this respect is resembles theory or hypothesis. But more properly we should regard the ecosystem model as an entire array of hypotheses. In evaluating the model, we must examine each of the assumptions. The model must be used to generate individual testable hypotheses. By testing these and accepting or rejecting in favor of alternative hypotheses, the model is changed in the direction of a truer representation of whatever part of the real world it was designed to represent. I emphasize the distinction between this process and 'fine tuning' a model by simply changing it in whatever manner will best enable it to conform to observed data. The former procedure leads to understanding and explanation; the latter leads nowhere.

Discussion of the differing approaches and objectives of each model flows best from consideration of organization. The organization of a model comprises both structure and function. This distinction of structure versus function has been used for some time as a useful way of categorizing certain characteristics or attributes of ecosystems. Unfortunately, these two terms, as is the case with other terms in the growing science of ecology, came into general use without the benefit of critical examination or rigorous definition. By *structure* the ecologist often means the things in the ecosystem, species, members of individuals, biomass, etc. (Curiously, there is seldom explicit inclusion of the abiotic 'things' like soil, nutrients, etc.) *Function* is usually used to describe both the pathways of matter and energy transfer and the variation in flux with time. For example, nutrient cycling and energy flow are considered functional attributes of ecosystems. Function seems to be synonymous with system behavior. Hill and Wiegert (in press) pointed out the serious conflict of this view of structure vs. function as used in ecology and the definitions used in engineering.

In systems science, structure describes the state variables (boxes) in the system and the pathways (arrows) of interaction. Thus the structure of the system is an abstraction, existing in the absence of matter or energy. Structure is an organizational property that changes relatively slowly. In the case of ecosystems, structure changes only as a result of major evolutionary change or permanent physical change. The trophic web is a good example of the structure of an ecosystem. The food web does not change with change in number, biomass, etc. Such changes define the *behavior* of the system, an attribute that results from the interaction of structure (empty niches and pathways of transfer) with function (species and matter/energy flows), that is, the filling of the boxes with numbers and biomass and of the arrows with actual matter or energy transfers. The structure *of* a system is quite distinct from the structures *in* the system. The latter are equivalent to the system structure of ecology. If one is asked to enumerate the structures in a city, a list of numbers and kinds of houses, factories, stories, lamp posts, etc., will suffice nicely. If, on the other hand, you are asked to describe the structure of the city, the answer includes the various categories of things the city can manufacture, sell, import, and transform as well as the ways these materials and energy are transformed and transported. Thus, structure *of* a system can be separated from the individual structures *in* the system. The latter, however, have individual functions and therefore the function of the entire system depends directly on the function of all the individual structures. Function logically includes the numbers, biomass and concentrations (the contents of the boxes) as well as the magnitudes of the flows associated with the arrows.

3. MODEL SYNOPSES

I discuss and compare seven models, four dealing with coastal marine communities or the estuary proper and three concerned specifically with salt marshes. The first, Steele (1974) concerns fisheries production in the North Sea and is one of the earliest attempts to construct a model of an ecosystem combining a phyto-synthesis model with a model of grazing food chain (phytoplankton-zooplankton-fish). Kremer and Nixon (1978) used the same approach in constructing a model of an estuary (Narragansett Bay). Their model is considerably more detailed than Steele's but remains basically a linear phytoplankton-zooplankton-fish food chain model. The role of microorganisms other than algae was not included in either the Steele or Kremer-Nixon models. Pomeroy (in press), in a conceptual model of the southeastern shallow-water coastal marine community, forcefully argues for an emphasis on the detrital part of the food web. Pomeroy presents a static box and arrow flow model (a simulation is under development, Pomeroy and Hagner, in preparation). McKellar (1977) described a unique model designed to predict the effect on an estuary of thermal pollution by a power plant.

The Salt Marsh models include: (1) a successional model of salt marsh in Chesapeake Bay, Va. (Zieman and Odum, 1977), (2) the important question of man-marsh interactions in the Louisiana Gulf coast salt marshes by Hopkinson and Day (1977), and (3) versions of a *Spartina* marsh model (Wiegert *et al.*, 1975; Wiegert and Wetzel, 1979; and Wiegert, unpublished).

4. METHODS OF COMPARISON

To facilitate understanding and to emphasize certain important aspects of structure and function in the different models, I have redrawn the flow diagram of each model using a simple notation. Variables of state, and boxes, tanks, etc. of model diagrams are represented by a simple irregular closed contour. A flow of energy or material is indicated by a solid arrow. A gate across this solid arrow indicates control of the flow by *other than the donor state variable*. Arrows with no gates are linear donor-controlled flows; the flow is the product of a rate coefficient times the value of the donor variable. The source of the control for flows with a gate is indicated by dotted lines going from the control variable to the gate. This Forrester notation is sufficient to show those characteristics of each model that I want to emphasize in this paper.

In addition to redrawing the flow diagram of each model, I have simplified all the models by eliminating variables of state, flows, or controls wherever these were not needed to illustrate the essentials of the model.

Wherever possible I consider in order the topics: (1) general description of the model, (2) objectives and/or philosophy, (3) the structure of the model, (4) the type of control(s) employed, (5) data needed and how acquired, (6) conclusions, and (7) special comments or criticisms.

5. THE STEELE NORTH SEA MODEL

Steele (1974) published a small book describing the step-by-step construction of a model of the North Sea fishery involving a food chain from phytoplankton to calanoid copepods. The model tracks carbon flow, but the source is always expressed in carbon equivalents of the nitrogen transferred. Steele deliberately chose to model a simple food chain with complex trophic interactions as opposed to a diverse food web with simplified equations of interaction.

The three major objectives are summarized in the following questions:

1) Can the model predict conditions in the North Sea phytoplankton and zooplankton and the dynamics of populations without incorporating stochastic processes or patchiness in distributions?

2) How can behavioral responses by organisms influence or regulate the system productivity?

3) How does the North Sea pelagic system compare with terrestrial ecosystems?

The structure of the Steele model is simple (Figure 1). Three compartments represent the source of carbon, the phytoplankton, and the calanoid copepods (both weight and numbers are tracked). The effect of the predatory fish (pelagic species, mainly herring and mackrel) is represented in the model as loss from a variable mortality to the copepods.

Nitrogen is implicitly a control of primary productivity. Mixing of N from the lower water levels into the upper photic zone was modeled as the product of the water exchange times the relative difference in N concentration in the two layers. Light and temperature were constant over the ranges of simulation explored. Thresholds were used at various points in the model. A resource level threshold regulated uptake of N by phytoplankton. This group was in turn provided with a refuge threshold preventing overgrazing by copepods. Two different refuge thresholds, numbers and biomass, protected the copepods against predatory overgrazing.

Steele discusses some of the problems encountered in the experimental measurement of parameters, most notably with respect to copepod respiration rates. To obtain an adequate measure of oxygen uptake, the copepod density in the respiration bottles had to be increased from 10-100 times the usual density in the open sea. Because food was scarce, feeding soon stopped. Indeed, no feeding

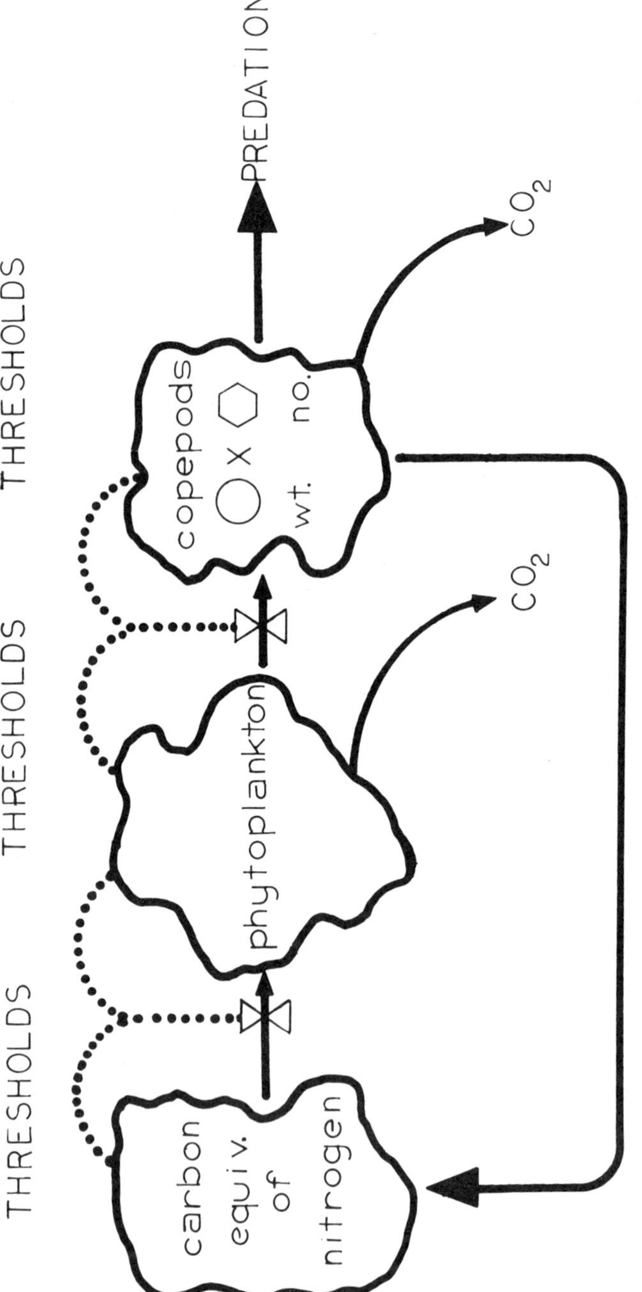

FIG. 1: Steele model of the North Sea trophic web.

was possible in those experiments where filtered seawater was used. Steele's data showed diminished food or no food caused a lowering in the respiratory rate of *Calanus*. In general, however, his discussion of parameterization concerns justification of the particular values.

Steele stressed in conclusion the inadequacies of the data, the lack of temperature and light as controls, and the failure to incorporate patchiness in the model. The defects, however, could be remedied. Given these limitations, the model simulations appeared as reasonable representations of conditions in the North Sea. Three conclusions with respect to stability of the system were: (1) A relationship between nutrient concentration and nutrient uptake was necessary for stability, but the form of this relationship was unimportant. (To what extent this conclusion is forced by the control functions actually used was not discussed.) (2) The system survives but in an unrealistic mode when the copepod refuges against predation were removed. (3) The major stabilizing factor was resource dependent control of grazing by the copepods. If one accepts terrestrial plant availability not limiting grazers, comparison of the North Sea to terrestrial ecosystems was difficult (see Hairston, Smith, and Slobodkin, 1960). This model of Steele's is, however, too simplistic and contains too many gaps to be used for such a complex comparison as that between terrestrial and open water marine communities. Steele discusses some of these problems. (See also the comparison of terrestrial vs. open water communities by Wiegert and Owen, 1971, in proposing an explanation for the questions posed by Hairston, Smith, and Slobodkin, 1960).

6. THE KREMER-NIXON NARRAGANSETT BAY MODE

Kremer and Nixon (1978) - see also Nixon and Kremer (1977) - described a large multi-simulation model of Narragansett Bay, a New England estuary. The several parts of the total model couple the ecological fluxes and interactions with submodels describing oil pollution, hydrodynamics, and changes in oxygen, salinity, and temperature. The model simulates energy flow modified by available nutrients. The phytoplankton-zooplankton-predator food chain is emphasized; the microorganism-detritus-benthos food web is given little attention because of lack of information. In this respect the model rationale differs from that of Steele (1974) who assumed detrital-based food chains were unimportant. The organization of the Kremer-Nixon model differs from that of the Steele model principally by being coupled to a detailed hydrodynamic model and by incorporating varying light and temperature and time delays.

The major objective was to develop a model able to advance basic understanding of the behavior of the estuary. The ecological model was to simulate spatial and temporal variation in the nutrients and plankton, a worthwhile objective in view of the important stabilizing effect of heterogeneity (Smith, 1972) and the variable effect of time delays on stability.

The structure of the ecological model is outlined in Figure 2. Only those pathways and compartments that seemed of fundamental importance to the phytoplankton-zooplankton food chain were retained. To facilitate simulation of spatial variation Kremer and Nixon divided the Bay into 8 different regions within which model parameters and functional controls could be determined.

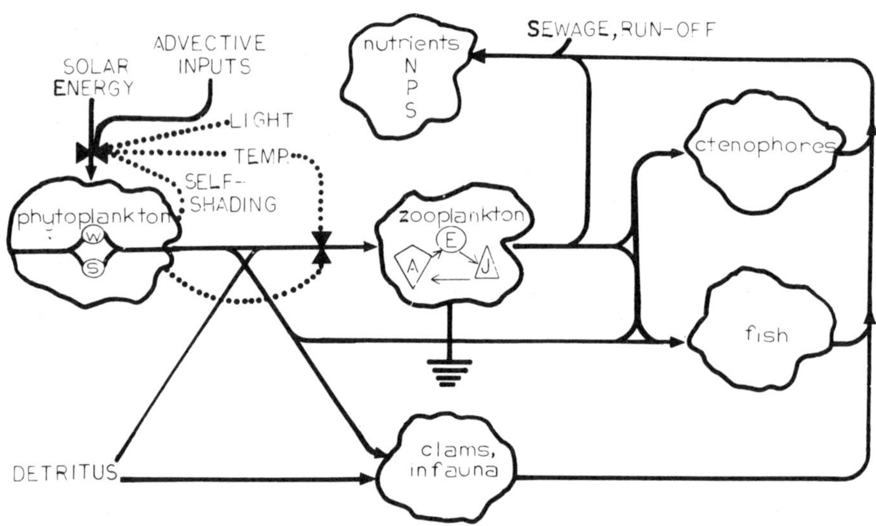

FIG. 2: Kremer-Nixon model of Narragansett Bay, Rhode Island.

A maximum rate of transfer for each pathway defined by temperature was reduced by unitless fractions as functions of the other regulatory factors. This is a realistic approach that can, indeed must, be taken when modeling the dynamic behavior of populations. It forced Kremer and Nixon to consider a severe shortcoming of the most common means of representing the effects of a limiting nutrient in aquatic population models, the Monod equation. The particular expression they discussed is

$$f(X_i) = X_i/(k_s + X_i) \tag{1}$$

where

X_i = the value of some limiting factor (light, nutrient, temperature, etc.)
k_s = the half saturation value of X_i.

The half saturation value is that value of X_i where $f(X_i) = 0.5$ and thus the realized transfer coefficient is 0.5 the maximum. But $f(X_i)$ cannot equal 1 for $X_i < \infty$. Kremer and Nixon noted that several such functional controls, multiplied together, would return a realized rate far below the maximum even when the several resources (values of X) were all optimum. They thus rejected the multiplicative coupling of limiting nutrients and used the Monod equation to calculate a limiting factor only for the nutrient in shortest supply at a particular time. The effects of light and temperature were modeled with specific functions derived from experimental data. A refuge for phytoplankton is implied by the exponential evaluation of the rate functions (as opposed to the finite difference method). Kremer and Nixon, with Steele, recognized the vital stabilizing role of density-dependent response by grazers to phytoplankton availability. The predation drain on the zooplankton was obtained from field data. Predator dynamics were not included. Direct interference, i.e., competition for space caused by overcrowding, was not modeled. The regeneration of nutrients by the clams and infauna were decoupled from the N, P, Si compartment, further reducing the dynamic impact of any detritus or benthic dynamics on the overall behavior of the system.

The dynamics of 6 state variables were simulated: phytoplankton, zooplankton, NH_3, NO_2-NO_3, SiO_3, and PO_4.

Simulation runs using the Bay model convinced Kremer and Nixon that most of the primary and secondary production was used (grazed or degraded) within the confines of Narragansett Bay. Little organic material was exported to Rhode Island Sound.

Despite its limitations the model seemed to give a reasonable simulation of behavior by the Narragansett ecosystem. The entire model continues to serve as an adequate tool of management for the Bay. It should also prove useful as a device for suggesting hypotheses as well as stimulating thought about the problems of modeling itself. The concern of Kremer and Nixon about the use of the Monod function is an example of the latter. The validity of using this particularly hyperbolic function must be decided in each case and has been discussed elsewhere (see Wiegert, 1975). Of concern here is how to solve the difficulty perceived by Kremer and Nixon. Continuous mathematical functions such as the Monod (which was developed as a model of the kinetics of enzymes) will often fail when used to simulate the dynamics of populations where physiology and behavior can change the rules at any given point. The fact of a maximum rate of uptake implies the existence of at least an upper or satiation threshold at some resource concentration short of infinity. Similarly, one might reasonably expect some lower level short of zero where the resource is unavailable or where it is so difficult to obtain that the consumer abandons the effort! Why not incorporate these biological realities into the equation, thereby making possible the simultaneous limitation by more than one material resource. Certainly this is the case in many real situations. Rewrite the Monod function in the form

$$f(X_i) = 1 - k'_s/k'_s + X_i \quad ; \tag{2}$$

then a modification of the second term gives

$$f(X_i) = \left[1 - \frac{k'_s - \gamma_{ij}}{k'_s + X_i - 2\gamma_{ij}}\left(\frac{\alpha_{ij} - X_i}{\alpha_{ij} - \gamma_{ij}}\right)_+\right]_+ \tag{3}$$

where $(\cdot)_+ = \begin{matrix} 0 \text{ if } (\cdot) \leq 0 \\ (\cdot) \text{ if } (\cdot) > 0 \end{matrix}$,

α_{ij} = satiation threshold concentration,

γ_{ij} = refuge threshold concentration.

Note that as $\alpha_{ij} \rightarrow \infty$, an assumption of the Monod function, $f(X_i) \rightarrow 1 - k'_s/k'_s + X_i$ and $k'_s \rightarrow k_s$. Otherwise k'_s is found by solving the equation with $X_i = k_s$. For densities of $X_i \leq \gamma_{ij}$, $f(X_i) = 0$. Using this function, any number of limiting factors can be employed multiplicatively.

7. THE POMEROY CONTINENTAL SHELF MODEL

Pomeroy (in press) developed a conceptual model of the marine food web of the continental shelf off the southeastern coast of the U.S.A. The published model is a static box and arrow structural diagram with function represented only by the annual magnitude of fluxes of energy. A dynamic simulation model of this system is to be based on Figure 3 (Pomeroy and Hagner, unpublished). The simulation model is being constructed following the principles of choosing biologically realistic, definable, and measurable parameters as suggested by Wiegert *et al.* (1975) and Wiegert and Wetzel (1979) for salt marsh models.

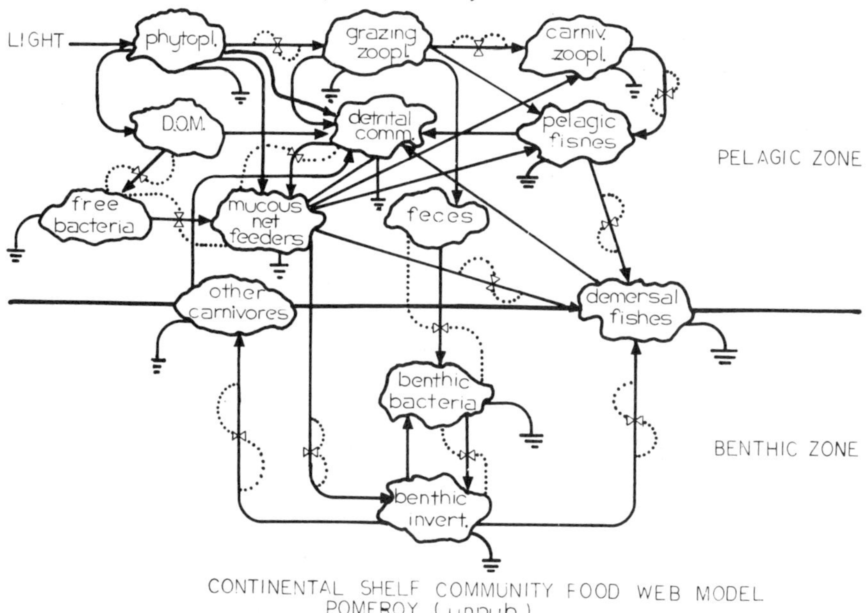

FIG. 3: Pomeroy model of the continental shelf trophic web off the southeastern coast of the United States.

Pomeroy's major objective in constructing both the static and dynamic models was to investigate the consequences of varying the importance of the detrital-benthic part of the food web as opposed to the open water grazing-zooplankton trophic pathways. Pomeroy's studies of the continental shelf community invalidated the idea of a minor detrital food web. Pomeroy found only 25-50% of the production by phytoplankton eaten by grazers; the remainder entered the detrital food chains. Because a portion of even the grazer intake also ends up as detritus, through egestion, production of the latter accounts for far more than half the annual net primary production. The model represents all of the important feeding groups in the community (Figure 3) has controls of both donor and recipient densities wherever needed. These controls, in contrast to the Steele and Kremer-Nixon models include not only density-dependent competition for limited material resources (exploitative competition) but also competition for space when overcrowding occurs (interference competition). The data and conclusions considered here are based on simple manipulations of the stati model (Pomeroy, in press).

Net primary production of the continental shelf community (measured with 14-C methods) approximated 1000 kcal m^{-2} yr^{-1}. There are four pathways whereby this production is transferred to the other groups in the ecosystem (Figure 3): (1) grazing by zooplankton, (2) cpature by mucous net feeders, (3) secretion as dissolved organic matter (DOM), or (4) transformation into non-living particulate organic matter and associated bacteria (detrital community). An important and novel aspect of this model is the separation of the bacteria into three spatially and operationally distinct groups: (1) those associated with the sediments (benthic), (2) those living free in the water column, and (3) those living on detritus particles in the water column.

In one set of experimental manipulations of the model, Pomeroy used efficiencies of energy transfer similar to those employed by Steele (1974) in the model of the North Sea. By varying the proportion of the 100 kcal net primary production among the four pathways, Pomeroy generated three different examples of behavior by the system. In Case I only small amounts (.100 kcal each) were allotted to detritus, DOM, and mucous net feeders; the bulk of the net primary production (700 kcal) went to grazing zooplankton. This was the nominal case against which the other examples were compared. No matter how energy was directed from the detritus or the DOM components, production at the carnivore level changed little. In case 2 production was assumed to be mostly nanoplankton available to net feeders (700 kcal to this group, only 100 each to the others). A major result was a severe reduction in production by the benthos. In case 3, a late stage phytoplankton bloom was modeled with the bulk (500 kcal) of the net production going directly to detritus, 300 kcal to DOM, and only 100 kcal each to zooplankton and net feeders. This favored production of the benthos and demersal fishes. An interesting feature of all three examples was the failure of these rather major shifts in trophic transfer of net primary production to eliminate any group or even reduce it below what might have seemed just a bad year class. However, Pomeroy warned that the efficiencies used were minimal. Further reduction such that all are equal to the carnivore 10% renders the community energy insufficient to support all groups.

Pomeroy argued that some efficiencies employed by Steele may be too low rather than too high. A second set of manipulations with the model was generated using gross growth efficiencies of 30% for grazers, net-feeders, benthic invertebrates, and carnivorous zooplankton. Bacterial efficiency remained at 50% and efficiency of fishes and other carnivores at 10%. Dividing the production as 700 kcal to grazers and 100 each to the remaining groups gave increases in the production of the terminal groups with carnivorous zooplankton and pelagic fishes making the greatest gains. Dividing the primary production as 450 kcal to grazing zooplankton, 150 kcal to net feeders, and 200 kcal each to detritus and DOM gave a more equable increase in pelagic and demersal fishes. Finally late bloom conditions were modeled; the bulk (500 kcal) of production goes to detritus, 300 kcal to DOM, and only 100 each to grazing zooplankton and net-feeders. However, this time the benthic and demersal groups declined instead of increasing, the result of a much higher proportion of the detritus going to the pelagic chains via the increased efficiency of the mucous net feeders.

Simple models of this kind do not tell what conditions occur in the real world, but can suggest the kind of conditions that might result in large changes in the marine community purely because of shifts in available energy. Pomeroy feels the addition of the capability to simulate a wealth of population behaviors with time lags, as well as nutrient and space control will further increase the utility of the model as a tool to formulate testable hypotheses about the marine ecosystem.

8. THE McKELLAR CRYSTAL RIVER ESTUARY MODEL

McKellar (1977) described a simulation model of the Crystal River Estuary, Florida that couples the estuarine bay ecosystem with a coastal power plant. The Crystal River Estuary, on the Gulf coast of Florida north of Tampa Bay, has a salinity range of 17-30°/oo and an annual ambient temperature range of 14-30°C. The estuary consists of an inner bay with sea grass beds often exposed at low tide. The outer bays are covered to 2m at low tide and phytoplankton production is important. This model had as the major objective the prediction of the effect of a specific perturbation. The bay ecosystem was to be used to dispose of a pollutant, hot waste water from the power plant. The plant was already in operation, pumping 3.5×10^6 m^3 of water day^{-1} raising the temperature of the thermal plume 5-6°C above ambient. The question McKellar asked of the model was prediction of the effects on the community of doubling the power output (and waste water production) of the plant and the consequent increase in temperature of the thermal plume of an additional 1°C above ambient (to 6-7°C).

The model is an energy flow model with the additional capability of tracking the flow and storage of available phosphorus (Figure 4). Differential equations were developed for each flow. External driving forces of sunlight water temperature and imports of phytoplankton, zooplankton, and phosphorus were given sinusoidal functions representing seasonal variations. The nominal model was tested by comparing simulated behavior (total energy flow and component standing stock) with observed values from the outer control bays protected from thermal discharges.

Figure 4 shows the variety of controls used in the model. A Monod type approach to a maximum rate of primary production was dependent on available phosphorus and affected by light and water temperature. Control of ingestion by zooplankton was accomplished

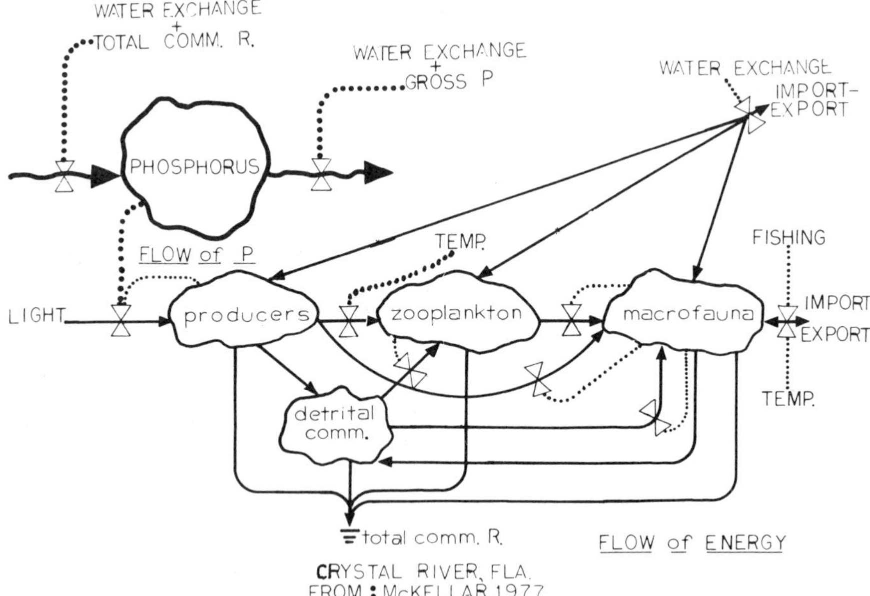

FIG. 4: McKellar model of the Crystal River estuary, Florida.

by a novel combination of availability of phytoplankton times the respiration of the zooplankton. The respiration of the latter was in turn made a function of zooplankton standing stock and of water temperature. The major shortcoming of control in the model was the absence of any thresholds.

As with most ecosystem models of this degree of complexity, the data were gathered from a number of sources, literature as well as experiments, with varying degrees of precision. Initial conditions were obtained from measurements on the outer control bays. Effects of the existing power plants were simulated with the increased water exchanges and temperatures (15-35°C) caused by the discharge. Total respiration was higher throughout the year with a maximum 20% in summer. Predicted producer biomass was lower, but the model simulation did not reach the observed 15-20% reduction in the discharge bay. Zooplankton were lower in spring and summer due to entrainment mortality. Benthic invertebrates and fish were about 40% lower in summer and 60-70% higher in winter due to temperature-caused migration.

The new nuclear powered power plant was expected to double the volume of heated water pumped into the estuary and increase the average temperature of the plume by 1°C. The amount of unmixed water that reached the outer bay would increase greatly and was calculated to be capable of causing a 6-7°C increase above ambient in the outer bay. The simulation was therefore programmed for a summer maximum of 37°C and a winter minima of 19°C. Total water exchange increased one-third. Comparing the new simulation with the simulated effects of the pre-existing power plant effects showed: (1) there was a stimulation of gross primary production by an average of 30% during winter and spring; (2) total respiration was increased by 30% in winter and 10% in summer; (3) Zooplankton biomass suffered a further 40% reduction in summer caused by increased entrainment mortality. McKellar discusses the various possible effects of increased temperature depending on the availability of auxiliary sources of energy (wave action, tidal exchange) and nutrient stocks. When these are large, temperature increases usually increase the normal ecosystem biological processes. The model in this instance proved a useful tool to evaluate the effect of changes in the thermal additions to the bay and permitted a rational discussion of the advantages and disadvantages of alternative means of waste disposal.

9. THE ZIEMAN-ODUM MODEL OF MARSH SUCCESSION

Zieman and Odum (1977) presented a model of ecological succession and production in two areas of salt marsh along Chesapeake Bay. The marshes consist of three zones: a low marsh area fringing the tidal creek, an ecotone, and a high marsh area where the frequency of tidal flooding is markedly less frequent. Mean tidal amplitude at these two sites was 1 m. Salinity ranged from 0-15°/oo. Four species of vascular plants were sampled in the low marsh: dominant, smooth cordgrass (*Spartina alterniflora*), the two high marsh dominants *Spartina patens* and *D st chilis spicata*, and *Aster tenuifolius*. All four species occur in the ecotone. The aster was a minor component.

The successional model was used to predict the conditions and time sequence necessary for the development of a salt marsh, and the kind of spoil bank that must be constructed if a marsh was to form.

A point model of each species was constructed and coupled with other species point models to give simulated dynamics in both space and time (Figure 5). The model is empirical, developed directly from time series data on colonization and growth correlated with measurement of the physical and biological regulating factors. The resulting model(s) is (are) site specific and nonexplanatory. The selection of parameters for inclusion in the model was based

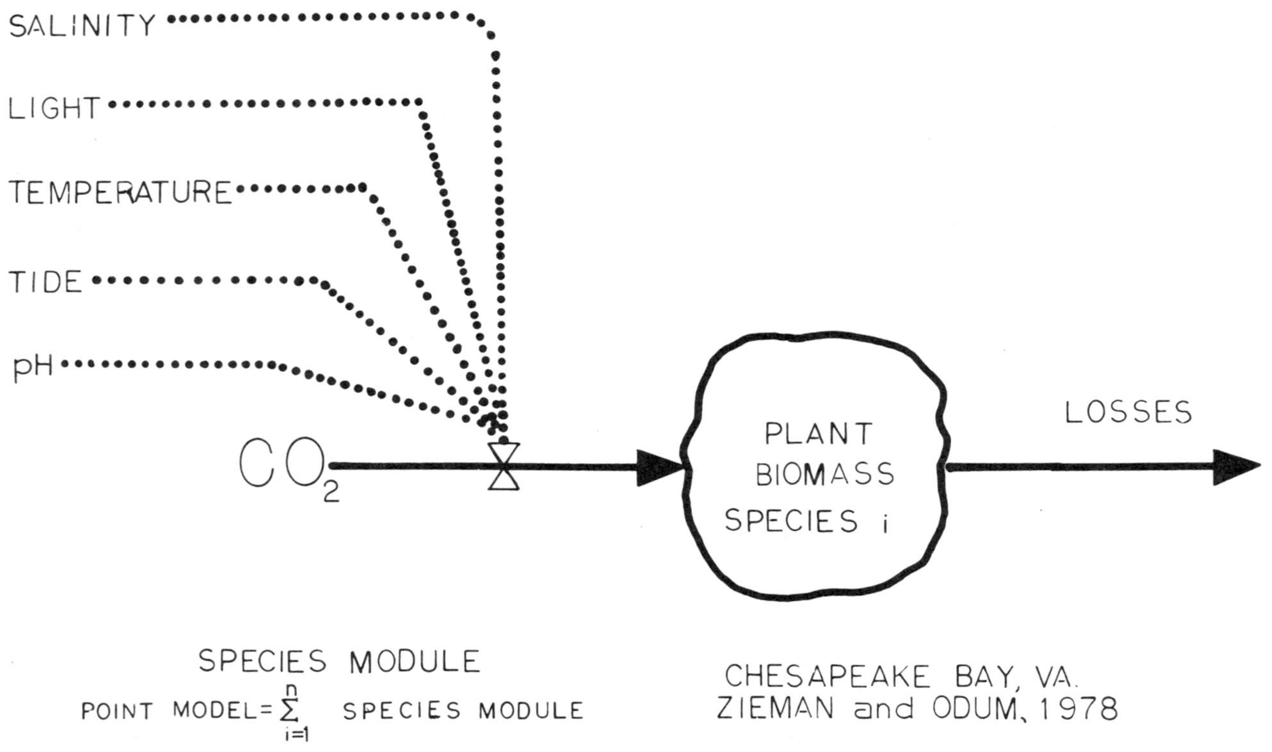

FIG. 5: The species module of the Zieman-Odum model of succession.

solely on the degree of contribution to prediction obtained. By studying intensively two different sites and surveying many additional sites outside the Chesapeake Bay area, Zieman and Odum were able to obtain some idea of the range of applicability of the model(s).

Controls in the usual sense were not used. Figure 5 shows that total biomass increase in a given species of plant is a function of salinity, light, temperature, tide, and pH. In addition, available Fe was found to exert an effect, and at any particular point, elevation was a factor because of its relation to tidal inundation. Growth of *S. alterniflora* was correlated positively with number of tidal inundations and height (although the analysis did not extend seaward far enough to detect the presumed negative effect of submersion too often). Growth was inversely related to sediment interstitial salinity.

Data were obtained by continuous recording of variables correlated with measurements of standing stock and growth rate.

Conclusions emphasized the differences between the successional trends on the two intensive study sites. One marsh was eroding with *Spartina* invading landward. The second may be accreting, but the direction of succession was uncertain. The site characters elevation and organic content were most important in determining the spatial limits of the *S. alterniflora* marsh. Accurate prediction of growth and succession became more difficult as the elevation gradient was traversed from low marsh landward.

This type of model has no explanatory power and thus little relevance to the advancement of ecological theory. It is an example of the proper and valuable use of correlation analyses to devleop a management tool, one that can be used by persons with little training in either ecology or modeling. Indeed, our knowledge of and progress in modeling *Spartina* salt marshes is still so rudimentary that simulation models that do attempt explanation do not include succession. The Zieman-Odum model would be a good point from which to develop a more fundamental explanatory model of succession.

10. THE HOPKINSON-DAY SALT MARSH MODEL

Hopkinson and Day (1977) published a simulation model of the salt marsh ecosystem of Barataria Bay on the Gulf Coast of Louisiana. In Barataria Bay, *Spartina alterniflora* produces 2-3 times more energy than the phytoplankton. The water and the marsh immediately adjacent to the shore are more productive than areas further inland. The ecosystem is the result of the interaction of climate, delta formation (Mississippi River), and physical gradients. The authors wished to model the productivity and hydrology of the ecosystem such that evaluation of selected parameters for maintaining marsh function was possible. The large-scale ecosystem model coupled the marsh and offshore systems and incorporated certain aspects of man's impact on the system, including dredging, preventing flooding by levee construction, and draining. A simplified diagram of the model structure is given in Figure 6. Carbon and nitrogen flow were modeled jointly. Figure 6 shows only carbon flow with nitrogen used as a regulatory factor; 'minor' flows of carbon have also been eliminated from Figure 6.

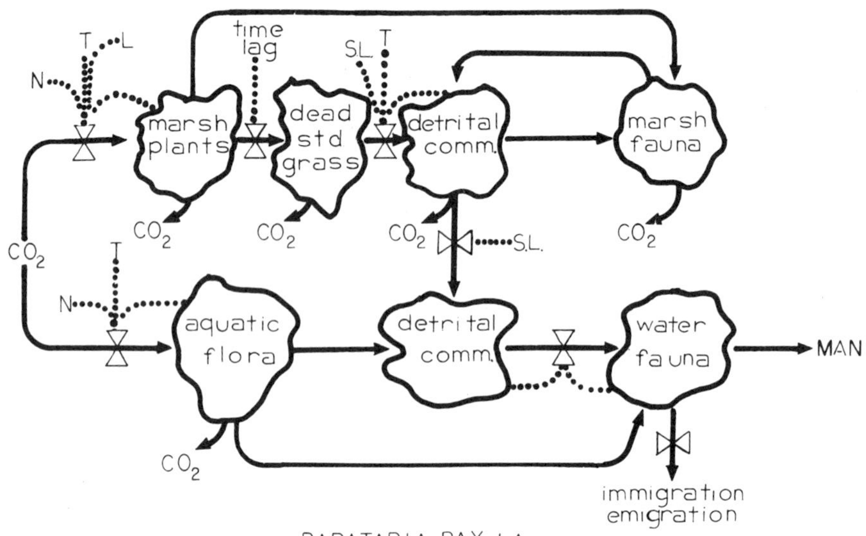

FIG. 6: Hopkinson-Day model of Barataria Bay, Louisiana.

A number of realistic controls were used. Nutrients, temperature, and light determined the realized specific rate of net photosynthesis which was then multiplied by standing stock to obtain the instantaneous flux. A time lag was employed to simulate the mortality of marsh plants. Annual variation in sea level and temperature were the factors governing the transformation of dead, standing *Spartina* into detritus. One unrealistic treatment was modeling the transfer from the detritus. One unrealistic treatment was modeling the transfer from the detrital community and the marsh plants to the marsh fauna as a linear donor-controlled flux. Variations in marsh fauna could thus have no direct effect on their resources. This eliminated what could have been an interesting type of perturbation simulation. A sensitivity analysis of the model showed temperature to be by far the most important variable in the control of primary production. Nutrients and light were limiting only during blooms of phytoplankton. Sea level was instrumental in determining the timing and magnitude of flushing of organic matter into the estuary.

Simulations with the model included runs where primary production was eliminated alternately from the marsh or from the water. The latter, despite the smaller standing stock and net production of the phytoplankton, had the greatest effect on the aquatic fauna, probably because N cycling by the phytoplankton is much faster than that of the marsh plants. The phytoplankton not only produce organic carbon, but also help enrich the organic food resources of the aquatic fauna. The aquatic fauna and not the marsh fauna is of major interest to man. Barataria Bay, for example, accounts for almost half of Louisiana's commercial fisheries catch. Hopkinson and Day feel that the modeling effort gave them insight into the operation of the ecosystem that they could have obtained in no other way. Specifically some new areas of research were suggested by the model results (this is an implication that they developed new hypotheses on the basis of model simulations): (1) Efforts are needed to find out why soil N was such a poor predictor of productivity; (2) the importance of sea-level variation pointed to the need for more concentration in hydrologic studies; (3) the model used constant coefficients and was thus representative of average conditions. Measurement of the degree of variability in some of the more sensitive coefficients would permit inclusion of variable coefficients.

11. THE WIEGERT-WETZEL SALT MARSH MODEL

A second-generation carbon flow simulation model of the Duplin River *Spartina alterniflora* marsh was described by Wiegert and Wetzel (1979). During the past year, additions to the model have included the provision of representing spatial heterogeneity and the addition of a simple complementary nitrogen flow model. The Duplin River marsh is located on Sapelo Island, Georgia. These are intertidal marshes in which *Spartina alterniflora* is the dominant higher plant. Salinity ranges normally from 20 to 30°/oo, although following periods of heavy rain it may be close to zero on the marsh. Annual water temperature varies from near freezing to 30°C.

The objectives of the model construction were to combine knowledge of biotic and abiotic ecological interactions in the ecosystem into a simulation model that would help generate hypotheses. These could in turn guide current research and suggest new avenues of exploration. No management objectives were originally included. As the modeling and field research has progressed, however, certain predictive capabilities of the model have become apparent and the third generation model should function as a management as well as a research tool. A central question is now prediction of the fate and effect of the surplus carbon produced each year by the marsh. This necessitates a much closer coupling of the marsh model to the estuary, sound, and nearshore environments. This requires, in turn, more information on the movements and feeding of the aquatic macrofauna.

A simplified diagram of the model is given in Figure 7. The marsh is divided into three distinct spatial regions: air, water, and sediment. In addition the marsh (air + sediment portions) can be divided into creek bank (high production of *Spartina*) and high marsh (lower *Spartina* production). The model is process-oriented in scope rather than species-oriented, but future development will concentrate on increasing the resolution with respect to certain important species groups in the macrofauna.

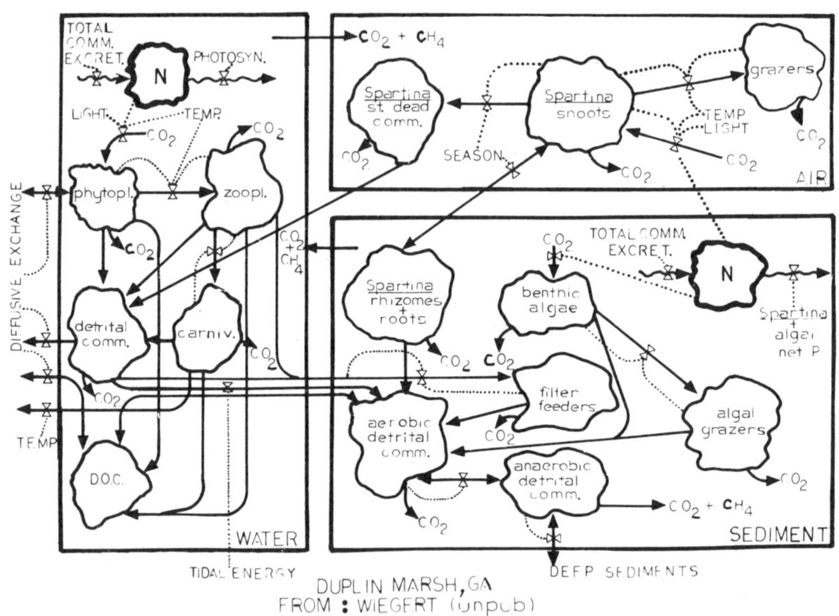

FIG. 7: Wiegert-Wetzel model of the Georgia salt marsh.

Many of the model equations are nonlinear, as indicated in Figure 7. All flows between two biotic compartments have the potential of being regulated by the density of either the donor or the recipient. Regulation is accomplished by reducing a given transfer coefficient from some maximum value. Constraints prohibited reduction of this specific rate below zero. Both explotative and direct interference competition is included, as are numerous thresholds, both upper satiation densities and lower refuge densities are used. Time lags were not included in the Wiegert and Wetzel (1979) model but are in the unpublished version shown in Figure 7. Physical factors (light, temperature, nutrients) were included implicitly, in early versions of the model by seasonal changes in maximum transfer coefficients. In the latest version many (such as N) are explicitly included.

Data came from original experiments and observations by investigators working at the Sapelo Island Marine Institute, from the literature on other *Spartina* marshes, or represent best guesses. Parameters were seldom derived from field measurements of standing crop. Thus seasonal changes in the latter, compared with predicted changes, are a good way of testing model simulations. A number of simulations have been run with the model(s), first as tests of sensitivity and corroboration; later for formulate hypotheses. Many of these are discussed in Wiegert and Wetzel (1979) and Imberger *et al.* (in press). The most interesting and latest use of the model has been to investigate the various possible ways that surplus carbon is used or transported out of the marsh. Simulations suggest (and the observations confirm) that the difference between carbon fixed and carbon-degraded *in situ* on or in the marsh sediments leaves a net of several hundred gC m^{-2} yr^{-1} (out of a total net fixation from all sources of about 1500 gC m^{-2} yr^{-1}). Model simulations of the effects of heavy rainstorms showed that the normal complement of storms, with accompanying freshwater flushing of the Duplin River can account for much of the surplus carbon. Reasonable assumptions about bacterial degradation and the assimilation of bacteria into food chains, leading to large, mobile aquatic macrofauna could account for the remaining 200-300 gC m^{-2} yr^{-1} (Imberger *et al.*, in press).

12. CONCLUSIONS

This examination of 7 different models of coastal, estuarine, and marsh models has emphasized the different scope, objectives, and structure that can be accommodated by ecosystem simulation efforts. Models may serve variously as management tools, value assessors or guides to research, and hypotheses generators. Given the need for improvement in the speed and efficiency with which we gather ecological information and develop both management procedures and basic theory, the effort of developing and *using* computer models concurrent with field research appear from this brief survey to be well worthwhile. No other way to organize, integrate, and actually use the large amounts of data coming from studies of ecosystems presents itself. Increased emphasis is needed on the best techniques for translating ecological knowledge into useful models.

ACKNOWLEDGMENTS

This paper constitutes as contribution 393 from the University of Georgia Marine Institute, Sapelo Island, Ga. Preparation of the paper was supported by N.S.F. Grant OCE75-20842 A03. For critical comments on the manuscript I thank C. Hopkinson, J. Kremer, C. Montague, S. Nixon and W. Wiebe.

REFERENCES

Hairston, N. G., Smith, F. E., and Slobodkin, L. B. (1960). Community structure, population control, and competition. *American Naturalist*, 94, 421-425.

Hopkinson, C.S.,Jr. and Day, J. W., Jr. (1977). A model of the Barataria Bay salt marsh ecosystem. In *Ecosystem Modeling in Theory and Practice*, C. Hall and J. Day, eds. Wiley Interscience, New York.

Imberger, J., Berman, T., Christian, R., Haines, E., Hanson, R., Pomeroy, L., Whitney, D., Wiebe, W., and Wiegert, R. The influence of water motion on the spatial and temporal variability of chemical and biological substances in a salt marsh estuary. *Limnology and Oceanography*. (In press).

Kremer, J. N. and Nixon, S. W. (1978). *A Coastal Marine Ecosystem--Simulation and Analysis*. Springer Verlag, New York.

McKellar, H. N., Jr. (1977). Modeling effects of coastal power plant thermal plume on the outer Crystal River estuary ecosystem. *Ecological Modelling*, 3, 85-118. Elsevier, Amsterdam.

Nixon, S. and Kremer, L. (1977). Narragansett Bay - the development of a composite simulation model for a New England estuary. In *Ecosystem Modeling in Theory and Practice*, C. Hall and J. Day, eds. Wiley, New York. 621-673.

Pomeroy, L. R. Secondary production mechanisms of continental shelf communities. (In press).

Smith, F. E. (1972). Spatial heterogeneity, stability, and diversity in ecosystems. *Transactions of the Connecticut Academy of Arts and Sciences*, 44, 307-335.

Steele, J. H. (1974). *The Structure of Marine Ecosystems*. Blackwells, Oxford.

Wiegert, R. G. (1975). Simulation models of ecosystems. In *Annual Review of Ecology and Systematics*, 6, 311-338.

Wiegert, R. G. and Owen, D. F.(1971). Trophic structure, available resources, and population density in terrestrial vs. aquatic ecosystems. *Journal of Theoretical Biology*, 30, 69-81.

Wiegert, R. G. and Wetzel, R. (1979). Simulation experiments with a 14-compartment salt marsh model. In *Estuarine Simulation and Analysis*, R. Dame, ed. University of South Carolina Press, Columbia.

Zieman, J. C. and Odum, W. E. (1977). Modelling of ecological succession and production in estuarine marshes. Technical Report D-77-35, Office, Chief of Engineers, Washington, D. C.

[*Received June* 1979. *Revised July* 1979]

SECTION IV

STATISTICAL METHODOLOGY
AND
SAMPLING

G. P. Patil and M. Rosenzweig, (eds.),
Contemporary Quantitative Ecology and Related Ecometrics, pp. 345-359.

FIELD ESTIMATES OF INSECT COLONIZATION, II

JANICE A. DERR

Department of Zoology
University of Iowa
Iowa City, Iowa 52242 USA

J. KEITH ORD

Department of Statistics
University of Warwick
Coventry, CV4 7AL, England

SUMMARY. Existing life table studies of insect populations assume constant survival rates over time, which precludes any check for declining survival due to deterioration of the environment. A simple version of the Kalman filter is introduced and modified so that it may be applied to life table data. A study of a population of the Central American cotton strainer bug, *Dysdercus bimaculatus* (Pyrrhocoridae; Heteroptera) shows that the proportion of adults who survive and stay to reproduce does indeed decline as food resources are depleted.

KEY WORDS. *Dysdercus*, forecasting, Kalman filter, life tables, moisture stress, mortality rates, Poisson distribution, survival rates, time-series.

1. INTRODUCTION

The dynamics of insect populations have been studied in various ways, of which life table methods form a major part. If N_j survivors enter the j*th* stage of the life cycle and N_{j+1} survive until the next stage, then the survival rate from stage j to $j+1$, $a_{j,j+1}$, may be estimated by

$$\hat{a}_{j,j+1} = N_{j+1}/N_j \quad . \tag{1.1}$$

In laboratory studies, a single cohort can be monitored throughout the life-cycle and the numbers surviving at each stage recorded (cf Southwood, 1966, chap. 10). However, in the field, several problems arise: (a) individuals cannot be tracked so that the available data consist of sample counts for each larval stage; (b) times of hatching and entry into successive stages are unknown; and (c) conditions are not homogeneous over time and so survival rates may vary.

Several approaches to time-specific life tables (i.e. using data described in (a)) have been given in the literature. The best methods currently available use unimodal probability distributions to describe the hatching and duration times (cf. Read and Ashford, 1968; Manly, 1974; Birley, 1977; and Kempton, 1979). A fuller review of these methods is given by Derr and Ord (1979), which we refer to as FE1 (field estimates, paper 1).

The approach in FE1 differed from that of earlier authors in that we were concerned about problem (c) above. The particular insect under study in FE1 was the central American cotton stainer bug, *Dysdercus bimaculatus* (Pyrrhocoridae; Heteroptera), which feeds upon the seeds of the *Sterculia* tree (*Sterculia apetala*). The field problem is described in greater detail in Section 2.

When a female reaches the adult stage, it may either reproduce at the same tree, or migrate to a different tree before laying eggs. The wings muscles degenerate after egg development so that any migration must precede reproduction. Derr (1977) has hypothesize that changes in the microhabitat (under a particular tree) act as stimuli to encourage migration. For a study of this question, it is essential that we be able to monitor changes in the 'survival rate' of adults, where we now redefine the 'survival rate' to mean the proportion of individuals surviving from the last (fifth) larval instar to maturity *and* staying at the same location. 'Mortality' then includes the proportion of individuals not surviving from the fifth instar to maturity *and* that proportion leaving the site of maturation.

1.1 *Estimation of Time-Dependent Survival Rates.*

Let $N_j(t)$ denote the numbers observed in the j*th* stage of the life cycle at time t and let $a_{j,j+1}(t)$ denote the time-dependent survival rate at time t. We assume that observations are made regularly (every 'week') so that time $t=1,2,\cdots,$ is measured in 'weeks'. Henceforth, a week will be used to denote the interval between successive observations, but this is a convenient description of a unit of time rather than any restriction upon the method. Further, let L_j denote the mean duration of the j*th* stage in 'days', where there are w 'days' in a 'week.' Then, in FE1, the authors showed that the survival rate at time t could be estimated by

$$\hat{a}_{j,j+1}(t) = N^*_{j+1}(t)/N^{**}_j(t) \quad , \tag{1.2}$$

where $N^*_{j+1}(t)$ and $N^{**}_j(t)$ represent appropriate averages of recent counts for the $(j+1)th$ and jth stages.

These averaged counts are built up as follows. First, we define the backward shift operator B such that

$$B\,N_j(t) = N_j(t-1), \quad B^2N_j(t) = N_j(t-2) \text{ and so on.}$$

Next, let $g_j(B)$ denote a polynomial in B , so that

$$g_j(B) = c_j(0) + B\,c_j(1) + \quad + B^{u_j}\,c_j\,(u_j)$$

where u_j denotes the number of complete weeks in the jth stage of the life cycle. In this paper we assume that the stage durations are known and fixed. Given these definitions, the authors showed in FE1 that the appropriate averages are

$$N^*_{j+1}(t) = g_j(B)\,N_{j+1}(t) \tag{1.3}$$

$$\text{and} \quad N^{**}_j(t) = g_{j+1}(B)\,B^{u_j}[(1-fr_j) + fr_jB]N_j(t) \quad , \tag{1.4}$$

where $L_j = wu_j + r_j$ and $f = w^{-1}$.

We note that primary interest focuses upon the changes in the 'survival rate' over time, since we would expect the rate to be very high initially (high immigration) but then to decline and level out (end of immigration, stable situation) before there is a further drop (emigration). In FE1, we estimated the rates using equation (1.2) and found effects of this kind. However, sampling variation in the estimates was rather high and the purpose of the present study is to smooth the estimates so that a clearer picture emerges. To do this, we use the Kalman filter, which is outlined in Section 3.

Some adaptation of the Kalman scheme is necessary for our present purposes, and these changes are outlined in Section 4. We then go on to obtain the estimates $\hat{a}_{j,j+1}(t)$ for each t and to provide one step ahead predictions which help validate the model and show how the population is progressing. The results are presented and discussed in Section 5. The smoothed estimates confirm the earlier analysis in FE1 that there is a marked drop in the 'survival' rate during the later weeks of the study.

2. THE FIELD PROBLEM

Dysdercus bimaculatus is an opportunistic species of insect which colonizes a series of 'temporary' habitats throughout the year on an annual cycle. The major type of 'temporary' habitat in the Pacific savannahs of Costa Rica is provided by the crops of seeds from *Sterculia apetala* trees. A *Sterculia* tree produces a crop of seed pods in the early dry season, around December. Adults *D. bimaculatus* are attracted to the maturing pods, and fly to a tree, possibly from great distances. When the seed pods mature, the seeds fall to the ground, where the bugs can feed on them. The females feed, mate, and lay eggs in the dirt near the seeds. Nymphs feed on the seeds and develop through five wingless instar morphs, a process which takes about 30 days, before eclosing to the winged adult form. It is at this stage when the adult female has the 'option' to remain at the same tree to reproduce, or to fly to a different tree.

Depending on the size of the *Sterculia* crop and the synchrony of pod maturation, seeds may be available to the bugs at one site from about 4 to 20 weeks (Derr, 1977). The tree studied here, 'L.S.' *Sterculia,* had a large seed crop in 1974-1975 and supported a 20 week colonization. Derr (1977) hypothesized that two major factors contributed to the deterioration of the habitat and hence to the tendency of females to migrate away from a *Sterculia* tree: 1) the availability of food and 2) the degree of environmental 'moisture stress,' a collection of biotic and abiotic factors which decrease the ability of bugs to utilize the seeds. Both these factors become less favorable through time at one tree due to the synchronous maturation of the crop and the progression of the tropical dry season. Hence, Derr (1977) hypothesized that the proportion of migrants would increase through time at a tree which could support more than one generation of bugs. Over a wide range of laboratory conditions, eclosion mortality remained relatively constant. Therefore, we have assumed that changes in 'survival rates' in the field reflect changes in migration. A second assumption is that the duration of the different instar periods remains the same throughout colonization; this seems reasonable given the constant daily temperature regime in this tropical area.

Weekly samples of insects were taken from 48 15cm × 15cm quadrats randomly located under the 'L.S.' *Sterculia.* Insects were identified to nymphal stage. For further details, see Derr (197

3. THE KALMAN FILTER

Linear regression models with time dependent parameters were first developed by Kalman (1960, 1963); the general methodology has become known as *The Kalman filter*. Harrison and Stevens (1976) give a full account of the Kalman filter and our description here deals with a special case of their work.

Let the random variable X_t be observed at regular intervals $t=1,2,\cdots$ and suppose that

$$X_t = z_t \beta_t + \varepsilon_t \quad , \tag{3.1a}$$

$$\beta_t = \beta_{t-1} + \delta_t \quad , \tag{3.1b}$$

where z_t is known, while ε_t and δ_t denote independent, identically distributed error terms with zero means. That is,

$$E[\varepsilon_t] = E[\delta_t] = 0 \quad , \quad \text{for all } t \quad ,$$

$$\text{Var}(\varepsilon_t) = \sigma^2 \quad , \quad \text{Var}(\delta_t) = \omega^2 \quad , \quad \text{for all } t \quad ,$$

$$\text{Cov}(\varepsilon_t,\varepsilon_s) = \text{Cov}(\delta_t,\delta_s) = 0 \quad , \quad \text{for all } s \neq t \quad ,$$

and $\text{Cov}(\varepsilon_t,\delta_s) = \text{Cov}(\varepsilon_t,z_s) = \text{Cov}(\delta_t, z_s) = 0$ for all s and t.

Equations (3.1) are known as the *state equations* of the system. Once β_0, σ^2, and ω^2 are specified, it is possible to estimate β_t at time t, given the past history and present state of the system. We denote the data record at time t by

$$D_t = (X_1,X_2,\cdots,X_t) \quad .$$

Parameter estimation may be carried out sequentially using an updating algorithm which we describe below. Although other assumptions are possible, we shall assume that the error terms are normally distributed. Further, we assume that values for σ^2 and ω^2, or at least their ratio $\tau^2 = \omega^2/\sigma^2$, are available and that prior information on β_0 is also available. In the work in Section 5, we tried several values for β_0 and the ratio τ^2 and it can be seen that the exact choice is not critical. Harrison and Stevens (1976) present their results in a Bayesian setting but it is equally reasonable to specify a prior likelihood for β_0 and then to obtain the estimators by maximum likelihood; the estimating equations assume the same form.

Turning back to equations (3.1), we see that when $\omega^2 = 0$, $\beta_t = \beta$ for all t and we are back at the classical linear regression model with a single regressor variable. Clearly, further terms may be included if desired. Conversely, when $\omega^2 >> \sigma^2$, we have a situation where there is very little carry-over from one time period to the next and we approach the estimators

$$\hat{\beta}_t = X_t/z_t \quad , \quad t=1,2,\cdots \quad .$$

The estimators (1.1), used in FE1, correspond to this situation.

3.1 The Updating Equations. Given that the errors are normally distributed and that the ratio $\tau^2 = \omega^2/\sigma^2$ is known, the log-likelihood at time t, given D_t, is

$$\ell_t(\beta_t|D_t) = \text{const} - \frac{1}{2\sigma^2}(X_t - z_t\beta_t)^2 - \frac{1}{2\omega^2}(\beta_t - \beta_{t-1})^2 + \ell_{t-1}(\beta_{t-1}|D_{t-1}) \quad . \tag{3.2}$$

Solving sequentially, we obtain

$$\hat{\beta}_t = \frac{\hat{\beta}_{t-1} + \tau^2 X_t z_t}{1 + \tau^2 z_t^2}$$

or

$$\hat{\beta}_t = \alpha_t \hat{\beta}_{t-1} + (1-\alpha_t)(X_t/z_t) \quad , \tag{3.3}$$

where

$$\alpha_t = 1/(1 + \tau^2 z_t^2) \quad . \tag{3.4}$$

The trade-off between past and current information is clearly seen in equation (3.4). When τ^2 is large, $\hat{\beta}_t$ will track the current ratio X_t/z_t, while when τ^2 is small successive $\hat{\beta}_t$ will vary little and are similar to the least squares estimators for constant β. Let the variance of estimator $\hat{\beta}_t$ in equation

(3.3) be $V_t = \text{Var}\,(\hat{\beta}_t | D_t)$. Since $\hat{\beta}_{t-1}$ and X_t are independent, it follows from equation (3.4) that

$$V_t = \alpha_t^2 V_{t-1} + [(1 - \alpha_t)^2 / z_t^2] \text{Var}(X_t | \beta_t)$$

$$= \alpha_t^2 \{V_{t-1} + \sigma^2 \tau^4 z_t^2\} \quad . \tag{3.5}$$

Thus, once the initial values have been specified, we can evaluate the variance of the estimator at each stage and construct confidence statements. However, these confidence intervals will not narrow as time increases, since new variability is introduced at each stage through equation (3.1b). For example, when $z_t \equiv 1$ for all t, so that β_t represents the mean of X_t at time t, the variance of β_t for large t approaches the limit

$$\text{Var}(\hat{\beta}_t) \simeq \sigma^2 \tau^2 / (2 + \tau^2) \quad .$$

Given $\hat{\beta}_t$, we may forecast the level of the process at time $(t+1)$ by

$$\hat{X}_{t+1} = z_{t+1} \hat{\beta}_t(1) \tag{3.6}$$

where $\hat{\beta}_t(1) \equiv \hat{\beta}_t$ is the one-step ahead predictor; that is, the value predicted for β_{t+1} at time t. From (3.1a) and (3.6) we see that

$$X_{t+1} - \hat{X}_{t+1} = z_{t+1} \{\beta_{t+1} - \hat{\beta}_t(1)\} + \varepsilon_{t+1}$$

$$= z_{t+1} \{\beta_t - \hat{\beta}_t + \delta_{t+1}\} + \varepsilon_{t+1}$$

so that the variance of the forecast is

$$E[(X_{t+1} - \hat{X}_{t+1})^2] = z_{t+1}^2 \text{Var}(\hat{\beta}_t) + z_{t+1}^2 \text{Var}(\delta_{t+1}) + \text{Var}(\varepsilon_{t+1})$$

$$= z_{t+1}^2 \text{Var}(\hat{\beta}_t) + \omega^2 z_{t+1}^2 + \sigma^2 \quad , \tag{3.7}$$

$$= \sigma_p^2(t) \quad , \text{ say.}$$

Thus, we may use an estimate of the variance, based upon (3.7), to develop prediction intervals of the form

$$\Pr[|X_{t+1} - \hat{X}_{t+1}| > k_\alpha\, \sigma_p(t)] = \alpha \quad , \tag{3.8}$$

where k_α is the $100(1-\alpha)\%$ percentage point of the standard normal distribution (two-tailed). Since we must replace $\sigma_p(t)$ by an estimate, the actual probability will be somewhat greater than the nominal value α.

The details given in this section are essentially a brief summary of the properties of the Kalman filter which will be used later; for further details, consult Harrison and Stevens (1976).

4. A KALMAN MODEL FOR INSECT POPULATIONS

In Section 1 we developed the model

$$N_{j+1}^{*}(t) = a_{j,j+1}(t) N_j^{**}(t)$$

to describe developments from one stage to the next in the life cycle. It was argued in FE1 that the individual counts followed independent Poisson distributions, so that $N_{j+1}^{*}(t)$ and $N_j^{**}(t)$ are weighted averages of Poisson variates; see equations (1.3) and (1.4). Thus the variance of $N_{j+1}^{*}(t)$ is unlikely to be constant as is required by model (3.1). However, it is well known (cf. Ord, 1972, p. 188) that if the random variable Y is Poisson with mean λ then $(Y + 3/8)$ has near constant variance, equal to $\frac{1}{4}$. Also, the distribution of $(Y + 3/8)$ is somewhat closer to the normal than is that of Y.

Thus, we let

$$X_t = [N_{j+1}^{*}(t) + 3/8]^{\frac{1}{2}} \quad , \tag{4.1}$$

$$z_t = [N_j^{**}(t) + 3/8]^{\frac{1}{2}} \quad , \tag{4.2}$$

$$\beta_t = [a_{j,j+1}(t)]^{\frac{1}{2}} \quad ,$$

and use the Kalman filter model as in equations (3.1). Using equations (3.3) and (3.4) we obtain the recurrence relations

$$\hat{a}_{j,j+1}(t) = \hat{\beta}_t^2$$

where $\hat{\beta}_t = \alpha_t \hat{\beta}_{t-1} + (1-\alpha_t)(X_t/z_t)$ and $\alpha_t = 1/(1 + \tau^2 z_t^2)$, X_t and z_t being given by equations (4.1) and (4.2). The variance of the estimator is, approximately,

$$\text{Var}\{\hat{a}_{j,j+1}(t)\} \simeq 4\,\beta_t^2\,\text{Var}(\hat{\beta}_t) \quad . \qquad (4.3)$$

Using equation (3.5), this may be estimated by $4\,\hat{\beta}_t^2\,\hat{V}_t$ where $\hat{V}_t = \alpha_t^2\,\{\hat{V}_{t-1} + \sigma^2\,\tau^4\,z_t^2\}$, once the initial value V_0 is given.

Since we are interested in the square of the one step ahead predictor, our development now deviates from that in Section 3, although the principles remain the same. If we set the observed $N_j^{**}(t+1) = n$, our one-step ahead forecast may be written as

$$\hat{N}_{j+1}^{*}(t+1) = n\,\hat{\beta}_t^2 \quad .$$

Taking the conditional distribution of $N_{j+1}(t+1)$ given n to be approximately Poisson with parameter $n\hat{\beta}_2^t$, the variance of the forecast is

$$\text{Var}\{N_j^{*}(t+1)|n\} + \text{Var}(n\,\hat{\beta}_t^2|n)$$

$$= n\,\beta_t^2 + 4n^2\beta_t^2\,\text{Var}(\hat{\beta}_t) \quad , \qquad (4.4)$$

using equation (4.3). This expression may be estimated by

$$n\hat{\beta}_t^2(1 + 4n\,\hat{V}_t) \quad ; \qquad (4.5)$$

approximate confidence intervals may then be established as in Section 3.

We now turn to an analysis of the 'mortality' rates for the population of cotton stainer bugs described in Section 2. In FE1, we noted that a question of major interest for *Dysdercus bimaculatus* is the extent to which the adults opt for migration rather than immediate reproduction as the quality of their habitat deteriorates. So, in this paper we concentrate upon the 'survival' rates from the fifth instar (stage 5) to adult (stage 0), written as $a_{50}(t)$ in our earlier notation.

From laboratory trials, the mean times spent in the fifth and 'adult' (including egg and first instar) stages were $L_5 = 11$ days and $L_0 = 14$ days. All adults are assumed to survive (and commence the next reproductive cycle) or not to survive at that site (either through death or migration). Using equations (1.3) and (1.4) we obtain (see FE1 for details)

$$N_0^*(t) = [7N_0(t) + 4\ N_0(t-1)]/7 \quad ,$$

$$N_5^{**}(t) = [3N_5(t-1) + 7\ N_5(t-2) + 4\ N_5(t-3)]/7 \quad .$$

The data collected by J.A.D. under the *Sterculia* tree known as 'L.S.' are given in Table 1, together with the values of $N_0^*(t)$ and $N_5^{**}(t)$ for each time period.

Given this background, we now proceed to determine initial values for the parameters σ^2, τ^2, β_0, and V_0. These were set as follows:

σ^2: since the square root transformation induces a constant variance for X_t, we set $\sigma^2 = ¼$.

τ^2: the changes in $a_{50}(t)$ tend to be very high in the early stages when most of the adults are immigrants, but to fall off later. Thus, the square root transformation is likely to help in stabilizing the variance of β_t also as the model requires $\text{Var}(\beta_t) = \omega^2$. We felt that ω^2 was likely to be in the range 0 - 0.08 (implying a standard error of not more than 0.40 at $a_{50}(t) = 0.5$). In turn this suggests a range of 0 - 0.32 for τ^2. We tried the values $\tau^2 = 0.001$, $\tau^2 = 0.05$, and $\tau^2 = 0.30$ in our calculations, for comparative purposes.

β_0: the initial value $a_{50}(0) = \beta_0^2$ is likely to be high because of immigration, although the numbers involved are small. We tried the two values $\beta_0 = 0.45$ and $\beta_0 = 1$, corresponding to $a_{50}(0) \simeq 0.2$ and $a_{50}(0) = 1$, recognizing that the choice will affect the early

TABLE 1: Original counts and weighted averages for the D. bimaculatus *population under the* Sterculia *tree 'L.S.'*

Week number	Adults: Count $N_0(t)$	Adults: Weighted average $N_0^*(t)$	Fifth stage instars: Count $N_5(t)$	Fifth stage instars: Weighted average $N_5^{**}(t)$
1	130	130	0	0
2	61	135.1	0	0
3	86	120.8	1	0
4	139	188.0	21	0.4
5	55	134.2	121	10.0
6	72	103.4	365	73.6
7	69	110.0	383	289.9
8	85	124.3	234	598.7
9	51	99.5	99	691.7
10	54	83.1	145	494.9
11	24	54.8	43	294.7
12	32	45.7	128	219.9
13	14	32.2	174	180.7
14	12	20.0	108	227.3
15	24	30.8	151	293.4
16	32	45.7	47	272.1
17	28	46.2	21	232.8
18	14	30.0	5	142.1
19	4	12.0	11	49.9
20	2	4.3	5	21.7

estimates, but should have little impact as the population grows. Since β_0^2 denotes the survival rates, it seems reasonable to select a value near the median of the unsmoothed $\hat{a}_{50}(t)$ values given in Table 2. The second value corresponds to the maximum possible in the absence of migration. Since the initial β_0 is determined almost entirely by immigration, these choices represent a downward biassing of the results, but a legitimate one given our desire to reproduce the broad trends in the 'survival' rates.

V_0: this initial value will not affect the results very much unless it is large relative to the $N_j^{**}(t)$ values. Indeed, there is something to be said for making the analysis conditional upon β_0 and setting $V_0 = 0$. However, we adopted the alternative of allowing β_0

TABLE 2: Estimated survival rates (fifth larval instar to adult) using ratios and using smoothed estimates.

<table>
<tr><th rowspan="3">Week</th><th rowspan="3">Original ratios</th><th colspan="6">Estimated survival rate, $\hat{a}_{50}$</th></tr>
<tr><th colspan="2">τ^2=0.001</th><th colspan="2">τ^2=0.05</th><th colspan="2">τ^2=0.30</th></tr>
<tr><th>β_0=0.45</th><th>β_0=1.0</th><th>β_0=0.45</th><th>β_0=1.0</th><th>β_0=0.45</th><th>β_0=1.0</th></tr>
<tr><td>4</td><td>437.2</td><td>0.21</td><td>1.02</td><td>1.05</td><td>2.41</td><td>11.14</td><td>14.30</td></tr>
<tr><td>5</td><td>13.4</td><td>0.24</td><td>1.08</td><td>3.63</td><td>5.08</td><td>12.51</td><td>13.29</td></tr>
<tr><td>6</td><td>1.40</td><td>0.29</td><td>1.10</td><td>1.79</td><td>1.99</td><td>1.65</td><td>1.67</td></tr>
<tr><td>7</td><td>0.38</td><td>0.31</td><td>0.90</td><td>0.44</td><td>0.45</td><td colspan="2">0.39</td></tr>
<tr><td>8</td><td>0.21</td><td>0.27</td><td>0.59</td><td colspan="2">0.21</td><td colspan="2">0.21</td></tr>
<tr><td>9</td><td>0.14</td><td>0.21</td><td>0.37</td><td colspan="2">0.15</td><td colspan="2">0.14</td></tr>
<tr><td>10</td><td>0.17</td><td>0.20</td><td>0.29</td><td colspan="2">0.17</td><td colspan="2">0.17</td></tr>
<tr><td>11</td><td>0.19</td><td>0.20</td><td>0.27</td><td colspan="2">0.19</td><td colspan="2">0.19</td></tr>
<tr><td>12</td><td>0.21</td><td>0.20</td><td>0.26</td><td colspan="2">0.21</td><td colspan="2">0.21</td></tr>
<tr><td>13</td><td>0.18</td><td>0.20</td><td>0.24</td><td colspan="2">0.18</td><td colspan="2">0.18</td></tr>
<tr><td>14</td><td>0.09</td><td>0.17</td><td>0.21</td><td colspan="2">0.10</td><td colspan="2">0.09</td></tr>
<tr><td>15</td><td>0.11</td><td>0.16</td><td>0.18</td><td colspan="2">0.11</td><td colspan="2">0.11</td></tr>
<tr><td>16</td><td>0.17</td><td>0.16</td><td>0.18</td><td colspan="2">0.16</td><td colspan="2">0.17</td></tr>
<tr><td>17</td><td>0.20</td><td>0.17</td><td>0.18</td><td colspan="2">0.20</td><td colspan="2">0.20</td></tr>
<tr><td>18</td><td>0.21</td><td>0.17</td><td>0.19</td><td colspan="2">0.21</td><td colspan="2">0.21</td></tr>
<tr><td>19</td><td>0.24</td><td>0.18</td><td>0.19</td><td colspan="2">0.24</td><td colspan="2">0.24</td></tr>
<tr><td>20</td><td>0.20</td><td>0.18</td><td>0.19</td><td colspan="2">0.22</td><td colspan="2">0.22</td></tr>
</table>

to be one standard deviation from the origin, so that $V_0 = \beta_0^2$.

It needs to be recognized that our 'prior information' on these parameters is not pure prior information since we looked at the data to guide our thinking. Nevertheless, since the primary purpose of the analysis is to obtain smoothed estimates rather than to forecast, this seems a reasonable procedure. If we wished to estimate τ^2 then this might be done by minimizing $\Sigma(X_t-\hat{X}_t)^2$ with respect to τ^2; however, this is not pursued further in this paper.

5. DISCUSSION

The estimates of $a_{50}(t)$ are given in Table 2, both the unsmoothed estimates derived in FE1 and the smoothed estimates for the six different parameter configurations $\tau^2 = 0.001, 0.05, 0.30$ together with $\beta_0 = 0.45, 1.0$.

TABLE 3: Standard errors for $\hat{a}_{50}(t)$ *for different* τ^2 *and* β_0 *values.*

<table>
<tr><th></th><th colspan="2">τ^2=0.001</th><th colspan="2">τ^2=0.05</th><th colspan="2">τ^2=0.30</th></tr>
<tr><th>Week</th><th>β_0=0.45</th><th>β_0=1.0</th><th>β_0=0.45</th><th>β_0=1.0</th><th>β_0=0.45</th><th>β_0=1.0</th></tr>
<tr><td>4</td><td>0.45</td><td>1.0</td><td>0.43</td><td>0.96</td><td>0.38</td><td>0.81</td></tr>
<tr><td>5</td><td>0.45</td><td>0.99</td><td>0.29</td><td>0.64</td><td>0.15</td><td>0.23</td></tr>
<tr><td>6</td><td>0.41</td><td>0.92</td><td>0.08</td><td>0.14</td><td colspan="2">0.06</td></tr>
<tr><td>7</td><td>0.32</td><td>0.71</td><td colspan="2">0.03</td><td colspan="2">0.03</td></tr>
<tr><td>8</td><td>0.20</td><td>0.45</td><td colspan="2">0.02</td><td colspan="2">0.02</td></tr>
<tr><td>9</td><td>0.12</td><td>0.26</td><td colspan="2">0.02</td><td colspan="2">0.02</td></tr>
<tr><td>10</td><td>0.08</td><td>0.18</td><td colspan="2">0.02</td><td colspan="2">0.02</td></tr>
<tr><td>11</td><td>0.06</td><td>0.14</td><td colspan="2">0.03</td><td colspan="2">0.03</td></tr>
<tr><td>12</td><td>0.05</td><td>0.11</td><td colspan="2">0.03</td><td colspan="2">0.03</td></tr>
<tr><td>13</td><td>0.04</td><td>0.10</td><td colspan="2">0.03</td><td colspan="2">0.04</td></tr>
<tr><td>14</td><td>0.04</td><td>0.08</td><td colspan="2">0.03</td><td colspan="2">0.03</td></tr>
<tr><td>15</td><td>0.03</td><td>0.06</td><td colspan="2">0.03</td><td colspan="2">0.03</td></tr>
<tr><td>16</td><td>0.02</td><td>0.05</td><td colspan="2">0.03</td><td colspan="2">0.03</td></tr>
<tr><td>17</td><td>0.02</td><td>0.04</td><td colspan="2">0.03</td><td colspan="2">0.03</td></tr>
<tr><td>18</td><td>0.02</td><td>0.03</td><td colspan="2">0.04</td><td colspan="2">0.04</td></tr>
<tr><td>19</td><td>0.02</td><td>0.03</td><td colspan="2">0.05</td><td colspan="2">0.07</td></tr>
<tr><td>20</td><td>0.02</td><td>0.03</td><td colspan="2">0.06</td><td colspan="2">0.09</td></tr>
</table>

When $\tau^2 = 0.001$, the estimates are very strongly smoothed and the starting value $\beta_0 = 1.0$ is, perhaps, more appropriate than $\beta_0 = 0.45$ since it allows for immigration in the early weeks. However, it is apparent that this value for τ^2 is much too small since it completely flattens out the trends in $\hat{a}_{50}(t)$. The inappropriateness of this value of τ^2 is shown clearly in Table 3, which gives the standard errors of the estimates. The poor assumption made in setting τ^2 so low is reflected in the high estimated standard errors, particularly over weeks 6-12. Further, the standard error fails to rise in weeks 18-20 when the sample numbers are small.

The values $\tau^2 = 0.05$ and $\tau^2 = 0.30$ represent more reasonable prior estimates in the light of our comments in Section 4 and the results in Table 3. For both of these values of τ^2 the choice of β_0 affects only the first three weeks when immigration is high. The general picture from all four initial parameter configurations is of high $a_{50}(t)$ as far as week 7, a steady period over weeks 8-13, a sharp dip in weeks 14 and 15 and then steady values thereafter (although the smoothing is more apparent here as the sample numbers diminish.

We now compare this analysis with independent evidence on the quality of the micro-habitat.

In FE1 we divided the 20-week colonization period into 3 stages of habitat favorability, based on an independent assessment of seed quality and environmental moisture stress. Stage 1, weeks 1 through 7, represented the most favorable period: both food and water were plentiful. After week 4, the occurrence of immigration was quite low (cf. FE1; most seed pods had matured by week 4). Taking into account the influence of immigration in weeks 1-4, the trend in values for a_{50} is consistent with the hypothesis that increased migration occurs with the 'stage-wise' deterioration of the habitat. The pronounced dip in weeks 13, 14, and 15 corresponds to the transition from Stage 2 (weeks 8-12) to Stage 3(weeks 13-20). Stage 2 was a period of high food availability and moderate moisture stress; Stage 3 was a period of low food availability and high moisture stress.

The only minor discrepancy between our analysis and the independent assessment of the habitat arises in week 13. It is at this point that sample numbers decline and the approximations inherent in the construction of our estimates are most susceptible to error. Nevertheless, the general accord is good and we feel that the study illustrates the potential advantages of the method over the traditional model with fixed survival rates.

ACKNOWLEDGEMENT

While this research was in progress, J.A.D. held a Post-doctoral Fellowship from the National Institutes of Health at Iowa and J.K.O. was partially funded by a Senior Visiting Fellowship from the Science Research Council. Our thanks go to these organizations for their support.

REFERENCES

Birley, M. (1977). The estimation of insect density and instar survivorship functions from census data. *Journal of Animal Ecology*, 46, 497-510.

Derr, J. A. (1977). *Population movements of* Dysdercus bimaculatus *(Pyrrhocoridae; Heteroptera) in relation to moisture stress and fruiting cycles of its different host plants*. Ph.D. thesis, Washington University, St. Louis.

Derr, J. A. and Ord, J. K. (1979). Field estimates of insect colonization. *Journal of Animal Ecology*, 48 (to appear).

Harrison, P. J. and Stevens, C. L. (1976). Bayesian forecasting (with discussion). *Journal of the Royal Statistical Society, Series B*, 38, 205-247.

Kalman, R. E. (1960). A new approach to linear filtering and prediction problems. *Journal of Basic Engineering*, 82, 35-45.

Kalman, R. E. (1963). New methods in Wiener filtering theory. In *Proceedings of First Symposium on Engineering Applications of Random Function Theory and Probability*, J. L. Bogdanoff and F. Kozin, eds. Wiley, New York.

Kempton, R. A. (1979). Statistical analysis of frequency data obtained from sampling an insect population grouped by stages. In *Statistical Distributions in Ecological Work*, J. K. Ord, G. P. Patil and C. Taillie, eds. Satellite Program in Statistical Ecology, International Co-operative Publishing House, Fairland, Maryland.

Manly, B. F. J. (1974). Estimation of stage-specific survival rates and other parameters for insect populations developing through several stages. *Oecologia*, 15, 277-285.

Ord, J. K. (1972). *Families of Frequency Distributions*. Griffin, London.

Read, K. L. Q. and Ashford, J. R. (1968). A system of models for the life cycle of a biological organism. *Biometrika*, 55, 211-221.

Southwood, T. R. E. (1966). *Ecological Methods*. Methuen, London.

[*Received March* 1979. *Revised May* 1979]

G. P. Patil and M. Rosenzweig, (eds.),
Contemporary Quantitative Ecology and Related Ecometrics, pp. 361-397.

ANALYSES OF SPECIES OCCURRENCES IN COMMUNITY, CONTINUUM, AND BIOMONITORING STUDIES

JOHN A. HENDRICKSON, JR.

Division of Limnology and Ecology
Academy of Natural Sciences of Philadelphia
Philadelphia, Pennsylvania 19103 USA

SUMMARY. Many conceptual approaches to ecological questions can be stated in terms of the distribution of occurrences of a large number of (often unspecified) species, rather than strictly in terms of the abundance of those species. Examples considered in this paper concern continuum studies, community studies, and biomonitoring studies. Some recent methodological developments are considered with respect to the evaluation of specific ecological predictions. Some common ecological applications of the Jaccard coefficient are considered objectionable.

KEY WORDS. ecology, environment, similarity, association, community, continuum, "River Continuum," Jaccard coefficient, clustering, Cochran Q-test, McNemar test, cophenetic correlation, environmental impact, aquatic insects.

1. INTRODUCTION

It seems as though one of the simplest possible observations in ecology is noting the occurrence of a species at a site: systematists often quibble (generally justifiably) that an identifiable stage of the species must be present and clearly recognized, but conceptually this is not a very abstract level of observation. Much of the history of quantitative ecology is devoted to the comparison of species lists from several samplings.

This paper presents some ecological settings in which these comparisons may be relevant. and suggests that a variety of quantitative approaches might be appropriate in different studies, depending on the types of comparisons considered to be important. The paper also provides formal support for statistical approaches advocated earlier (Hendrickson, 1978). One particular comparative index, developed by Jaccard (1901), is rather extensively considered in various settings.

2. SOME ECOLOGICAL MODELS

In this section, I have attempted a rather brief and cursory view of certain ecological constructs and applications. I grew up on the Pacific Coast of North America in an era when life zones and communities/associations were in vogue from alpine tundra to the lowest limit of low tide. My graduate work was done on the Great Plains, when biomes were the prevalent approach, and my working career as a biometrician/ecologist has been in a culture which emphasizes continua and gradients. I learned succession as a dominant ecological principle, and while I no longer hold to its full tenets, I am fascinated by colonization and subsequent succession; therefore island biogeography interests me. As a final point, I have long been aware of ecological perturbations due to man's activities.

As I try to interrelate my own experiences, I find that I do have many intellectual forefathers such as Gleason (1926), Clausen (1965), and Filice (1959). I also have learned to accept models as merely approximations which serve to explain certain phenomena. We seem only to get into trouble when models are proclaimed with the force of dogma.

2.1 Community Approaches. In a community study, sample sites are usually selected to represent maximal comparability in habitat and stage of development. Typical examples might be the study of ten alpine meadows by Jaccard (1901, parts III and IV), Gleason's (1925) study of a Michigan woodlot, Needham and Usinger's (1956) study of stream insects.

In such studies, one tends to begin with such assumptions as that most sampled units will have similar number of species and that the cumulative number of species will tend to increase with increasing numbers of samples taken (e.g., Jaccard, 1901; Arrhenius, 1921, 1923; Gleason, 1925).

If one is sampling within such a community or association, one generally finds very few species common to all the samples. Thus, Jaccard (1901) found only three species in common to his ten sites, and, by 1926, Gleason had concluded that the plant association was an artifact created by our human categorizing tendencies.

Goodall (1969) quantified one aspect of some community studies, namely the detection of outlying samples, by examining whether a sample contained an unusual combination of species. For the two-sample case, he tabulated a statistic which is equivalent to the MacNemar (1947) test, which in turn is the two-sample analog of Cochran's Q-statistic (see below). Essentially, Goodall assumed that there should not be strong associations among the species within a community. Pielou (1971, 1972) also took up the question of whether or not a community displays 'net positive association,' and her measure, ν, is likewise a function of Cochran's Q-statistic. Hendrickson (1978) has also used the Q-statistic and developed a related statistic, M, based on pairwise matches between samples (see below). Both Q and M have simple distributions, well approximated by χ^2-distributions, for a null hypothesis of no 'net positive association.' The M-statistic tests a more stringent version of no association. (Goodall's multi-sample case is not precisely the same hypothesis as the Q tests, since his criterion is essentially the joint probability of having excessive numbers of rare species in the same sample. Moreover, although the probabilities are measured in terms that are appropriate for *a priori* comparisons, the example he used was phrased as if it were an *a posteriori* comparison.) (A sample which contains an excessive number of species which are rare in one 'association' is not likely to have been drawn from the same community as the remainder of the samples.)

2.2 Continuum Approaches. Rivers are now being viewed under continuum theory, particularly with respect to the changing patterns of processes and how these changes determine the kinds of flora and fauna present (Vannote, manuscript; Sweeney, 1976; Cummins, 1977).

Independently planned studies relating to rivers as continua include Horwitz (1978) and an older group of papers (Ide, 1935, 1940; Sprules, 1947) on the response to physical gradients.

Most of the experience in data analysis of biological responses to continua is based on vegetation studies, although Terborgh (1971) (also Terborgh and Weske, 1975) has examined distributions of birds along continua with a view to examining the role of interspecific competition in determining distributional limits. Certainly the experience of aquatic biologists has been that the reliability of abundance data on stream organisms is substantially poorer than the reliability of qualitative (presence-absence) data, particularly for stream insects.

Thus we tend very much to be restricted to looking at data on distributional limits, although with certain known complications -

for instance, a single stream will tend to have a more-or-less repeated sequence of pools and riffles, at least near the headwaters, but pools and riffles present very different habitats for stream animals.

Secondly, a river tends to be the product of repeated junctions of two streams - unless one of the streams is very much smaller than the other, there will be a detectable downstream reach where one bank shows the influence of the larger stream while the other tends to show the influence of the tributary - such a 'link effect' (Vannote, manuscript) is particularly notable even to the casual observer, wherever a tributary brings a different color of water, which can often be seen to hug the bank for some distance before being very diffused into the main stream. It appears that many mainstream distributional irregularities can be traced to samples taken just downstream from such a confidence.

Another small source of complications is the directional geography of rivers. It is much easier for an organism to move downstream than upstream; however, for many fish and some aquatic insects, active instream movement against the current has been documented (Hynes, 1970). For many stream insects, the post-emergence adult aerial flight tends to be back up the stream course, which serves to avoid a progressive downstream movement of the entire population over a series of generations.

Although some aspects of the river continuum are necessarily monotonic functions of the gradient in most natural streams (e.g., base flow, sometimes channel width), others have been found to be non-linear functions of the gradient in many natural streams (e.g., diurnal temperature variation) (Sweeney, 1976, 1978).

Here, then, in a continuum study we could hypothesize a chaining pattern in the samples, or, more likely, a collection of chains, connected, at least on a topographic map, in a partial ordering, or, in some cases, only in a quasi-ordering (*sensu* Barlow *et al.*, 1972).

If the responses of the species may be non-linear in the continuum and if partial ordering (or worse) is also anticipated, the topographic pattern is unlikely to be recoverable from most forms of commonly used multivariate techniques, such as cluster analysis of similarity coefficients or even non-metric scaling based on sample 'distances.' This leads us to consider other techniques, with various reservations. First, we need techniques which begin with our hypothesized structure and test for departure from it, while bearing in mind that we need a measure of unsuitability of our hypothesized model, analogous to the stress measure in non-metric scaling and to the kind of rejection region recommended by Hogg (1965).

The assumption of unbroken distribution between known observation points on a river continuum (e.g., Horwitz, 1978) could sometimes be false, either due to link effects or to local extinctions, analogous in space to the situation reported on by Diamond and May (1977) regarding the underreporting of species turnover in time on infrequently censused islands.

2.3 Biomonitoring Studies. Biomonitoring studies usually involve a predicted site (or sites) where an impact will be detected if there is any, and additional sites which, ideally, differ only in lacking the potential adverse impact. Consequently, the simplest design for analysis is based on a community-theory style null hypothesis formulated in terms of the impact site(s) being different in having unusual combinations of species.

Sites are selected *a priori*. Unusual combinations usually appear either as fewer species or as fewer matches than expected--often pollution-intolerant species are considered likely to drop out at impact sites. Robson (in Pielou, 1972) takes up the issue of independence of two sets of species on the assumption of no association.

The Kaesler *et al.* (1974) paper attempts to interpret the correlation coefficient, rather than its square, in terms of the fraction of one variable 'explained' by another. Thus, even if the correlation between two sets of Jaccard coefficients is readily interpretable as due to common sources of variation (which I doubt), they have certainly overstated the case for substantially reducing the taxonomic efforts undertaken in biomonitoring surveys.

Moreover, in one of the studies they use as a basis for their recommendations, there is little evidence that any structure exists in the data within surveys, and the comparison of several surveys over time at one station is of doubtful use.

(Jaccard coefficients are defined, and vary rather widely, even in no association is detectable in the data matrix. In addition, as I have pointed out elsewhere, the widely used cophenetic correlation has no direct information about whether there was structure in the original data matrix.)

2.4 Other Ecological Studies. It is worthwhile to note that parallel problems in the analysis of species occurrence data, as well as parallels in the models, exist between the three topics we're considering (communities, continua, and biomonitoring) and topics arising in island biogeography.

Thus, for a continuum with a defined beginning, such as a headwater stream, there is a gradient of complexity and community development which is analogous to the process of recolonization of a denuded island over time (e.g., Simberloff, 1976; Diamond, 1975). Similarly, for islands at the same stage of (re)-colonization, one might expect a 'community' model to be well approximated, with little (or no) positive association among species across the group of islands. Further, in certain kinds of 'experimental' studies in island biogeography, it could be predicted that one group of islands might differ from others after the same temporal period of recolonization because of differences in relative isolation from sources of immigrant species.

I find these analogies between island biogeography and the kinds of systems with which I work to be useful for me. However, unless I begin actively to work on projects which were specifically conceived to test hypotheses arising from the theory of island biogeography, I don't anticipate being in a position to evaluate whether my approaches would be useful in answering current or future questions in island biogeography. They do, however, have certain parallels with the approaches taken by Simberloff and Connor (1979).

3. MATHEMATICAL AND STATISTICAL DEVELOPMENTS

As we established in the introduction, biologists often collect (or observe) samples of organisms, which may be identified to species (or other appropriate taxonomic level). Two or more such samples are often compared on the basis of their species lists to evaluate whether they were drawn from the same biological 'community' (e.g., Patrick, 1949; Williams, 1964; Goodall, 1969; Crossman *et al.*, 1974).

Such taxonomic lists (or other attribute lists where one state is labeled positive) can be combined in an occupancy table, Y, in which the t rows represent the taxa, s columns represent the samples, and the entires, Y_{ij}, are 1 for the presence or 0 for the absence of taxon i in sample j. Such occupancy tables are implicit in the committee problem (Mantel, 1974, and references therein), and Feller (1968, p. 41) introduces them as tables of Fermi-Dirac statistics.

In some cases the comparison of rows has been of interest (Cole, 1957), in other the comparison of columns (Goodall, 1969), but the general class of problems involves a $t \times s$ binary matrix Y, with $\sum_{i=1}^{t} \sum_{j=1}^{s} Y_{ij}$ fixed. Most of the questions in this paper concern the homogeneity of the s columns (or samples) or of subsets of these samples.

As has been noted by Barton and David (1959), Pielou (1972), Mantel and Fleiss (1975), and Bishop *et al.* (1975), a $t \times s$ binary occupancy tables can be collapsed into a 2^t or 2^s contingency table depending on which marginal sums are of interest.

The marginal sums of the rows, $Y_{i.}$, have been regarded as fixed in several independent papers, including Cochran (1950), Goodall (1967, 1969), and Mantel (1974). Pielou (1971, p. 129) used this model to simulate an approximation to Barton and David's approximating binomial. She was interested in evaluating association among the t species and used this model to develop a distribution of the expected numbers of samples with n species, $0 \leq n \leq t$, if the t species were distributed independently.

Cochran's Q-test compares the sum of squares of the column sums with its expectation under independence of the rows. Goodall (1967) considered the distribution of the simple matching coefficient (defined below), while his 1969 paper approached the question of the detection of atypical samples by examining the occurrence or cooccurrence of species found in very few samples.

In this paper I also introduce the model in which both the row and column sums are fixed, and the constraints on the distributions of positive and negative matches and mismatches and of coefficients of association derived therefrom are developed. This permits the consideration of whether additional structure is present which is not retained by the marginal sums.

These last two models (which are conditioned on the row sums) will also be considered for cases in which the s samples are ordered in time or space (or both); such orderings occur in real-world applications at least in ecology.

It should be noted that instead of fixed row sums, Sokal and Sneath (1963, p. 125-141) considered the column sums fixed and computed the expected values of the numbers of positive matches (both columns show a one in the same row), negative matches (both have zeroes in the same row) and mismatches (one column has a zero, the other a one in the same row) on the assumption of conditional independence of positive entries in the two columns. These expectations were used to estimate the expected values of the coefficients of association among columns. One consequence of their approach will be considered below. (Mantel (1974) arrived at this case with rows and columns reversed (his formulas (5)).)

3.1 Matches and Coefficients of Similarity.[1] Consider two columns, j and k, of Y. In a given row, i, the pair of elements, Y_{ij} and Y_{ik}, can both be ones, termed a positive match, both can be zeroes, termed a negative match, or one of the pair is a one and the other a zero, terms a mismatch. In the notation of a 2×2 contingency table, the number of positive matches is A, negative matches are counted in D, and mismatches are counted in B if Y_{ij} is positive or in C if Y_{ik} is positive. When $s > 2$ columns, A, B, C, and D must be subscripted to indicate the pair of samples being compared. The following equations hold:

$$A_{jk} = \sum_{i=1}^{t} Y_{ij} Y_{ik}, \tag{1}$$

$$B_{jk} = \sum_{i=1}^{t} Y_{ij} (1-Y_{ik}), \tag{2}$$

$$C_{jk} = \sum_{i=1}^{t} (1-Y_{ij})Y_{ik}, \tag{3}$$

$$D_{jk} = \sum_{i=1}^{t} (1-Y_{ij})(1-Y_{ik}), \tag{4}$$

$$Y_{\cdot j} = A_{jk} + B_{jk}, \tag{5}$$

$$Y_{\cdot k} = A_{jk} + C_{jk}, \tag{6}$$

$$t = A_{jk} + B_{jk} + C_{jk} + D_{jk} \tag{7}$$

A wide variety of coefficients of similarity based on A, B, C, and D have been used in biology (see, for example, Sokal and Sneath, 1963; Goodall, 1973; Sneath and Sokal, 1973); the basic properties of most of these can be understood by considering only three coefficients which differ in the treatment of negative matches. The simple matching coefficient (Sokal and Michener,

---

[1] 'Similarity' is taken as including at least measures of agreement and some measures of association as those words are used by Bishop, Fienberg, and Holland (1975, chap. 11). Usage seems to vary among subject matter fields.

1958) uses both positive and negative matches in the numerator;

$$S_{jk} = (A_{jk} + D_{jk})/t. \tag{8}$$

Russell and Rao (1940) used a coefficient with only positive matches in the numerator;

$$R_{jk} = A_{jk}/t. \tag{9}$$

Jaccard (1901) introduced a coefficient with positive matches in the numerator, but which excludes negative matches from the denominator;

$$J_{jk} = A_{jk}/(t-D_{jk}) \quad \text{if} \quad t > D_{jk}, \quad = 0 \quad \text{otherwise} \tag{10}$$

(Gower, 1971). Note that

$$0 \leq R_{jk} \leq J_{jk} \leq S_{jk} \leq 1. \tag{11}$$

The Jaccard coefficient has enjoyed wide currency in ecology, including a presentation of dendrograms based on clustering a similarity matrix of such coefficients. This paper had its beginnings in an effort to interpret some results presented solely as Jaccard coefficients; it was generalized when other problems were recognized as related.

Consider as an example Configuration A of Table 1. The three samples are scored for the presence of each of six taxa. Table 2 presents the elements A, B, C, and D, and the three sample similarity coefficients for each of the three (unordered) pairs of samples in configuration A. Thus, samples 1 and 2 have a positive match for taxon 1; they are mismatched on taxa 2-5; they have a negative match for taxon 6.

$$J_{12} = 1/(6-1) = 1/5; \quad R_{12} = 1/6; \quad S_{12} = (1+1)/6 = 2/6 = 1/3.$$

3.2 Expectations Conditioned on Row Sums. For a binary occupancy table, Y, the row sums, $Y_{i.}$, can range from 0 to s. Invariant rows do not always affect calculations, sometimes for algebraic reasons (as in the Cochran Q-test) and in other cases for subject matter reasons. Thus, in numerical taxonomy invariant attributes are usually deleted from the analyses (Sneath and Sokal, 1973); however, in ecology one cannot reasonably exclude ubiquitous species, while little is gained from comparing the absence of non-existent species. Conceivably in other cases the total absence of an attribute may be interesting.

TABLE 1: Four (of many configurations for 3 samples and 6 taxa.

Configuration A

	Sample 1	2	3	$Y_{i\cdot}$
Taxon 1	1	1	0	2
2	1	0	1	2
3	0	1	1	2
4	1	0	0	1
5	0	1	0	1
6	0	0	1	1
$Y_{\cdot j}$	3	3	3	9

Configuration B

	Sample 1	2	3	$Y_{i\cdot}$
Taxon 1	1	1	0	2
2	1	1	0	2
3	1	0	1	2
4	0	1	0	1
5	0	0	1	1
6	0	0	1	1
$Y_{\cdot j}$	3	3	3	9

Configuration C

	Sample 1	2	3	$Y_{i\cdot}$
Taxon 1	1	1	0	2
2	1	1	0	2
3	1	1	0	2
4	0	0	1	1
5	0	0	1	1
6	0	0	1	1
$Y_{\cdot j}$	3	3	3	9

Configuration D

	Sample 1	2	3	$Y_{i\cdot}$
Taxon 1	1	0	1	2
2	0	1	1	2
3	0	1	1	2
4	1	0	0	1
5	0	1	0	1
6	0	0	1	1
$Y_{\cdot j}$	2	3	4	9

TABLE 2: Computation of sample similarities for the configurations of data presented in Table 1.

Configuration	Sample Pairs	A	B	C	D	J	R	S
A	1-2	1	2	2	1	1/5	1/6	2/6
	1-3	1	2	2	1	1/5	1/6	2/6
	2-3	1	2	2	1	1/5	1/6	2/6
	Σ	3	6	6	3	3/5	3/6	1
	Σ(B+C)		12					
B	1-2	2	1	1	2	2/4	2/6	4/6
	1-3	1	2	2	1	1/5	1/6	2/6
	2-3	0	3	3	0	0/6	0/6	0/6
	Σ	3	6	6	3	7/10	3/6	1
	Σ(B+C)		12					
C	1-2	3	0	0	3	3/3	3/6	6/6
	1-3	0	3	3	0	0/6	0/6	0/6
	2-3	0	3	3	0	0/6	0/6	0/6
	Σ	3	6	6	3	1	3/6	1
	Σ(B+C)		12					
D	1-2	0	2	3	1	0/5	0/6	1/6
	1-3	1	1	3	1	1/5	1/6	2/6
	2-3	2	1	2	1	2/5	2/6	3/6
	Σ	3	4	8	3	3/5	3/6	1
	Σ(B+C)		12					

Where algebraic results may differ under restrictions on the admissibility of row sums, they will be developed for $0 \leq Y_{i\cdot} \leq s$.

For $m = 0,1,\cdots,s$, let f_m denote the number of rows with row sums equal to m, for convenience in collecting terms. Then

$$\sum_{m=0}^{s} f_m = t, \quad \text{and} \quad \sum_{m=0}^{s} mf_m = Y_{\cdot\cdot} \quad . \tag{12}$$

3.2.1 Expectations independent of column sums. Under independence of rows, the expectation of an observation is $E[Y_{ij}] = Y_{i\cdot}/s$.

The expectation of a column sum is

$$E[Y_{\cdot j}] = E\left[\sum_{i=1}^{t} Y_{ij}\right] = \sum_{i=1}^{t} E[Y_{ij}] = \sum_{i=1}^{t} Y_{i\cdot}/s = Y_{\cdot\cdot}/s$$

$$= (1/s) \sum_{m=0}^{s} mf_m = (1/s) \sum_{m=1}^{s} mf_m \quad . \tag{13}$$

Following Feller (1968, p. 231ff), the expected variance of $Y_{\cdot j}$ is obtained by summing the variances of the Y_{ij}, where $\mathrm{Var}(Y_{ij} = (1/s^2)Y_{i\cdot}(s-Y_{i\cdot})$, so that

$$\mathrm{Var}(Y_{\cdot j}) = (1/s^2) \sum_{i=1}^{t} Y_{i\cdot}(s-Y_{i\cdot}) = (1/s^2) \sum_{m=0}^{s} m(s-m)f_m$$

$$= (1/s^2) \sum_{m=1}^{s-1} m(s-m)f_m \quad . \tag{14}$$

However, following Cochran (1950, p. 259), we note that since we have conditioned on the row totals, $Y_{i\cdot}$, the row sum has zero variance, and the covariance of Y_{ij} with Y_{ik} is found to be

$$\mathrm{Cov}(Y_{ij}, Y_{ik}) = \frac{Y_{i\cdot}(s-Y_{i\cdot})/s^2}{(s-1)} \quad \text{for} \quad j \neq k \quad . \tag{15}$$

Hence, if the rows are independent, the s column totals, $Y_{\cdot j}$, should be distributed such that Q has an asymptotic χ^2 distribution with $s-1$ degrees of freedom, where

$$Q = \frac{(s-1) \sum_{j=1}^{s} (Y_{\cdot j} - Y_{\cdot\cdot}/s)^2}{\sum_{m=0}^{s} f_m \, m(s-m)/s} \tag{16a}$$

or,

$$Q = \frac{(s-1) \sum_{j=1}^{s} (Y_{\cdot j} - Y_{\cdot\cdot}/s)^2}{\sum_{i=1}^{t} Y_{i\cdot} - \sum_{i=1}^{t} Y_{i\cdot}^2/s}, \tag{16b}$$

which is familiar as the Cochran Q-statistic.

Comparing any two columns of Y, $j \neq k$, if the rows are independent, the probability of a positive match in row i (if there are at least two ones in row i) can be computed as the probability of a one in column j, $Y_{i\cdot}/s$, times the probability of another one in column k, $(Y_{i\cdot}-1)/(s-1)$; thus,

$$\Pr(Y_{ij}Y_{ik} = 1) = Y_{i\cdot}\ (Y_{i\cdot} - 1)/(s(s-1)). \tag{17}$$

Let $a_{ijk} = Y_{ij}Y_{ik}$, $1 \leq j \leq k \leq s$; (18)

this equals one for a positive match, zero otherwise.

Then $E[a_{ijk}] = Y_{i\cdot}\ (Y_{i\cdot} - 1)/(s(s-1))$, (19)

and $\mathrm{Var}(a_{ijk}) = E[a_{ijk}^2] - (E[a_{ijk}])^2$. (20)

Let $A_{jk} = \sum_{i=1}^{t} a_{ijk}$. (21)

From the independence assumption of the t rows, the expected number of positive matches between columns is

$$E[A_{jk}] = \sum_{i=1}^{t} E[Y_{ij}Y_{ik}] = \sum_{i=1}^{t} Y_{i\cdot}(Y_{i\cdot} - 1)/(s(s-1))$$

$$= \sum_{m=0}^{s} f_m m(m-1)/(s(s-1))$$

$$= \sum_{m=2}^{s} f_m m(m-1)/(s(s-1)), \tag{22}$$

since no matches can occur for $Y_{i\cdot} < 2$.

Similarity, $$\mathrm{Var}(a_{ijk}) = E[a_{ijk}^2] - (E[a_{ijk}])^2 \tag{23}$$

and $a_{ijk} \equiv a_{ijk}^2$, since only 0 and 1 can occur.

Note that $$\sum_{j=1}^{s-1} \sum_{k=j+1}^{s} a_{ijk} = Y_{i\cdot}(Y_{i\cdot}-1)/2; \tag{24}$$

thus, the total of the $s(s-1)/2$ values of a_{ijk} from a row has zero variance. Hence, for suitable constraints (e.g., $s \geq 3$; g, h, $k \neq j$; $g \neq h$), the covariance of a_{ijk} and a_{igh} can be given as

$$\mathrm{Cov}(a_{ijk}, a_{igh}) = -\frac{Y_{i\cdot}(Y_{i\cdot}-1)/\{s(s-1)\} - [Y_{i\cdot}(Y_{i\cdot}-1)/\{s(s-1)\}]^2}{s(s-1)/2 - 1} \tag{25}$$

From the independence assumption, the variances and covariances of the A_{jk} values should sum properly over the rows, and, for t sufficiently large, the joint distribution should be approximately multivariate normal, with a common variance, σ^2, and covariance, $\rho\sigma^2$; then $\sum\sum_{j<k} (A_{jk} - \bar{A})^2$ should be distributed as $\chi^2\sigma^2 \cdot (1-\rho)$ with $(s-1)(s-2)/2$ degrees of freedom, following Cochran (1950) and Fleiss (1965) and noting that $(s-1)$ degrees of freedom are lost in eliminating dependencies among the pairs of columns. Summing over the $s(s-1)/2$ terms and some rearrangement yields

$$M = \frac{(s-2)\left\{s(s-1)\sum_{j=1}^{s-1}\sum_{k=j+1}^{s} A_{jk}^2 - 2\left[\sum_{j=1}^{s-1}\sum_{K=j+1}^{s} A_{jk}\right]^2\right\}}{s\sum_{i=1}^{t} Y_{i\cdot}(Y_{i\cdot}-1) - \{1/(s-1)\}\sum_{i=1}^{t}\{Y_{i\cdot}(Y_{i\cdot}-1)\}^2} \tag{26}$$

which should be distributed as χ^2 with $(s-1)(s-2)/2$ degrees of freedom. The small sample distribution is discussed in Section 3.2.3.

In closing this subsection, let us examine the effects of conditioning on the row sums on the similarity coefficients.

Similarly to the results for A_{jk}, the expected number of mismatches is

$$E[B_{jk}+C_{jk}] = E[B_{jk}] + E[C_{jk}]$$

$$= \sum_{i=1}^{t} E[(1-Y_{ij})Y_{ik} + Y_{ij}(1-Y_{ik})]$$

$$= 2 \sum_{m=0}^{s} f_m m(s-m)/\{s(s-1)\}$$

$$= 2 \sum_{m=1}^{s-1} f_m m(s-m)/\{s(s-1)\}. \qquad (27)$$

The expected number of negative matches is

$$E[d_{jk}] = \sum_{i=1}^{t} E[(1-Y_{ij})(1-Y_{ik})]$$

$$= \sum_{m=0}^{s-2} f_m (s-m)(s-m-1)/(s(s-1)). \qquad (28)$$

If $s \geq 4$, the covariance of A_{jk} with D_{jk} may be non-zero.

$$\mathrm{Cov}(A_{jk},D_{jk}) = \sum_{i=1}^{t} E[Y_{ij}Y_{ik}(1-Y_{ij})(1-Y_{ik})]$$

$$- \sum_{i=1}^{t} E[Y_{ij}Y_{ik}]\ E[(1-Y_{ij})(1-Y_{ik})]$$

$$= 0 - \sum_{m=2}^{s-2} f_m m(m-1)(s-m)(s-m-1)/(s^2(s-1)^2). \qquad (29)$$

Since $A_{jk}+B_{jk}+C_{jk} = t - D_{jk}$,

$$\mathrm{Cov}(A_{jk},A_{jk}+B_{jk}+C_{jk}) = -\mathrm{Cov}(A_{jk},D_{jk}). \qquad (30)$$

Also, $$f_s \leq A_{jk} \leq \sum_{m=2}^{s} f_m, \tag{31}$$

$$f_0 \leq D_{jk} \leq \sum_{m=0}^{s-2} f_m, \text{ and} \tag{32}$$

$$0 \leq B_{jk} + C_{jk} \leq \sum_{m=1}^{s-1} f_m. \tag{33}$$

The expectations of two of the similarity coefficients clearly follow in closed form

$$E[R_{jk}] = E[A_{jk}]/t = \sum_{m=2}^{s} f_m m(m-1)/(st(s-1)), \tag{34}$$

and $$E[S_{jk}] = E[A_{jk} + D_{jk}]/t$$

$$= \sum_{m=0}^{s} f_m(m(m-1) + (s-m-1))/(st(s-1)). \tag{35}$$

However, $$E[J_{jk}] = E[A_{jk}/(t-D_{jk})] \tag{36}$$

cannot generally be expressed in closed form since the numerator and denominator are both random variables and may not be independent.

If $\bar{R} = [2/(s(s-1))] \sum\sum_{j<k} R_{jk}$, the arithmetic mean of the similarity values for the $s(s-1)/2)$ unordered pairs of columns,

then $$E[R_{jk}] = \bar{R}. \tag{37}$$

Likewise $$E[S_{jk}] = \bar{S}. \tag{38}$$

Note that in the examples of Table 1 all the row sums, $Y_{i.}$, are either $[s/2]$ or $s - [s/2]$, and $\bar{S} = 1/3$ (where the square brackets indicate 'the integer part of'). This represents a minor class of exceptions to Goodall's (1967) assertion that $E[S_{jk}] \geq 0.5$, which presumed sampling with replacement.

A comparison of configurations A-D in Tables 1 and 2 will indicate that $\bar{J}$ for a particular set of samples is not defined by conditioning on the row sums.

3.2.2 Conditional on both row and column sums. For a particular pair of columns, $j \neq k$, conditioning the column sums (ignoring, briefly, conditioning on the row sums in this section) provides the following constraints on numbers of matches and mismatches, which follow, for A_{jk}, from

$$\binom{Y_{\cdot j}}{A_{jk}}\binom{t - Y_{\cdot j}}{Y_{\cdot k} - A_{jk}}:$$

$$\max(Y_{\cdot j} + Y_{\cdot k} - t,\ 0) \leq A_{jk} \leq \min(Y_{\cdot j},\ Y_{\cdot k}), \tag{39}$$

$$\max(t - Y_{\cdot j} - Y_{\cdot k},\ 0) \leq D_{jk} \leq \min(t - Y_{\cdot j},\ t - Y_{\cdot k}), \tag{40}$$

$$|Y_{\cdot j} - Y_{\cdot k}| \leq B_{jk} + C_{jk} \leq t - \{\min(A_{jk}) + \min(D_{jk})\}. \tag{41}$$

Further, since $Y_{\cdot j} = A_{jk} + B_{jk}$ and $Y_{\cdot k} = A_{jk} + C_{jk}$, a one unit increase (decrease) in A_{jk} requires a one unit decrease (increase) each in B_{jk} and C_{jk} and a one unit increase (decrease) in D_{jk}. Thus, for any pair of columns with known sums, the possible values of A_{jk} can be readily enumerated, and to each will correspond the unique corresponding values, B_{jk}, C_{jk} and D_{jk}.

Conditioning on the row sums as is presumed in the remainder of this section will further constrain the possible values of A_{jk}, etc. The combined constraints follow.

$$\max(Y_{\cdot j} + Y_{\cdot k} - t,\ f_s) \leq A_{jk} \leq \min(Y_{\cdot j},\ Y_{\cdot k},\ \sum_{m=2}^{s} f_m), \tag{42}$$

$$\max(t - Y_{\cdot j} - Y_{\cdot k},\ f_0) \leq D_{jk} \leq \min(t - Y_{\cdot j},\ t - Y_{\cdot k},\ \sum_{m=0}^{s-2} f_m), \tag{43}$$

$$|Y_{\cdot j} - Y_{\cdot k}| \leq B_{jk} + C_{jk} \leq t - \{\min(A_{jk}) + \min(D_{jk})\}. \tag{44}$$

Consider configuration D of Table 1. Here, $s = 3$, $t = 6$, $Y_{\cdot 1} = 2$, $Y_{\cdot 2} = 3$, $Y_{\cdot 3} = 4$, $f_0 = f_3 = 0$, and $f_1 = f_2 = 3$. From the constraints, it is apparent, for example, that $0 \leq A_{12} \leq 2$ and $1 \leq A_{23} \leq 3$. The full set of possible values is presented in Table 3.

TABLE 3: Possible values of matches, mismatches and similarity coefficients for the sample pairs of configuration D *of Table 1.*

Pair	A	B	C	D	R_{jk}	J_{jk}	S_{jk}
1,2	0	2	3	1	0/6	0/5	1/6
	1	1	2	2	1/6	1/4	3/6
	2	0	1	3	2/6	2/3	5/6
1,3	0	2	4	0	0/6	0/6	0/6
	1	1	3	1	1/6	1/5	2/6
	2	0	2	2	2/6	2/4	4/6
2,3	1	2	3	0	1/6	1/6	1/6
	2	1	2	1	2/6	2/5	3/6
	3	0	1	2	3/6	3/4	5/6

TABLE 4: Possible combinations of A_{jk} *values (from Table 3) for the sample pairs of configuration* D.

A_{12}	A_{13}	A_{23}	Probability this combination
0	0	3	1/20
0	1	2	6/20
0	2	1	3/20
1	0	2	3/20
1	1	1	6/20
2	0	1	1/20

Moreover, conditional on the row sums, $\sum_{j<k} A_{jk} = \sum_{m=2}^{s} f_m m(m-1) = 3$ in this example, and it is possible to enumerate the admissible combinations of A_{12}, A_{13}, and A_{23}. Thus, if $A_{23} = 3$, $A_{12} = A_{13} = 0$. Likewise, once A_{jk}, $Y_{\cdot j}$, and $Y_{\cdot k}$ are specified, B_{jk}, C_{jk}, and D_{jk} are known. The admissible combinations of the A_{jk} values of Table 3 are presented in Table 4.

The combinations shown in Table 4 exhaust the points with positive probability. It is now necessary to assign probabilities to these combinations.

Just as the admissible combinations of matches are identifiable and enumerable, the data configurations leading to any one combination of matches are identifiable and countable, which leads to a basis for computing and probability of a given combination of matches conditional on row and column sums. These computations have been performed for configuration D and the results are shown in Table 4.

The distribution of M implied by this model, as illustrated from the data of Table 4, constitutes a subset of the distribution developed in the last subsection. For such a subset, both the expectation and the upper-tail probabilities will ordinarily be changed. Obviously, the Q-statistic is of no interest in this case.

3.2.3 Distribution of the M-statistic. It was asserted earlier that M should be distributed under the null hypothesis of independent rows as χ^2 with $(s-1)(s-2)/2$ degrees of freedom when conditioned only on the row sums, $Y_{i\cdot}$. In the cases investigated thus far $(t=2,6(1), s=3, 4, \text{ or } 6)$, the expectation of M is identically its degrees of freedom, and the upper tail probabilities are fairly closely approximated by χ^2 for $t \geq 4$, if only rows with positive matches and non-zero variance $(2 \leq Y_{i\cdot} < s)$ are included.

However, when both row and column sums are fixed (e.g., Section 3.2.2), the variance and covariance terms developed in Section 3.2.1 are not applicable. Presumably a revised variance structure could be developed for the doubly-conditioned case, and this might be useful in continuum studies. Simberloff and Connor (1979) have attacked this problem by Monte Carlo simulation studies.

3.2.4 Hypothesis testing with the Q-statistic. Cochran (1950) showed that Q could be treated as approximating a χ^2-variate with s-1 degrees of freedom. He also proposed that if one were interested in dividing the sample into exactly two (non-overlapping) groups, the sample Q statistics for each group could be computed and their sum subtracted from the original Q statistic to yield a Q difference, which would be distributed as χ^2 with one degree of freedom. With real data, this method for estimating Q difference occasionally gives rise to a negative value (usually treated as 0.00).

Fleiss (1965, 1973) suggested recomputing only the numerator of Q for each of the two groups as well as directly for Q difference, a procedure which assures additivity (within rounding error). Thus if the first s_1 sites were in group 1 and the remaining s_2 sites in group 2, $s_1 + s_2 = s$, following (16b) we arrive at

$$Q_{group\ 1} = \frac{(s-1)}{s_1} \cdot \frac{s_1 \sum_{j=1}^{s_1} Y_{\cdot j}^2 - (\sum_{j=1}^{s_1} Y_{\cdot j})^2}{\sum_{i=1}^{t} Y_{i\cdot} - \sum_{i=1}^{t} Y_{i\cdot}^2/s}, \tag{45}$$

$$Q_{group\ 2} = \frac{(s-1)}{s_2} \cdot \frac{s_2 \sum_{j=s_1+1}^{s_1} Y_{\cdot j}^2 - (\sum_{j=s_1+1}^{s_1} Y_{\cdot j})^2}{\sum_{i=1}^{t} Y_{i\cdot} - \sum_{i=1}^{t} Y_{i\cdot}^2/s}, \tag{46}$$

$$Q_{difference} = \frac{(s-1)}{ss_1s_2} \cdot \frac{(s_2 \sum_{j=1}^{s_1} Y_{\cdot j} - s_1 \sum_{j=s_1+1}^{s} Y_{\cdot j})^2}{\sum_{i=1}^{t} Y_{i\cdot} - \sum_{i=1}^{t} Y_{i\cdot}^2/s}. \tag{47}$$

There are s_1-1 degrees of freedom for $Q_{group\ 1}$, s_2-1 degrees of freedom for $Q_{group\ 2}$, and 1 degree of freedom for $Q_{difference}$.

Unfortunately, the sample computations given in Hendrickson (1978) followed the procedure in Cochran (1950) rather than Fleiss (1973).

3.2.5 Hypothesis testing with the M-statistic. The M-statistic is defined for $s \geq 3$ columns. Section 3.2.3 presented some evidence that both the mean and the upper-tail probabilities are in accord with the appropriate χ^2 distribution. If M exceeds the selected critical value from the χ^2 distribution, one can reject the null hypothesis of independence of the rows (taxa). I presume that the potential for standard simultaneous test procedures exists in this setting, but my further comments will be restricted to certain classes of *a priori* hypotheses.

If $s = 3$, there is only a single degree of freedom for the M-statistic, and no further testing is possible. Consequently, *a priori* tests can only be designed for $s \geq 4$, since otherwise no partitioning is possible. It is, then, apparent that in any partition, at least one group must contain at least 3 columns, or there is no subset against which to define the M-statistic for the difference between the two groups, which is the natural test statistic for a difference between the groups.

Analogously to the partitioning of Q, we can define $M_{group\ 1}$ and $M_{group\ 2}$, and calculate $M_{difference}$ by subtraction. Again for ease in notation, we assume that the first s_1 sites are group 1, and the remaining s_2 sites are group 2. From (26)

$$M_{group\ 1} = \frac{s(s-1)}{s_1(s_1-1)} \cdot F_1, \tag{48}$$

$$M_{group\ 2} = \frac{s(s-1)}{s_2(s_2-1)} \cdot F_2, \tag{49}$$

$$M_{difference} = M - M_{group\ 1} - M_{group\ 2}, \tag{50}$$

where

$$F_1 = \frac{(s-2)\{s_1(s_1-1)\sum_{j=1}^{s_1-1}\sum_{k=j+1}^{s_1} A_{jk}^2 - 2(\sum_{j=1}^{s_1-1}\sum_{k=j+1}^{s_1} A_{jk})^2\}}{[s\sum_{i=1}^{t} Y_{i.}(Y_{i.}-1) - \{1/(s-1)\}\sum_{i=1}^{t}\{Y_{i.}(Y_{i.}-1)\}^2]}$$

$$F_2 = \frac{(s-2)\{s_2(s_2-1)\sum_{j=s_1+1}^{s-1}\sum_{k=j+1}^{s} A_{jk}^2 - 2(\sum_{j=s_1+1}^{s-1}\sum_{k=j+1}^{s} A_{jk})^2\}}{[s\sum_{i=1}^{t} Y_{i.}(Y_{i.}-1) - \{1/(s-1)\}\sum_{i=1}^{t}\{Y_{i.}(Y_{i.}-1)\}^2]}$$

In Hendrickson (1978), I had underestimated the need for preserving the original denominator in partitioning M; additivity is not guaranteed unless the same expected variance is used for all the computations.

If M is partitioned into three components, $M_{group\ 1}$, $M_{group\ 2}$, and $M_{difference}$, both $M_{difference}$ and one of the two group statistics must have positive degrees of freedom in order to permit a test of the difference, hence, at least larger of group 1 and group 2 must include at least 3 columns. The design implications of this are discussed elsewhere (Hendrickson, 1978). There is, of course, no requirement that both groups include at least 3 columns.

Unless some way is developed for computing $M_{difference}$ directly, analogous to $Q_{difference}$ (equation 47), the present course of computing $M_{difference}$ by subtraction will need to be continued.

If s_1 and s_2 are the numbers of columns in groups 1 and 2 respectively, the residual degrees of freedom for $M_{difference}$ are $s_1 s_2 - 1$, subject to the preceding restrictions on the partitioning of M and its degrees of freedom; obviously

$$(s-1)(s-2)/2 \equiv (s_1-1)(s_1-2)/2 + (s_2-1)(s_2-2)/2 + s_1 s_2 - 1, \quad (51)$$

for s_1 and s_2 both positive integers, $s_1 + s_2 = s$.

Lastly in this section, I wish to note that Barlow *et al.* (1972) provide a number of tests in which the alternative hypothesis is a strict ordering of the expectation at various points along a gradient. One version of this is a non-parametric test which can be applied to a series of points for which only single observations are available. In essence, the procedure consists just in smoothing the data into strict (but weak) order, using the Pool Adjacent Violators algorithm. Once the data fit strict order, the number of distinct values is counted, and compared with the tabulated probability of finding at least that many distinct values, in order, in a series of that length under the null hypothesis of no order. If that probability is small enough, the alternative hypothesis of strict order can be selected. This is applied in several of the examples.

4. SOME ECOLOGICAL EXAMPLES

As far as feasible, I have tried to select some examples in which I have a personal ecological interest and which also permitted consideration of at least one of the kinds of study described in Section 2. The presentation of each example is rather abbreviated in this paper, primarily because I am not, generally, stating fully new ecological conclusions.

4.1 Sampling a Series of Alpine Meadows. If what is hypothesized is a strict ordering on a gradient, the statistical tools exist which will both force the data to fit and also evaluate whether too much force was required to admit the hypothesis of strict order. Hence, if a direct ordination is of interest, it is often possible to smooth the data to conform to the stated order, and then to examine how much smoothing was necessary. In the published tables for a test for simple order (Barlow *et al.*, 1972) the probability tabled is for the null hypothesis of no order when considered against the specific alternative hypothesis that strict simple order obtains. This is a more stringent test than simply a test for a positive correlation between sample value and sample position.

Even though I'm an entomologist by training, a botanical example, from a classic study, would suffice to illustrate this point. Paul Jaccard and two other botanist surveyed 10 Alpine meadows in the course of one summer. They recorded, for purposes of Jaccard's (1901) paper, only those species of flowering plants which were recognizable because of buds, flowers or fruits. On examining the itinerary and the sampling techniques, it appears that a minimum of twelve days, and probably about twenty days elapsed between the first and last meadows. Jaccard had already completed work on the general floras of these areas and felt that some of the low values of species in common were unexplained by his knowledge of the general flora. Jaccard, however, argued that the data were comparable.

My own knowledge of Alpine flora may be weak, but a few ideas strike me. At least in North America, the floral expression is not phenologically unified at any one site - some plants bud, bloom, and fruit much earlier than others. Secondly, I don't tend to notice every kind of plant in the first visit to any meadow, so there might be a learning curve involved. Both of these ideas suggest that these three botanists should have tended to record more species as time passed.

Interestingly, at least for this paper, whether one can draw this conclusion depends on how the alternate hypothesis is phrased. If the alternative hypothesis is simple order, we cannot reject the null hypothesis of no order. However, if the alternative hypothesis is merely that a positive relationship exists between the order of sampling and the number of species recorded, the alternative hypothesis is selected. Such a result might be plausible if some early blooming plants tended to lose their fruits before the botanists reached the last few sites, while other plants were coming into bud or bloom during the study (See Tables 5-7.) I would concur in selecting the phenological arguments as opposed to the learning curve anyway, since Jaccard felt that he could not explain why some sites he regarded as having similar floras had such low indices of overlap; however, I must leave the question to some future botanical study, if it is of any interest at all. (One cannot restudy the original ten meadows, since several have been extensively manipulated, either by regular 'haying' or by inundation by the waters stored in reservoirs for hydro-electric generation.)

4.2 A Midstream Riffle 'Community'. The microhabitats in a stream riffle tend to be more variable along the margins than in

TABLE 5: Localities in Vaud, Valais, or Oberland, Switzerland, of the 10 *alpine meadows studied by Jaccard* (1901) *and colleagues. Number of species of flowering plants recorded in the* 100 *meter-wide transect from* ca. 1900-2300 *meters in elevation is listed for each locality.*

Locality Number	Name	Number of species
1	Plan la Chaud (val Ferret)	101
2	La Peulaz (val Ferret)	107
3	Col Ferret	106
4	Alpes de Tsessettaz (combe de La, Entremont)	99
5	Alpage des Vingt-Huit (Bagnes)	140
6	Barberine (Trient)	114
7	Luisin (Emaney)	173
8	Gagnerie (Salanfe)	165
9	Iffigen (Wildhorn)	147
10	Kúh Dungel (Wildhorn)	150

TABLE 6: Number of species recorded from exactly n *meadows, after Jaccard* (1901). *See Table 5 for explanation.*

n sites	number of species	n sites	number of species
1	108	6	20
2	73	7	22
3	43	8	19
4	33	9	17
5	32	10	3

TABLE 7: Numbers of species in common between pairs of meadows, from Jaccard (1901). *Site numbers refer to localities listed in Table 5.*

Sites	1	2	3	4	5	6	7	8	9	10
1	X									
2	53	X								
3	59	57	X							
4	57	42	58	X						
5	66	68	69	65	X					
6	38	50	50	45	71	X				
7	48	65	59	53	73	66	X			
8	58	60	60	61	84	62	69	X		
9	64	48	53	69	76	60	69	74	X	
10	60	62	56	64	53	66	82	86	89	X

midstream. If samples are to be gathered that would give the appearance of being from a single community, midstream would be the most likely habitat. As an example, I've selected the two midstream samples from each of the ten transects sampled by Needham and Usinger (1956) in Prosser Creek, California. Samples 5 and 6 were from the middle of the first (most downstream) transect, while samples 95 and 96 were the middle elements of the tenth transect. These samples are presented in Tables 8-9. The M and Q tests were applied to these data.

The observed Q-statistic, 30.44, is slightly greater than $\chi^2_{.05}(19) = 30.14$, which suggests looking at whether there are distinct outliers. The observed distribution of the C_j terms is more platykurtic than heavy-tailed or long-tailed, and hence without other evidence I see no basis for regarding these samples as heterogeneous in terms of number of taxa. We do in this instance, have no evidence that might lead us to suspect that 4 of these samples might tend to have more species than the other 16 samples.

Needham and Usinger (1956) had used a "modified Latin square design" in order to detect and statistically remove any effects due to differences in sampling techniques among the five individuals who made the collections. They found that one individual (designated as 'H') had misinterpreted instructions and included rocks which lay mainly outside the sample perimeter, effectively sampling 1.57 times the area sampled by the other four investigators, which increased the total number of individuals collected. Chutter (1972) has reviewed other aspects of the analyses of these data. Following Sanders (1968) (among many possible authors) we could note that it is usually observed that if two samples which are otherwise similar have different numbers of individuals, one would expect the larger one to contain more species. On this basis we have at least an *ex post* hypothesis that one of the five investigators should have collected more species. 'H' collected samples 6, 45, 56, and 95 among those we are considering. If we treat 'H''s samples as one group and the rest as the other group, we obtain $Q_H = 2.56$ with 3 degrees of freedom, $Q_{\text{not H}} = 21.90$ with 15 degrees of freedom, and $Q_{\text{difference}} = 5.98$ with 1 degree of freedom. Of these, only Q_{diff} is even suspiciously large ($\Pr\{\chi^2(1) > 5.98\} = 0.0145$) and it does appear that 'H' sampled about 2.5 more species the other ivestigators, on average.

The observed M statistic, 211.24, is significant ($0.025 > P > 0.01$) when compared with a χ^2-distribution for 171

TABLE 8: Insect taxa occurring in twenty midstream samples from a riffle in Prosser Creek, California. (Extracted from Needham and Usinger (1956: Table 3,6). A '1' indicates the occurrence of this taxon in this sample. R_i is the number of occurrences of the ith taxon. C_j is the number of insect taxa recorded for the jth sample.

	5	6	15	16	25	26	35	36	45	46	55	56	65	66	75	76	85	86	95	96	R_i
Alloperla	1	1	0	1	0	1	1	0	1	0	1	1	1	1	1	0	1	1	0	1	14
Isogenus	0	0	0	0	0	0	1	0	1	0	0	0	0	0	0	0	0	0	0	0	2
Nemoura	0	0	0	0	1	1	0	0	0	0	0	0	0	0	0	0	0	0	0	0	2
Isoperla	0	1	0	0	0	0	0	0	0	0	0	0	0	0	0	0	1	0	1	0	3
Baetis	1	1	1	1	1	1	1	1	1	1	1	1	0	0	1	1	1	1	1	1	18
Ameletus	0	0	0	0	0	0	0	0	0	0	0	1	0	0	0	0	0	0	0	0	1
Cinygmula	1	1	1	1	1	1	1	1	1	1	1	1	1	1	1	0	1	1	1	0	18
Rhithrogena	1	1	1	1	1	1	1	1	1	1	1	1	1	1	1	1	1	1	1	1	20
Iron	1	1	1	1	1	1	1	1	1	1	1	1	1	1	1	0	1	1	1	1	19
Ephemerella	1	1	1	1	1	1	1	1	1	1	1	1	1	0	1	1	1	1	1	1	19
Sericostoma	0	0	0	0	0	0	1	0	0	0	0	1	0	1	0	0	1	1	1	0	6
Glossosoma	1	1	1	1	1	1	1	1	1	1	1	1	1	1	1	1	1	1	1	1	20
Hydropsyche	0	0	0	0	1	0	0	0	1	1	1	0	0	1	0	0	0	0	0	0	5
Brachycentrus	1	1	0	0	1	0	0	1	1	0	1	0	0	0	0	0	0	0	1	1	8
Lepidostoma	0	1	1	1	0	0	0	0	1	0	1	1	0	0	1	1	1	1	1	0	11
Rhyacophila	0	0	1	0	0	0	0	0	0	1	1	1	0	0	1	1	1	0	0	0	7
Limnephilidae	0	0	0	0	0	0	0	0	1	0	0	0	0	0	0	0	0	0	0	0	1
Dytiscidae	0	0	0	0	0	0	1	0	0	1	1	0	0	0	0	0	0	0	1	0	4
Helophorus	0	0	0	0	1	0	0	0	0	0	0	0	0	0	0	0	0	0	0	0	1
Elmus	0	0	0	0	0	0	0	0	0	0	0	1	0	0	0	0	1	0	0	0	2
Psephenidae	1	1	0	0	1	1	1	1	0	1	1	1	0	1	0	0	1	1	1	1	14
Heleidae	0	0	0	0	0	0	0	0	0	1	0	0	0	0	0	0	0	0	0	0	1
Chironomidae	0	0	1	1	1	1	1	1	1	1	1	1	1	1	1	1	0	1	1	1	17
Limnophila	0	0	0	0	0	0	0	0	0	0	0	0	0	0	0	1	0	0	0	0	1
Hexatoma	0	0	1	1	0	0	0	0	0	0	0	0	0	0	0	0	0	0	0	1	3
Eriocera	0	0	0	0	0	0	0	0	1	0	0	0	0	0	0	0	0	0	0	0	1
Blepharoceridae	0	0	0	0	0	0	0	0	0	0	0	0	0	0	0	0	0	0	1	0	1
Antocha	0	0	0	1	0	1	0	0	1	1	1	0	0	0	0	0	0	0	0	1	6
Tabanidae	0	0	0	0	0	0	0	1	0	0	0	0	0	0	0	0	0	0	0	0	1
Nostoc Chironomidae	0	0	0	0	0	0	0	0	0	1	1	0	0	0	0	0	0	1	0	0	3
C_j	9	11	10	11	12	11	12	10	15	14	16	14	7	9	10	8	13	12	14	11	

TABLE 9: Numbers of species in common for pairs of samples from Needham and Usinger data given in Table 8.

Samples	5	6	15	16	25	26	35	36	45	46	55	56	65	66	75	76	85	86	95	96
5	X																			
6	9	X																		
15	6	7	X																	
16	7	8	9	X																
25	8	8	7	7	X															
26	8	8	7	9	9	X														
35	8	8	7	8	8	9	X													
36	8	8	7	7	9	8	8	X												
45	8	9	0	10	9	9	9	8	X											
46	7	7	8	8	9	9	9	8	9	X										
55	9	10	9	10	10	10	10	9	12	13	X									
56	8	9	9	9	8	9	10	8	9	9	11	X								
65	6	6	6	7	6	7	7	6	7	6	7	7	X							
66	6	6	5	6	7	7	8	6	7	7	8	8	6	X						
75	7	8	9	9	7	8	8	7	9	8	10	10	7	6	X					
76	4	6	7	6	5	5	5	5	6	6	7	7	4	3	7	X				
85	8	10	8	8	7	8	9	7	8	8	10	12	6	7	9	6	X			
86	8	9	8	9	8	9	10	8	9	9	11	11	7	8	9	6	10	X		
95	8	10	8	8	9	8	9	9	9	9	11	10	6	7	8	6	10	10	X	
96	8	8	7	9	8	9	8	8	9	8	10	8	6	6	7	5	7	8	8	X

degrees of freedom. Hence, there is some indication that we should consider dividing this group of samples on the basis of M-alone. Since we have divided the samples between those taken by 'H' and by others, we should estimate M_H, $M_{not\ H}$, and $M_{difference}$. $M_H = 0.63$ (3 degrees of freedom), $M_{not\ H} = 133.07$ (105 degrees of freedom; $0.05 > P > 0.025$), and $M_{difference} = 77.54$ (63 degrees of freedom; $P > 0.10$). These do not suggest that much of the variation in matches is due to the 'extra species' collected by 'H'; a plausible explanation is, in part, that 'H' was more efficient at retaining material in deep-water samples, where the Surber sampler is least effective (Chutter, 1972).

4.3 Mayflies in Ontario Streams. Ide (1935) sampled both riffles and pools in his study of this distribution of mayflies along a thermal gradient in Ontario streams. His species occurrence data are presented in Table 10, and the pairwise species-in-common values are shown in Table 11.

As in many of the River Continuum studies now in procress, Ide's gradient is formed from riffles occurring in two similar watersheds. Stations 1, 2, and 3 occur in descending order along a small tributary, while stations 4, 5, and 6 occur in descending order on a much larger stream, but the two segments can be considered as joined in forming a series of samples along a riffle continuum. The number of species of mayflies increases monotonically downstream from Station 1 to 6 as would be predicted, and most of the pairwise matches among riffles also occur in strict order of decreasing overlap with increasing number of intervening stations. There is heterogeneity as shown by $Q = 39.8$ and $M = 100.9$.

The four pool samples, A, B, C, D, form a downstream sequence ending just upstream from station 4, but since pools are a very different habitat, serving essentially for the storage of detritus, they are not part of the riffle continuum, and do not even seem to follow strict order. This is consistent with the silence of the river continuum hypothesis on the relationship between stream order and pool community function.

4.4 Stoneflies and Caddisflies in Ontario Streams. Sprules (1947) reported on a number of stations in the Madawaska River basin, Ontario. As did Ide, he sampled several riffles, and his data for Plecoptera are presented in Table 12 (and the species in common for pairs of stations in Table 13), and for Trichoptera in Table 14 (species in common in Table 15).

TABLE 10: Ephemeroptera recorded by Ide (1935) *from three riffles in a tributary of the Pine River* (1, 2, 3), *and three riffles* (4, 5, 6) *and four pools* (A, B, C, D) *in the Mad River, Ontario.*

Stations	1	2	3	4	5	6	A	B	C	D
Arthroplea bipunctata							1		1	
Baetis brunneicolor	1	1	1			1				
B. cingulatus		1	1	1	1	1				
B. frondalis								1		
B. intercalaris		1	1	1	1	1				
B. parvus		1	1	1	1	1				
B. pygmaeus								1	1	1
B. vagans	1	1	1	1	1	1				
Blasturus nebulosus							1	1	1	1
Caenis sp.								1		1
Callibaetis americana							1			
Centroptilum bellum (?)				1				1	1	
C. convexum							1	1	1	1
Cloeon rubropicta									1	1
C. simplex									1	1
C. sp.										1
Ephemera guttulata						1				
E. simulans								1	1	1
Ephemerella aurivilli	1	1								
E. deficiens				1	1	1				
E. depressa		1	1		1					
E. excrucians						1				
E. fuscata					1	1				
E. invaria			1	1	1	1				
E. needhami					1	1				
E. serrata				1		1				
E. subvaria		1	1	1	1	1				
E. temporalis						1	1	1		1
Epeorus humeralis		1	1	1	1	1				
Habrophlebiodes americana				1	1	1				
Heptagenia hebe				1	1	1			1	
H. pulla	1	1	1	1	1	1				
Hexagenia viridescens								1	1	1
Iron pleuralis	1	1	1	1	1					
Isonychia bicolor				1	1	1				
Leptophlebia adoptiva	1	1	1	1	1	1				
L. debilis	1	1	1	1		1	1	1	1	
L. guttata					1	1				
L. mollis			1	1	1	1				
Pseudocloeon carolina		1	1	1	1	1				
P. sp.								1	1	1
Siphlonurus alternatus							1		1	1
S. quebecensis							1		1	1
Stenonema canadense				1	1	1	1	1	1	
S. fuscum		1	1	1	1	1	1	1	1	
S. (fuscum group)						1				
S. heterotarsale						1				
S. nepotellum						1				
S. tripunctatum							1	1	1	1
Tricorythodes atrata								1		1
	7	15	16	21	22	29	11	15	17	15

TABLE 11: Numbers of species in common for pairs of samples for Ide's (1935) *mayfly data given in Table 10.*

Sites	1	2	3	4	5	6	A	B	C	D
1	X									
2	7	X								
3	6	14	X							
4	6	12	14	X						
5	4	12	14	18	X					
6	5	12	14	19	20	X				
A	1	2	2	3	2	4	X			
B	1	2	2	4	2	4	7	X		
C	1	2	2	5	3	4	9	11	X	
D	0	0	0	0	0	1	6	10	11	X

TABLE 12: Plecoptera recorded by Sprules (1947) from three riffle stations (2, 3a, 3) in Mud Creek and at a riffle station (4) in the Madawaska River, Algonquin Park, Ontario

	Station			
	2	3a	3	4
Acroneuria abnormis		1	1	1
A. lycorias				1
Allocapnia pygmaea	1	1		
Alloperla imbecilla			1	
A. mediana	1			
Hastaperla brevis			1	1
Hydroperla subvarians				1
Isoperla montana	1	1	1	
I. sp. 1		1	1	
I. sp. 2				1
I. transmarina				1
I. truncata		1	1	1
Leuctra biloba	1			
L. decepta	1	1	1	
L. hamula	1	1	1	
L. sara	1			
L. sibleyi	1	1	1	
L. tenuis		1	1	
Nemoura punctipennis	1	1	1	
N. serrata	1			
N. trispinosa	1			
N. venosa	1	1		
	12	11	11	7

TABLE 13: Numbers of Plecoptera species in common to pairs of samples in Sprules' (1947) data presented in Table 12.

Stations	2	3a	3	4
2	X			
3a	8	X		
3	6	16	X	
4	2	10	14	X

TABLE 14: Trichoptera recorded by Sprules (1947) *from three riffle stations* (2, 3a, 3) *in Mud Creek and a riffle station* (4) *in the Madawaska River, Algonquin Pack, Ontario.*

	Station					Station			
	2	3a	3	4		2	3a	3	4
Agapetus sp.		1	1		*L. ontario*	1			
Agraylea costello			1	1	*Macronema zebratum*				1
Athripsodes alces				1	*Micrasema sprulesi*	1			
A. angustus		1	1	1	*M. wataga*				1
A. dilutus		1	1	1	*Mystacides sepulchralis*		1		1
A. wetzeli				1	*Neophylax autumnus*	1	1		
Cheumatopsyche campyla			1	1	*Neureclipsis crepuscularis*				1
C. gracilis				1	*N. parvulus*				1
C. minisculа				1	*Nyctiophylax vestitus*		1	1	1
C. pettiti		1	1		*Oecetis avara*				1
Chimarrha aterrima		1	1	1	*Osythira* sp.			1	1
C. lucia				1	*Parapsyche apicalis*	1			
C. socia				1	*Philopotamus distinctus*	1	1	1	
Dolophilus moestus	1				*Phylocentropus placidus*			1	1
Goera stylata	1				*Plectrocnemia cinerea*		1	1	1
Hydropsyche betteni		1	1		*Polycentropus confusus*	1	1	1	1
H. dicantha				1	*Psychomyella flavida*			1	1
H. morosa				1	*Psychomyia diversa*	1	1	1	1
H. recurvata				1	*Rhyacophila carolina*	1	1	1	
H. sparna		1	1	1	*R. fuscula*		1	1	1
H. ventura	1	1			*R. vibox*	1			
Hydroptila sp.	1				*R. vuphipes*				1
Lepidostoma grisea	1	1	1		*Stenophylax guttifer*	1	1	1	
						15	19	21	30

TABLE 15: Numbers of Trichoptera species in common to pairs of samples in Sprules' (1947) *data presented in Table 14.*

Stations	2	3a	3	4
2	X			
3a	7	X		
3	5	9	X	
4	0	2	3	X

TABLE 16: Combined Plecoptera and Trichoptera results from Tables 12-15.

a) Combined numbers of species (Tables 12 and 14).

Station	Plecoptera and Trichoptera
2	27
3a	30
3	32
4	37

b) Combined numbers of species in common to pairs of samples (Table 13 and 15).

Station	2	3a	3	4
2	X			
3a	15	X		
3	11	25	X	
4	2	12	17	X

Although the number of species of Plecoptera does decline as would be expected from the gradient in dissolved oxygen, $Q = 2.49$ is not significant. However, $M = 22.64$ is highly significant, indicating that the species in common are not independent, and it can be seen from Table 13 that the number of species in common declines uniformly as a function of distance.

The number of species of Trichoptera, on the other hand, increases uniformly going downstream, and here both $Q = 10.13$ and $M = 31.17$ are significant. Again the distribution of species is consistent with the stream gradient.

When data for Trichoptera and Plecoptera are treated in the same Table (16), $Q = 2.49$ is no longer significant, although $M = 21.37$ certainly is, and the strict ordering of species in common is preserved.

4.5 Insects in a Biomonitoring Study. Hendrickson (1978) presented data on non-insect macroinvertebrates from Wurtz and Dolan (1960), finding that Station 10 was significantly poorer than the group of stations 1, 2, 3 and 5, using both $Q_{difference}$ and $M_{difference}$. However, when the insect data (Tables 17 and 18) from Wurtz and Dolan (1960) are analyzed, the same pattern is not found.

This again serves to question the generality of the conclusions of Kaesler *et al.* (1974) that fewer taxonomic groups need to be sampled in biomonitoring studies. This subject warrants further consideration, which may become possible if Q and M are accepted as appropriate test statistics in the biomonitoring setting.

5. COMMENTS ON THE USE OF THE JACCARD COEFFICIENT IN ECOLOGY

As I mentioned in Section 3, most of the mathematical treatment in that section arose because I was asked to provide guidance on the interpretation of a particular, published triangular matrix of Jaccard coefficients and of the dendrogram which resulted from clustering those coefficients using the unweighted pair-group method with arithmetic averages (Sokal and Sneath, 1963). I have since concluded that the particular example is uninterpretable as it stands, and that conclusion would extend to most published ecological examples. This conclusion does not rest solely on any single point raised so far in this paper.

TABLE 17: List of insects recorded by Wurtz and Dolan (1960) *from five stations in the Schuylkill River, Pennsylvania*

	Station					
	1	2	3	5	10	R_i
Pseudocloeon sp.	1	0	0	0	0	1
Boyeria vinosa	1	0	0	0	0	1
Macromia sp.	0	0	1	0	0	1
Gomphus spiniceps	0	0	1	0	0	1
Perethemis tenera	0	0	1	0	1	2
Plathemis lydia	0	0	0	0	1	1
Libellulinae g. sp.	0	0	1	0	0	1
Argia sp. 1	1	1	1	1	1	5
Argia sp. 2	1	1	0	0	1	3
Enallagma sp. 1	0	0	1	0	0	1
Ischnura sp. 1	1	0	1	0	1	3
Ischnura sp. 2	0	0	1	0	0	1
Ischnura sp. 3(?)	0	0	1	0	0	1
Coenagrionidae g. sp. 1	0	0	1	0	0	1
Hydroptilidae g. sp. (pupa)	0	1	0	0	0	1
Stenelmis sp. 1	1	1	0	0	0	2
Stenelmis sp. 2	1	0	0	0	0	1
Stenelmis sp. 3	1	0	0	0	0	1
Stenelmis sp. 4	0	1	0	0	0	1
Helichus sp.	1	0	0	0	0	1
Dubiraphia sp.	1	0	0	0	0	1
Psephenus prob. *lecontei*	0	1	0	0	0	1
Hydrophilidae g. sp.	0	0	1	1	0	2
Berosus prob. *peregrinus*	1	0	0	0	0	1
Parargyractis sp.	1	1	0	0	0	2
Simulium vittatum	1	1	0	0	0	2
Pentaneura cf. *basalis*	0	0	1	1	0	2
Pentaneura sp.	0	0	1	1	1	3
Cricotopus bicinctus	1	1	1	0	1	4
Cricotopus nr. *trifasciatus*	0	0	1	0	0	1
Psectrocladius sp.	0	0	1	1	1	3
Calopsectra sp.	1	0	0	0	0	1
Polypedilum illinoense	1	1	1	0	1	4
Glyptotendipes senilis	1	0	0	1	0	2
C_i	17	10	17	6	9	59

TABLE 18: Number of insect species (below diagonal) and total macroinvertebrate species (above diagonal) in common between pairs of Schuylkill River stations. After Wurtz and Dolan (1960) *and Hendrickson* (1978).

Stations	1	2	3	5	10
1	x	19	16	13	14
2	7	x	12	10	11
3	4	3	x	16	13
5	2	1	5	x	10
10	5	4	7	3	x

The Jaccard coefficient for a pair of samples can be stated as the number of species in the intersection of the two species lists divided by the number of species in the union. It is clear that Jaccard (1901) intended this intersection/union approach to apply to larger aggregates than pairs, and that he computed the index in this fashion for groups of four, six, eight, or all ten sites. In original usage, then, the coefficients were not to be clustered, but rather the samples.

Gower (1971) does suggest that the Jaccard coefficient might be suitable for ecological data solely on the grounds that D, the number of jointly absent species, cannot reasonably be estimated in many cases. He then provides an infinite series expression involving D as part of his formula for the expectation of the Jaccard coefficient, where D is once again defined in the usual way based on the available species list. This may qualify at least as inconsistent with the rationale for selecting the coefficient.

As the Jaccard coefficient has been used in ecology, species which are invariably present (at least in the samples under consideration) enter into all the pairwise coefficients. Since this tends to inflate the estimated similarity of a species-poor station with either a species-rich or a species-poor station, as opposed to excluding invariant species, or using a constant denominator (as in R_{jk}), this use of the Jaccard coefficient seems to particularly objectionable in biomonitoring studies.

As Cormack (1971) and other authors have pointed out, clustering methods are not hypothesis-testing procedures. In particular (Hendrickson, 1978) the clustering of a matrix of Jaccard coefficients may obscure clear differences between impact and control stations. Many clustering algorithms (particularly those commonly used in ecology) will cluster coefficients based on independently distributed observations without providing any indication that the imposed structure is merely 'noise.' Related to this is the common practice of deciding that the last sample to enter the main cluster must be an 'outlier.' Certainly we should have learned enough in the past fifteen years about the unanticipated, but axiomatic, behavior of clustering algorithms given particular data configurations to avoid jumping to such conclusions.

ACKNOWLEDGEMENTS

By the time a paper has grown to this size, it's a difficult task to thank everyone who has helped in significant ways. The sheer assembly of the paper was tremendously assisted by our excellent and unflappable Clerical Services staff, headed by Mrs. Eleanor Thomas.

Various aspects of the aquatic applications were based on suggestions or information provided by C. E. Goulden, J. W. Richardson, Jr., B. W. Sweeney, R. Horwitz, R. L. Vannote, G. W. Minshall, R. C. Petersen, R. Patrick, C. E. Cushing, and K. W. Cummins. Portions of this work were funded by a National Science Foundation grant, number BMS-75-07333, to the Academy of Natural Sciences. This paper, is issued as Number 7 in the River Continuum Contribution series. Additional research funds were provided by the Division of Limnology and Ecology of the Academy of Natural Sciences of Philadelphia.

On the statistical side, my thanks go to N. Mantel, D. K. Hildebrand, M. I. Artuz, G. P. Patil, R. H. Green, and R. L. Kaesler. S. Zeger, B. J. Wilson, and J. N. Hansen provided some early feedback.

Among many biologists who have commented on the biomonitoring applications, I particularly recall R. Giles, S. L. H. Fuller, R. Patrick, S. Halterman, S. S. Roback, C. B. Wurtz, and J. Cairns, Jr.

I apologize for my oversights to those others whom I should have included. Last, and perhaps most, I thank my wife for putting up with all the false leads and long hours I've devoted to discovering the right way to add zero or multiply by one.

REFERENCES

Arrhenius, O. (1921). Species and area. *Journal of Ecology*, 9, 95-99.

Arrhenius, O. (1923). On the relation between species and area. A reply. *Ecology*, 4, 90-91.

Barlow, R. E., Bartholomew, D. J., Bremner, J. M., and Brunk, H. D. (1972). *Statistical Inference Under Order Restrictions. The Theory and Application of Isotonic Regression*. Wiley, New York.

Barton, D. E. and David, F. N. (1959). The dispersion of a number of species. *Journal of the Royal Statistical Society, Series B*, 21, 190-194.

Bishop, Y. M. M., Fienberg, S. E., and Holland, P. W. (1975). *Discrete Multivariate Analysis*. M.I.T. Press, Cambridge, Massachusetts.

Chutter, F. M. (1972). A reappraisal of Needham and Usinger's data on the variability of a stream fauna when sampled with a Surber sampler. *Limnology and Oceanography*, 17, 139-141.

Clausen, J. (1965). Population studies of Alpine and Sub-Alpine races of conifers and willows in the California High Sierra Nevada. *Evolution*, 19, 56-68.

Cochran, W. (1950). The comparison of percentages in matched samples. *Biometrika*, 37, 256-266.

Cole, L. C. (1957). The measurement of partial interspecific association. *Ecology*, 38, 226-233.

Cormack, R. M. (1971). A review of classification. *Journal of the Royal Statistical Society, Series A*, 134, 321-367.

Crossman, J. S., Kaesler, R. L. and Cairns, J., Jr. (1974). The use of cluster analysis in the assessment of spills of hazardous materials. *American Midland Naturalist*, 92, 94-117.

Cummins, K. W. (1977). From headwater streams to rivers. *American Biology Teacher*, 39, 305-312.

Diamond, J. M. (1975). Assembly of species communities. In *Ecology and Evolution of Communities*, M. L. Cody and J. M. Diamond, eds. Belknap Press, Cambridge, Massachusetts, 342-444.

Diamond, J. M. and May, R. M. (1977). Species turnover rates on islands: Dependence on census interval. *Science*, 197, 266-270.

Feller, W. (1968). *An Introduction to Probability Theory and Its Applications, Vol. I*, 3rd edition. Wiley, New York.

Filice, F. P. (1959). The effect of wastes on the distribution of the bottom invertebrates in the San Francisco Bay estuary. *Wasmann Journal of Biology*, 17, 1-17.

Fleiss, J. L. (1965). Estimating the accuracy of dichtomous judgments. *Psychometrika*, 30, 469-479.

Fleiss, J. L. (1973). *Statistical Methods for Rates and Proportions*. Wiley, New York.

Gleason, H. A. (1925). Species and area. *Ecology*, 6, 66-74.

Gleason, H. A. (1926). The individualistic concept of the plant association. *Bulletin of the Torrey Botanical Club*, 53, 237-266.

Goodall, D. W. (1967). The distribution of the matching coefficient. *Biometrics*, 23, 647-656.

Goodall, D. W. (1969). A procedure for recognition of uncommon species combinations in sets of vegetation samples. *Vegetatio*, 18, 19-35.

Goodall, D. W. (1973). Sample similarity and species correlation. In *Handbook of Vegetation Science*. R. Tuxen (ed.) Dr. W. Junk, The Hague. Part V, 105-156.

Gower, J. C. (1971). A general coefficient of similarity and some of its properties. *Biometrics*, 27, 857-874.

Hendrickson, J. A., Jr. (1978). Statistical analysis of the presence-absence component of species composition data. In *Biological Data in Water Pollution Assessment: Quantitative and Statistical Analyses, ASTM STP 652*. K. L. Dickson, J. Cairns, Jr., and R. L. Livingston, eds. American Society for Testing and Materials, Philadelphia. 113-124.

Hogg, R. V. (1965). On models and hypotheses with restricted alternatives. *Journal of the American Statistical Association*, 60, 1153-1162.

Horwitz, R. J. (1978). Temporal variability patterns and distributional patterns of stream fishes. *Ecological Monographs*, 48, 307-321.

Hynes, H. B. N. (1970). *The Ecology of Running Waters*. University of Toronto Press, Toronto.

Ide, F. P. (1935). The effect of temperature on the distribution of the mayfly fauna of a stream. *University of Toronto Studies, Biological Series*, 39, 9-73.

Ide, F. P. (1940). Quantitative determination of the insect fauna of rapid water. *University of Toronto Studies, Biological Series*, 47, 5-20.

Jaccard, P. (1901). Distribution de la flore alpine dans le Bassin des Dranses et dans quelques regions voisines. *Bulletin de la Societe vaudoise des Sciences naturelles*, 37, 241-272.

Kaesler, R. L., Cairns, J., Jr., and Crossman, J. S. (1974). Redundancy in data from stream surveys. *Water Research*, 8, 637-642.

Mantel, N. (1974). Approaches to a health research occupancy problem. *Biometrics*, 30, 355-362.

Mantel, N. and Fleiss, J. L. (1975). The equivalence of the generalized McNemar tests for marginal homogeneity in 2^3 and 3^2 tables. *Biometrics*, 31, 727-729.

McNemar, Q. (1947). Note on the sampling error of the difference between correlated proportions or percentages. *Psychometrika*, 12, 153-157.

Needham, P. R. and Usinger, R. L. (1956). Variability in the macrofauna of a single riffle in Prosser Creek, California, as indicated by the Surber sampler. *Hilgardia*, 24, 383-409.

Patrick, R. (1949). A proposed biological measure of stream conditions, based on a survey of the Conestoga Basin, Lancaster County, Pennsylvania. *Proceedings of the Academy of Natural Sciences of Philadelphia*, 101, 277-341.

Pielou, E. C. (1971). Measurement of structure in animal communities. In *Ecosystem Structure and Function*, J. A. Weins, ed. Oregon State Universite Press, Corvallis. 113-135.

Pielou, E. C. (1972). 2^k contingency tables in ecology with an appendix by D. S. Robson. *Journal of Theoretical Biology*, 34, 337-352.

Russell, P. F. and Rao, T. R. (1940). On habitat and association of species of anopheline larvae in southeastern Madras. *Journal of the Malaria Institute of India*, 3, 153-178.

Simberloff, D. (1976). Experimental zoogeography of islands: Effects of island size. *Ecology*, 57, 629-648.

Simberloff, D. and Connor, E. F. (1979). Q-mode and R-mode analyses of biogeographic distributions: Null hypotheses based on random colonization. In *Contemporary Quantitative Ecology and Related Ecometrics*, G. P. Patil and M. Rosenzweig, eds. International Cooperative Publishing House, Fairland, Maryland.

Sneath, P. H. A. and Sokal, R. R. (1973). *Numerical Taxonomy*. Freeman, San Francisco.

Sokal, R. R. and Michener, C. D. (1958). A statistical method for evaluating systematic relationships. *University of Kansas Science Bulletin*, 38, 1409-1438.

Sokal, R. R. and Sneath, P. H. A. (1963). *Principles of Numerical Taxonomy*. Freeman, San Francisco.

Sprules, W. M. (1947). An ecological investigation of stream insects in Algonquin Park, Ontario. *University of Toronto Studies, Biological Series*, 56, 1-81.

Sweeney, B. W. (1976). A diurnally fluctuating thermal system for studying the effect of temperature on aquatic organisms. *Limnology and Oceanography*, 21, 758-763.

Sweeney, B. W. (1978). Bioenergetic and developmental response of a mayfly to thermal variation. *Limnology and Oceanography*, 23, 461-477.

Terborgh, J. (1971). Distribution on environmental gradients: Theory and a preliminary interpretation of distributional patterns in the avifauna of the Cordillera Vilcabamba, Peru. *Ecology*, 52, 23-40.

Terborgh, J. and Weske, J. S. (1975). The role of competition in the distribution of Andean birds. *Ecology*, 56, 562-576.

Vannote, R. L. The river continuum: A theoretical construct for analysis of river ecosystems. (manuscript) (Being revised for *Bioscience*.)

Williams, C. B. (1964). *Patterns in the Balance of Nature*. Academic Press, New York.

Wurtz, C. B. and Dolan, T. (1960). A biological method used in the evaluation of effects of thermal discharge in the Schuylkill River. *Proceedings of the Tenth Industrial Waste Conference* Purdue, Indiana, 461-472.

[*Received* *August* 1978. *Revised* *March* 1979]

G. P. Patil and M. Rosenzweig, (eds.),
Contemporary Quantitative Ecology and Related Ecometrics, pp. 399-420.

INTERPOLATION IN THE PLANE: The Robustness to Misspecified Correlation Models and Different Trend Functions

RUTH SHESHINSKI

Department of Statistics
The Hebrew University of Jerusalem
Jerusalem, Israel

SUMMARY. We deal with spatially autocorrelated random processes in the plane. Optimal linear interpolation between discrete observations in R^2 depends on the random function model assumed to generate the two-dimensional field. Sensitivity of the optimal solution to errors in specification of the model is studied. Both misspecifications of the autocorrelation function and misspecifications of the trend are considered.

We have found that the optimal linear estimator is grossly insensitive to various misspecifications of the autocorrelation function (provided isotropic models are considered). However, if the misspecified autocorrelation function is used in the calculation of the squared interpolation error, then such error calculations can be grossly misleading.

The additional misspecification is an error in modeling the trend function (i.e., the phenomena measured is not necessarily a second order stationary process). Such misspecifications are examined in terms of the strength of the underlying autocorrelation model. For interpolation, in general, the MSE's tend to be close for the 'true' and misspecified trends except for high underlying correlation models. For extrapolation, however, the MSE values can be different.

As an application, we interpolated 16 air pollution monitoring stations, measuring Oxidant levels in the San Francisco Bay Area, using different correlation and trend models.

KEY WORDS. random functions, autocorrelation, misspecified spatial correlation, trend misspecification, interpolation in the plane, isopleth maps, oxidant levels, Kriging.

1. INTRODUCTION

The intent of the following work is to map pollutant concentrations, using a random-function approach, to estimate the needed values for the interpolation scheme. The correlations between various monitoring stations are needed in the estimation procedure and have to be modeled. The question that naturally arises is how robust the interpolation scheme is to misspecifications in the correlation model.

The application has been carried out on ozone measurements. Oxidants and PAN, as pollution measures, are especially important in ecological studies, as their harmful influence on plants and microorganisms has long been recognized. See National Academy of Sciences (1977).

The problem in hand is to map an area, with its ozone levels, under the following circumstances:

(1) The spatial distribution of the measured pollution process is unknown and it is by no means normal.

(2) Distributional assumptions, such as stationarity, are far too strict and are not representative of the data.

(3) We have no knowledge of the underlying spatial or time dependent correlations between the various monitoring stations. As far as we could judge from studying the ozone data, these correlations should probably be described by an anisotropic model, with an imbedded 'nugget' effect. The correlation, denoted $\rho(d)$ at distance d between two stations, is less than one for $d = 0$.

(4) We have a small sample, i.e., there is a discrete number of monitoring stations. (In our case $n = 16$, for the San Francisco Bay Area). No replications are available in the space-domain but, in air pollution measurements, there is a great deal of replication in the time-domain. For full generality one should deal with a time-space correlation function. However, in the present work, we consider only a spatial correlation model with $n(n-1)/2 = 120$ observed correlations between the various pairs of stations.

We have used a stochastic-process approach, called Kriging, which has been developed extensively by other authors, notably

Matheron (1965, 1973) and Delfiner (1976), and which takes its name from Krige (1951). The following is an elementary special case of the more general theory.

2. THE KRIGING METHOD

Let us have a stochastic process $Z(\underset{\sim}{X})$ at point $\underset{\sim}{X} \in R^2$. We take a sample of size n (monitoring stations), located at $s = (\underset{\sim}{X}_1, \underset{\sim}{X}_2, \cdots, \underset{\sim}{X}_n)$ and let Z_s denote $(Z(\underset{\sim}{X}_1), Z(\underset{\sim}{X}_2), \cdots, Z(\underset{\sim}{X}_n))'$.

$$Z(\underset{\sim}{X}) = M(\underset{\sim}{X}) + Y(\underset{\sim}{X}) \tag{1}$$

where $M(\underset{\sim}{X})$ is the trend function and $Y(\underset{\sim}{X})$ is an error function, which is a stochastic process with a smaller scale variation (random function with pattern). Let us define the estimator of Z as the linear estimator Z^*.

$$Z^*(\underset{\sim}{X}) = \sum_{i=1}^{n} C_i(\underset{\sim}{X}) \, Z(\underset{\sim}{X}_i). \tag{2}$$

We interpolate at point $\underset{\sim}{X}$.

The assumptions (under which we look for an optimal estimator Z^*) concerning the stochastic process Z are:

(1) $E[Z(\underset{\sim}{X}) | \underset{\sim}{X}] = M(\underset{\sim}{X})$ (i.e., the random function $Y(\underset{\sim}{X})$ has a zero expectation);

(2) $\frac{1}{2}E[Z(\underset{\sim}{X}+\underset{\sim}{h}) - Z(\underset{\sim}{X})]^2 = \gamma(\underset{\sim}{h})$ depends only on $\underset{\sim}{h}$.

$\gamma(\underset{\sim}{h})$ is called the semivariogram (or variogram) function. $\gamma(\underset{\sim}{h})$ does not necessarily have to be finite (Brownian motion, for example.)

At first, we simplify the model, and deal with simple Kriging. Further on, we add the trend function to the model.

2.1 Simple Kriging. Assume (1) $E[Z(\underset{\sim}{X})] = M$, a constant, and (2) $\gamma(\underset{\sim}{h}) < \infty$, i.e., the process $Z(\underset{\sim}{X})$ has a finite variance (WLOG=1). The second assumption is equivalent to the assumption that $Z(\underset{\sim}{X})$ is a second order weak stationary process with correlation function

$$\rho(\underset{\sim}{h}) = \frac{\text{Cov } (Z(\underset{\sim}{X}+\underset{\sim}{h}) \text{ , } Z(\underset{\sim}{X}))}{\text{Var } Z(\underset{\sim}{X})} = \text{Cov } (Z(\underset{\sim}{X}+\underset{\sim}{h}), Z(\underset{\sim}{X})), \text{ i.e.,}$$

$$\gamma(\underset{\sim}{h}) = \rho(\underset{\sim}{0}) - \rho(\underset{\sim}{h}) \quad \text{where} \quad \rho(\underset{\sim}{0}) = 1. \tag{3}$$

Under these assumptions we find an optimal unbiased linear estimator Z^* such that we have minimum mean squared error

$$\min_{C} E[Z^*(\underset{\sim}{X}) - Z(\underset{\sim}{X})]^2 = \min_{C} \text{Var}[Z^*]$$

under the constraint $E[Z^*(\underset{\sim}{X})] = Z(\underset{\sim}{X})$, i.e.,

$$\sum_{i=1}^{n} C_i = 1. \tag{4}$$

Applying Lagrange multipliers to minimize the MSE under the constraint given in formula (4) results in Z^* , the known BLUE of Z (Regression line) under the regressors Z_s (Goldberger, 1962).

$$Z^*(\underset{\sim}{X}) = \hat{M} + \Sigma'_{sx} \, \Sigma^{-1}_{ss} \, (Z_s - \hat{M} \, \underset{\sim}{1}) \tag{5}$$

where

$$\Sigma_{sx} = (\rho(\underset{\sim}{X}_1 - \underset{\sim}{X}), \rho(\underset{\sim}{X}_2 - \underset{\sim}{X}), \cdots, \rho(\underset{\sim}{X}_n - \underset{\sim}{X}))' \; ; \; n \times 1$$

$$\Sigma_{ss} = (\rho(\underset{\sim}{X}_i - \underset{\sim}{X}_j)) \; ; \; n \times n$$

$$\underset{\sim}{1} = (1, 1, \cdots, 1)' \; ; \; n \times 1$$

$$\hat{M} = \frac{\underset{\sim}{1}' \, \Sigma^{-1}_{ss} \, Z_s}{\underset{\sim}{1}' \, \Sigma^{-1}_{ss} \, \underset{\sim}{1}}$$

The mean squared error is

$$[\text{MSE}] = \Sigma'_{sx} \, \Sigma^{-1}_{ss} \, \Sigma_{sx} - \frac{\underset{\sim}{1}' \, \Sigma^{-1}_{ss} \, \Sigma_{sx} - 1)^2}{\underset{\sim}{1}' \, \Sigma^{-1}_{ss} \, \underset{\sim}{1}} \; ; \tag{6}$$

see Matheron (1973) or Delfiner (1976).

3. MISSPECIFICATION OF PARAMETERS IN $\rho(d)$

3.1 Optimal and Suboptiman Variance Comparisons. Let us assume that the spatial correlations between n monitoring stations $\rho_{ij} = \rho(d_{ij}) = \rho(\underset{\sim}{X}_i, \underset{\sim}{X}_j)$ is of an exponential form. $\rho(d) = \rho(|d|)$ is a function of the distance d_{ij} between $\underset{\sim}{X}_i$ and $\underset{\sim}{X}_j$:

$$\rho(d) = \exp\{-ud^2\} , \ u > 0. \tag{7}$$

The larger u is, the closer we are to the case of independent observations in which case we have $\rho(d) = \begin{cases} 0 & \text{if } h > 0 \\ 1 & \text{if } h = 0 . \end{cases}$

This case is of course the worst from the point of view of relevant information that we are able to extract in estimating Z out of the given Z_s .

We simulated a situation of having three (or four) points (monitoring stations) in the plane, trying to interpolate and extrapolate estimates for $Z(\underset{\sim}{X})$ with some underlying correlation model assumption. We decided on the exponential model in (7) with parameter u_o to be the 'true' correlation model and denoted the misspecified parametric value of u_o, by u. We simulated over the whole range of u_o $(0 \leq u_o \leq 1)$ and u $(0 \leq u \leq 1)$ (Results and tables are available from the author upon request). As a measure we defined the ratio R_1 between the mean squared error of the estimation error under the misspecification to the MSE under the 'true' parametric value.

$$R_1 = \left[\frac{MSE_{u_o}\ [Z^*_u\ (\underset{\sim}{X}) - Z(\underset{\sim}{X})]}{MSE_{u_o}\ [Z^*_{u_o}\ (\underset{\sim}{X}) - Z(\underset{\sim}{X})]}\right]^{\frac{1}{2}} \geq 1. \tag{8}$$

This ratio may be looked upon as the square root of suboptimal to optimal variance of estimation error.

It seems evident from the simulations that the linear interpolation procedure is not very sensitive to choice of coefficients C_i in Formula (2). Even if coefficients are calculated using a grossly misspecified parameter in the correlation model, the estimation error is not severely influenced.

By and large if the misspecified model is of stronger underlying correlation than the true model, the suboptimal coefficients are almost unchanged in comparison to the optimal ones. The errors are more noticeable (but still small) when the estimation point $\underset{\sim}{X}$ is close to an observation point.

The worst error appeared when we in fact had a very high correlation model $\rho_{u_o}(1) = 0.90$ and misspecified it by $\rho_u(1) = 0$ in the estimate Z*. (Misspecifying correlated observations by uncorrelated ones.) The most extreme case of misspecification is when $\rho_u(1) = 0$ $(u=\infty)$. This again illuminates the well-known fact that in the case where we deal with correlated observations, we should by no means treat them as independent ones.

For the more likely to appear situations, (where in practice we know that we deal with correlated observation but we misspecify the parameter of the model either by over or under estimating it) the errors are much smaller and may be considered in most practical applications as negligible.

If the true situation has independent observations, any misspecified model of correlation does not (for practical purposes) influence the precision of the estimate, even for locations close to an observation point.

As an examples of simulation, we assume the four stations located in the corners of a unit square as described in Figure 1 and interpolate the random function $Z(\underset{\sim}{X})$ at the points $\underset{\sim}{X} = \underset{\sim}{X}_0$ in the square.

The three 'true' underlying correlation models that we consider are:

$$\rho_{u_o}(1) = 0 \ (u_o = \infty) , \quad \text{Figure 2A}$$

$$\rho_{u_o}(1) = 0.3679 \ (u_o = 1) , \quad \text{Figure 2B}$$

$$\rho_{u_o}(1) = 0.9048 \ (u_0 = 0.1) , \quad \text{Figure 2C.}$$

At all three estimation points that we considered, the worst ratio is 3.88 at $\underset{\sim}{X}_o = (0.1, 0.1)$ for the extreme misspecification $\rho_u(1) = 0$ instead of $\rho_{u_o}(1) = 0.37$.

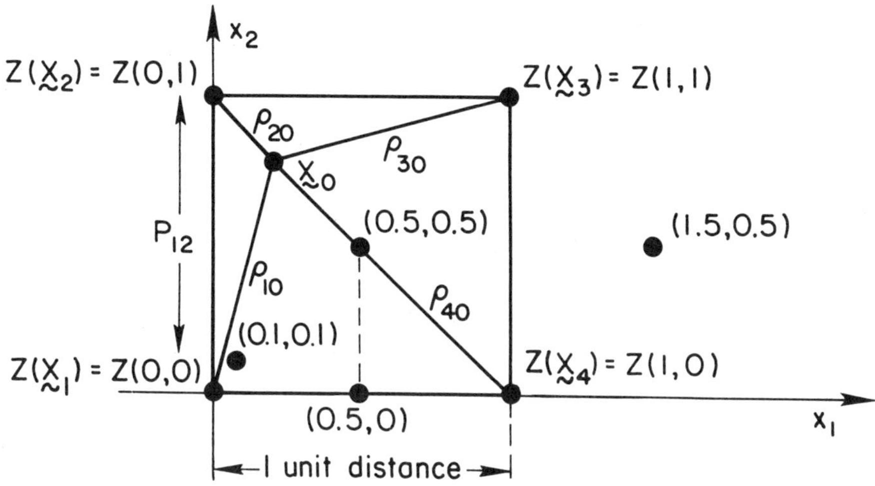

FIG. 1: Geometrical setup for 4 observations.

By choosing the above three mentioned estimation locations X_o, under three different 'true' correlation models, we tried to get some feeling how a misspecified parameter u would influence the performance ratio in equation 8. See Figure 1 for locations of the various X_o relative to the observations and consult Table 1 for results. We plot the performance ratios as a function of $\rho_u(1)$ in Figures 2A, 2B, and 2C. The range of the misspecified correlation model is: $0 \leq \rho_u(1) \leq 0.90$.

The preformance ratios corresponding to different misspecified correlations $\rho_u(1)$ are not going up monotonically as a function of the parameter u going down.

We noticed that whenever misspecified association $\rho(1)$ is weaker than the true model, the performance ratio is increasing. On the other hand, assuming stronger association $\rho(1)$ than the true one, does not cause any substantial deviation from the optimal estimation variance.

3.2 Calculated Versus Experienced Variance of Estimation Error. We refer to *calculated* MSE of the estimation error as $MSE[Z_u^* - Z]$, i.e., we use the misspecified parameter u in the calculation of the coefficients C_i (in (2) as well as in the calculation of the minimal MSE, in (6). We are interested in the

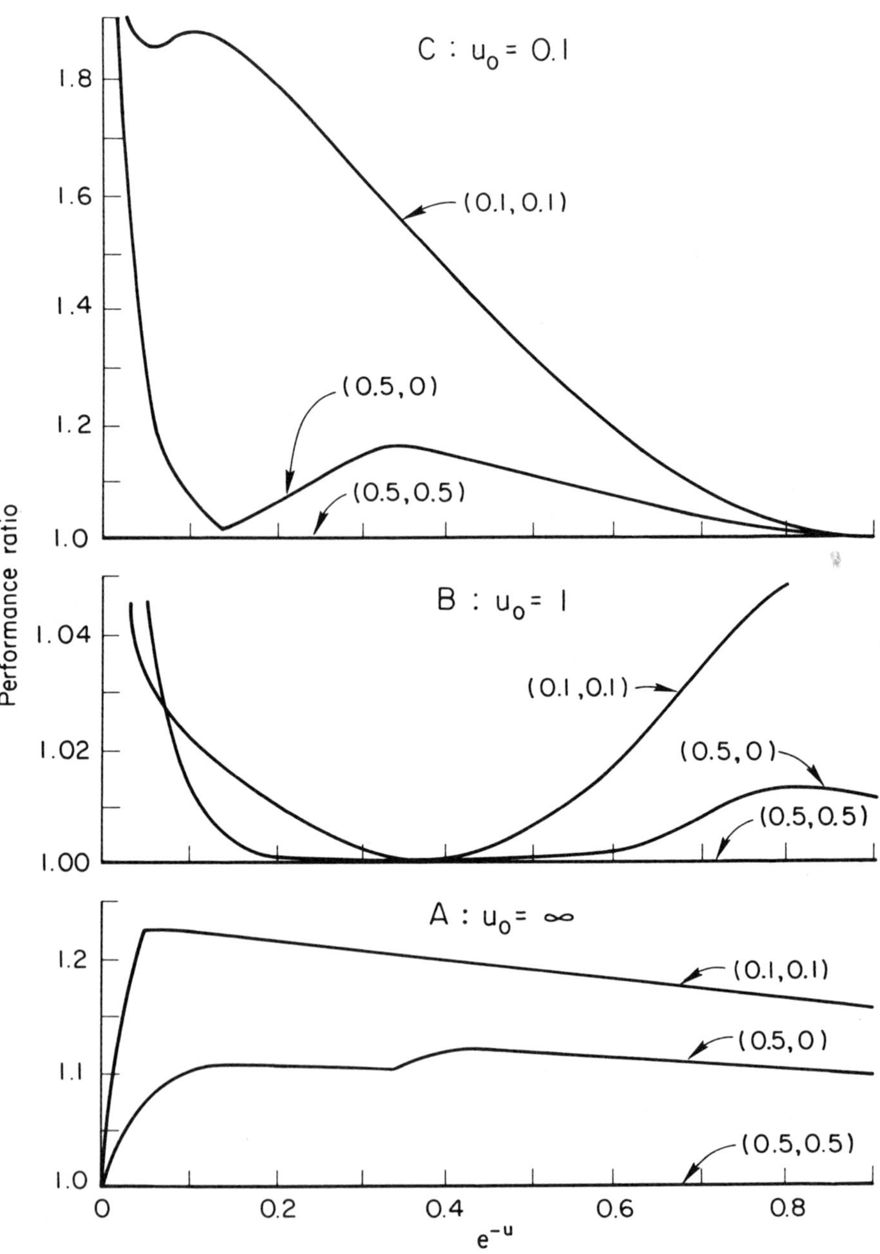

FIG. 2: Performance ratio R_1 *vs. misspecified correlation.* $n = 4$.

TABLE 1: Three different correlation models for three interpolation locations.

Misspecified $\rho_u(1)$	u	True $\rho_{u_o}(1) = 0$, $u_o = \infty$ (0.5,0.5)	(0.1,0.1)	(0.5,0.)	True $\rho_{u_o}(1) = 0.3679$, $u_o = 1$ (0.5,0.5)	(0.1,0.1)	(0.5,0.)	True $\rho_{u_o}(1) = 0.9048$, $u_o = 0.1$ (0.5,0.5)	(0.1,0.1)	(0.5,0.)
0.0	∞	5/4	5/4	5/4	0.2547	0.4383	0.4025	0.0046	0.06161	0.7808
		1.	1.	1.	1.	3.8792	1.8150	1.	3.2315	6.6418
0.0498	3.	5/4	1.8743	1.4525	0.2547	0.0310	0.1345	0.0046	0.0020	0.0025
		1.	1.2245	1.0780	1.	1.0310	1.0493	1.	1.8660	1.1807
0.1353	2.	5/4	1.8582	1.5354	0.2547	0.0309	0.1232	0.0046	0.0021	0.0019
		1.	1.2192	1.1083	1.	1.0180	1.0040	1.	1.8742	1.0247
0.3679	1.	5/4	1.7976	1.5741	0.2547	0.0291	0.1222	0.0046	0.0014	0.0024
		1.	1.1992	1.1012	1.	1.	1.	1.	1.5178	1.1558
0.4066	0.90	5/4	1.7880	1.5723	0.2547	0.0292	0.1222	0.0046	0.0012	0.0023
		1.	1.1959	1.1215	1.	1.0005	1.0000	1.	1.4534	1.1476
0.4724	0.75	5/4	1.7723	1.5670	0.2547	0.0293	0.1222	0.0046	0.0011	0.0022
		1.	1.1907	1.1196	1.	1.0035	1.0001	1.	1.3494	1.1252
0.6065	0.50	5/4	1.7427	1.5517	0.2547	0.0301	0.1225	0.0046	0.0008	0.0020
		1.	1.1807	1.1142	1.	1.0172	1.0015	1.	1.1704	1.0686
0.7788	0.25	5/4	1.7091	1.5289	0.2547	0.0319	0.1250	0.0046	0.0006	0.0182
		1.	1.1693	1.1059	1.	1.0462	1.0116	1.	1.0300	1.0132
0.9084	0.10	5/4	1.6874	1.5122	0.2547	0.0335	0.1250	0.0046	0.0006	0.0018
		1.	1.1619	1.0999	1.	1.0730	1.0114	1.	1.	1.

First entry in the Table is $MSE_{u_o}(Z^*_u - Z)$, Second entry is R_1, eq, (8), $n = 4$; $\rho_u(d) = \exp[-ud^2]$.

ratio

$$R_2 = \left[\frac{MSE^*_u[Z^*_u(\cdot) - Z(\cdot)]}{MSE_{u_o}[Z^*_u(\cdot) - Z(\cdot)]}\right]^{\frac{1}{2}} \tag{9}$$

The calculated variance will be larger than the true-experienced variance if $\rho_u(1) < \rho_{u_o}(1)$, $(u > u_o)$, i.e., using correlation which are too weak, and smaller if $\rho_u(1) > \rho_{u_o}(1)$, $(u < u_o)$, i.e., using correlations which are too strong.

We would hope that the ratio (9) from our simulations is close to one. However, misspecification of the correlation model in the variance calculation may lead to large errors. An appreciable error appears even for modest misspecifications, especially when misspecification is in the direction of overestimating correlations. This conclusion contrasts with the conclusion of Section 3.1.

It is very important to have the precise knowledge of correlations when calculating the estimation variance $MSE_u(Z^*_u(\cdot) - Z(\cdot))$, which is heavily dependent on the parameter value. The estimate $Z^*_u(\cdot)$ is not very sensitive to misspecified models, but the variance of estimation error is very sensitive to change of parameters in the model.

Similar calculations have been done for interpolation between four observations on a square as described for the previous ratio R_1, equation (8). The results are summed up in Table 2 and Figure 3, for the interpolation point $\underset{\sim}{X}_o = (0.5,0.5)$. This specific case being the center of the unit square has all four coefficients identical (= 1/4) for any underlying correlation model. But yet, using a misspecified parameter in the correlation model, when calculating the variance estimate, results in noticeable overestimate of $MSE(Z^*_u(\cdot) - Z(\cdot)$ if the misspecification was of the form $\rho_u(1) > \rho_{u_o}(1)$ and results in an underestimate of the MSE if $\rho_u(1) > \rho_{u_o}(1)$. The standard deviation ratio formula (9) for this configuration (in either direction of misspecification, underestimation as well as overestimation) is as large as 16.

For the interpolation points $\underset{\sim}{X}_o = (0.1,0.1)$ and $\underset{\sim}{X}_o = (0.5,0.)$ at three different underlying models ($\rho_{u_o}(1)$ assumes values 0.90,

TABLE 2:

Misspecified		True $\rho_{u_o}(1) = 0$ $u_o = \infty$			True $\rho_{u_o}(1) = 0.3679$ $u_o = 1$			True $\rho_{u_o}(1) = 0.9048$ $u_o = 0.1$		
$\rho_u(1)$	u	(0.5,0.5)	(0.1,0.1)	(0.5,0.)	(0.5,0.5)	(0.1,0.1)	(0.5,0.)	(0.5,0.5)	(0.1,0.1)	(0.5,0.)
0.0	∞	5/4	5/4	5/4	0.2547	0.4383	0.4025	0.0046	0.0616	0.7808
		1.	1.	1.	2.2153	1.6888	1.7623	16.4845	4.5043	1.2653
0.0498	3	...	1.8743	1.4525	...	0.0310	0.1345	...	0.0020	0.0025
		0.8145	0.2424	0.6306	1.8043	1.8864	2.0724	13.3649	7.3436	15.2907
0.1353	2	...	1.8582	1.5354	...	0.0309	0.1232	...	0.0021	0.0019
		0.6850	0.1947	0.4798	1.5174	1.5275	1.6935	11.2396	5.8456	13.7772
0.3679	1	...	1.7976	1.5741	...	0.0291	0.1222	...	0.0013	0.0024
		0.4514	0.1273	0.2786	1.	1.	1.	7.4071	4.6419	7.1822
0.4066	0.9	...	1.7880	1.5723	...	0.0292	0.1222	...	0.0012	0.0023
		0.4189	0.1185	0.2552	0.9280	0.9284	0.9156	6.8737	4.5026	6.6233
0.4724	0.75	...	1.7723	1.5670	...	0.0293	0.1222	...	0.0011	0.0022
		0.3659	0.1043	0.2188	0.8106	0.8111	0.7836	6.0046	4.2497	5.7814
0.6065	0.5	...	1.7427	1.5517	...	0.0301	0.1225	...	0.0008	0.0020
		0.2648	0.0770	0.1539	0.5865	0.5853	0.5475	4.3447	3.5840	4.2592
0.7788	0.25	...	1.7091	1.5289	...	0.0319	0.1250	...	0.0006	0.0018
		0.1443	0.0432	0.0819	0.3197	0.3162	0.2881	2.3683	2.2630	2.3749
0.9048	0.1	5/4	1.6874	1.5122	0.2547	0.0335	0.1250	0.0046	0.0006	0.0018
		0.0609	0.0186	0.0342	0.1350	0.1323	0.1191	1.	1.	1.

First entry in the Table is $MSE_u(Z_u^*-Z)$; Second entry is R_2, eq. (9), $n = 4$, $\rho(d) = \exp(-ud^2)$.

... indicates that the MSE value is the same as above.

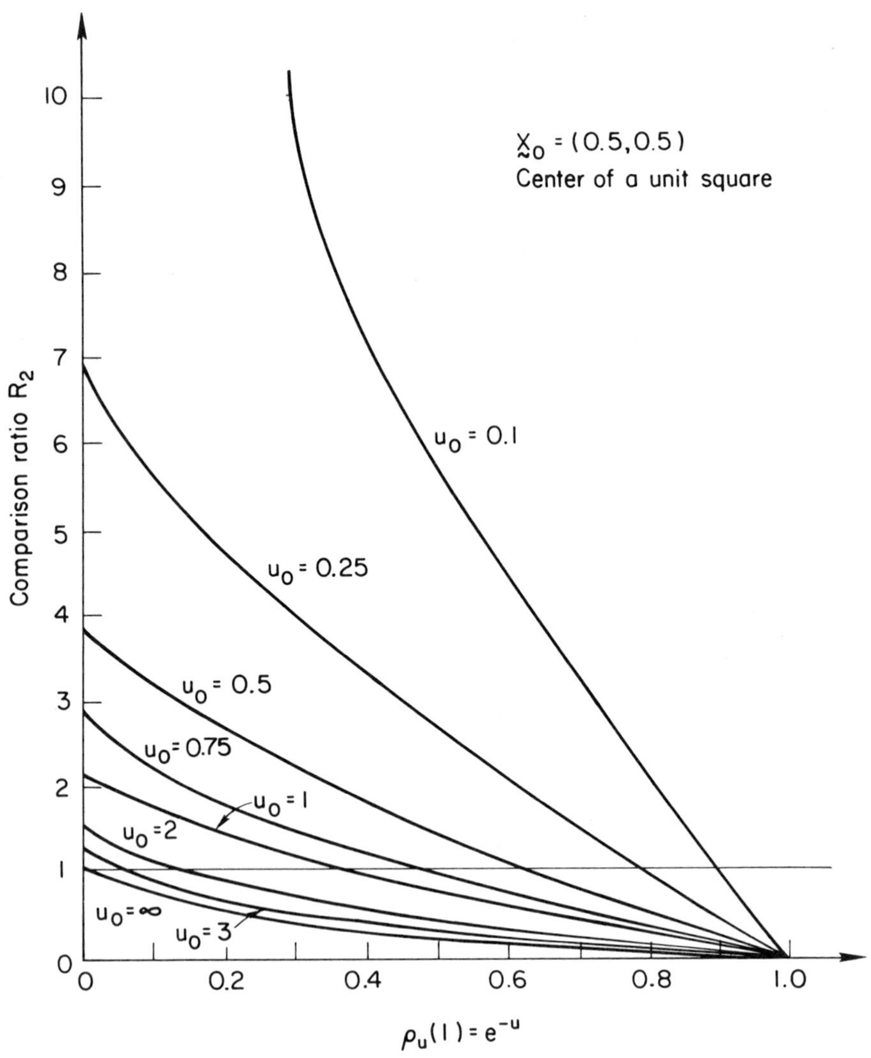

FIG. 3: Comparison ratio R_2 *vs. misspecified correlation model at unit distance.* $n = 4$.

0.37, and 0), we get the results that are summarized in Table 2. The graphs of the ratios versus $\rho_u(1)$ are similar in shape to Figure 3, and therefore omitted. At these locations the discrepancies are even higher, than those achieved at the center of the square.

To sum up the results of the last two sections, we can conclude that:

(1)
Optimal Variance
$(MSE_{u_o}\ (Z^*_{u_o}(\cdot) - Z(\cdot))$

(2)	(3)
Calculated Variance	Experienced Variance
$MSE_u\ (Z^*_u(\cdot) - Z(\cdot))$	$MSE_{u_o}\ (Z^*_u(\cdot) - Z(\cdot))$

(1) is very different from (2)
(1) is approximately equal to (3)
(2) is very different from (3).

where u_o defines the 'true' parametric value of the underlying correlation model, u is the misspecified parameter, and $Z^*(\cdot)$ is the linear estimate (interpolation value) for $Z(\cdot)$.

4. ROBUSTNESS OF INTERPOLATION IN THE PLANE TO TREND MISSPECIFICATIONS

The trend $M(\underset{\sim}{X})$ in (1) was taken as a constant (or zero) in Station 3 and as a result we dealt with a weak second order stationary stochastic process.

Now, we deal with processes having a trend, dependent on $\underset{\sim}{X}$. Minimizing the MSE of estimation error we have to add to constraint (4) some additional constraints, resulting from the request that the estimator $Z^*(\underset{\sim}{X})$ is unbiased over some class of mean functions $M(\underset{\sim}{X})$.

Constrained estimators are described in Matheron (1973) where the trend is given as an analytic expression

$$M(\underset{\sim}{X}) = \sum_{i=0}^{k} a_i f_i(\underset{\sim}{X}), \tag{10}$$

where $\{f_i(\underset{\sim}{X})\}$ are specified functions and $\{a_i\}$ are unspecified constants. For example (10) can be used to represent a polynomial trend in the mean function. Then the conditions on the weights $\{C_i\}$ in (1), (so that $Z^*(\underset{\sim}{X})$ is unbiased for any $\{a_i\}$) are:

$$E[Z^*(\underset{\sim}{X})] = \sum_{j=1}^{n} C_j \; E[Z(\underset{\sim}{X}_j)] = \sum_{j=1}^{n} C_j \sum_{i=0}^{k} a_i \, f_i(\underset{\sim}{X}_j)$$

$$= \sum_{i=0}^{k} \sum_{j=1}^{n} a_i \, C_j \, f_i(\underset{\sim}{X}_j). \qquad (11)$$

As shown in the above reference, the necessary and sufficient condition for unbiasedness of $Z^*(\underset{\sim}{X})$ is

$$\sum_{j=1}^{n} C_j \, f_i(\underset{\sim}{X}_j) = f_i(\underset{\sim}{X}) , \qquad i = 0,1,2,\cdots,k. \qquad (12)$$

When assuming a misspecified trend other than $M(\underset{\sim}{X}) = M$ (constant), we actually miscalculate the estimates for the autocorrelation functions. The term

$$\sigma_{ij} = E[(Z(\underset{\sim}{X}_i) - M(\underset{\sim}{X}_i))(Z(\underset{\sim}{X}_j) - M(\underset{\sim}{X}_j))] \qquad (13)$$

that enters in ρ_{ij} will be different for misspecified $M(\underset{\sim}{X}_i)$. These errors in estimating autocorrelation functions imply wrong coefficients in the estimator. For an arbitrary trend (nonconstant) the location of the estimation point will certainly be a factor in determining the severity of the discrepancy.

The BLUE $Z^*(\underset{\sim}{X})$ is that unbiased linear estimator of $Z(\underset{\sim}{X})$ which minimizes the mean squared error of $(Z^* - Z)$. The application of Lagrange multipliers with constraints (12) shows that the MSE is minimized by

$$Z^*(\underset{\sim}{X}) = \hat{M} + \gamma'_{sx} \cdot \gamma_{ss}^{-1} \cdot (Z_s - \hat{M}\,\underset{\sim}{1}) , \qquad (14)$$

where

$$Z_s = (Z(\underset{\sim}{X}_1) \; , \; Z(\underset{\sim}{X}_2)\cdots Z(\underset{\sim}{X}_n))', \qquad n \times 1;$$

$$\gamma_{sx} = (\gamma(\underset{\sim}{X}_1 - \underset{\sim}{X})\cdots\gamma(\underset{\sim}{X}_n - \underset{\sim}{X}))' \; , \qquad n \times 1;$$

$$\gamma_{ss} = (\gamma(\underset{\sim}{X}_i - \underset{\sim}{X}_j)) \; , \qquad n \times n;$$

$$\underset{\sim}{1} = (1,1,\cdots,1)' , \qquad n \times 1$$

and
$$\hat{M} = \frac{\underset{\sim}{1}'\gamma_{ss}^{-1} Z_s}{\underset{\sim}{1}'\gamma_{ss}^{-1} \underset{\sim}{1}} .$$

with $\gamma(\underset{\sim}{h})$ and Z_s as defined in Section 2, and the minimum MSE of $(Z^* - Z)$ equals

$$\min \text{MSE} = \gamma_{sx}' \gamma_{ss}^{-1} \gamma_{sx} - \frac{(\underset{\sim}{1}' \gamma_{ss}^{-1} \gamma_{sx} - 1)^2}{\underset{\sim}{1}' \gamma_{ss}^{-1} \underset{\sim}{1}} . \tag{15}$$

As a straightforward example, we simulate a linear trend represented by the mean function.

$$E[Z(\underset{\sim}{X})] = M(\underset{\sim}{X}) = \alpha_o + \alpha_1 U + \alpha_2 V , \tag{16}$$

where (U,V) are the coordinates of the point $\underset{\sim}{X}$ in the plane, and α_o, α_1, α_2 are fixed constants. Let $Z^*(\underset{\sim}{X})$ be of the form (1) , i.e., a linear combination of the observations used to estimate $Z(\underset{\sim}{X})$ at an interpolation point X.

If the trend constants α_o, α_1, α_2 are not specified and we want $E[Z^*(\underset{\sim}{X})] = M(\underset{\sim}{X})$ for all α_o, α_1, α_2 , then the constraints on the weighting coefficients, according to equation (12) are:

$\sum_{i=1}^{n} C_i^* = 1$, $\sum_{i=1}^{n} C_i^* U_i = U$, $\sum_{i=1}^{n} C_i^* V_i = V$, where (U,V) are the coordinates of $\underset{\sim}{X}$. The kind of misspecification are to assume unknown constant trend versus specified known trend, or constant trend instead of a linear one, or to assume wrong parametric values in the functional form of the trend or to assume only one constraint $(\Sigma C_i = 1)$, which results in a biased estimator, and compare it to a fully constrained solution.

We performed all these simulations. As an example, we shall list a few results of interpolation and extrapolation, using the random function approach and assuming the misspecifications in the trend.

$$r = (\sigma_c^2/\sigma^2 \text{ opt.})^{\frac{1}{2}} , \tag{a}$$

where σ^2opt. indicates the MSE of the optimal *unconstrained* linear estimator, where all the trend parameters α_o, α_1, α_2 are fully, correctly specified. σ_c^2 is the optimal MSE for a *constrained* estimator which is unbiased in all coefficients α_o, α_1, α_2 . In ratio (a) we assume σ_c^2 to describe the MSE of the constrained estimator under a constant trend, i.e., $\alpha_o = \text{constant}$, $\alpha_1 = \alpha_2 = 0$.

$$r_1 = (\sigma_c^2/\sigma^2\text{opt.})^{\frac{1}{2}} , \tag{b}$$

where the underlying trend is a linear one, with coefficients α_o, α_1, α_2 .

$$r_2 = (\sigma_b^2/\sigma_c^2)^{\frac{1}{2}} , \tag{c}$$

where σ_c^2 is as defined above in (b) and σ_b^2 indicates that in the algorithm of minimizing MSE we used only one constraint $\sum_{i=1}^{4} C_i = 1$, and as a result deal with a *biased* estimator.

Each of the above defined ratios will depend on the parameters α_o, α_1, α_2 of the linear trend function

$$M(\underset{\sim}{X}) = \alpha_o + \alpha_1 U + \alpha_2 V \quad \text{where} \quad \underset{\sim}{X} = (U,V).$$

To simplify comparisons between r_1 and r_2 we set $\alpha_o = \alpha_2 = 0$ throughout and α_1 takes values 0.2, 0.5, 1.0, 10.0, which respond to an angle θ between the U axis and the linear trend of about 12, 30, 45, and 85 degrees, respectively.

See Table 3 for the ratio r_2 , with the above four mentioned α_1 parameters, as well as r and r_1 , using the autocorrelation model $\rho(d) = \exp(-ud^2)$, where d is distance and $u > 0$ is a parameter, in the range $\rho(1) = 0.9$ till $\rho(1) = 0$.

TABLE 3: Interpolation and extrapolation points. $r = [\sigma_c^2/\sigma_{opt}^2]^{1/2}$ *, constant mean ;* $r_1 = [\sigma_c^2/\sigma_{opt}^2]^{1/2}$*, linear trend with* α_o*,* α_1*,* α_2*;* $r_2 = [\sigma_b^2/\sigma_c^2]^{1/2}$*; single constraint* σ_b^2 *versus linear trend with* α_o*,* α_1*,* α_2*.*

					r_2			
$\rho_u(1)$	u	X_0	r	r_1	α_1=0.2	α_1=0.5	α_1=1.	α_1=10.
0.9048	.1		1.1915	1.2002	0.9927	0.9927	0.9927	0.9927
0.6065	.5		1.0982	1.1210	0.9797	0.9797	0.9797	0.9797
0.3679	1	(0.5,0.)	1.0390	1.0565	0.9834	0.9834	0.9834	0.9834
0.0498	3		1.0024	1.0046	0.9978	0.9978	0.9978	0.9978
0	∞		1.1180	1.2247	0.9129	0.9129	0.9129	0.9129
0.9048	.1		1.3626	1.3707	0.9947	0.9982	1.0103	2.0598
0.6065	.5		1.2058	1.2343	0.9777	0.9815	0.9952	2.1352
0.3679	1	(0.75,0.25)	1.0916	1.1273	0.9691	0.9730	0.9868	2.1296
0.0498	3		1.0012	1.0131	0.9885	0.9897	0.9942	1.4647
0	∞		1.1180	1.1726	0.9544	1.0063	1.0247	2.4496
0.9048	.1		1.3473	1.4669	0.9278	0.9754	1.1292	6.6319
0.6065	.5		1.1992	1.5578	0.7781	0.8204	0.9565	5.7290
0.3679	1	(1.5,-0.5)	1.1274	1.5946	0.7152	0.7564	0.8883	5.4239
0.0498	3		1.0881	1.6854	0.6546	0.6992	0.8421	5.4450
0	∞		1.1180	1.8028	0.6300	0.6793	0.8230	5.5815
0.9048	.1		1.0844	1.1022	0.9839	0.9839	0.9839	0.9839
0.6065	.5		1.0541	1.1340	0.9295	0.9295	0.9295	0.9295
0.3679	1	(0.5,-0.5)	1.0373	1.1783	0.8803	0.8803	0.8803	0.8803
0.0498	3		1.0523	1.3314	0.7904	0.7904	0.7904	0.7904
0	∞		1.1180	1.5000	0.7453	0.7453	0.7453	0.7453

These simulations indicated that the estimated interpolation value at $\underset{\sim}{X}$ is insensitive to changes of the underlying model of correlation, as well as to changes in the linear trend parameters. The MSE values for extrapolation, though, are very different.

Tables and detailed results are available from the author upon request.

5. APPLICATION - INTERPOLATING OXIDANT MEASUREMENTS IN THE PLANE

Data. Ozone measurements are used frequently, because they are relatively precise. Data base is large and quickly available, both in time and in geographical extense. The measurements are taken as hourly averages (calculated from five minutes' observations), 24 hours of each day, for the years 1971-1973, at 16 stations located in the San Francisco Bay area.

We used 1) Daily averages, 2) Weekly averages, 3) Averages from four summer months. Each approach is applicable for different sort of questions. Mapping of the three different data averages (daily, weekly, or monthly) indicates that the variability in the data is immense, and different patterns show on the maps.

We used log-Oxidant measurements (given in pphm). A zero observation was transformed into 0.1 pphm, and all the transformed data was shifted so that the original zero measurements should be zero again.

Isopleth Maps. Assuming a correlation model,

$$\rho_{ij} = \rho(\underset{\sim}{X}_i, \underset{\sim}{X}_j) = e^{-uh^2}, \quad h = |\underset{\sim}{X}_i - \underset{\sim}{X}_j| ,$$

we plotted the isopleth maps for different u parameters. The maps have a very similar pattern and support the conclusion drawn from the simulations that the interpolated values are insensible to changes in the parameter of the underlying correlation model. The data used were log-Oxidant means for one day in mid 1973. See Figure 4. The Kriging method estimates linear trend functions which differ from each other as a result of using different underlying correlation models.

Figure 5 shows three isopleth maps drawn for monthly averages of mid 1973 by using the correlation function $\rho(\underset{\sim}{h}) = \exp(-0.05h^2)$,

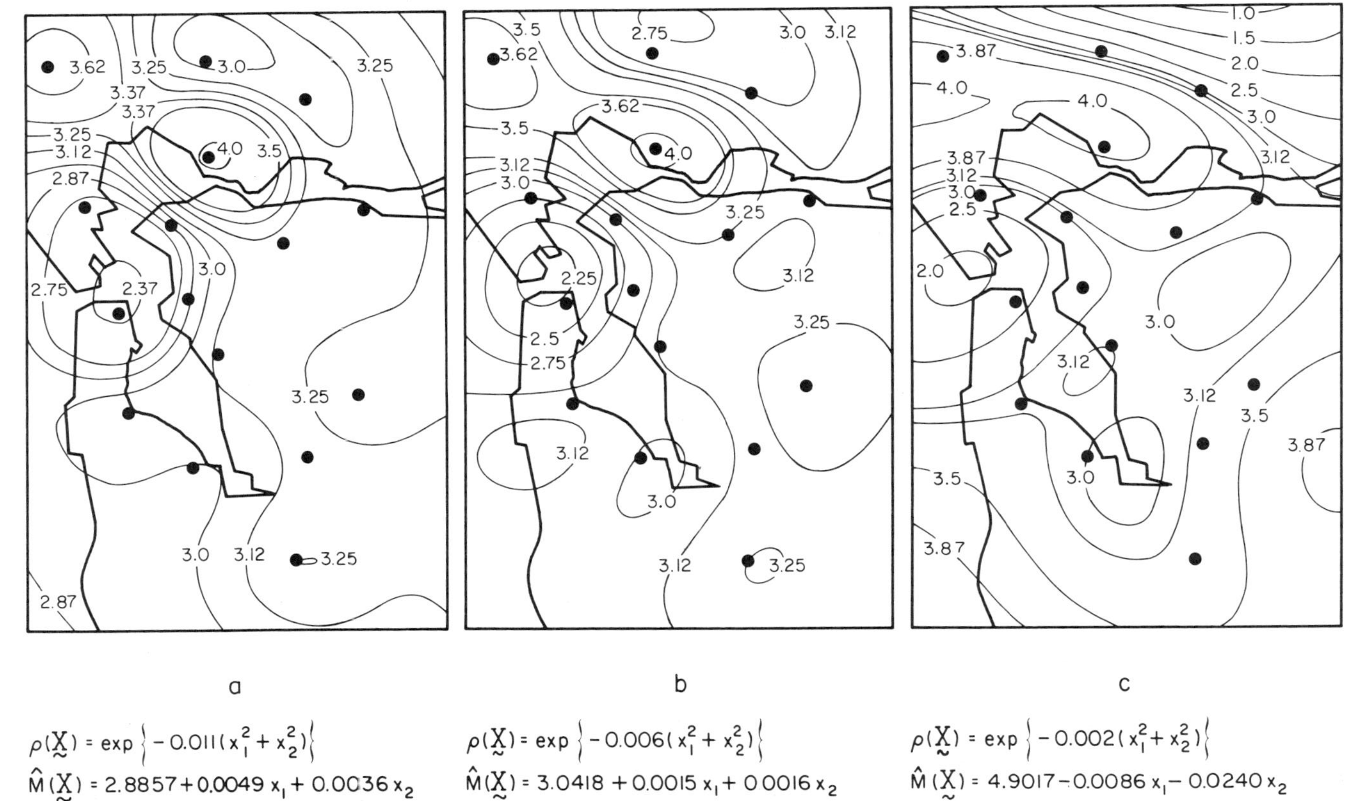

FIG. 4: Oxidant 32nd day mid 1973 *(pphm).*

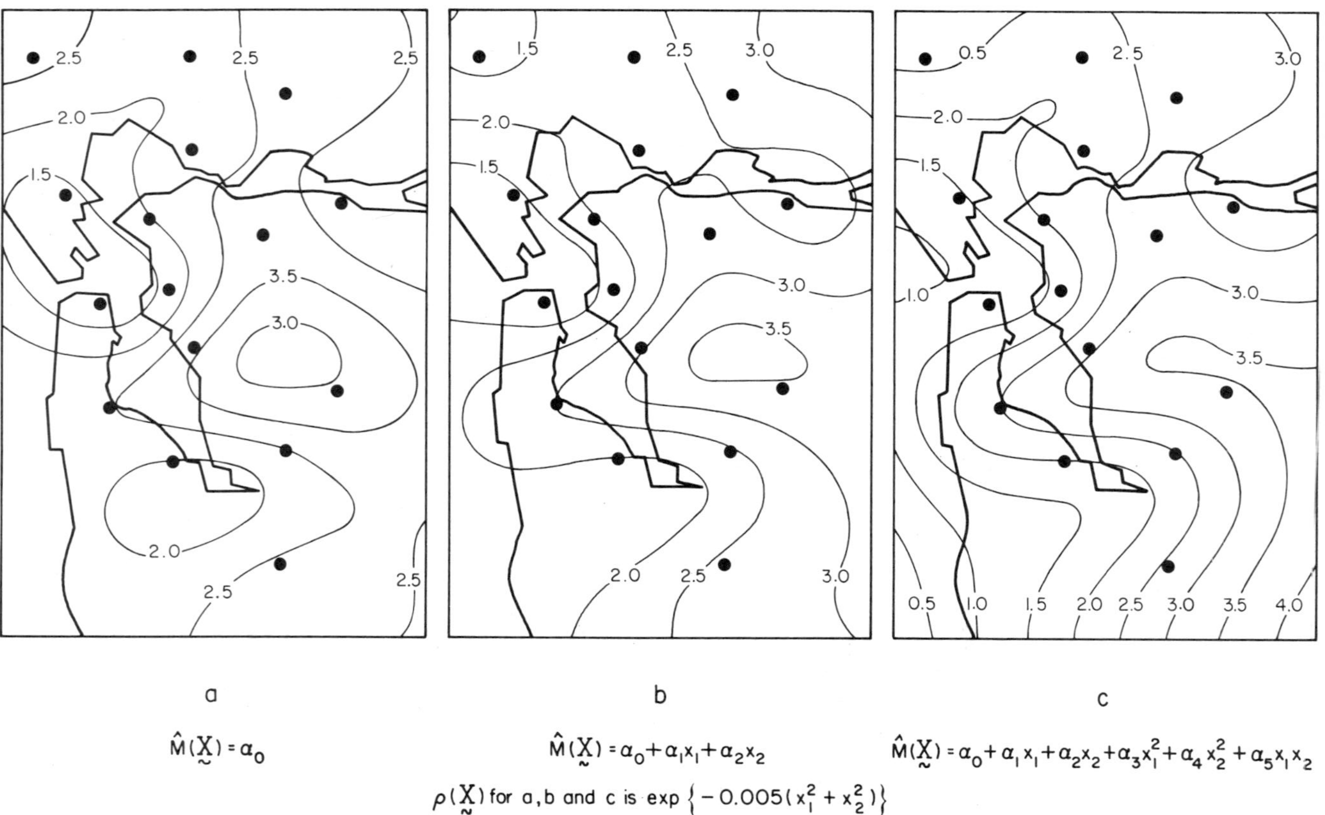

FIG. 5: Oxidant 1 month mid 1973 *(pphm).*

assuming different linear trends chosen arbitrarily (without any effort to fit the most suitable one, which eventually has to be done in a real problem situation):

$$M(\underset{\sim}{X}) = \alpha_o \text{ , } M(\underset{\sim}{X}) = \alpha_o + \alpha_1 X_1 + \alpha_2 X_2 \text{ , and}$$

$$M(\underset{\sim}{X}) = \alpha_o + \alpha_1 X_1 + \alpha_2 X_2 + \alpha_3 X_1^2 + \alpha_4 X_2^2 + \alpha_5 X_1 X_2 \text{ , respectively,}$$

where $\underset{\sim}{X} = (X_1, X_2)$.

Comparison of the maps supports again the assumption of robustness of the interpolated value to different trend models.

We looked into correlation models which were directionally dependent (Anisotropic models). Correlation function of the form

$$\rho_{ij} = \rho\ (\underset{\sim}{X}_1, \underset{\sim}{X}_2) = a \ \exp\{X_1^2 + \alpha X_2^2\} \text{ ,}$$

$0 < \alpha < 1$, gave rise to very different isopleth maps than the maps achieved by using isotropic models. The problem of estimating and fitting a good correlation model (spatial as well as temporal) to the given data is still not solved and has yet to be investigated.

ACKNOWLEDGMENTS

I am indebted to Ray Faith for letting me use his data sets, Kriging and plotting programs. These results are part of my Ph.D. thesis, written under Paul Switzer whom I owe my warmest thanks. This work has been prepared under the auspices of SIAM Institute for Mathematics and Society SIMS, at Stanford University.

REFERENCES

Delfiner, P., (1976). Linear estimation of non-stationary spatial phenomena, Proceedings of NATO A.S.I. "Geostat 75", Rome, Italy, Reidel, Dordrecht, Netherlands.

Goldberger, A. S., (1962). Best linear unbiased prediction in the generalized linear regression model. *Journal of the American Statistical Association*, 57, 369-375.

Krige, D. G., (1951). A statistical approach to some basic mine value problems. *Journal of Chemical, Metallurgical and Mining Society of South Africa*, 52, 119-139.

Matheron, G. (1965). *Les Variables Regionalisees et Leur Estimation.* Masson et Cie, Paris.

Matheron, G. (1973). The intrinsic random functions and their applications. *Advances in Applied Probability*, 5, 439-468.

National Academy of Sciences (National Research Council) (1977). Medical and Biological Effects of Environmental Pollutants "*Ozone and other photochemical oxidants*". 437-585.

Sheshinski, R. (1976). *Robustness of linear estimators to misspecification of correlation models.* SIMS Technical Report No. 2, Department of Statistics, Stanford University.

[*Received December* 1978. *Revised February* 1979]

G. P. Patil and M. Rosenzweig, (eds.),
Contemporary Quantitative Ecology and Related Ecometrics, pp. 421-438.

THE EFFECT OF RANDOM SELECTIVE INTENSITIES ON FIXATION PROBABILITIES

WILLIAM C. TORREZ[1]

Department of Mathematical Sciences
New Mexico State University
Las Cruces, New Mexico 88003 USA

SUMMARY. An important factor determining fixation probabilities of a particular haplotype in a binomial selection model or a birth and death selection model is the effect of environmental randomness on selective intensity. Karlin and Levikson (1974) made a comprehensive investigation on the effect of stochastic temporal fluctuations in selection intensities for the Wright-Fisher binomial sampling model for small population size and their results show that for every initial frequency of a particular haplotype, the probability of fixation of that haplotype in the asymmetrical random case is smaller than the corresponding fixation probability of the non-random selection model. In this paper, we similarly show that for the birth and death selection model, variation in selection intensities due to stochastic environments diminishes the probability of fixation of a particular haplotype.

KEY WORDS. random selective intensities, birth and death process in a random environment, fixation probability.

1. INTRODUCTION

Presently there are two distinct methods of modeling genetic

---

[1]Supported in part by National Institutes of Health Grant No. RR-08136-04 and a Ford Foundation post-doctoral fellowship.

fluctuations due to ecological pressures such as selection, mutation, and migration in genetic populations of fixed size N. One is based on binomial or multinomial sampling and was first investigated by Fisher (1930), Wright (1931, 1939), and Feller (1951). The other is based on the intuitive concept of a birth and death process on the integers $0,1,\ldots,N$, and has been studied by Moran (1958a) and Karlin and McGregor (1962). In both types of models, an important factor determining fixation probabilities of a particular haplotype is the effect of environmental randomness on selective intensity. Karlin and Levikson (1974) made an elaborate investigation on the effects of stochastic temporal fluctuations in selection intensities for the Wright-Fisher binomial sampling model for small population size. In this paper, we consider a selection model based on a continuous-time birth and death process $Z(t)$, $t \geq 0$, with state space $\{0,1,\ldots,N\}$ evolving in a random environment controlled by a Markov process $Y(t)$, $t \geq 0$, with stationary transition matrix $\underset{\sim}{K}(t)$. When no mention is made of the environmental process $Y(t)$, $Z(t)$ is referred to as a birth and death process in a random environment. The mathematical details of this model including instability theorems, extinction theorems, and methods for calculating extinction probabilities are given in Torrez (1979c).

In this study, we consider the following question: does variation in selection intensities due to stochastic environments diminish the probability of fixation of a particular haplotype? In Section 2, we describe in detail the selection model to be considered, and in Section 3, we investigate the case that $Z(t)$ evolves in the environment controlled by $\underset{\sim}{K}(t)$, and we compare the probability of fixation for an asymmetrical selection model in the environment $\underset{\sim}{K}(t)$ with the corresponding probability of fixation for the selection model in constant environment.

The work of Karlin and Levikson (1974) shows that for binomial sampling, the probability of fixation is reduced in stochastic environments chosen by an independent sequence of random variables, when compared to a non-random asymmetrical selection model. Our main result shows that the probability of fixation is diminished when there is temporal randomness in selection intensities as compared to the case when selection intensities are constant over time. Thus, the result of Karlin and Levikson for the binomial sampling model holds true for the birth and death model as well.

2. DESCRIPTION OF THE WRIGHT-FISHER SELECTION MODEL AND ITS GENERALIZATION INCORPORATING RANDOM ENVIRONMENTS

2.1 The Classical Wright-Fisher Selection Model for Small Population Size. Consider a population of M diploid monoecious organisms reproducing in discrete generations, in a fashion to be described below, and suppose that at a particular locus there are two, and only two alleles, A and a, present in the population. Let the number of A genes in the adult population of the *tth* generation be $X(t)$, $t = 0,1,2,\ldots$. Then our reproductive scheme requires that the number $X(t+1)$ of zygotes of the $(t+1)$*st* generation is given as a binomial variate with index $2M$ and parameter $X(t)/2M$. Hence, the sequence $X(t)$, $t = 0,1,2,\ldots$, is Markovian; explicitly, given that $X(t) = i$, the probability that $X(t+1) = j$ is given by

$$p_{ij} = \binom{2M}{j}(i/2M)^j\{1 - (i/2M)\}^{2M-j} \;,\; j = 0,1,\ldots,2M \;. \qquad (2.1)$$

This model is due implicitly to Fisher (1930) and explicitly to Wright (1931). The one-step transition probabilities (2.1) together with the distribution of $X(0)$, the initial number of A alleles, determine the entire process. Note that we have two absorbing states 0 and $2M$ for the Markov chain $X(t)$; in biological terms, absorption into these boundary states is called *fixation*. For the model (2.1) and its generalization incorporating mutation and selection forces, the calculations of fixation probabilities, rates to fixation, the expected time to fixation, and other quantities of interest are exceedingly difficult to obtain directly. Classically, approximations to these quantities are obtained by diffusion approximation to the discrete process. The use of diffusion methods in genetics originates with Fisher (1922); for particular important uses, see Wright (1931, 1945), Feller (1951), and a review article by Kimura (1964). See also Crow and Kimura (1970), Chapters 8 and 9.

The kind of model that we consider in this paper is a generalization of (2.1) allowing for the influence of selection pressures (for a generalization of (2.1) incorporating mutation pressures, see Feller (1951)). This model may be described as follows (cf. Ewens, 1969): Suppose that among the $2M$ genes which form the mature zygotes in generation t , there are exactly i A genes and thus $2M-i$ a genes. Then the expected frequencies of AA , Aa , and aa zygotes at the time of conception of generation $t+1$, assuming random mating, are $(i/2M)^2$, $2(i/2M)(1-i/2M)$, and $(1-i/2M)^2$, respectively. If the zygotes survive to reach maturity and in turn produce genes for the next generation in the ratios $1+s_1:1+s_2:1$, then the expected frequency of A genes at the time of formation of the zygotes of generation $t+1$ is

$$\pi_i = \frac{(1+s_1)i^2 + (1+s_2)i(2M-i)}{(1+s_1)i^2 + 2(1+s_2)i(2M-i) + (2M-i)^2} . \tag{2.2}$$

Suppose that, at maturity, there will be exactly M individuals in any generation. As before, let $X(t)$ be the number of A genes in this population. Then in analogy with (2.1), we postulate that $X(t)$, $t = 0,1,2,...$ is a Markov process with one step transition probabilities given by

$$p_{ij} = \binom{2M}{j}\pi_i^j(1-\pi_i)^{2M-j} , \tag{2.3}$$

π_i being defined by (2.2). For subsequent analysis, we consider our model as a population of $N=2M$ haploid individuals (cf. Moran, 1958a) and give fitness values s_1 and s_2 to the haplotypes A and a, respectively. The random fluctuation of allele types is still postulated Markov (as in the Wright-Fisher diploid formulation reflected by (2.3)), but the expected frequency π_i of A genes at the time of formation of the $(t+1)st$ generation, given that this frequency was i/N in the previous generation t, is now

$$\pi_i = (1+s_1)i/[(1+s_1)i + (1+s_2)(N-i)] . \tag{2.4}$$

2.2 Random Drift and Random Fluctuation of Selection Intensity. It has long been recognized that in nature, the process of evolution may not be quite deterministic (for a classical deterministic approach, see the work of Haldane, 1924a,b,c,d, 1927) because of the existence of factors which produce *random* fluctuations in gene frequencies, of which two different types have been underscored (Wright, 1948; Kimura, 1954). One is the random sampling of gametes in reproduction and the other is the random fluctuation of selection intensity over time. The process of change in gene frequency which is due solely to the first factor is called *random drift* and was first considered by Fisher (1922, 1930) and Wright (1931) and later by Malécot (1944). The relations (2.1)-(2.4) serve to model the fluctuations of gene frequencies in small populations due to random sampling of gametes by superimposing a binomial sampling scheme where the average changes are determined by (2.2) and (2.4) in the diploid and haploid cases, respectively. (One can easily generalize (2.2) for the case of polyploidy by imposing multinomial sampling.)

The other factor (affecting both large and small populations) consists of random fluctuations in what Wright (1949) called the systematic evolutionary pressures, of which random fluctuation of selection intensity is especially important. This process of change was recognized by Fisher and Ford (1947) and Wright (1948). (For recent work incorporating randomly varying parameters into some standard models of theoretical ecology not strictly in population genetics, see, for example, Lewontin and Cohen (1969), Levins (1969), May and MacArthur (1972), May (1973, Chapter 5), Capocelli and Ricciardi (1974), Feldman and Roughgarden (1975), Keiding (1975), Levikson (1976), and Turelli (1977, 1978). The interested reader should consult these works and their bibliographies for further references.) The effect of random fluctuation of selection intensity has been investigated in several contexts. We begin with a review of some important studies for the case of large populations. Gillespie (1972) provides a description of large haploid populations subject to a stationary Gaussian selection process. He assumes that the rate of change in the relative frequency of a haplotype is governed by the differential equation $dx/dt = s(t)x(1-x)$, where $s(t)$ is a stochastic process with continuous sample paths describing the process of selection. He restricts attention to the case when $s(t)$ is of the form $m(t) + \xi(t)$ where $m(t)$ is a non-random continuous function and $\xi(t)$ is a stationary Gaussian process with continuous covariance function $r(t) = \int_0^\infty \cos \lambda t \, g(\lambda) \, d\lambda$, where $g(\lambda)$ is the spectral density of $r(t)$. He obtains necessary and sufficient conditions on $g(\lambda)$ for fixation, and gives an example which shows that fixation need not occur in autocorrelated environments, in contrast to Kimura's (1954) result that random fluctuations in the fitnesses of two haploid genotypes will ultimately lead to fixation when the fitnesses in successive environments are uncorrelated. Gillespie (1973) obtains conditions for the maintenance of variation (polymorphism) for both uncorrelated and autocorrelated environments. Tuckwell (1974), using transformation methods, and the Stratonovich calculus, solves the stochastic differential equation $dN(t)/dt = f(N(t)) + g(N(t))n(t)$, where $N(t)$ is the size of the population at time t , f and g are specified functions, and $n(t)$ is taken to be a stationary Gaussian process. He offers his results as justification for the use of the Stratonovich integral rather than the Ito integral since they agree qualitatively (and, in some cases, quantitatively) with some previous findings of Kimura (1954, 1964), Lewontin and Cohen (1969), Levins (1969), and Gillespie (1972). Tuckwell (1976), again employing the Stratonovich calculus, concludes, among other findings that, under random selection quasifixation (the phenomenon of gene frequency near one for large populations; cf. Kimura (1954)) of an allele which has a mean selective advantage is certain to occur independent of initial frequency and selection variance. An excellent study on the uses of the Ito and Stratonovich calculi in population growth models is Turelli (1977). In particular, he discusses the accuracy of some of Tuckwell's (1974, 1976) results.

The first study of an infinite haploid discrete generation population with selection pressures was that of Haldane and Jayakar (1963). They discussed a number of cases involving deterministic temporal selection changes for a dominant trait and seasonal variations in fitness coefficients. An earlier paper of Dempster (1955) dealt briefly with this topic. Hartl and Cook (1973), considering an infinite diploid population with stochastic selection changes, rationalize a discovery of Haldane and Jayakar (1963) that a recessive genotype that is favored from time to time but disfavored otherwise will be fixed in a population if the geometric mean of its number of progeny is greater than one. The work of Karlin and Lieberman (1974), among other findings, conclude that the Hartl-Cook (1973) model is not a case of average neutral effects, but rather represents in its mean fitness impact either overdominance, underdominance, or directional selection, depending on the specific magnitude of the statistical variances associated with the fitness values of the alternative genotypes. Their work also finds that variance selection mitigates the deterrent mean effects against polymorphism. Follow-up papers to this work, including precise mathematical analyses and extensions to continuous time processes, are Karlin and Lieberman (1975) and Levikson and Karlin (1975), respectively. Cook and Hartl (1975) consider the problem of characterizing the nature of polymorphism and gene frequency equilibria in stochastic selection models they formulated in their earlier paper (Hartl and Cook, 1973).

Wright (1948) first enunciated the importance of the joint effects of random drift of gene frequencies in small populations and temporal variation in selection intensity and derived a distribution of gene frequencies in steady state for a special case. In a series of papers commencing in 1952, Kimura undertook a quantitative analysis to study the effects of these two factors, first considering them separately (Kimura, 1952, 1954, 1955a,b,c). In these studies he was concerned mainly with the case of a single diallelic locus without dominance, and considered neutral variation in the sense that allelic fitnesses fluctuate randomly from generation around a mean value of zero. Later Kimura (1957) extended these results to include any level of dominance. Kimura (1962) then considered the joint effects of random drift and random selection intensity and settled a conjecture of his (Kimura, 1955c) in the affirmative that there is positive probability that a gene will be lost from the population by random fluctuation in selection intensity (plus random sampling of gametes when the frequency of the gene is near zero) even if the gene is advantageous on the average. Jensen and Pollak (1969) also studied the process of change for a population exhibiting both effects of randomness (and interestingly, do not cite the earlier work of Kimura (1962)). They consider the case when selection advantage of a gene is zero on the average. Ohta (1972) uses diffusion formulae of Kimura (1962) and considers the concept of what she termed nearly neutral mutations when taking into account random fluctuation of selection intensity. More precisely if p denotes the initial frequency of a gene A in a haploid population of constant size N, then the ultimate probability $u(p)$ that A becomes fixed in the population satisfies (Kimura, 1962):

$$\frac{1}{2} v(\Delta p)u''(p) + m(\Delta p)u'(p) = 0 , \quad 0 \le p \le 1 ,$$

with $u(0) = 0$, $u(1) = 1$, and the functions $m(\Delta p)$ and $v(\Delta p)$ are respectively, the first and second moments of the change in gene frequency in one generation. Ohta (1972) assumed a random variable s with mean μ/N and variance σ^2/N as describing the selective advantage of gene A over gene a and she investigated the probability of fixation for different values of the two parameters μ and σ^2, considering selection fluctuations 'nearly neutral' when μ in some sense is small. One of her findings was that variability in s reduces the chance of fixation of a rare gene. However, Jensen (1973) elaborated on the small population model incorporating 'neutral' random selection intensity formulated in Jensen and Pollak (1969) and derived an expression for $u(p)$ using

$$m(\Delta p) = (\theta/2N)p(1-p)(1-2p) \quad \text{and}$$

$$v(\Delta p) = (1/N)p(1-p) + (\theta/N)p^2(1-p^2) .$$

Here the selection variable s is assumed to have second moment equal to θ/N, $\theta > 0$, with the two genotypes A and a having adaptive values $1+s:1$ or $1:1+s$ each chosen in any generation with probability $1/2$. Jensen (1973) found that as

$\theta \to +\infty$, $u(p) \uparrow 1/2$ for $0 < p < 1/2$ and $u(p) \downarrow 1/2$ for $1/2 < p < 1$, thus indicating that variability in selection increases the chance of fixation of a rare gene, in contrast to Ohta's (1972) findings. Gillespie (1973b), althouth considering a different random selection model than Jensen (1973), independently derived $m(\Delta x)$ and obtained some results almost identical to Jensen's. Follow-up work can be found in Gillespie (1974) and Ohta and Kimura (1972). The model of Karlin and Levikson (1974), postulated that the selection intensities (s_1, s_2) in (2.4) vary over generations t as a stochastic process $\underset{\sim}{S}(t) = (s_1(t), s_2(t))$ and that $\{X(t)\}$, conditional on a realization of $\underset{\sim}{S}(t)$ is Markov, that is,

$$\begin{aligned} &P[X(t+1)=j \mid X(t)=i,\ \underset{\sim}{S}(\tau),\ 0 \le \tau \le t] \\ &\quad = P[X(t+1)=j \mid X(t)=i,\ \underset{\sim}{S}(t)] \\ &\quad = \binom{N}{j} \pi_i^j(s_1(t), s_2(t))[1-\pi_i(s_1(t), s_2(t))]^{N-j}, \end{aligned} \tag{2.5}$$

where

$$\begin{aligned} &\pi_i(s_1(t), s_2(t)) \\ &\quad = [1+s_1(t)]i/\{[1+s_1(t)]i + [1+s_2(t)](N-i)\} \end{aligned} \tag{2.6}$$

for the haploid case; analogous relations hold for the diploid case (compare with (2.2)-(2.4)). Hence their scheme requires that $X(t+1)$ be a binomial variate with index N and parameter $\pi_i(\underset{\sim}{S}(t))$. Thus, binomial sampling is maintained to account for random drift simultaneously allowing the selection intensities to vary as a stochastic process reflecting the influence of random environmental factors. Mathematically, the bivariate process $\{X(t), \underset{\sim}{S}(t)\}$ is known as a Markov chain in a random environment (cf. Cogburn, 1979). In the Karlin-Levikson model $\underset{\sim}{S}(t)$ is taken to be a sequence of i.i.d. random variables.

3. PROBABILITY OF FIXATION IN RANDOM ENVIRONMENTS

By specifying the distribution of $\underset{\sim}{S}(t)$ in (2.5) and (2.6) and its autocorrelation properties as a stochastic process, the variation affecting the changes in the population numbers of A genes over successive generations due to random selective intensities can be studied. For the model of Karlin and Levikson (1974), a number of conclusions are drawn concerning the extent to which randomness in the selection coefficients influences the probability of fixation and the expected time to fixation in the case $s_2(t)=0$ (asymmetrical case) and $\{s_1(t)\}$ is a sequence of independent, identically distributed random variables. Their main emphasis is contrasting qualitatively and quantitatively the findings on these problems for the case of selective intensities in random environments with the corresponding case of constant environment selective intensities. Their methods involve in all circumstances approximations by appropriate diffusion processes, and they show, for example, that for every initial frequency x of gene A, the probability of fixation on A, for the asymmetrical haploid case (2.5)-(2.6), is smaller than the corresponding probability of fixation for a non-random selection model. We would like to ask whether the same conclusion can be drawn about the selection model (2.5)-(2.6) viewed within the framework of a birth and death process in a random environment (Torrez, 1979c). (For a discrete-time formulation and examples of the birth and death chain in a random environment, see Torrez, 1978, 1979a, 1979b.) The use of a continuous-time birth and death process to study the Wright-Fisher process with mutation pressures is due to Moran (1958a). Moran (1958b) also dealt with this model incorporating selectivity factors. See his papers for some biological justification of the birth and death scheme as well as analytic methods for finding the probability of fixation without resorting to diffusion approximations. Later, Karlin and McGregor (1962) gave a complete mathematical analysis of the Moran process, unifying the treatments of Wright (1951), Feller (1951), and that of Kimura (1957); the latter considered a slight variant of the Wright-Fisher model using hypergeometric sampling (sampling without replacement) rather than binomial sampling (sampling with replacement).

To formulate the birth and death model, we begin with a description of the process excluding temporal variation in selection intensities. The number of A gametes at time t will be a birth and death process $Z(t)$ on the state space $S_N = \{0,1,\ldots,N\}$ with birth rates $\lambda_z = \alpha(z/N)P_z$ and death rates $\mu_z = \alpha(1-z/N)Q_z$ where $\alpha^{-1} > 0$ is the mean of the (exponential) waiting time until a change in the population occurs (independent of the values of $Z(t)$), and

$$P_z = (1+s_2)(N-z)/[(1+s_1)z + (1+s_2)(N-z)] ,$$

and

$$Q_z = 1 - p_z = (1+s_1)z/[(1+s_1)z + (1+s_2)(N-z)] \ .$$

The rationale of this model is based on the following structure (cf. Karlin and Taylor (1975)): at the (random) times $T_1 < T_2 < \cdots$, one individual dies and is replaced by another of type A or a. If just before a replacement time T_n, there are z A genes and N-z a genes present, we postulate that the probability that an A individual dies is $(1+s_1)z/B_z$ and that an a individual dies is $(1+s_2)(N-z)/B_z$ where $B_z = (1+s_1)z + (1+s_2)(N-z)$. Generally the A- and a-type individuals have chance $(1+s_1)/(2+s_1+s_2)$ and $(1+s_2)/(2+s_1+s_2)$ of dying at time T_n, respectively. Thus, $(1+s_1)/(1+s_2)$ can be interpreted as the selective advantage of A-types over a-types. Now let Y(t) be an irreducible Markov pure jump process with finite state space $Y = \{1,\ldots,m\}$. Let the jump times of Y(t) be denoted by $\tau_n \uparrow +\infty$ and consider the embedded chain $(Y(\tau_n))$. Let $(i_1,i_2,\ldots)$ be a realization of this process and consider the sequence $(s_1^{(i_n)}, s_2^{(i_n)})$ of selection intensities. These intensities are now temporally varying and in the time block $w_n = \tau_n - \tau_{n-1}$, $n \geq 1$, $\tau_0 = 0$, the Z(t) process evolves according to a birth and death process with birth rates

$$\lambda_z^{(i_n)} = \alpha^{(i_n)}(z/N)P_z^{(i_n)}$$

and death rates

$$\mu_z^{(i_n)} = \alpha^{(i_n)}(1-z/N)Q_z^{(i_n)} \tag{3.1}$$

where $1/\alpha^{(i_n)}$ is the mean of the exponential waiting time for the birth and death process in environment i_n and

$$P_z^{(i_n)} = (1+s_2^{(i_n)})(N-z)/[(1+s_1^{(i_n)})z + (1+s_2^{(i_n)})(N-z)],$$

and

$$Q_z^{(i_n)} = 1 - p_z^{(i_n)} = (1+s_1^{(i_n)})z/[(1+s_1^{(i_n)})z + (1+s_2^{(i_n)})(N-z)].$$

We have thus formulated $Z(t)$ as a birth and death process in a random environment. In this paper, we consider only the asymmetrical case that $s_1^{(i_n)} = 0$ and $s_2^{(i_n)}$ will fluctuate between real values $s^{(1)},\ldots,s^{(m)}$.

For the discussion in this section, we refer to the theorem on extinction probabilities for the birth and death process with continuously temporal environments (cf. Torrez (1979c)). Let $\{\lambda_z^{(i)},\mu_z^{(i)}\}$, $(i,z) \varepsilon \{1,\ldots,m\} \times \{0,1,\ldots N\}$ (where m and N are finite positive integers) be the infinitesimal parameters for $Z(t)$. Define quantities $q_z^{(i)} = \mu_z^{(i)}/(\lambda_z^{(i)} + \mu_z^{(i)})$,

$\underline{q}_z = \inf_{i\varepsilon Y} \mu_z^{(i)}/(\lambda_z^{(i)}+\mu_z^{(i)})$ and $\bar{p}_z = \sup_{i\varepsilon Y} \lambda_z^{(i)}/(\lambda_z^{(i)}+\mu_z^{(i)})$.

Analogously define quantities $\bar{q}_z$ and $\underline{p}_z$.

Theorem. Assume the quantities $\bar{p}_z$ and $\underline{p}_z$ are positive for all $z \varepsilon \mathbb{Z}^+ = \{1,2,\ldots\}$ and that $q_z^{(i)} \downarrow$ as a function of z for all $i \varepsilon Y$.

(1) If $\sum_{z=1}^{\infty} \prod_{k=1}^{z} (\underline{q}_k/\bar{p}_k) = +\infty$, then $P_{(i_0,z_0)}[Z(t) \to 0] = 1$ for any initial state $(i_0,z_0) \varepsilon Y \times \mathbb{Z}^+$.

(2) If $\sum_{z=1}^{\infty} \prod_{k=1}^{z} (\bar{q}_k/\underline{p}_k) < +\infty$, then $P_{(i_0,z_0)}[Z(t) \to 0] < 1$ for any initial state $(i_0,z_0) \varepsilon Y \times \mathbb{Z}^+$. □

Let ${}_AZ(t)$ and ${}_aZ(t)$ denote the number of A genes and a genes at time t, respectively. Let $f_A(i_0,z_0)$ be the probability of fixation of A genes for the process with initial A type population z_0 in initial environment i_0. Then

$$f_A(i_0,z_0) \equiv P_{(i_0,z_0)}[{}_AZ(t) = N \text{ for some } t]$$

$$= P_{(i_0,N-z_0)}[{}_aZ(t) = 0 \text{ for some } t]$$

$$\equiv u_a(i_0,N-z_0) = \text{probability of extinction of}$$

a genes for the process ${}_aZ(t)$ with initial a type population $N-z_0$, environment i_0, and birth and death parameters

$${}_a\lambda_z^{(i)} = {}_A\mu_z^{(i)} \quad \text{and} \quad {}_a\mu_z^{(i)} = {}_A\lambda_z^{(i)}$$

where ${}_A\lambda_z^{(i)}$ and ${}_A\mu_z^{(i)}$ are given in (3.1). From the above theorem, we may assert

$$u_a(i_0,N-z_0) \leq \bar{u}_a(N-z_0) \equiv P_{N-z_0}[{}_a\bar{Z}(t) = 0 \text{ some } t]$$

$$= \text{probability of extinction of}$$

a genes for the process with initial a type population $N-z_0$, environment i_0 and birth and death parameters

$${}_a\underline{p}_z = \min_{1 \leq i \leq m} {}_a\lambda_z^{(i)} / ({}_a\lambda_z^{(i)} + {}_a\mu_z^{(i)}) ,$$

$${}_a\bar{q}_z = \max_{1 \leq i \leq m} {}_a\mu_z^{(i)} / ({}_a\lambda_z^{(i)} + {}_a\mu_z^{(i)}) .$$

But this probability is just

$$P_{z_0}[{}_A\bar{Z}(t) = N \text{ for some } t]$$

= probability of fixation of A genes for the process with initial A type population z_0, environment i_0, and birth and death parameters

$${}_Ap_z = {}_a\bar{q}_z , \; {}_Aq_z = {}_a\underline{p}_z .$$

But

$$ {}_A p_z = \max_{1 \le i \le m} {}_A\lambda_z^{(i)} / ({}_A\lambda_z^{(i)} + {}_A\mu_z^{(i)}) $$

$$ = (1+s)/(2+s) , $$

and

$$ {}_A q_z = \min_{1 \le i \le m} {}_A\mu_z^{(i)} / ({}_A\lambda_z^{(i)} + {}_A\mu_z^{(i)}) $$

$$ = 1/(2+s) \quad \text{where} \quad s = \max_{1 \le i \le m} s^{(i)} . $$

But from these parameters we obtain (see p.430) the same fixation probability as for the birth and death selection model with one-step transition probabilities

$$ p_z = (1+s)(N-z)z/B_z N , \; q_z = z(N-z)/B_z N , \; r_z = 1 - {}_A p_z - {}_A q_z , $$

where $B_z = z + (1+s)(N-z)$, and $s = \max_{1 \le i \le m} s^{(i)}$.

REFERENCES

Capocelli, R. and Ricciardi, L. (1974). A diffusion model for population growth in random environment. *Theoretical Population Biology*, 5, 28-41.

Cogburn, R. (1978). Markov chains in random environments: The case of Markovian environments. *Annals of Probability*. To appear.

Cook, R. and Hartl, D. (1975). Stochastic selection in large and small populations. *Theoretical Population Biology*, 7, 55-63.

Crow, J. and Kimura, M. (1970). *An Introduction to Population Genetics Theory*. Harper and Row, New York.

Dempster, E. (1955). Maintenance of genetic heterogeneity. *Cold Spring Harbor Symposium*, 20, 25-32.

Ewens, W. (1969). *Population Genetics*. Methuen, London.

Feldman, M. and Roughgarden, J. (1975). A population's stationary distribution and chance of extinction in a stochastic environment with remarks on the theory of species packing. *Theoretical Population Biology*, 7, 197-207.

Feller, W. (1951). Diffusion processes in genetics. *Proceedings of the Second Berkeley Symposium on Mathematical Statistics and Probability*. Berkeley, California. 227-246.

Fisher, R. (1922). On the dominance ratio. *Proceedings Royal Society of Edinburgh*, 42, 321-341.

Fisher, R. A. (1930). *The Genetical Theory of Natural Selection*. Oxford University Press, Oxford.

Fisher, R. and Ford, E. (1947). The spread of a gene in natural conditions in a colony of the moth *Panaxia dominula* L. *Heredity*, 1, 143-174.

Gillespie, J. (1972). The effects of stochastic environment on allele frequencies in natural populations. *Theoretical Population Biology*, 3, 241-248.

Gillespie, J. (1973a). Polymorphism in random environments. *Theoretical Population Biology*, 4, 193-195.

Gillespie, J. (1973b). Natural selection with varying selection coefficients--a haploid model. *Genetical Research*, 21, 115-120.

Gillespie, J. (1974). Natural selection for within-generation variance in offspring number. *Genetics*, 76, 601-606.

Haldane, J. (1924a). A mathematical theory of natural and artificial selection. I. *Transactions Cambridge Philosophical Society*, 23, 19-41.

Haldane, J. (1924b). A mathematical theory of natural and artificial selection. II. *Proceedings, Biological Sciences, Cambridge Philosophical Society*, 1, 158-163.

Haldane, J. (1924c). A mathematical theory of natural and artificial selection. III. *Proceedings Cambridge Philosophical Society*, 23, 363-372.

Haldane, J. (1924d). A mathematical theory of natural and artificial selection. IV. *Proceedings Cambridge Philosophical Society*, 23, 235-243.

Haldane, J. (1927). A mathematical theory of natural and artificial selection. V. Selection and mutation. *Proceedings Cambridge Philosophical Society*, 28, 838-844.

Haldane, J. and Jayakar, S. (1963). Polymorphism due to selection of varying direction. *Journal of Genetics*, 58, 237-242.

Hartl, D. and Cook, R. (1973). Balanced polymorphism of quasi-neutral alleles. *Theoretical Population Biology*, 4, 163-172.

Jensen, L. (1973). Random selection advantages of genes and their probabilities of fixation. *Genetical Research*, 21, 215-219.

Jensen, L. and Pollak, E. (1969). Random selective advantages of a gene in a finite population. *Journal of Applied Probability*, 6, 19-37.

Karlin, S. and Levikson, B. (1974). Temporal fluctuations in selection intensities: case of small population size. *Theoretical Population Biology*, 6, 383-412.

Karlin, S. and Lieberman, U. (1974). Random temporal variation in selection intensities: case of large population size. *Theoretical Population Biology*, 6, 355-382.

Karlin, S. and Lieberman, U. (1975). Random temporal variation in selection intensities: one locus two allele model. *Journal of Mathematical Biology*, 2, 1-17.

Karlin, S. and McGregor, J. E. (1962). On a genetics model of Moran. *Proceedings Cambridge Philosophical Society*, 58, 229-311.

Karlin, S. and Taylor, H. (1975). *A First Course in Stochastic Processes*. Second edition. Academic Press, New York.

Keiding, N. (1975). Extinction and exponential growth in random environments. *Theoretical Population Biology*, 8, 49-63.

Kimura, M. (1952). Process of irregular change of gene frequencies due to the random fluctuation of selection intensities. *Annual Report of the National Institute of Genetics of Japan*, 1, 45-47.

Kimura, M. (1954). Process leading to quasi-fixation of genes in natural populations due to random fluctuations of selection intensities. *Genetics*, 39, 280-295.

Kimura, M. (1955a). Solution of a process of random genetic drift with a continuous model. *Proceedings National Academy of Sciences*, 41, 144-150.

Kimura, M. (1955b). Random genetic drift in multi-allelic locus.

Evolution, 9, 419-435.

Kimura, M. (1955c). Stochastic processes and distribution of gene frequencies under natural selection. *Cold Spring Harbor Symposium*, 20, 33-53.

Kimura, M. (1957). Some problems of stochastic processes in genetics. *Annals of Mathematical Statistics*, 28, 882-901.

Kimura, M. (1962). On the probability of fixation of mutant genes in a population. *Genetics*, 47, 713-719.

Kimura, M. (1964). Diffusion models in population genetics. *Journal of Applied Probability*, 1, 177-232.

Levikson, B. (1976). Regulated growth in random environments. *Journal of Mathematical Biology*, 3, 19-26.

Levikson, B. and Karlin, S. (1975). Random temporal variation in selection intensities acting in infinite diploid populations: diffusion and analysis. *Theoretical Population Biology*, 8, 292-300.

Levins, R. (1969). The effect of random variations of different types on population growth. *Proceedings National Academy of Sciences*, 62, 1061-1065.

Lewontin, R. and Cohen, D. (1969). On population growth in a randomly varying environment. *Proceedings National Academy of Sciences*, 62, 1056-1060.

Malecot, G. (1944). Sur une probleme de probabilites en chaine que pose la genetique. *Comptes Rendus de l'Academie des Sciences, Paris*, 219, 379-381.

May, R. (1973). *Stability and Complexity in Model Ecosystems.* Princeton University Press, Princeton, New Jersey.

May, R. and MacArthur, R. (1972). Niche overlap as a function of environmental variability. *Proceedings National Academy of Sciences*, 69, 1109-1113.

Moran, P. A. P. (1958a). Random processes in genetics. *Proceedings Cambridge Philosophical Society*, 54, 60-72.

Moran, P. A. P. (1958b). The effect of selection in a haploid genetic population. *Proceedings Cambridge Philosophical Society*, 54, 463-467.

Ohta, T. (1972). Fixation probability of a mutant influenced by random fluctuation of selection intensity. *Genetical Research*, 19, 33-38.

Torrez, W. (1978). The birth and death chain in a random environment: instability and extinction theorems. *Annals of Probability*, 6, 1026-1043.

Torrez, W. (1979a). On a genetics model of Moran evolving in random environments. *Rocky Mountain Journal of Mathematics*, 9. To appear Winter 1979.

Torrez, W. (1979b). Calculating extinction probabilities for the birth and death chain in a random environment. *Journal of Applied Probability*, 16. To appear.

Torrez, W. (1979c). Birth and death processes with continuously temporal environments. *Journal of Applied Probability*. Submitted.

Tuckwell, H. (1974). A study of some diffusion models of population growth. *Theoretical Population Biology*, 5, 345-357.

Tuckwell, H. (1976). The effects of random selection on gene frequency. *Mathematical Biosciences*, 30, 113-128.

Turelli, M. (1977). Random environments and stochastic calculus. *Theoretical Population Biology*, 12, 140-178.

Turelli, M. (1978). A re-examination of stability in randomly varying versus deterministic environments with comments on the stochastic theory of limiting similarity. *Theoretical Population Biology*, 13, 244-267.

Wright, S. (1931). Evolution in Mendelian populations. *Genetics*, 16, 97-159.

Wright, S. (1939). Statistical genetics in relation to evolution. *Actualites Scientifiques et Industrielles*. No. 802. Herman et Cie, Paris.

Wright, S. (1945). The differential equation of the distribution of gene frequencies. *Proceedings National Academy of Sciences*, 31, 382-389.

Wright, S. (1948). On the role of directed and random changes in gene frequencies in the genetics of populations. *Evolution*, 2, 279-294.

Wright, S. (1949). Adaptation and selection. In *Genetics, Paleontology, and Evolution.* Jepsen *et al.*, eds. Princeton University Press, Princeton, New Jersey.

Wright, S. (1951). The genetical structure of populations. *Annals of Eugenics*, 15, 323-354.

[*Received June* 1978. *Revised March* 1979]

G. P. Patil and M. Rosenzweig, (eds.),
Contemporary Quantitative Ecology and Related Ecometrics, pp. 439-452.

A GRAPH THEORETICAL TEST TO DETECT INTERFERENCE IN SELECTING NEST SITES

RICHARD F. GREEN

Department of Statistics
University of California
Riverside, California 92521 USA

SUMMARY. Birds nesting in boxes may interfere with other members of the same species in the sense that nearby boxes are unlikely to be occupied by members of the same species. Interspecific interference may also occur. In this paper simple methods are described to test for intraspecific and interspecific interference in selecting nest sites. These tests are based on looking at neighboring pairs of nest boxes to see whether they are occupied by the same species, different species, or if one or both boxes are empty. A similar procedure is suggested to test for nest site interference in digger wasps. The tests are applied to some data on Great and Blue Tits nesting in Marley Wood, near Oxford, England and on Great Golden Digger Wasps digging their nest burrows in a planter on the campus of the University of Michigan, Dearborn. The wasp data was not sufficient to show interference but intraspecific interference was shown in both Great and Blue Tits. Interspecific interference was not shown for the tits.

KEY WORDS. England, graph theory, intra- and interspecific interference, Michigan, nest sites, *Parus caeruleus*, *Parus major*, *Sphex ichneumoneus*, statistical tests.

1. INTRODUCTION

Individual plants and animals may interfere with each other in the sense that the presence of one individual repels others or interferes with their growth. Plants may retard the growth of neighbors in several ways including shading and chemical interference. Animals often exclude others from their vicinity by territorial behavior. Interference may result in individuals spacing themselves out. Interference may be within species, between species or both.

A number of procedures exist to test whether points are distributed at random in space or whether several types of points are distributed at random with respect to each other. Many of these procedures of importance in ecology are described by Pielou (1969, Chapters 7-16), who is mainly concerned with plants, which have fixed locations. Many of the procedures test whether the individuals are distributed completely at random, that is, according to a Poisson distribution. For animals seeking sites for nests or burrows, only certain locations may be possible and these possible locations may not be randomly distributed. In this paper a method is described to test whether nest sites are chosen at random from among a number of fixed locations.

The method is illustrated with two examples.

(1) In a study of two species of birds nesting in boxes, the locations of the boxes are known and it is known which species nest in which boxes and which boxes are empty. There are two questions here. (a) Are boxes chosen at random within species? (Is there intraspecific interference?). (b) Are boxes chosen at random between species? (Is there interspecific interference?).

The question of intraspecific interference has been considered by Krebs (1971) who used a method that is not quite satisfactory statistically. Krebs investigated nest site interference by considering nearest neighbor distances. Analyzing Great Tit data from Wytham Wood, Krebs measured the distance from each nest box occupied by Great Tits to the nearest box also occupied by Great Tits. Years were broken into high and low density years and the observed yearly distributions of nearest neighbor distances were compared with simulated distributions obtained by choosing boxes at random to be occupied. Simulations were done by using one number for low density years and a different number of boxes for high density years. The observed nearest neighbor distances (obtained by combining nearest neighbor distances for low density years in one group and distances for high density years in another group) were compared with the theoretical nearest neighbor distances by using a Chi-squared test. There are several reasons why this procedure is unsatisfactory. Using a simulation to obtain the theoretical distribution and combining several sets of observations both tend to produce high values for a Chi-squared statistic and the nearest neighbor distances are not independent of each other since if two nearby nests are both occupied by Great Tits then each box is likely to have the other as its nearest neighbor and the distance is counted twice. This violates the assumption in the Chi-squared test that observations are independent. Krebs' procedure is more likely to show a significant difference when none actually exists than the stated significance level suggests.

(2) Female digger wasps dig their burrows, provision them and then fill them in. The locations of the burrows and the time during which they are active may be observed. The question is whether burrows that are active at the same time are less likely to be close together than those that are not active at the same time.

The method described in this paper to test for nest site interference is similar to the method used by Knox (1964) in his epidemiological study of childhood leukemia.

2. METHOD

2.1 Leukemia Cases and Digger Wasp Burrows. In his study of childhood leukemia Knox (1964) used data on 96 cases of childhood leukemia recorded in northern England from 1951 to 1960. Knox considered the number of pairs of cases that occurred close together in space (1 km.) and in time (60 days). Knox's method has been given a general formulation in terms of graph theory by Barton and David (1966).

In graph theory a graph is a collection of points and lines which connect some of the points. In the case of Knox's leukemia data the 96 cases used to test for epidemicity could be considered as points with lines connecting all pairs of cases that occurred within 1 km. distance of each other. This would give a graph which we could call the space graph. Again, the 96 cases could be considered as points with lines connecting all pairs of cases whose dates of onset were within 60 days of each other. This would give a graph we could call the time graph. The statistic Knox suggests to test for epidemicity is the number of pairs of cases close together both in space and time. Barton and David (1966) interpret Knox's statistic in terms of the intersection of the time graph and the space graph. (The intersection of two graphs consists of the points and the lines that the two graphs have in common.)

The observed value of Knox's statistic can be compared with the expected value calculated under the assumption that there is no space-time interaction. Using the notation of David and Barton (1966):

X = the observed number of cases that are close in both space and time (Knox's statistic)

N_{1S} = the number of pairs of cases close in space (the number of lines in the space graph),

N_{1T} = the number of pairs of cases close in time (the number of lines in the time graph), and

n = the number of cases (the number of points in each graph),

we have the expected value of X given by

$$E[X] = 2N_{1S}N_{1T}/n^{(2)} \tag{1}$$

where $n^{(2)} = n(n-1)$ is the second factorial power of n. The *rth* factorial power of n is $n^{(r)} = n(n-1)(n-2)\cdots(n-r+1)$.

To calculate the variance of X the idea of the degree of a point is needed. The degree of a point in a graph is the number of lines to that point. If we denote the degree of the *ith* point in a graph by ℓ_i then we define

$$N_2 = \sum_{i=1}^{n} \ell_i(\ell_i - 1)/2. \tag{2}$$

Notice that $N_1 = \sum_{i=1}^{n} \ell_i/2$ since N_1 is the total number of lines in a graph and each line connects two points.

All that need be observed in order to calculate the mean and variance of Knox's statistic is the degree of each point in the space graph and in the time graph. That is, we must see how many neighbors each point has in space and in time.

For the variance we have

$$\begin{aligned} \mathrm{Var}(X) &= 2N_{1S}N_{1T}/n^{(2)} + 4N_{2S}N_{2T}/n^{(3)} - (2N_{1S}N_{1T}/n^{(2)})^2 \\ &+ 4(N_{1S}^2 - N_{1S} - 2N_{2S})(N_{1T}^2 - N_{1T} - 2N_{2T})/n^{(4)}. \end{aligned} \tag{3}$$

Formulas (1) and (3) give the mean and variance of X under the null hypothesis that the space graph and the time graph are independent. This is equivalent to assigning the observed locations and times to the points at random. It is not necessary to assume that the locations or times themselves are random. In fact, in Knox's leukemia study the cases are more frequent in cities than in the country and are more frequent in summer than in winter.

Exactly the same method as described for leukemia can be used to test whether digger wasps interfere with each other in the choice of nest sites. For the wasps the time graph will have lines connecting each pair of burrows that are active at the same time. For both leukemia cases and wasp burrows the null hypothesis is that the space and time graphs are independent. The only difference is that the alternative hypotheses predict more leukemia cases close together in space and time than under the null hypothesis (epidemicity) and fewer wasp burrows close together in space and time than under the null hypothesis (interference). To test whether the observations are sufficiently different from the expected we can calculate

$$z = (X - E[X])/[\mathrm{Var}(X)]^{\frac{1}{2}} \tag{4}$$

which will be approximately normal with mean 0 and variance 1 under the null hypothesis if the number of burrows is large enough.

2.2 Bird Nests: Intraspecific Interference. The method suggested in this paper to test for interference between birds in the selection of nest boxes is quite similar to that used in the case of wasp burrows. If we only consider one species of bird, for example, the Great Tit, then the nest boxes are the points and we will have a space graph with lines connecting boxes within some specified distance of each other (say 50 m.). However, in place of a time graph we will have a graph with lines between each pair of nest boxes with Great Tits nesting in them. The test statistic will again come from taking the intersection of two graphs. In this case X will be the number of "neighbor" nest box pairs with Great Tits using each box. This model has been used by Cliff and Ord (1973) in their work on statistical geography.

Under the null hypothesis that the Great Tits choose their nest boxes at random the mean and variance of the number of Great-Great neighbor box pairs, X, may be obtained from (1) and (3) where the time graph is replaced by a species graph with lines connecting boxes occupied by Great Tits. The calculations will be a bit simpler using the expressions

$$E[X] = N_{1S} g^{(2)}/n^{(2)}, \quad \text{and} \tag{5}$$

$$\text{Var}(X) = N_{1S} g^{(2)}/n^{(2)} + 2N_{2S} g^{(3)}/n^{(3)} + (N_{1S}^2 - N_{1S} - 2N_{2S}) g^{(4)}/n^{(4)} - (N_{1S} g^{(2)}/n^{(2)})^2, \tag{6}$$

where g is the number of nest boxes occupied by Great Tits. Again, if $E[X]$ is reasonably large the value of z calculated from (4) will be approximately standard normal under the null hypothesis and may be used to test for intraspecific interference in choosing nest sites.

The same procedure may be followed for Blue Tits as well, merely substituting the number of Blue-Blue neighbor box pairs, Y, for X in (4) and the number of nest boxes occupied by Blue Tits, b, for g in (5) and (6).

2.3 Bird Nests: Interspecific Interference. Interspecific interference may be tested for by using the number of neighbor nest box pairs with Great Tits occupying one box and Blue Tits occupying the other. Call this number U.

If there is no intraspecific interference then the null hypothesis of no interspecific interference may be tested by using (4) to compare z with its mean and variance where

$$E[U] = 2N_{1S} bg/n^{(2)}, \quad \text{and} \tag{7}$$

$$\text{Var}(U) = 2N_{1S} bg/n^{(2)} + 2N_{2S} bg(b+g-2)/n^{(3)} + 4\,(N_{1S}^2 - N_{1S} - 2N_{2S}) b^{(2)} g^{(2)}/n^{(4)} - (2N_{1S} bg/n^{(2)})^2. \tag{8}$$

If intraspecific interference exists it must be taken into account when testing for interspecific interference. If there is intraspecific interference but not interspecific interference and if a given box is occupied by Great Tits, say, then a nearby box would be more likely to be occupied by Blue Tits than by chance simply because Great Tits tend to be excluded. Thus if there is intraspecific but not interspecific interference then Great-Blue neighbor box pairs would be more likely than if all boxes are occupied completely at random.

In order to calculate the mean and variance of U under the null hypothesis of no interspecific interference but conditioned on the observed number of Great-Great and of Blue-Blue neighbor box pairs, X and Y, respectively, it is necessary to calculate the covariances: Cov(X,Y), Cov(X,U), and Cov(Y,U). These are given by

$$\mathrm{Cov}(X,Y) = (N_{1S}^2 - N_{1S} - 2N_{2S})b^{(2)}g^{(2)}/n^{(4)}$$

$$- N_{1S}^2 b^{(2)}g^{(2)}/(n^{(2)})^2, \quad \text{and} \tag{9}$$

$$\mathrm{Cov}(X,U) = 2N_{2S}bg^{(2)}/n^{(3)} + 2(N_{1S}^2 - N_{1S} - 2N_{2S})bg^{(3)}/n^{(4)}$$

$$- 2N_{1S}^2 bg^2(g-1)/(n^{(2)})^2. \tag{10}$$

The expression for Cov(Y,U) will be the same as that for Cov(X,U) given in (10) but with b and g interchanged.

To test the null hypothesis of no interspecific interference conditional on the values of X and Y observed, we compare the observed value of U with its conditional mean $E[U|X,Y]$ and variance $\mathrm{Var}(U|X,Y)$. If we denote the means of the variables X, Y, U by μ_x, μ_y, μ_u, their variances by a_{xx}, a_{yy}, a_{uu}, and their covariances by a_{xy}, a_{xu}, a_{yu} then we have

$$E[U|X,Y] = \mu_u + [a(X-\mu_x) + b(Y-\mu_y)]/c, \quad \text{and} \tag{11}$$

$$\mathrm{Var}(U|X,Y) = a_{uu} + [a(a_{xu}) + b(a_{yu})]/c, \tag{12}$$

where $a = (a_{xu}a_{yy} - a_{yu}a_{xy})$, $b = (a_{xx}a_{yu} - a_{xy}a_{xu})$, and $c = (a_{xx}a_{yy} - a_{xy}^2)$.

3. RESULTS

The methods described in the preceding section are illustrated using two sets of data.

3.1 Marley Wood Tit Data for 1977. Since 1947 a population study of Great Tits, *Parus major*, and Blue Tits, *Parus caeruleus*, has been carried on by members of the Edward Grey Institute of Field Ornithology in Marley Wood, a part of Wytham Wood, near Oxford,

England. The birds nest in boxes which have been provided for them. Each year it is observed which boxes are occupied by which birds and nesting success is recorded. Much of this work has been described by Lack (1966).

The locations of the nest boxes in Marley Wood have recently been accurately mapped by Ed Minot (personal communication, 1978) who found that earlier maps were in error. In 1977 there were 214 nest boxes in Marley Wood. Thirty pairs of Great Tits nested there and 86 pairs of Blue Tits (I have not counted boxes used by the five pairs of Blue Tits that renested in the area and I have only counted once the box that was used twice). The calculations have been done four times, successively defining nest boxes within 20, 30, 40 and 50 meters of each other as neighbors. The results are shown in Table 1.

Testing the data shows significantly too few Blue-Blue neighbor pairs for all four distances considered. There are also too few Great-Great neighbor pairs for the greater distances 40 and 50 meters and possibly for 30 meters as well. Since only 30 boxes were occupied by Great Tits the expected number of Great-Great neighbor pairs within 20 m. is only about 1 and the fact that no such pairs were observed is not statistically significant.

TABLE 1: 1977 Marley Wood Tit Data and Analysis.

n = 214 boxes

g = 30 Great Tit nesting pairs

b = 86 Blue Tit nesting pairs

N_{1S} = the number of neighbor pairs of boxes (unordered)

N_{2S} = half the sum of the degree × (degree − 1) of the points

Distance (meters)	20	30	40	50
N_{1S}	49	170	341	561
N_{2S}	22	255	1059	2900
X	0	0	1	3
Y	0	12	38	72
U	8	20	41	64
E(X)	.94	3.24	6.51	10.71
E(Y)	7.86	27.26	54.69	89.97
E(U)	5.55	19.24	38.60	63.51
Var(X)	.92	3.11	6.20	10.15
Var(Y)	6.59	20.78	43.73	71.96
Var(U)	4.91	16.50	33.02	54.13
Cov(X,Y)	-.15	-.44	-.91	-1.54
Cov(X,U)	-.10	-.50	-.95	-1.44
Cov(Y,U)	-.89	-3.18	-6.28	-10.14
z(X)	-.98	-1.84	-2.21	-2.42
z(Y)	-3.06	-3.27	-2.53	-2.12
z(U|X,Y)	.55	-.51	-.17	-.45

These significance tests indicate that intraspecific interference in the selection of nest boxes exists for both Great Tits and Blue Tits. Tests of interspecific nest site interference do not show significance for any neighbor distance. Although further analyses may yield more definitive results, the 1977 data suggest that if interspecific interference does occur its effect is small compared to that of intraspecific interference.

3.2 Michigan Digger Wasp Data for 1974. The Great Golden Digger Wasp, *Sphex ichneumoneus* provides another example of possible interference in the choice of nest site. The female wasp digs a burrow which she provisions with prey. She lays her eggs on the prey and fills in the burrow. In some cases burrows are abandoned before the process is completed.

Brockmann (1976) has studied digger wasps for several years. In one study she observed wasps digging their burrows in a planter on the campus of the University of Michigan, Dearborn. She observed a total of 79 burrows being dug over the six week study period. At any one time some of the burrows were active, some were abandoned, and some had been filled in completely. The location of each burrow and the time during which it was active were recorded. The question here is whether the position of active nests influenced where a wasp dug a new nest.

Brockmann's 1974 Michigan wasp data are tested for nest site interference using formula (4) where the mean and variance are calculated using formulas (1) and (3). The calculations have been done four times, successively defining burrows within 4, 6, 8, and 10 cm. of each other as neighbors in space. Burrows active at the same time are considered to be neighbors in time. The results are shown in Table 2.

The observed numbers of burrows close together in space and simultaneous in time are not significantly different from those expected by chance. Since the total number of burrows is small the differences would have to be striking in order to be statistically significant.

Several burrows were excavated by the same individual wasps. It might be interesting to see whether each wasp tended to clump or to space out her own burrows. This could be tested by using the same general method used here with neighbor burrows defined in the same way but with the other graph connecting burrows dug by the same individual rather than burrows active at the same time. I have not looked at the data in this way.

TABLE 2: 1974 Dearborn Digger Wasp Data and Analysis

n = 79 burrows

N_{1S} = number of neighbor pairs (within chosen distance)

N_{1T} = number of simultaneous pairs of burrows

$N_{2S} = \sum_{i=1}^{n} \ell_i^{(2)}/2$, $N_{2T} = \sum_{i=1}^{n} k_i^{(2)}/2$, where

ℓ_i = local degree of the *ith* burrow in space

k_i = local degree of the *ith* burrow in time

X = number of simultaneous neighbor pairs.

Distance (cm.)	4	6	8	10
N_{1S}	8	17	26	42
N_{1T}	800	800	800	800
N_{2S}	3	11	26	74
N_{2T}	18513	18513	18513	18513
X	3	3	4	8
EX	2.08	4.41	6.75	10.91
Var(X)	1.57	3.35	5.18	8.66
z	.74	-.77	-1.21	-.99

4. DISCUSSION

4.1 The Results. The 1974 wasp burrow data suggest there may be nest site interference but the differences observed were not statistically significant. It would take several times as much data as analyzed here to detect interference if it is not stronger than these data suggest.

The 1977 nest data do show significant intraspecific interference for both Great Tits and Blue Tits but interspecific has not been shown. More useful than statistical tests, however, is the direct comparison of the observed with the expected number of intraspecific neighbor pairs. There are neither Great-Great nor Blue-Blue neighbor pairs for the shortest distance considered (20 m.). But while there are still no Great-Great neighbor pairs for 30 m. there are 12 Blue-Blue pairs for the distance (with about 27 expected under the null hypothesis). By the time we consider neighbors within 50 m. there are 72 such Blue-Blue neighbor pairs (with about 90 expected) while there are only 3 Great-Great neighbor pairs (with about 11 expected). Thus for a distance of 50 meters there are about 80% of the number of Blue-Blue neighbor pairs expected under the null hypothesis while there are fewer than 30% of the expected number of Great-Great neighbor pairs. This difference suggests that the effect of nest box interference extends further for Great Tits than for Blue Tits. This is certainly reasonable because the Great Tits are substantially larger birds.

Another way of looking at the data is to compare the observed and expected number of neighbors within a given distance interval. For example, looking at Table 1, we can see that for Blue Tits the number of neighbor pairs from 20 to 30 meters apart is 12 and the expected number is $27.26 - 7.86 = 19.40$. There are 26 Blue-Blue neighbor pairs from 30 to 40 meters apart with 27.43 expected if boxes are chosen at random. From 40 to 50 meters there are 34 Blue-Blue neighbor pairs with 35.28 expected if boxes are chosen at random. These observations suggest that nest site interference does not extend further than 30 meters for Blue Tits. For Great Tits the effect of interference seems to extend to 50 meters although there are too few Great-Great neighbor pairs expected in each interval to show statistical significance.

Krebs (1971) presented evidence that in years of low Great Tit density the interference effect extended as far as 50 meters. In years of high density the effect did not extend as far. By Krebs' criterion 1977 would be a low density year for Great Tits.

4.2 Knox's Method and Other Methods. The methods I have described and illustrated in this paper are based on that used by Knox (1964) in his study of childhood leukemia. Knox's method can be applied directly to test for nest site interference in digger wasps. For wasp burrows there will be a space graph (where neighbor burrows are joined) and a time graph (where burrows active at the same time are joined). For bird nests there will be a space graph (where neighbor nest boxes are joined) and a species graph (where boxes occupied by the same species are joined). In each case X will be the number of lines (pairs of nests or burrows) the two graphs have in common. For bird nests the variance of X under the hypothesis of random occupation of boxes will be easier to calculate using formula (6) than using formula (3). I have extended Knox's method to the case of interspecific interference when it is known that intraspecific interference may occur.

Methods similar to Knox's have been used by statistical geographers for some time. This work is discussed in the book by Cliff and Ord (1973). Cliff and Ord give conditions for the asymptotic normality of (4).

Several alternatives to Knox's method have been proposed.

1. Simulations may be used to produce an exact test for nest site interference. Such procedures are known as Monte Carlo tests (see Cliff and Ord, 1973, p. 50; or Hope, 1968). For example, to test for nest site interference among Great Tits using the 1977 Marley Wood data one could pick a test statistic, say the number

of Great-Great neighbor pairs within 50 meters. The observed value is: $X = 3$. Then a simulation is done, choosing 30 boxes at random from among the 214 and imagining these are occupied by Great Tits. The statistic is calculated for the simulation and the process is repeated until there are, say 99, simulated populations each with 30 boxes occupied by Great Tits. The statistic, X, is calculated for each of the 99 simulations. If the observed value $X = 3$ is one of the five smallest values when combined with those from the 99 simulations then the null hypothesis that the nest boxes were chosen at random is rejected.

Besag and Diggle (1977) have described the use of Monte Carlo tests for a number of spacial processes including Knox's problem. Monte Carlo tests have the advantage of not requiring knowledge of the distribution of the test statistic, X. If the simulation can be done (in this case by choosing 30 boxes at random) the test can be performed.

2. Morris (1975) tested whether the pattern of development of retinal cells was random or nonrandom. She constructed maps of labelled and unlabelled principal cones in the retinae of embryo chicks. The cones were labelled or not according to whether or not they were developing when they were exposed to a radioactive marker. The maps were tested for contagion by grid-analysis (Greig-Smith, 1964) or by graphical methods where neighboring pairs of cones were looked at to see whether they were of the same type (both labelled or both unlabelled).

Morris concluded that the graphical methods were generally more successful than grid-analysis in detecting contagion. One of the difficulties with grid-analysis is that the quadrat boundaries are arbitrary and contagion is likely to act across boundaries as well as within them. This is especially important when local effects are of interest and the grids must be fine.

3. The method of Knox has been generalized by Mantel (1967). Instead of simply counting the number of cases that occur close together in time and in space it is possible to weight the pairs of cases according to how close they are in space and by how close they are in time. If the proper weights are used such a procedure may be more powerful in detecting contagion or interference than Knox's method. Even with Knox's method the distance and time used to define neighbors are arbitrary. The advantage of Knox's method is its conceptual simplicity.

5. SUMMARY AND CONCLUSION

In this paper a simple method is described to test for interference in the choice of nest sites. Similar methods have been used for some time by epidemiologists and statistical geographers. The method is illustrated with two sets of data. Intraspecific interference in the choice of nest sites has been shown for both Great Tits and Blue Tits but interspecific interference has not been shown. If it exists it is almost certainly weaker than intraspecific interference.

Interference was not shown in the choice of burrow sites in the Great Golden Digger Wasp.

ACKNOWLEDGMENTS

Most of the work on this paper was done while I was on sabbatical from the University of California, Riverside, at the Department of Zoology, Oxford. I would like to thank Dr. C. M. Perrins, director of the Edward Grey Institute of Field Ornithology, for permission to use the Wytham tit nesting data, Mr. E. O. Minot for the use of his data on nest locations, and Dr. H. J. Brockmann for the use of her data on the digger wasps. The problem of detecting interspecific nest site interference was described to me by Dr. André Dhondt. I would like to especially thank Professor F. N. David, through whom I first heard of the problem. She suggested the method of solution and has given me a great deal of help and encouragement. An earlier draft of this paper was read and criticized by H. J. Brockmann, E. O. Minot, M. P. Sloan, and F. C. Vasek.

REFERENCES

Barton, D. E. and David, F. N. (1966). The random intersection of two graphs. In *Research Papers in Statistics*, F. N. David, ed. John Wiley & Sons, New York. 455-469.

Besag, J. and Diggle, P. J. (1977). Simple Monte Carlo tests for spatial pattern. *Applied Statistics*, 26, 327-333.

Brockmann, H. J. (1976). *The control of nesting behavior in the Great Golden Digger Wash,* Sphex ichneumoneus L. *(Hymenoptera, Sphecidae)*. Ph.D. dissertation, University of Wisconsin, Madison.

Cliff, A. D. and Ord, J. K. (1973). *Spatial Autocorrelation.* Pion Limited, London.

David, F. N. and Barton, D. E. (1966). Two space-time interaction tests for epidemicity. *British Journal of Preventive and Social Medicine*, 20, 44-48.

Greig-Smith, P. (1964). *Quantitative Plant Ecology. Second edition.* Butterworths, London.

Hope, A. C. A. (1968). A simplified Monte Carlo significance test procedure. *Journal of the Royal Statistical Society, Series B*, 30, 582-598.

Knox, G. (1964). Epidemiology of childhood leukemia in Northumberland and Durham. *British Journal of Preventive and Social Medicine*, 18, 17-24.

Krebs, J. R. (1971). Territory and breeding density in the great tit, *Parus major* L. *Ecology*, 52, 2-22.

Lack, D. (1966). *Populations Studies of Birds.* Clarenden Press, Oxford.

Mantel, N. (1967). The detection of disease clustering and a generalized regression approach. *Cancer Research*, 27, 209-220.

Morris, V. B. (1975). Non-randomness in the sequential formation of principal cones in small areas of the developing chick retina. *Journal of Comparative Neurology*, 164, 95-104.

Pielou, E. C. (1969). *An Introduction to Mathematical Ecology.* John Wiley & Sons, New York.

[*Received March* 1979. *Revised May* 1979]

G. P. Patil and M. Rosenzweig, (eds.),
Contemporary Quantitative Ecology and Related Ecometrics, pp. 453-472.

GRAPHICAL MODELS AND METHODS IN ECOLOGY

I. NOY-MEIR

Department of Botany
Hebrew University
Jerusalem, Israel

SUMMARY. Graphical models and methods are used in ecology both on their own and in association with mathematical methods. Being precise, yet simple and not too abstract, they are useful as didactic aids and often also as research tools. Their main limitation is dimensionality. The contribution of graphical methods to two areas in ecology is reviewed: community ordination and stability analysis.

KEY WORDS. graphical models, geometrical models, ordination, stability analysis, isoclines.

1. INTRODUCTION

It may be questioned whether graphical models and methods do properly and formally come under the heading of statistical and mathematical ecology. They are certainly very closely related to that subject, and relevant to both teaching and research in mathematical ecology. In this review, I would like first to discuss briefly the relationships between graphical and mathematical models; then to point out both the advantages and the limitations (or dangers) of graphical methods; and to demonstrate those by examples from two fields of ecology.

A graphical (or geometrical) model, here, is any representation of elements and relations in graphical form in a Euclidean space, usually a two-dimensional one (paper, screen, etc.).

2. RELATIONS BETWEEN GRAPHICAL AND MATHEMATICAL MODELS

The process of modelling can be considered as a process of mapping, or translation, between the world of real systems and the world of models. We can further subdivide the world of models into (at least) three worlds, each with a different language: conceptual-verbal, mathematical, and graphical (Figure 1).

The first step is usually an abstraction from the real system into a conceptual model of this system, which is normally expressed in words. This is a difficult and critical step, but it is not our concern here. Once the conceptual-verbal model has been properly formulated, we try to apply logic, common sense, or verbal arguments in order to derive from it new results, new conclusions which will be relevant to the real system.

Unless the model is very simple, this mental or verbal procedure rapidly becomes rather difficult and confusing, at least for the average human mind. Thus we turn to mathematical or graphical procedures to help us solve the problem. But first the conceptual-verbal model has to be translated into mathematical or geometrical language.

The conclusions from the mathematical or graphical operations on the model must then be translated back (interpreted) into the conceptual-verbal model, and through it into the real system. The translation and re-translation are the difficult and tricky parts of the whole process. The use of mathematical procedures on mathematical models, and of graphical procedures on graphical models, is relatively straightforward. The translation between mathematical and graphical languages is also usually not too difficult. But there are many pitfalls and dangers in the translation of concepts or words into either mathematical or graphical language.

There are several possible ways in which graphical models and methods can be used (Figure 2).

Case A: A mathematical model is formulated directly from concepts. Some problems may be difficult or impossible to solve by mathematical techniques. For these, we translate the mathematical model into a two-dimensional Euclidean model, and solve the problem by graphical methods. The results may be interpreted directly in concepts or words.

Even if the graphical model is not actually necessary to solve problems, it may still be useful as a didactic aid. The mathematical models and methods are at a level of abstraction which makes it difficult for most non-mathematicians to understand

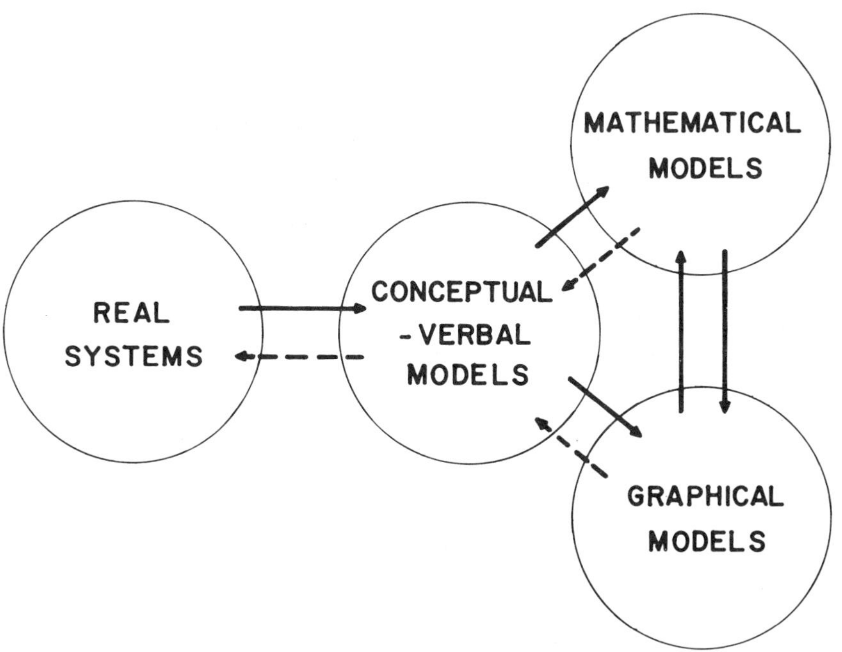

FIG. 1: A graphical model of the modelling process, as mapping between worlds, or translation between languages. Solid arrows indicate forward translation, broken arrows indicate re-translation (interpretation).

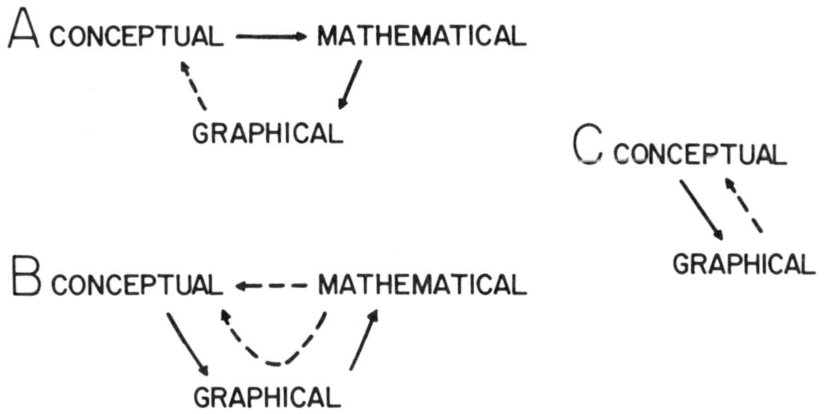

FIG. 2: Three possible ways of using graphical models in relation to conceptual and mathematical models.

what is happening, let alone explain this to others. The graphical model then helps us to visualize, understand, and explain what the mathematical model is doing, at least for the two-dimensional case.

Case B: In other cases, the first step is a translation of concepts into a geometrical model. Having understood and used this model, we then translate it into the appropriate two-dimensional mathematical model, and use mathematical methods to solve problems and get new results (beyond those obtainable from the graph). These results are interpreted in conceptual terms, either directly or through the graphic visualization. Often it is then easy to extend the mathematical model from 2 dimensions to many ($\underline{n}$), and solve problems which cannot be solved graphically.

Case C: Sometimes we use only a graphical model (without its mathematical counterpart) either to obtain new results and interpret them, or merely to visualize and explain the conceptual-verbal model.

3. ADVANTAGES OF GRAPHICAL MODELS

3.1 They make it possible to present problems and sometimes solve them, in a way which is simple yet precise. Graphs are usually simpler to work with than the corresponding mathematical models. Yet graphs are more precise than concepts and words, which tend to be vague and ambiguous.

3.2 Because they are simple, it is often easier, in terms of human time and effort, to solve problems and obtain new information by graphical rather than by mathematical methods. Thus, graphical models are useful as a research tool.

3.3 Because graphs are less abstract than mathematical expressions, it is usually easier with them to demonstrate and explain the principles of the model, the operations on it, and the results obtained, to students and users. Graphical models are useful as a teaching aid.

Taking an example from another field, one may see that an important part of economic theory has been developed by graphical methods. Much of the teaching of this theory is still done mainly by use of graphical models.

4. LIMITATIONS OF GRAPHICAL MODELS

4.1 They are effectively limited to two dimensions, though with some effort and ingenuity they can be extended to three dimensions.

In real ecological systems, more than three state variables are often important enough to be included in our conceptual models. It is possible to work with 2- or 3-dimensional graphs, and then extend the results, by analogy, to more dimensions. The problem with that is that in higher dimensional space qualitatively new phenomena may sometimes appear. Thus there is a danger that the graphical model will over-simplify the results and sometimes actually distort them seriously.

4.2 There is a danger of incorrect or imprecise translation between geometrical language on the one hand, and conceptual or mathematical language on the other. This danger also exists for direct translation between the conceptual and mathematical languages. But the introduction of a third language may increase the risk of mistranslation and misinterpretation.

5. EXAMPLES OF GRAPHICAL MODELS IN ECOLOGY

5.1 Patterns of Community Composition: Ordination and Classification. There is a set of graphical models and methods which is closely associated with the concepts of compositional similarity (distance), compositional gradients (ordination), and compositional groups (classification). It is hard to decide whether this is a case A or B (Figure 2), as in this field graphical and mathematical models developed hand in hand, one step here, one there.

The concepts of vegetation as a continuum and of vegetation gradients were from their beginning usually translated into both graphical and mathematical models. The models developed from rough to more precise (Ramensky, 1930; Brown and Curtis, 1952; Goodall, 1954, 1963; Whittaker, 1956).

On the other hand, measures of compositional similarity (or dissimilarity) between stands, species, or communities were first defined mathematically (Jaccard, 1901), and only later interpreted as distance measures and displayed graphically in roughly constructed plexus graphs (de Vries, 1956).

The Wisconsin school first used a graphical method of ordination to obtain site positions on axes from between-site similarities in composition (Bray and Curtis, 1957). But this method was soon supplemented with a precisely equivalent mathematical method of calculating the positions (Beals, 1960).

Goodall (1954), Dagnelie (1960), Groenewoud (1965), Orloci (1966), and subsequent users of principal component analysis used a purely mathematical procedure to obtain ordinations. But the results were always displayed and interpreted graphically. The

Euclidean model of "vegetation space" (Goodall, 1963; Williams and Dale, 1965), with species as dimensions and sites as points, or *vice versa* (Figure 3), was also always used to explain both the method or principal component analysis and the concepts of vegetation gradients and ordination scores.

Indeed, without the graphical description (Figure 4) it is very hard to explain these rather abstract concepts to non-mathematical biology students. With the aid of graphs they become much more plausible.

The concepts of classes and classification are intuitively more familiar. But even here geometrical analogies and representations are useful, as shown by the terms cluster, group centroid, nearest neighbor linkage, etc.

Some problems still remain. Mathematical ordination methods can obtain solutions in any number of dimensions, but the geometrical representation allows one to see only the configuration in two dimensions at a time. Many methods have been suggested in order to present and visualize a three-dimensional ordination on two-dimensional paper; some are rather ingenious and involve the use of stereoscopic pairs of plots (Cook, 1969; R. Gittins, unpublished).

Another problem is nonlinearity in ordination. This is not associated exclusively with the graphical representation, but is best demonstrated by it. Both the graphical model of Euclidean vegetation-space and most mathematical ordination models assume a linear relation between species quantities on the one hand (possibly after some transformation) and ordination axes and the gradients they represent on the other hand. This assumption is implicit in principal component analysis in all its variations, in reciprocal averaging or correspondence analysis (Hill, 1973) and polar ordination (Bray and Curtis, 1957). However, the original intuitive concept of vegetation gradients did not assume a linear species/gradient relation; if it assumed any relation, it was a unimodal (bell-shaped) one. Many ecologists tended to (mis-)interpret linear ordination axes as if they were nonlinear gradients. Since this inconsistency or mis-translation has been identified (Swan, 1970; Austin and Noy-Meir, 1971), there have been many efforts to develop mathematical methods of ordination which are not restricted by the assumption of linearity, or methods of unfolding nonlinear configurations in vegetation - or ordination-space (which is essentially the same thing).

Several such nonlinear ordinations (or catenation) methods are available (see Noy-Meir and Whittaker, 1977). They are generally computationally complex and time-consuming (even for

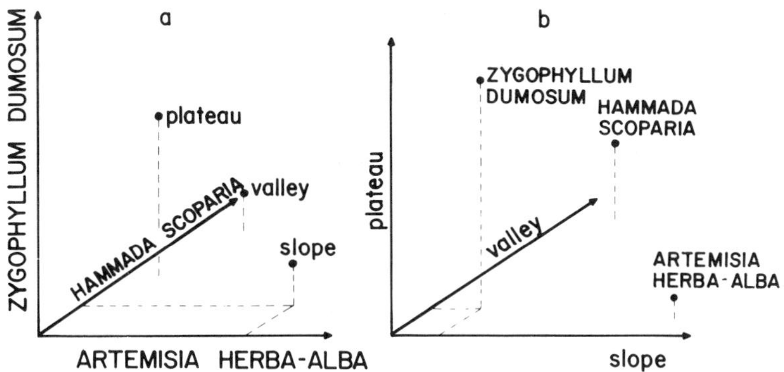

FIG. 3: The Euclidean vegetation space model, (illustrated with a 3 × 3 example from a desert community in Israel): a) composition-space, with species as dimensions, and sites as points; b) distribution-space, with sites as dimensions and species as points.

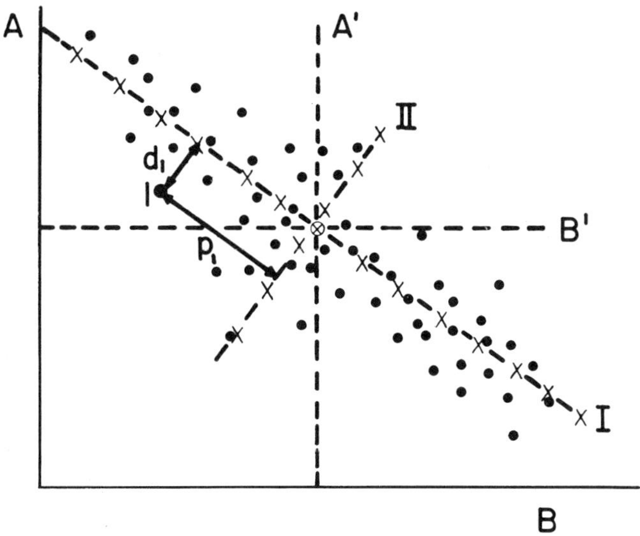

FIG. 4: Geometrical interpretation of centered principal component analysis. Composition space is defined by the quantities of species A and B. Centering by the means, transfers the origin to the centroid of the configuration of points-sites. Principal component analysis selects the first axis I through the configuration such that the sum of squared distances (d_i) of points from it is minimal, i.e., the sum of squared projections (p_i) of points on it is maximal.

large computers), and their effectiveness is limited in several senses.

The strange thing is that, at least in some two-dimensional cases (e.g. Figure 5), it is easy to solve graphically a mathematically difficult problem. What is required to obtain a non-linear ordination (or unfolding) of Figure 5 is to find a smooth curve which will pass through the configuration of points, such that the sum of squares of distances of points from the curve is minimized. Such a solution can be approached mathematically by several iterative algorithms, all rather complex, fairly slow and sensitive to the starting solution. But it is easy to draw by eye a curve in Figure 5 that should be quite close to the least squares solution. Maybe geometry can again come to the assistance of mathematics; for instance, it may be possible to obtain a good solution rapidly by displaying the configuration on a computer-linked screen, drawing curves with a light pencil and computing goodness-of-fit. With curved surfaces in three-dimensional space such an exercise becomes more difficult, but may still be feasible.

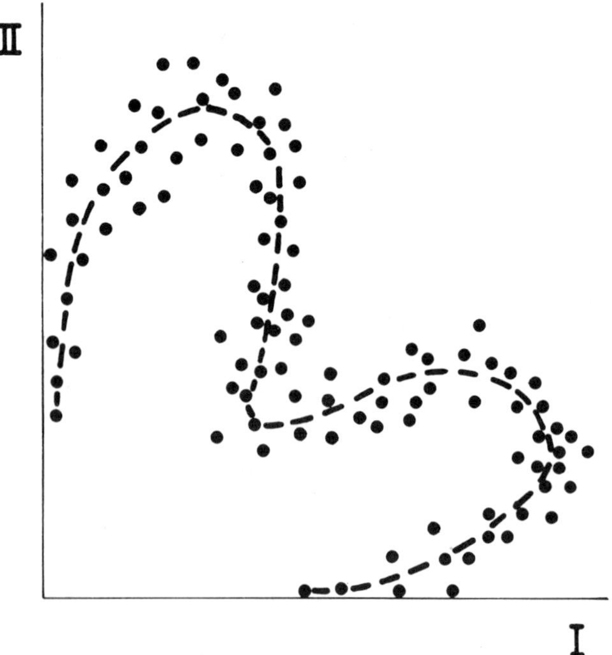

FIG. 5: A curvilinear configuration of points in ordination-space (or composition-space), and a hand-drawn curve through it.

5.2 Graphical Stability Analysis. The analysis of stability conditions of sets of differential or difference equations, representing growth rates of interaction populations has been one of the most active directions of research in theoretical and mathematical ecology recently. The standard mathematical technique is to obtain explicit solutions for the joint equilibrium of all species ($dN_i/dt = 0$ for all i) and to do an eigenvalue analysis near this equilibrium with a linearized interaction matrix (May, 1971, 1973; and many others). This methodology has been very productive in the development of ecological theory. However, some problems are associated with it:

a) in its abstract mathematical form, it is difficult to explain, to students and scientists in ecology, and it is difficult to convince them of its biological relevance;

b) unless the functions appearing in the differential equations are of linear or other simple forms, explicit solutions for the overall equilibrium may be impossible to obtain;

c) the very necessity to specify population growth equations as explicit functions of population densities limits the generality of the analysis. Often the biological information is sufficient to indicate the direction and general form of an effect, but not to determine a specific mathematical relationship.

The last difficulty may be to overcome by analysis of several alternative mathematical formulations (Rosenzweig, 1971; Noy-Meir, 1978). In some cases both of the last two problems can be partially solved by a general mathematical analysis, which does not require the explicit functions, but only some properties of these functions (e.g., May, 1973; p. 70-73, p. 86-89).

The graphical analysis of stability comes to the aid of the mathematical analysis and helps to overcome all three difficulties. The basic method involves the following steps (Figure 6):

1. Define the draw a phase space (or state space), the dimensions of which are the quantities of the interacting species; thus the dynamics of community composition (as determined by the population growth equations dN_i/dt) are represented as trajectories in this space.

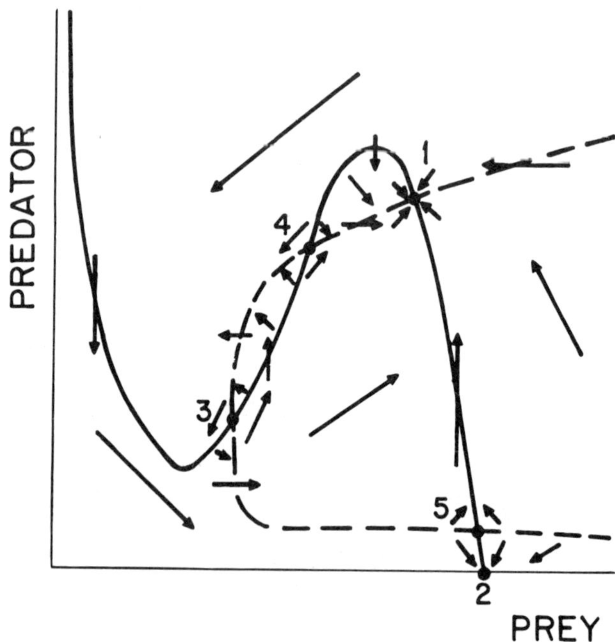

FIG. 6: Graphical stability analysis of a hypothetical predator-prey model, with a refuge for the prey, and a predator which shows both an Allee-effect at low density, and interference (or competition for another resource) at high density:——prey isocline, ---predator isocline. Arrows show the general direction of change in each region and near critical points: 1 *and* 2 *are stable equilibria (nodes),* 4 *and* 5 *are unstable saddle points,* 3 *is a focus of unstable (amplifying) oscillations (which will tend to a limit cycle around it).*

2. For each species, draw its zero-isocline in phase space, i.e., the line (in two dimensions) or surface (in three) on which its net growth rate is zero, $dN_i/dt = 0$.

3. Find the point or points at which the isoclines of all species intersect; these represent equilibrium points of the community.

4. For each of the regions of phase space (into which it has been divided by the isoclines) find the direction of trajectories.

5. For each equilibrium point, find its nature (stable or unstable node, saddle, or focus of oscillations) by considering the direction of trajectories around it.

This graphical method of isocline analysis was apparently introduced to ecology by Slobodkin (1961) for two-species competition systems, and by Rosenzweig and MacArthur (1963) (and in part also by Holling (1965)) for predation systems. Since then, it has been used frequently, in a number of different ways.

In some cases it is used to illustrate the stability properties of an explicit mathematical model, even though these properties can be fully derived analytically, e.g., the Lotka-Volterra competition model and the Lotka-Volterra predation model (in most ecology textbooks) and more sophisticated predation or herbivory models (e.g., Caughley, 1976). The graphical model is here merely a didactic tool which helps to explain and demonstrate the results obtained from the mathematical model. The word merely is perhaps inappropriate as it usually means the difference between less than 5% and more than 50% success in transmitting the information to an audience of students, biologists, resource managers, readers of ecological journals, or any other kind of non-mathematicians.

Often the graphical model enables us to carry on stability analysis where the mathematical model has to stop; this is a typical type A situation (Figure 2). Consider, for instance a plant-herbivore model, a modification of Caughley's (1976) equations 6.6, 6.7, which includes also density-dependent mortality of herbivores:

$$\dot{V} = \frac{dV}{dt} = gV(1 - V/K) - cH(1 - e^{-V/Q}), \tag{1}$$

$$\dot{H} = \frac{dH}{dt} = H[-a - bH + \varepsilon c(1 - e^{-V/Q})]. \tag{2}$$

The zero isoclines of the plant (V) and of the herbivore (H) can be obtained as explicit expressions of the form H(V):

$$H(\dot{V} = 0) = \frac{gV(1-V/K)}{c(1-e^{-V/Q})}, \tag{3}$$

$$H(\dot{H} = 0) = \frac{1}{b}\,[-a + \varepsilon c(1 - e^{-V/Q})]. \tag{4}$$

The joint equilibrium point or points are where $H(\dot{V} = 0) = H(\dot{H} = 0)$; but this equation cannot be solved to obtain explicit expressions for H and V at equilibrium.

However, the isoclines (equations 3, 4) can be drawn as graphs in phase space and for any given parameter values, the equilibrium values can be obtained graphically. Moreover, since the general shape of the curves described by equations 3 and 4 is known, a general qualitative graphical analysis of this model can be done, defining the possible situations in terms of number and nature of equilibrium points (Figure 7).

Finally, such a general qualitative graphical stability analysis can be carried out without specifying any explicit mathematical expressions for isoclines or population growth rates. It is only necessary to derive the general shape of isoclines directly from biological considerations (case C in Figure 2). In fact, this is how Rosenzweig and MacArthur (1963) first developed their isocline analysis of predator-prey models in which (in contrast to the Lotka-Volterra model) the prey isocline has a hump. Explicit mathematical models which have this realistic property were suggested and analyzed only much later (Rosenzweig, 1971; 1973a; May, 1973; Caughley, 1976). By that time, the graphical analysis on its own had already made a major impact on

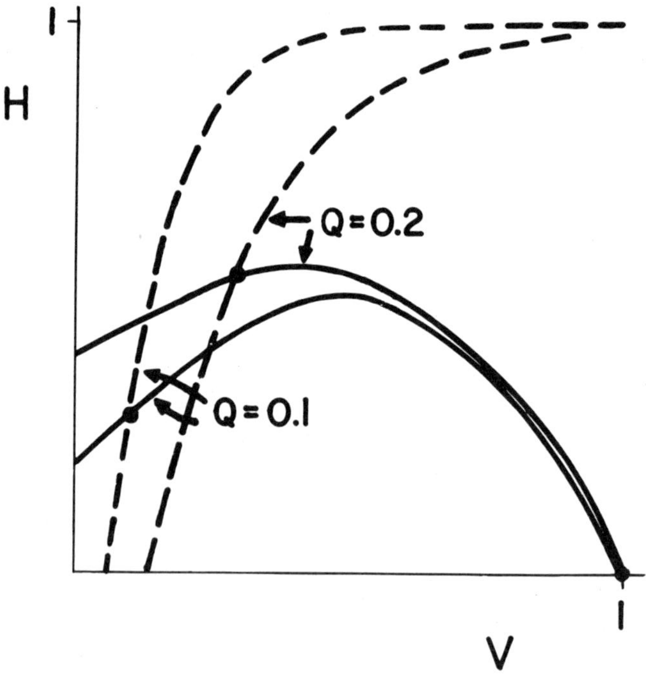

FIG 7: Isoclines of the vegetation V *(————) and the herbivore* H *(--------) and their intersection, for the model of equations* 1 *and* 2, *with two values of* Q (0.1, 0.2). *Other parameters:* $K = 1$, $c = 1$, $g = 1$, $\varepsilon = 0.1$, $a = 0.05$, $b = 0.1$.

ecological research and teaching and had become a standard textbook item (e.g., MacArthur and Connell, 1966; Ricklefs, 1973). The explicit mathematical models were then used to demonstrate in particular the results obtained in general by the graphical approach, rather than the other way round. Also, mathematical stability analysis was used to complement the graphical analysis (already in Rosenzweig and MacArthur, 1963; also in Rosenzweig, 1973b), in particular to determine stability near equilibrium points which turned out to be foci of oscillations. In terms of the relation between conceptual, graphical, and mathematical models, this is a type B situation.

The same may be said of my work on the stability of vegetation in grazing systems; a general graphical analysis (Noy-Meir, 1975) preceded analysis of specific mathematical models (Noy-Meir, 1978). However, in this case the general graphical solution emerged from experimenting with several rather simple mathematical functions (during a period of involuntary leisure somewhere in the Sinai Desert).

Graphical stability analysis has proved to be a most useful tool in the development and diffusion of ecological theory, at the level of two-population interactions. From a simple yet precise statement of assumptions, it allows one to derive by simple and easily explicable graphical operations, a set of conclusions or predictions about behavior and stability of population systems. The conclusions are sometimes judged intuitively reasonable in hindsight (though they were usually far from obvious before the analysis); sometimes the conclusions are surprising or counter-intuitive and provoke a reassessment of both our view of the system and the model. The graphical analysis (Noy-Meir, 1975; Figure 9) predicted that pastures under grazing would often be discontinuously stable (i.e. catastrophe prone), a view which had been vaguely expressed before, but never clearly explained. Rosenzweig (1971) used predator-prey graphs to demonstrate the enrichment paradox, i.e., destabilization by increased productivity which may be relevant to the effects of eutrophication in lake systems.

Using a simple phase diagram of an oscillating pest-predator system, Goh *et al.* (1974) showed that pest outbreaks can be controlled by several measures, including release of pest individuals into the system at certain times of the cycle. The graphical argument (Figure 9) turns what at first intuition seems a complete folly into a reasonable possibility (which in some situations is apparently of practical value).

In the last few years most papers on stability analysis of ecological systems have used purely mathematical methods rather than graphical techniques. However, several pioneering studies

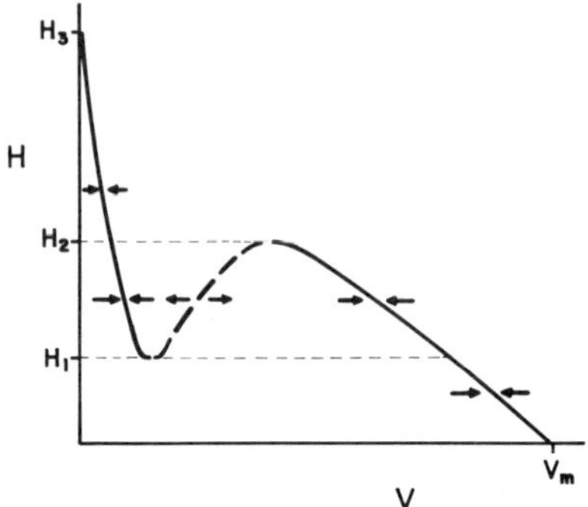

FIG. 8: The vegetation isocline and the stability of grazing systems. At each herbivore density H *below* H_1, *the vegetatiom has a single stable equilibrium at high biomass* V. *Between* H_1 *and* H_2, *if biomass passes below a threshold, it crashes to a low stable value. Between* H_2 *and* H_3, *only a low value is stable. Above* H_3, *the vegetation is totally grazed out.*

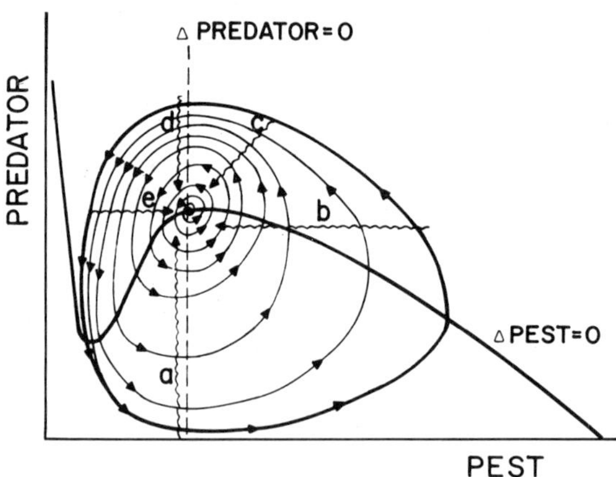

FIG. 9: Dynamics and control of a hypothetical pest-predator system (modified from Goh et al., 1974). *The system has a focus of weakly unstable oscillations, at a low pest density (economically acceptable damage). The oscillations slowly amplify towards a limit cycle, in which the pest population periodically reaches high and significantly damaging densities. A control policy aimed at dampling the outbreaks and returning the system to the focus region involves different actions at each phase of the cycle: a - addition of predators, b - specific biocides against prey, c - non-specific biocides against both species, d - specific biocide against predators, e - addition of prey.*

have used graphical analysis (some in conjunction with mathematical analysis). Thus, Vandermeer (1975) developed a graphical model of seed predation. Jones (1977) used a graphical model for a qualitative analysis of the spruce budworm/forest system in Canada in terms of catastrophe theory. The budworm outbreak cycle was described and explained first as a fold catastrophe in two dimensions (Figure 10), then in more detail, as a cusp catastrophe in three dimensions (Figure 11). The qualitative analysis was then supported by a quantitative simulation study using an extensive and long-term data set. A similar analysis was also applied to Pacific salmon fisheries (Jones and Walters, 1976; Peterman, Clark and Holling, 1978). In the further development and application of catastrophe theory to ecological systems, graphical models are likely to play an important role.

The possibilities of obtaining new and interesting results with the simple tools of graphical stability analysis are far from exhausted. Rosenzweig (1977) has come back to the basic predator-prey isocline graph and examined the effects of compensatory predation (the predator preferring individuals with a lower fitness) and of non-selective biocides. Graphical stability analysis of systems with two competing plant species under herbivore grazing has revealed an unexpectedly rich repertoire of behavior in such systems (Noy-Meir, in preparation).

The most severe limitation to the use of graphical analysis in developing community theory is dimensionality. Communities have many interacting species, while graphical analysis is most effective in two-dimensional phase-space. It is still possible to visualize and draw three-dimensional spaces, and look at the intersections of isoclines which are surfaces. In the case of three competitor systems, this is easy enough, particularly if planar isoclines are drawn (Ricklefs, 1973, p. 523).

In the case of a three-trophic-levels predation chain (Rosenzweig and MacArthur, 1963; Rosenzweig 1973b; Wollkind, 1976) a fair exercise of imagination is required to visualize the three surfaces in three-dimensional space, and the various possible ways in which they can intersect. A four-dimensional case seems to be beyond human powers of visualization.

It may be hopefully postulated that the essential qualitative behavior of many-species communities can be explained by the behavior of systems of two (or at most three) species in various modes of interaction (competition, predation, symbiosis etc.) The results of Rosenzweig (1973b) seem to support this to some extent, at least for exploitation chains. However, there are indications that in other cases, higher-order interactions can introduce qualitatively new effects and therefore cannot be

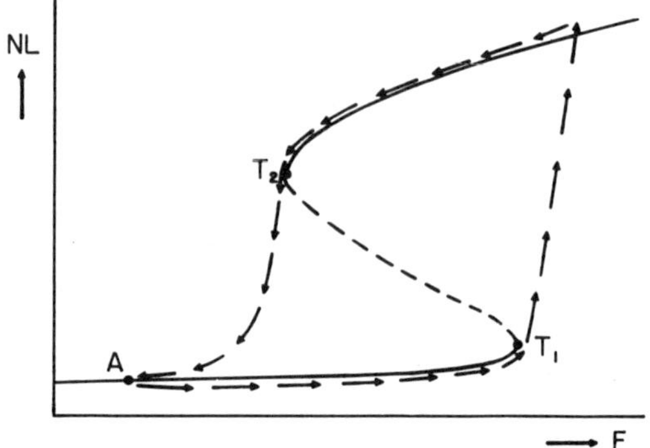

FIG. 10: The spruce budworm cycle as a fold catastrophe on the budworm isocline. The number of budworm larvae (NL) *varies much faster than the amount of available foliage* (F) *(after Jones,* 1977*).* A - *starting point in a young forest with (endemic) budworm density.* T_j - *budworm outbreak threshold.* T_2 - *budworm collapse threshold.*

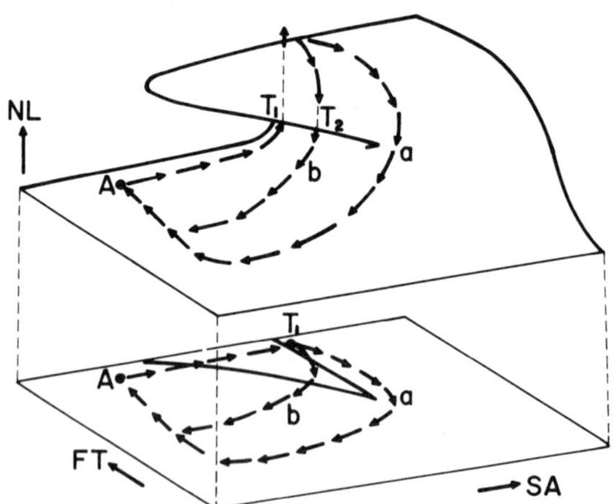

FIG. 11: The spruce budworm cycle as a cusp catastrophe on the budworm (NL) *isocline surface in relation to two slow variables: branch surface area* (SA) *and foliage per unit branch area* (FT). *The projection of the cusp and the trajectories on the* SA - FT *plane is also shown (after Jones,* 1977*).* A - *starting point: young forest, low budworm,* T_1 - *budworm outbreak threshold. Two possible paths of return: a) smooth budworm decrease b) budworm collapse at* T_2.

a priori neglected. For instance, oscillations and limit cycles do not occur in simple two-competitor systems (May, 1973), but can occur in three-competitor systems (Cockerham and Burrows, 1971; Gilpin, 1975) and in systems with two competitors under constant predation (Noy-Meir, in preparation).

Complex models (including some systems of four differential equations) can show extremely strange behavior, converging to strange attractors which are neither equilibrium points nor proper limit cycles (May and Oster, 1976). Apparently no ecologically relevant model has yet (fortunately?) been proved to have such behavior, but the possibility is disturbing. In any case, it would be wise to look for qualitatively new, higher-order effects in models of communities with four or more interacting species, before optimistically assuming them away. It would seem that qualitative graphical stability analysis cannot help us in this exploration into four-and-more-dimensional space. But perhaps the informal graphical method of isocline analysis can be translated into a formal mathematical method, which can then be extended into more than three dimensions.

REFERENCES

Austin, M. P. and Noy-Meir, I. (1971). The problem of non-linearity in ordination: experiments with two-gradient models. *Journal of Ecology*, 59, 217-227.

Bray, J. R. and Curtis, J. T. (1957). An ordination of the upland forest communities of southern Wisconsin. *Ecological Monographs*, 27, 325-349.

Brown, R. T. and Curtis, J. T. (1952). The upland conifer-hardwood forests of northern Wisconsin. *Ecological Monographs*, 22, 217-234.

Caughley, G. (1976). Plant-herbivore systems. In *Theoretical Ecology*, R. M. May, ed. Blackwell, Oxford, 94-113.

Cockerham, C. C. and Burrows, P. M. (1971). Populations of interacting autogenous components. *American Naturalist*, 105, 13-30.

Cook, B. G. (1969). STERPLOT: A computer program for plotting stereograms of the distribution of points in three dimensions. CSIRO Div. Land Research, Canberra, *Technical Memorandum*. 69/1.

Dagnelie, P. (1960). Contribution a l'étude des communautés végétales par l'analyse factorielle. *Bulletin du Service de la Carte phytogéographique Series B*, 5, 7-71 and 93-195.

Gilpin, M. E. (1975). Limit cycles in competition communities. *American Naturalist*, 109, 51-60.

Goh, B. S., Vincent, T. L., and Wilson, D. J. (1974). A method for formulating suboptimal policies for crudely modelled ecosystems. *Proceedings 1st International Congress of Ecology*. Pudoc, Wageningen. 405-408.

Goodall, D. W. (1959). Objective methods for the classification of vegetation. III. An essay in the use of factor analysis. *Australian Journal of Botany*, 2, 304-324.

Goodall, D. W. (1963). The continuum and the individualistic association. *Vegetatio*, 11, 297-316.

Groenewoud, H. van (1965). Ordination and classification of Swiss and Canadian coniferous forests by various biometric and other methods. *Berichte des Geobotanisches Institute Rübel Zürich*, 1964, 36, 28-102.

Hill, M. O. (1973). Reciprocal averaging: an eigenvector method or ordination. *Journal of Ecology*, 61, 237-249.

Holling, C. S.(1965). The functional response of predators to prey density and its role in mimicry and population regulation. *Memoirs of the Entomological Society of Canada*, 45, 1-60.

Jaccard, P. (1901). Distribution de la flore alpine dans le Bassin des Dranses et dans quelques régions voisines. *Bulletin de la Societé vaudoise des Sciences naturelles*, 37, 241-272.

Jones, D. D. (1977). Catastrophe theory applied to ecological systems. *Simulation*, 29, 1-15.

Jones, D. D. and Walters, C. J. (1976). Catastrophe theory and fisheries regulation. *Journal of the Fisheries Research Board of Canada*, 33, 2829-2833.

MacArthur, R. H. and Connell, J. H. (1966). *The Biology of Populations*. Wiley, New York

May, R. M. (1971). Stability in model ecosystems. *Proceedings of the Ecological Society of Australia*, 6, 17-56.

May, R. M. (1973). *Stability and Complexity in Model Ecosystems*. Princeton University Press, Princeton, New Jersey.

May, M. and Oster, F. (1976). Bifurcations and dynamic complexity in simple ecological models. *American Naturalist*, 110, 573-599.

Noy-Meir, I. (1975). Stability of grazing systems: an application of predator-prey graphs. *Journal of Ecology*, 63, 459-481.

Noy-Meir, I. (1978). Stability in simple grazing models: effects of explicit functions. *Journal of Theoretical Biology*, 71, 347-380.

Noy-Meir, I. (1979). Competition under grazing: graphical stability analysis of a two-species model. (manuscript).

Noy-Meir, I. and Whittaker, R. H. (1977). Continuous multivariate methods in community analysis: some problems and developments. *Vegetatio*, 33, 79-98.

Orloci. L.(1966). Geometric models in ecology. I. The theory and application of some ordination models. *Journal of Ecology*, 54, 193-215.

Peterman, R. M., Clark, W. C., and Holling, C. S. (1978). The dynamics of resilience: shifting stability domains in fish and insect systems. In *Population Dynamics*, R. M. Anderson and B. D. Turner, eds. Proceedings Symposium of the British Ecological Society. Blackwell, Oxford.

Ramensky, L. G. (1930). Zur Methodik der vergleichenden Bearbeitung und Ordnung von Pflanzenlisten und andere Objecten, die durch mehrere verschiedenartig wirkende Factoren bestimmt werden. *Beiträge zur Biologie der Pflanzen*, 18, 269-304.

Ricklefs, R. E. (1973). *Ecology*. Nelson, London.

Rosenzweig, M. L. (1971). The paradox of enrichment; destabilization of exploitation ecosystems in ecological time. *Science*, 171, 385-387.

Rosenzweig, M. L. (1973a). Evolution of the predator isocline. *Evolution*, 27, 84-94.

Rosenzweig, M. L.(1973b). Exploitation in three trophic levels. *American Naturalist*, 107, 275-294.

Rosenzweig, M. L. (1977). Aspects of biological exploitation. *Quarterly Review of Biology*, 52, 371-380.

Rosenzweig, M. L. and MacArthur, R. H. (1963). Graphical representation and stability conditions of predator-prey interactions. *American Naturalist*, 97, 209-223.

Slobodkin, L. B. (1961). *Growth and Regulation of Animal Populations*. Holt, Rinehart & Winston, New York.

Swan, J. N. A. (1970). An examination of some ordination problems by the use of simulated vegetational data. *Ecology*, 51, 89-102.

Vandermeer, J. H. (1975). A graphical model of seed predation. *American Naturalist*, 109, 147-160.

Vries, D. M., de (1953). Objective combinations of species. *Acta botanica reerlandica*, 1, 497-499.

Whittaker, R. H. (1956). The vegetation of the Great Smoky Mountains. *Ecological Monographs*, 26, 1-80.

Wollkind, D. (1976). Exploitation in three trophic levels: an extension allowing intraspecies carnivore interaction. *American Naturalist*, 110, 431.

[*Received January* 1979. *Revised April* 1979]

G. P. Patil and M. Rosenzweig, (eds.),
Contemporary Quantitative Ecology and Related Ecometrics, pp. 473-491.

THE EFFECTS OF SCHOOL STRUCTURE ON LINE TRANSECT ESTIMATORS OF ABUNDANCE

TERRANCE J. QUINN II

Center for Quantitative Science in
Forestry, Fisheries, and Wildlife
University of Washington
Seattle, Washington 98195 USA

SUMMARY. Of the many techniques available for the abundance estimation of animal populations, the line transect technique has become widely used. The basis of the method is to construct and estimate a sighting model from measurements of distance from the observer to the animal sighted. When animals congregate in schools, each school is considered as a single sighting. However, it is intuitively clear that the size of the school will affect the shape and scale of the sighting model. Empirical evidence from aerial and shipboard surveys of porpoise schools demonstrates this effect. Since the number of sightings from transect experiments is usually small, it is often not possible to stratify by school-size classes and it thus would be desirable to justify pooling the data. A theoretical development of school structure and transect methodology indicates that there is theoretical justification in pooling the data, because the sighting model that results is self-weighted by the relative abundances of the school-size classes. Of the selected parametric and non-parametric estimators evaluated through computer simulation, the Fourier series and GEM estimators are preferred.

KEY WORDS. line transects, school structure, abundance estimation, pooling data, porpoise schools, computer simulation.

1. INTRODUCTION

Of the many techniques available for the assessment of population abundance, the line transect technique has become increasingly popular. Transect sampling is direct, cost-efficient, and involves no killing or handling of individuals in contrast to mark-recapture and catch-effort methods. The method is especially useful for populations difficult to census due to logistic problems or legal restrictions. Furthermore, transect methodology provides the framework for the analysis of aerial, shipboard, or land-based surveys that are currently prevalent in environmental assessment.

In the line transect method, an observer travels along a transect line, which is usually placed randomly or systematically across the area of interest. All animals or groups of animals sighted or flushed are recorded along with some measure of distance from the animal to the transect line or to the observer. The most useful distance in the methods currently available is the perpendicular distance y from the animal to the transect line. The basis of the method is a postulated sighting model $g(y)$ which represents the non-increasing probability of sighting or flushing an animal as y increases, where it is assumed that all animals on the transect line are sighted (i.e., $g(0) = 1$).

There has been a plethora of transect-related papers since the advent of the statistical formulation of the method due to Gates *et al.* (1968) and Eberhardt (1968). Two general reviews of transect methods, sighting models, and assumptions are given by Seber (1973) and Gates (1979). The development of non-parametric transect estimators is based upon the general theory of probability density estimation (Burnham and Anderson, 1976). The interpretation and utility of transect methodology is discussed by Eberhardt (1978). A review of parametric estimation techniques and sighting models is given by Quinn and Gallucci (1979). Computer simulation studies have shown the importance of the choice of an appropriate sighting model in the estimation process and have determined the effect of relaxing some of the assumptions (Gates, 1969; Quinn, 1977).

Despite the extensive knowledge and application of the line transect method, it remains unclear if there exists a concise sighting model which can represent sighting phenomena over a range of different factors encountered in application of the method. During the course of the transect, a number of factors such as visibility, light intensity, sea state, spatial features, and animal response behavior can change and affect the sighting model. In addition, if the population is aggregated into schools, the size of a school, its species and stock composition, and the behavior of individuals can affect the sighting model. The question raised is whether or not a simple model $g(y)$ that is only a function of distance may be used validly to estimate abundance.

This paper seeks the solution of this problem with regard to the effect of school size. When animals congregate in schools, each school is considered as a single sighting and the assumptions of the method are applied to schools rather than individuals. Line transect methods are used to estimate the number of schools in the population, and the total number of individuals is estimated by multiplying the number of schools by some measure of the average school size. However, the size of a school should inherently affect the shape and scale of the sighting model, since the larger the school, the greater the probability of its sighting. Difficulties arise in the estimation of both the number of schools and the average school size due to the over-representation of the larger schools in the transect sample.

Empirical evidence supports the intuition. Aerial surveys of porpoise schools in the eastern tropical Pacific Ocean were carried out in 1977 by the Southwest Fisheries Center (SWFC) of the United States National Marine Fisheries Service (NMFS) (Dr. J. Powers, personal communication). A plot of the mean sighting distance as a function of the logarithms of school size demonstrated a linear increase (Dr. L. Eberhardt, personal communication). Supporting evidence is also found from similar surveys on bottle-nosed dolphins in the Gulf of Mexico (S. Leatherwood, personal communication).

The obvious solution would be to stratify by school size. However, for many populations, the number of schools is sufficiently scarce to result in few sightings (often less than 50). Furthermore, stratification is often necessary in the spatial-temporal domain which precludes further stratification by school size or other variables.

It would thus be desirable to develop methods which allow the pooling of data over school-size classes. This paper presents a theoretical development of transect methodology which incorporates school structure and determines criteria to be met for such pooling. Selected sighting models and corresponding estimators are then evaluated through computer simulation.

2. DEVELOPMENT

As mentioned earlier, the basis of the line transect method is a sighting model $g(y)$ which represents the probability of sighting a school as a function of its perpendicular distance y from the transect line. A parameter termed the effective width c is defined as

$$c = \int_0^\infty g(y)dy \quad ,$$

which represents the area under the sighting model curve (Seber, 1973). The parameter c may be interpreted as the distance where the probability of sighting greater than that distance is equal to ('balances') the probability of not sighting less than that distance (Gates, 1979). Furthermore, the inverse of c is equal to be the value of the probability density function $f(y)$ of sighting distances at the origin, i.e., $f(0) = c^{-1}$ (Burnham and Anderson, 1976). The general estimator of abundance N (the number of schools) is

$$\hat{N} = \frac{An}{2L}\hat{c}^{-1} = \frac{An}{2L}\hat{f(0)} \quad , \tag{1a}$$

with expected value

$$E[\hat{N}] = \frac{A}{2L}E[n]c^{-1} = \frac{A}{2L}E[n]f(0) \quad , \tag{1b}$$

where A is the population area, L is the transect length and n is the number of sightings. The estimator $\hat{N}$ is unbiased if the correct sighting model $g(y)$ is chosen. The six most used sighting models and their corresponding estimators are shown in Table 1. The theoretical variance estimate of $\hat{N}$ is

$$\hat{Var}(\hat{N}) = \hat{N}^2\{c.v.^2(n) + c.v.^2(c^{-1}|n)\} \quad , \tag{1c}$$

where $c.v.^2(\cdot)$ is the squared estimate of the coefficient of variation (Quinn and Gallucci, 1979). Thus the variation in $\hat{N}$ is decomposed into the variation in the number of sightings and variation in the effective width given the number of sightings. In the simulation study that follows where n is fixed, $c.v.^2(n)$ is set to 0. The term $c.v.^2(c^{-1}|n)$ follows from the specific sighting model (Table 1). The estimate of the total number of individuals T is $\hat{T}=\hat{N}\hat{\bar{S}}$, where $\hat{\bar{S}}$ is an estimate of the average school size in the population. Its variance is the variance of a product (Seber, 1973, p. 7-9).

In order to incorporate school size into this general framework, let there be t school-size classes with frequencies $N_1, \cdots, N_t$ in the population, and let $\Sigma N_i = N$. The *ith* school-size class represents S_i animals per school and has an associated sighting model $g_i(y)$. Define $c_i = \int_0^\infty g_i(y)dy$ as the effective width for the *ith* school-size class. It is now determined what criteria are necessary so that an estimator from the pooled data results in an approximately unbiased estimator of N and what interpretations the parameters of the pooled sighting model have.

TABLE 1: Common line transect abundance estimators and notes on their use in the current simulation study.

Estimator	Sighting Model $g(y)$	Estimator $\hat{c}^{-1}$	Theoretical Variance Estimate $\hat{Var}(\hat{c}^{-1}\|n)$	Notes on Use
Exponential Model (EM) (Gates *et al.*, 1968)	$\exp(-\lambda y)$	$\frac{n-1}{\Sigma y_i}$	$(\hat{c}^{-1})^2/(n-2)$	Easily computed
Half-normal Model (HNM) (Quinn and Gallucci, 1979)	$\exp\{-\frac{1}{2}(y/\sigma)^2\}$	$B(n) \times \sqrt{(2/\pi)(n/\Sigma y_i^2)}$	$(\hat{c}^{-1})^2(\frac{n}{n-2} B^2(n)-1)$	$B(n)$ (bias correction term = $\sqrt{\frac{2}{n}}\,\Gamma(\frac{n}{2})/\Gamma(\frac{n-1}{2})$
Cox-Eberhardt Model (COX) (Eberhardt, 1978)	Maclaurin series expansion of $g(y)$	$\frac{3f_1 - f_2}{2\Delta n}$	$(\hat{c}^{-1}) \frac{\left[3 + \frac{4f_2}{3f_1-f_2} - \frac{3f_1-f_2}{n}\right]}{2\Delta n}$	f_1 = frequency of observations in $[0,\Delta]$. f_2 = frequency of observations in $[\Delta,2\Delta]$. 2Δ = width that encompasses no more than 75% of the observations.
Fourier Model (FOURIER) (Crain *et al.*, 1978)	Fourier series expansion of $g(y)$	$\frac{1}{w} + \sum_{k=1}^{m} \hat{a}(k)^{\dagger}$	$\frac{1}{n^*-1} \sum\sum_{kj}\{\frac{\hat{a}(k+j)+\hat{a}(k-j)}{w} - \hat{a}(k)\hat{a}(j)\}$	w = truncation width chosen to encompass at least 90% of the observations, n^* = number of observations in $[0,w]$.
Generalized Exponential Model (GEM) (Quinn and Gallucci, 1979)	$\exp\{-\frac{1}{\beta}(y/\alpha)^{\beta}\}$	Iterative maximum likelihood solutions of the score functions		Data grouped into intervals to use multinomial theory
Kelker (strip model) (KELK) (Gates 1979)	$\begin{cases} 1 \text{ if } 0 \le y \le \Delta \\ 0 \text{ if } y > \Delta \end{cases}$	$f_1/n\Delta$	$f_1(n-f_1)/n^3\Delta^2$	f_1,Δ defined at in COX

$^{\dagger}\hat{a}(k) = \frac{2}{n^*w} \sum_{i=1}^{n^*} \cos(k\pi y_i^*/w)$. The Fourier abundance estimate $\hat{N}$ in (1a) replaces n with n^*.

The number of sightings n_i from each class is clearly multinomial, and its expected value from (1b) is

$$E[n_i] = \frac{2L}{A} N_i c_i \quad . \tag{2}$$

The expected proportion p_i^* of sightings of each school size class is then $p_i^* = E[n_i]/E[n]$, where $n=\Sigma n_i$, and simplifying,

$$p_i^* = N_i c_i / \Sigma N_i c_i \quad .$$

Thus the proportion of sightings is a function of the product of the abundance of the school-size class N_i and its effective width c_i.

The pooled sighting model $g_p(y)$ is constructed to examine the theoretical justification of examining the pooled data. Since each $g_i(y)$ is the contribution of one school to the pooled sighting model, the true relative abundance $p_i = N_i/N$ is the weighting factor of $g_i(y)$ to $g_p(y)$, i.e.,

$$g_p(y) = \Sigma p_i g_i(y) = \Sigma N_i g_i(y)/N \quad .$$

Define $c_p = \int_0^\infty g_p(y)dy$ as the pooled effective width, then $c_p = \Sigma p_i c_i = \Sigma N_i c_i/N$. Thus the pooled effective width is the sum of the individual c_i weighted by the relative abundance. Note that the observed relative frequency p_i^* is not equal to the true relative frequency p_i, in the population unless $c_i = c$ for all i.

The abundance estimator $\hat{N}_p$ of the pooled data follows from (1a) with c replaced by c_p, and $f(0)$ replaced by

$$f_p(0) = \frac{1}{c_p} = 1/\Sigma\{p_i/f_i(0)\} \quad . \tag{3}$$

Thus the pooled $f_p(0)$ is the inverse of the inverses of the individual $f_i(0)$'s weighted by the true relative abundance.

The critical question to be evaluated is: Will the estimator of abundance give a reasonable estimate of population size N? From (1b), the expected value of the population estimate of the *ith* school-size class is $N_i = \frac{A}{2L} E[n_i]f_i(0)$ and the expected value of the pooled estimate of N is $N_p = \frac{A}{2L} E[n]f_p(0)$ (Burnham and Anderson, 1976). It follows that

$$N_p = \frac{A}{2L} E[n]/\Sigma\{p_i/f_i(0)\}$$

$$= \Sigma N_i c_i/(\Sigma N_i c_i/N) = N.$$

Thus there appears to be theoretical justification in considering the data pooled over school-size classes.

However, the choice of an estimator of $f_p(0)$ or c_p is critical to the unbiased estimation of N. By equating N_p and ΣN_i, it is clear

$$E[n]f_p(0) = \Sigma E[n_i]f_i(0) \quad .$$

Thus an estimator that satisfies

$$n\hat{f}_p(0) = \Sigma n_i \hat{f}_i(0) \tag{4}$$

should give a reasonable estimate of abundance. Note from (4) that

$$\frac{An\hat{f}_p(0)}{2L} = \frac{A\Sigma n_i \hat{f}_i(0)}{2L} \quad ,$$

$$\hat{N}_p = \Sigma \hat{N}_i \quad .$$

Thus, if (4) is satisfied by an estimator, it can be said that the pooled population estimate is simply the sum of the individual population estimates considered separately. Thus the bias of a pooled estimator can be determined by the examination of each school-size class separately. The condition (4) is satisfied for many non-parametric estimators in Table 1, such as the Kelker, Cox-Eberhardt, and Fourier series estimators. It is in general not true for parametric estimators in Table 1 unless the $g_i(y)$ are all similar. As an example, consider the generalized exponential model (GEM) $g(y) = \exp\{-(y/\alpha)^{\beta}/\beta\}$ (Quinn and Gallucci, 1979) for fixed β, which represents a large class of commonly-used parametric estimators. If β=1, the exponential model (EM) results (Gates *et al.*, 1968); if $\beta=2$, the half-normal model (HNM) results.

The maximum likelihood estimator for the *ith* school-size class is

$$\hat{c}_i = d(\beta)(\sum_j y_{ij}^{\beta}/n_i)^{1/\beta}$$

where y_{ij} is the *jth* distance measurement and d is a proportionality constant for fixed β. The pooled estimator that results from combining the data is

$$\hat{c}_c = d(\beta)(\sum_{ij}\sum y_{ij}^{\beta}/n)^{1/\beta} = (\sum n_i \hat{c}_i^{\beta}/n)^{1/\beta}$$

which is asymptotically equal to

$$c_c = (\sum N_i c_i^{\beta+1}/\sum N_i c_i)^{1/\beta}$$

whereas the correct pooled version should be

$$c_p = \sum N_i c_i/\sum N_i \quad .$$

If and only if $c_i = c$ for all i is there equivalence between c_c and c_p.

For example, in the NMFS aerial surveys of porpoise schools, small schools are more abundant than large schools and are sighted closer to the transect. Suppose that we have four school-size classes with true relative abundances 40%, 30%, 20%, and 10% and effective widths of 1, 2, 3, and 4, respectively. Then $c_c = 2.7$ ($\beta=1$ and $\beta=2$) and $c_p=2.0$, which implies that the combined para-estimate would be 35% greater than the true value. Conversely, if the effective width were 4, 3, 2, 1, $c_c = 3.3$ ($\beta=1$) and 3.4 ($\beta=2$) and $c_p = 3.0$, which implies that the parametric pooled estimate would be 10% greater than the true value. If the relative abundances were equal (25%) and effective widths of 1, 2, 3, 4 occurred, then $c_c = 3.0$ ($\beta=1$) and 3.2 ($\beta=2$) and $c_p = 2.5$. Thus, for the three cases, expected overestimates occur from the combined estimate for fixed β.

However, if β is allowed to enter the estimation process, the interpretation becomes clouded. No closed-form estimators for α and β exist, so that estimates are produced from iterative maximum likelihood techniques (Quinn and Gallucci, 1979). It is possible that this estimation procedure may overestimate some classes and underestimate others so that the end result is an approximately unbiased result, which will be examined later through computer simulation.

The other approach is to use the parametric approach with school size S_i introduced as a covariate. For example, Eberhardt (personal communication) has proposed for the half-normal model (β=2) that α is a linear function of the logarithm of school size and thus $\alpha_i = b \ln S_i$ for the *ith* school size class. This procedure needs to be further explored for estimation properties.

Once the number of schools has been estimated, the estimate of the total number of individuals $\hat{T} = \hat{N}\hat{\bar{S}}$ requires an estimate of the average school size $\bar{S} = \Sigma N_i S_i / N$. The average school size from the transect sample $\bar{s} = \Sigma n_i S_i / n$ is a biased estimate of $\bar{S}$, because the true proportion of schools, p_i, is not equal to sampled proportion p_i^*. An unbiased estimate of $\bar{S}$ using (1b) is

$$\hat{\bar{S}}_1 = \Sigma n_i S_i \hat{c}_i^{-1} / \Sigma n_i \hat{c}_i^{-1} \quad , \tag{5}$$

which weights each school size observed by the inverse of its effective width. If the data are pooled over school size, then c_i is not estimated, so that (5) is not estimable. One method to alleviate this problem is to assume that c_i is proportional to the logarithm of school size as above. Then the estimate $\hat{\bar{S}}_2 = \Sigma(n_i S_i / \ln S_i) / \Sigma(n_1 / \ln S_1)$ is unbiased and weights each school size by the inverse of its logarithm. Another method is to use $\bar{s}$ with only the sighting data in a specified interval about the transect line where all schools are likely to be seen.

3. SIMULATION STUDY

3.1 Methodology. Because of the complexity of the estimation process for several estimators in Table 1, it is necessary to examine non-parametric and parametric estimators for finite samples as well as the earlier asymptotic arguments. A computer simulation study of the six estimators in Table 1 is undertaken to further explore the justification of pooling data when the data come from different school-size classes. The focus is on the estimation of the total number of schools N in the population.

The total number of sightings n is fixed at 50. The number of sightings from each of four school-size classes is generated from the multinomial distribution with expectations given by (2) for the three cases of relative abundance p_i in the total population: (I) p_i decreases linearly; (II) p_i is constant; (III) p_i increases linearly. The four school-size classes are of sizes 5, 25, 125, and 625. The effective widths c_i are calculated from the following empirical evidence from the 1977 NMFS surveys or porpoise schools. The half-normal sighting model is assumed for each school-size class. A regression equation from Eberhardt's covariate model yields $b = 0.08$, which determines α_i from the empirical relationship $\alpha_i = b \ln S_i$. The effective width c_i is related to α_i by $c_i = \sqrt{\pi/2}\,\alpha_i$ (Quinn and Gallucci, 1979). Distances are generated with the use of α_i in standard techniques for generating half-normal variates. The parameters of the simulation are summarized in Table 2. Plots of the individual and pooled sighting models for Case I show the heavier weighting of the more abundant classes in the pooled model (Figure 1).

The resultant distance data are subjected to the six estimators in Table 1, according to the criteria specified. Fifty replications of each procedure provide empirical means and standard errors to compare with the known values. Also, since theoretical variance estimates are available for each estimator, the theoretical standard error of the 50 replications is compared with the empirical standard error.

3.2 Results. The results of the simulation are shown in Tables 3(a), (b), (c) for the three cases I, II, III of relative abundance in Table 2, respectively. The parameters considered are the effective width c_p, its inverse $f_p(0)$, and finally, the product of the number of sightings n and $f_p(0)$, which is proportional to abundance in (1a). Since the number of sightings is fixed, the relation of the estimates and also the standard errors between $f_p(0)$ and $nf_p(0)$ is proportional, but both are reported for clarity.

The most difficult case to estimate is Case I, where the effective width and relative abundance are negatively correlated, which unfortunately is the most likely case to occur in practice. The estimates in Table 3(a) have larger biases and variances than in Tables 3(b) and (c), because the presence of the largest school-size classes is more variable. Although the results for Case I are more pronounced, the same trends are observed for the other two cases.

TABLE 2: Parameters used in the computer simulation for three cases of relative abundance.

School size class i	1	2	3	4
Common to all cases				
s_i	5	25	125	625
α_i	0.1288	0.2575	0.3863	0.5150
c_i	0.1614	0.3227	0.4841	0.6455
Case I: Relative abundance (p_i) *decreases as a function of school size*				
p_i	0.40	0.30	0.20	0.10
p_i^*	0.20	0.30	0.30	0.20
	c_p = 0.3227	$f_p(0)$ = 3.098		$nf_p(0)$ = 154.9
Case II: Relative abundance (p_i) *is constant as a function of school size*				
p_i	0.25	0.25	0.25	0.25
p_i^*	0.10	0.20	0.30	0.40
	c_p = 0.4034	$f_p(0)$ = 2.479		$nf_p(0)$ = 123.9
Case III: Relative abundance (p_i) *increases as a function of school size*				
p_i	0.10	0.20	0.30	0.40
p_i^*	0.03	0.13	0.30	0.53
	c_p = 0.3227	$f_p(0)$ = 2.066		$nf_p(0)$ = 103.3

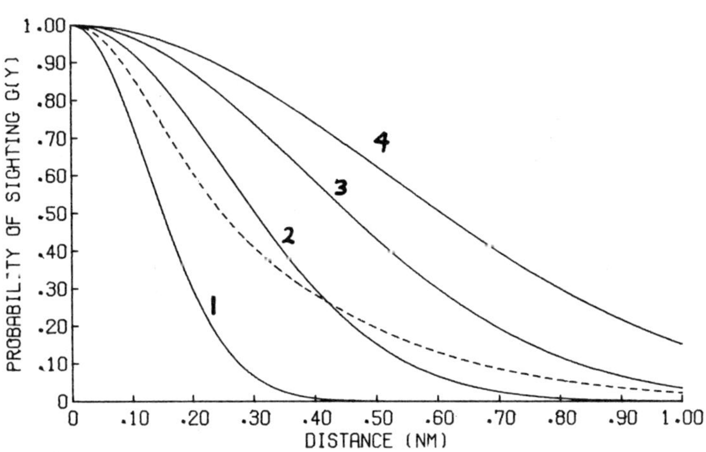

FIG. 1: Plots of the sighting model $g_i(y)$ *as a function of distance for each school-size class (solid lines) and the resultant pooled sighting model* $g_p(y)$ *(dashes) for Case I, Table 2.*

TABLE 3(a): Estimates of c_p, $f_p(0)$, *and* $nf_p(0)$ *for the estimators in Table 1 (Case I:* p_i *decreasing).*

Parameter	Estimator	Simulation Average	Theoretical Standard Error	Empirical Standard Error
	EM	0.2565	0.0052	0.0047
	HNM	0.4315	0.0062	0.0082
c_p	COX	0.3617	0.0359	0.0249
(0.3227)	FOURIER	0.2976	0.0107	0.0106
	GEM	0.3186	0.0117	0.0138
	KELK	0.3739	0.0113	0.0122
	EM	3.883	0.080	0.071
	HNM	2.335	0.034	0.044
$f_p(0)$	COX	3.174	0.139	0.140
(3.098)	FOURIER	3.547	0.095	0.113
	GEM	3.466	0.201	0.172
	KELK	2.808	0.080	0.086
	EM	194.2	4.0	3.5
	HNM	116.7	1.7	2.2
$nf_p(0)$	COX	158.7	7.0	7.0
(154.9)	FOURIER	150.4	4.0	4.6
	GEM	173.3	10.1	8.6
	KELK	140.4	4.0	4.3

TABLE 3(b): Estimates of c_p, $f_p(0)$, *and* $nf_p(0)$ *for the estimators in Table 1 (Case II:* p_i *constant).*

Parameter	Estimator	Simulation Average	Theoretical Standard Error	Empirical Standard Error
	EM	0.3068	0.0062	0.0053
	HNM	0.5057	0.0072	0.0090
c_p	COX	0.4254	0.0380	0.0273
(0.4034)	FOURIER	0.3641	0.0150	0.0138
	GEM	0.4056	0.0137	0.0151
	KELK	0.4473	0.0139	0.0140
	EM	3.241	0.067	0.056
	HNM	1.988	0.029	0.036
$f_p(0)$	COX	2.666	0.117	0.115
(2.479)	FOURIER	2.908	0.085	0.093
	GEM	2.634	0.112	0.099
	KELK	2.338	0.068	0.069
	EM	162.0	3.3	2.8
	HNM	99.4	1.4	1.8
$nf_p(0)$	COX	133.3	5.8	5.8
(123.9)	FOURIER	121.3	3.5	3.9
	GEM	131.7	5.6	4.9
	KELK	116.9	3.4	3.4

TABLE 3(c): Estimates of c_p, $f_p(0)$, *and* $nf_p(0)$ *for the estimators in Table 1 (Case III:* p_i *increasing).*

Parameter	Estimator	Simulation Average	Theoretical Standard Error	Empirical Standard Error
	EM	0.3420	0.0069	0.0059
	HNM	0.5515	0.0079	0.0097
c_p	COX	0.5325	0.0562	0.0369
(0.4841)	FOURIER	0.4282	0.0173	0.0167
	GEM	0.5050	0.0146	0.0165
	KELK	0.5286	0.0170	0.0164
	EM	2.907	0.060	0.050
	HNM	1.822	0.027	0.032
$f_p(0)$	COX	2.147	0.102	0.093
(2.066)	FOURIER	2.473	0.067	0.077
	GEM	2.086	0.073	0.068
	KELK	1.972	0.059	0.054
	EM	145.3	3.0	2.5
	HNM	91.1	1.3	1.6
$nf_p(0)$	COX	107.4	5.1	4.7
(103.3)	FOURIER	104.0	2.8	3.1
	GEM	104.3	3.7	3.4
	KELK	98.6	2.9	2.7

In general, the three parametric models (EM, HNM, GEM) estimate the effective width c_p better than $f_p(0)$ or $nf_p(0)$ and the non-parametric estimators (COX, FOURIER, KELK) estimate $f_p(0)$ better than c_p. The EM underestimates c_p in the three cases, which leads to overestimates of abundance. Conversely, the HNM overestimates c_p and thus underestimates abundance, which may seem surprising since the data from each school-size class come from a half-normal model. However, this effect was theoretically demonstrated earlier for fixed β in the GEM. In contrast to the EM and HNM, the GEM estimates c_p correctly and overestimates abundance very slightly, which demonstrates the compensatory effect of including β in the estimation process.

Of the non-parametric estimators, the Fourier and Cox models produce reliable abundance estimates. The Fourier estimator has a significantly lower standard error than the Cox estimator for all three cases. The Kelker estimator underestimates abundance slightly and has a small standard error, but the small bias tends to be an artifact of the HNM (Quinn, 1977). The Fourier estimator may only be compared with the simulation parameter of $nf_p(0)$, because the Fourier series estimates a truncated representation of the sighting model in Figure 1 (Crain *et al.*, 1978). In the simulation, the Fourier estimator produced excellent abundance estimates with low standard errors, which may be attributed to the use of a stopping rule which selected only a single Fourier term in many instances.

The theoretical and empirical standard errors are quite close for all estimates in Tables 3(a), (b), (c), which demonstrates the utility of the theoretical variance formulae in Table 1. Since the variance of the abundance estimator in (1c) is partitioned into the variability in the number of sightings and the variability in the sighting distances, the latter component may be estimated with confidence from the formulae in Table 1, regardless of the spatial distribution of schools or the distribution of school sizes.

In order to explore the estimators in greater detail, the sampling distributions of the six estimators of $nf_p(0)$ are plotted in Figures 2, 3, and 4 for cases I, II, and III, respectively. The sampling distributions of the EM and HNM are tight and approximately normally distributed, but are not centered about the true simulation parameter. The distribution of the GEM estimator is positively skewed but there is a strong tendency for the distribution to be centered correctly. The Cox estimator shows a quite variable sampling distribution with some clustering about the correct central value. The Kelker estimator has a rather tight,

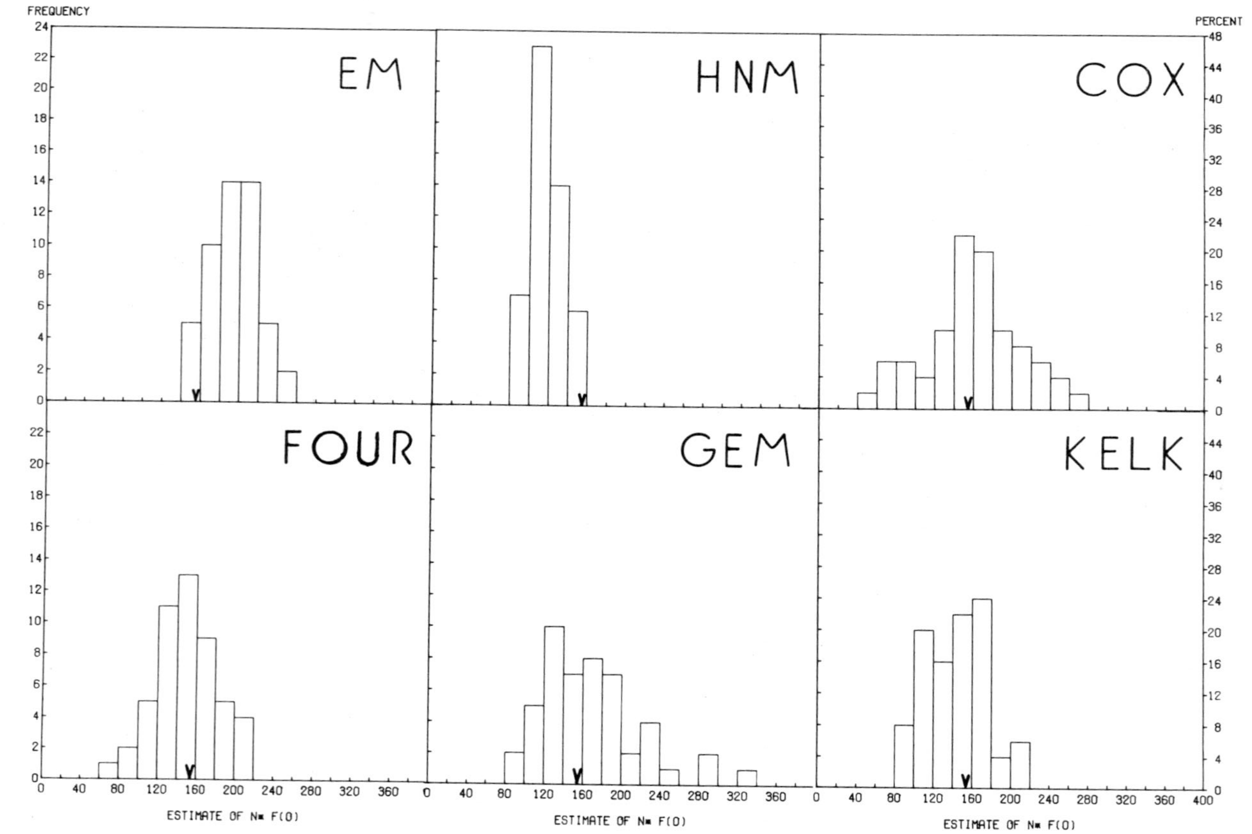

FIG. 2: Histograms of the sampling distributions of the estimators in Table 1 (Case I: p_i *decreasing).*

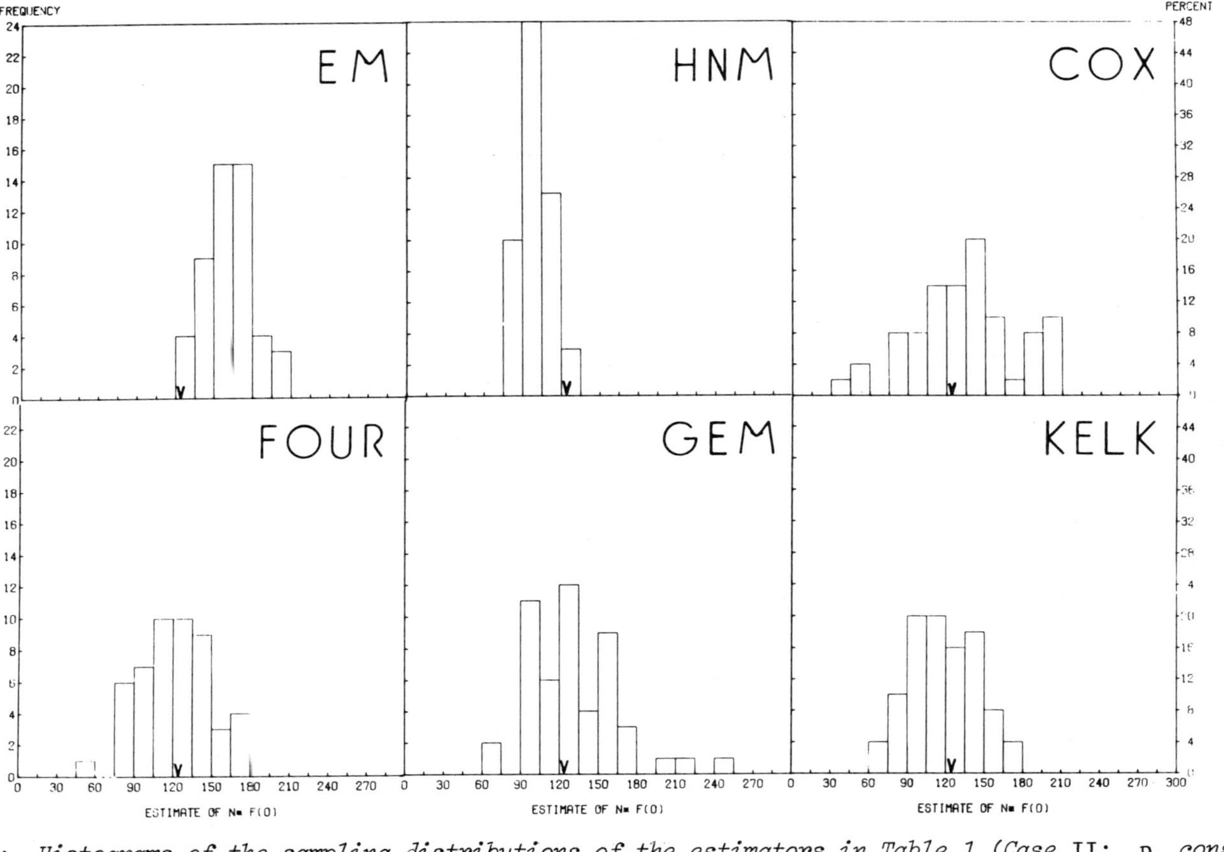

FIG. 3: Histograms of the sampling distributions of the estimators in Table 1 (Case II: p_i constant).

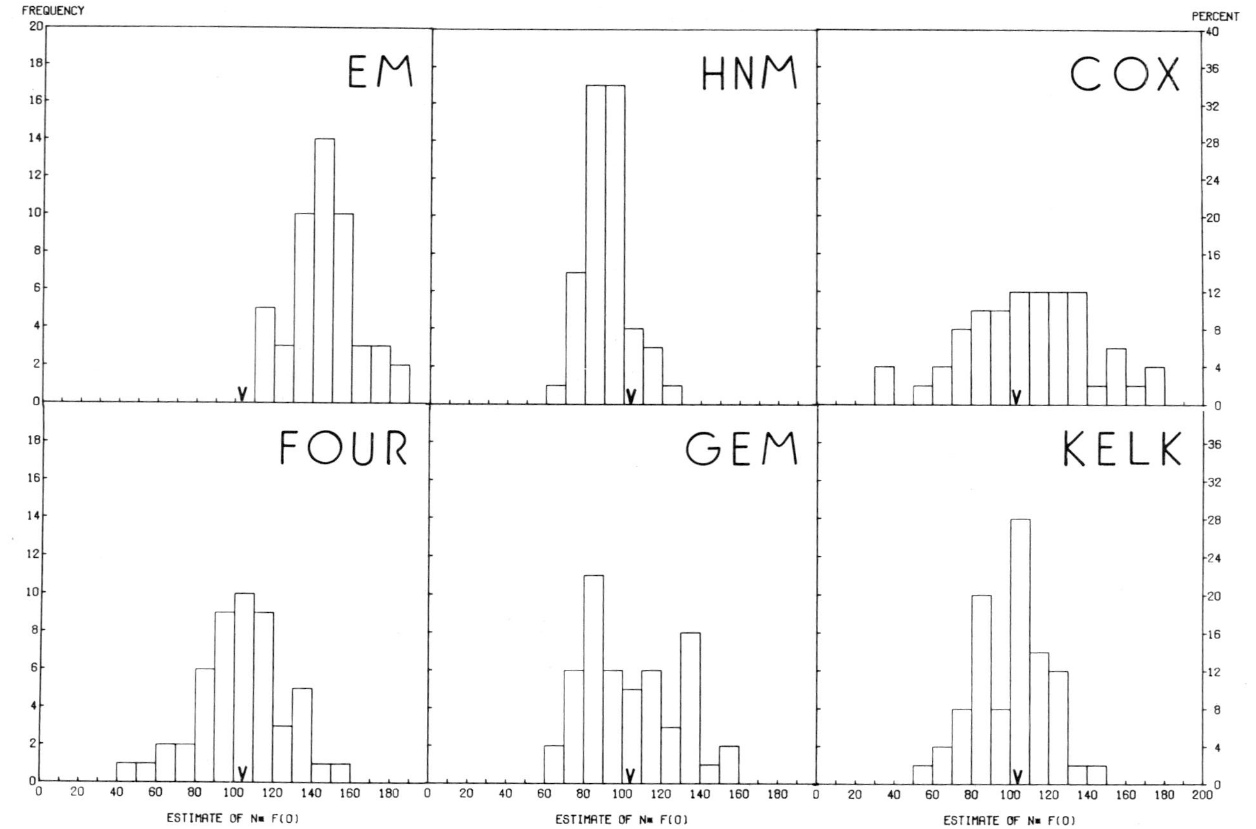

FIG. 4: *Histograms of the sampling distributions of the estimators in Table 1 (Case III:* p_i *increasing).*

normally distributed sampling distribution close to the true parameter, but this tendency is probably true for data coming from distributions rounded near the origin. The distribution of the Fourier estimator is approximately normal and properly centered around the true simulation parameter.

4. DISCUSSION

Of the estimators considered, the parametric and non-parametric estimators that perform the best are the GEM estimator and the Fourier estimator, respectively. The one-parameter EM and HNM parametric models do not have sufficient flexibility to account for differences in sighting models for different school-size classes. The Cox estimator is only slightly biased but its sampling distribution is quite variable. It should be used only when better non-parametric techniques are not available. The Kelker estimator can also be used for rounded sighting distributions, provided the choice of Δ is small enough to prevent bias.

The theoretical methodology shows that the sighting model for the pooled data is properly self-weighted by the true relative abundances in the population, so that estimation of parameters for abundance can be validly performed on the pooled data. The simulation study demonstrates the veracity of the theoretical development for sample sizes as small as 50, provided a flexible estimator such as GEM, Fourier, or perhaps Cox is chosen.

Further estimation possibilities include the examination of a covariate model where the actual school size enters the estimation process or a more general treatment of school size as a contaminating random variable of a size-biased distribution of distances. The latter procedure would follow from recent developments discussed by Patil and Rao (1978), where perhaps the lognormal distribution could be used for school size.

The results presented here for school size apply generally to other factors in the structure of schools. As long as a school may be accurately classified according to a given attribute or combination of attributes, such as stock grouping, species composition, spatial configuration, or behavior of individuals, each class is associated with an unknown sighting model $g_i(y)$ and the theoretical development above applies. Thus, the total number of schools may be estimated accurately using the pooled data. However, the estimate of the total number of individuals requires an unbiased estimate of the average school size over all attributes considered, which may require additional information depending on the population studied.

The more general question of whether or not a single sighting model that is a function solely of distance can fully account for sighting phenomena cannot be answered here. Ongoing work by others (K. P. Burnham, D. Burdick, personal communication) suggests the correct self-weighting tendency of the line-transect method for other variables such as visibility, length of transect, sea state, etc. as has been demonstrated in this paper for school size.

ACKNOWLEDGEMENTS

This research was motivated by a series of workshops held at the Southwest Fisheries Center of the National Marine Fisheries Service of the United States, convened by Dr. J. Powers and Dr. E. Barham. The exchange of ideas with these people and Dr. K. Burnham, Dr. D. Anderson, Dr. B. Crain, Dr. D. Burdick, Dr. L. Eberhardt, Dr. V. Gallucci, Dr. D. Chapman, and others greatly influenced this paper. This paper could not have been presented without generous support from the Graduate School of the University of Washington, Seattle, Washington, USA.

REFERENCES

Burnham, K. P. and Anderson, D. R. (1976). Mathematical models for non-parametric inferences from line-transect data. *Biometrics*, 32, 325-336.

Crain, B. R., Burnham, K. P., Anderson, D. R., and Laake, J. L. (1978). A Fourier series estimator of population density for line transect sampling. Utah State University Press, Logan, Utah.

Eberhardt, L. L. (1968). A preliminary appraisal of line transects. *Journal of Wildlife Management*, 32(1), 82-88.

Eberhardt, L. L. (1978). Transect methods for population studies. *Journal of Wildlife Management*, 42, 1-37.

Gates, C. E. (1969). Simulation study of estimators for the line transect sampling method. *Biometrics*, 25(2), 317-328.

Gates, C. E. (1979). Transect and related issues. In *Sampling Biological Populations*, R. M. Cormack, G. P. Patil, and D. S. Robson, eds. Satellite Program in Statistical Ecology, International Co-operative Publishing House, Fairland, Maryland.

Gates, C. E., Marshall, W. H., and Olson, D. P. (1968). Line transect method of estimating grouse population densities. *Biometrics*, 24(1), 135-145.

Patil, G. P. and Rao, C. R. (1978). Weighted distributions and size biased sampling with applications to wildlife populations and human families. *Biometrics*, 34(2), 179-189.

Quinn, II, T. J. (1977). *The effects of aggregation on line transect estimators of population abundance with application to marine mammal populations*. M.S. thesis, University of Washington, Seattle, Washington.

Quinn II, T. J. and Gallucci, V. F. (1979). Parametric models for line transect estimators of abundance. *Ecology*, to appear.

Seber, G. A. F. (1973). *The Estimation of Animal Abundance and Related Parameters*. Griffin, London.

[*Received September* 1978. *Revised May* 1979]

G. P. Patil and M. Rosenzweig, (eds.),
Contemporary Quantitative Ecology and Related Ecometrics, pp. 493-503.
International Co-operative Publishing House, Fairland, Maryland.

LINE INTERSECT SAMPLING OF FOREST RESIDUE

JOHN W. HAZARD

Pacific Northwest Forest and
Range Experiment Station
Portland, Oregon 97208 USA

STEWART G. PICKFORD

College of Forest Resources
University of Washington
Seattle, Washington 98195 USA

SUMMARY. Line intersect sampling of forest residue is simulated to determine the effect of changing piece diameter, length, and taper on the population variance and indirectly on the sample size. In addition, the question of optimal line length for specific populations is studied. It was found that the amount of sampling effort required for specified levels of precision varies considerably with the characteristics of the pieces in the population. Also, the precision of estimates of residue volume is a function only of total line length in random populations.

KEY WORDS: simulation, sampling methods, forest residue, forest mensuration.

1. INTRODUCTION

Simulation studies of sampling techniques have value only if the simulated populations have the same or similar characteristics as the natural populations to which the techniques will be applied. Without a link between the simulations and reality, there is no way of knowing whether the statistical properties of the sampling techniques or sampling guidelines generated by the simulations are useful. There are several ways one can insure that the results of simulations of residue populations meet this condition: one way is to map characteristics of residue populations from aerial photographs into the computer.

A second approach is to start with a uniform population, and introduce variability until a population is created which possesses all of the important natural characteristics of the population. In this way, the effects of individual characteristics on the statistical properties of the sampling technique can be determined as they are added. In these studies, we chose this second approach.

The line intersect technique (LIS) is used quite extensively in the Pacific Northwest area of the United States. LIS is currently used by several federal agencies which inventory slash or wildfire fuels for policy decisions or management planning, and for work such as forest survey (Howard, 1971), fire management (Dell and Ward, 1971), and timber management. In addition, research uses LIS to assess differences in the level of residue before and after treatments are applied. Even though the technique is frequently used, there are still only limited guidelines for field use.

Our objectives were:

(1) to determine the effect on the residue population variance of changing piece diameter, length, and taper and to indirectly determine the effect on the sample size required to achieve specified levels of precision for randomly distributed and oriented populations of residue pieces and (2) to explore the question of optimum line length for these same random populations of residue. Results of these two objectives are discussed in this paper.

A third objective, currently under investigation, is to examine the statistical properties of estimates of volume of residue per acre, the efficiency of the various sampling unit designs, and the optimum line length and number of lines for a variety of population characteristics and sampling techniques, including (a) several types of non-random spatial distributions of residue; clumped, parallel rows, and converging rows; (b) selected non-random piece orientations; (c) systematic arrangements of line locations; and, (d) various line cluster sampling units. The final product of these studies will be a set of guidelines for field use.

One additional study will generate cost functions to be

used to find optimum line lengths and numbers of lines. This study will generate cost functions for various patterns of line clusters. The total cost function for the line intersect sampling procedure will depend upon the average slope of the unit sampled, the density of residue in the population, and the total length and arrangement of sample lines.

Throughout this discussion, we assume readers have a basic understanding of the LIS sampling rules and estimation procedures (Warren and Olsen, 1964; Van Wagner, 1968; Bailey, 1969, 1970; Brown, 1970, 1974; Brown and Roussopoulos, 1974; De Vries, 1973, 1974). There are several terms used repeatedly which need defining:

Population refers to a well defined area containing residue. For the purpose of simulation, characteristics of populations were mapped into the computer. Sampling units are located via a coordinate system created on the population.

A *line* is a sampling unit located on the population by a well-defined sampling rule. Lines may be clustered or single segments, all of specified length. This definition creates a problem for LIS because a line has no width; thus, it is not possible to enumerate the total of all sampling units in the population in order to obtain the population characteristics. Instead, the residue volume must arise from summing all the piece volumes in the population. The population variance of residue volume per acre must be estimated by repeated sampling of the population (i.e., Monte Carlo sampling).

When random sampling, there are an infinite number of potential line locations within the population. Also, there are an infinite number of systematic samples if located with random starts.

Residue refers to solid pieces of wood on the ground in the field. It can result from commercial logging operations or simply from natural or artificial pruning or thinning.

A *piece* is an individual geometric solid item of residue existing in the population. Residue pieces resemble cylinders and frustums, the two forms used in this study.

Sample length, or total length of a sample, is the total length of all lines in a sample. It can be a continuous straight line or broken into disconnected segments.

2. LIS SIMULATIONS

The volume of residue on a given area can be estimated by various LIS estimators (Van Wagner, 1976). That is, volume can be estimated by measuring the midpoint diameter and length of intersected pieces, by both end diameters and piece length or by piece diameter at the point of intersection. This latter method is the one used in the Pacific Northwest. There are controlling practical reasons why the methods involving the measurement of piece length are not used. If it were simply a matter of the increased cost to measure these additional piece characteristics, there would definitely be times when another method would be more efficient; however, cost isn't the only consideration. Length, in many instances, is impossible to measure because of brush, debris, or piled residue. Also, the end or midlength diameter will frequently be difficult to measure. By taking the diameter at the point of intersection, one incurs much more variability; however, the amount of inaccuracy in diameter measurements and the cost are much less. Also, residue in the Northwest frequently occurs on very steep slopes, and the cost of taking additional measurements on steep slopes can become quite large. Therefore, the simulations of this study involve only estimates based upon diameters at the point of intersection.

We will briefly describe the simulation methods. There are actually two stages of simulation. First, characteristics of residue (i.e., number of pieces, assigned total volume, piece diameter and length distribution, piece shape, spatial distribution, and orientation) are mapped into the computer. Second, the sampling rule is applied to the population repeatedly to generate sample estimates for repeated random samples of the population.

Computer programs were written to perform these two operations. A program called SLASH generates simulated residue populations of any size possessing any residue characteristics. A program called INTRSCT samples these populations with a specified number of lines, line length, and design by a specified set of sampling rules. (These programs are currently available only through the Pacific Northwest Forest and Range Experiment Station, U. S. Forest Service, Portland, Oregon, USA.) INTRSCT computes the volume estimate for individual samples and computes the estimated expected values of sample estimates from repeated trials.

The results of simulations discussed in this paper satisfy objectives (1) and (2). And, as mentioned earlier, we have approached the problem by beginning with a very uniform set of population characteristics, observing the population variance and computing the sampling effort required to satisfy specified levels of precision. Then, by introducing new piece characteristics such as varying piece diameter, length, and taper, we observed the effect on the variance of residue volume and sample size. The end product is a set of guidelines for determining the sample size and length of lines for satisfying a specified level of precision for an artificially created population possessing nearly all of the natural characteristics which cause sample size requirements to change. In addition, we achieve a better understanding of the influence on population variance of individual population characteristics.

We began by constructing and sampling 5-acre (2.02-ha) populations in the order which follows:

(1) Cylinders of constant diameter and length with total volume of 3,562 cubic feet per acre (249.23 m^3/ha).

(2) Cones of constant base diameter and length with total volume of 3,652 cubic feet per acre (249.23 m^3/ha).

(3) Frustums of constant end diameters and length with total volume of 800 cubic feet per acre (55.98 m^3/ha).

(4) Frustums of constant end diameters and length with total volume of 3,122 cubic feet per acre (218.45 m^3/ha).

(5) Frustums of varying diameters and length with total volume of 2,341 cubic feet per acre (163.80 m^3/ha).

These simulations produced interesting results, the most noteworthy of which concerns the amount of sampling (sample size) required to satisfy particular levels of precision in these five populations. Also important is the result that, when pieces are truly randomly spaced and oriented, the length of an individual line is not a controlling factor in estimating precision. The total length of all lines in the sample establishes the level of precision of the estimate. This is true of any population regardless of the geometric configuration of the pieces. What this means is that, if the cost per foot of line decreases with length of line, it is more efficient to use longer lines. These results point out the need for accurate cost functions. Thus, one needs to be concerned only with the total length of line in a sample, and not with the length of segments.

Another point of interest is that estimates of residue volume per acre were unbiased for all populations. This, of course, was expected since we created populations which were randomly distributed and oriented, and we located lines randomly on the population. Additional results are given for specific populations.

Cylinders. One thousand cylinders 12 inches (30.48 cm) in diameter and 20 feet (6.10 m) long were created on the area of 470 by 470 feet (143.26 m). Under these conditions, it takes approximately 5,500 feet (1676.40 m) of line to satisfy a 10-percent margin of error for a 95-percent confidence interval on a mean residue volume of 3,562 cubic feet per acre (249.23 m^3/ha). This translates to 100 lines each 55 feet (16.76 m) long, 55 lines 100 feet (30.48 m) long, or any combination of number of lines and feet of line which result in 5,500 feet (1676.40 m) of line. We will use this 10-percent margin of error as a reference point for comparing the sampling effort required for various populations, even though practical situations may require more or less precision.

Cones. We created a population of cones having the same number of pieces, same spatial distribution, and same piece length and volume as the cylinders. Here, we wanted to estimate the increase in sampling effort for a specified level of precision when taper was introduced into the population. Also, we wanted to demonstrate that the presence of taper in pieces does not introduce a bias into the estimate of average residue per acre. The estimates of average residue were unbiased, illustrating that taper does not introduce bias; however, the population variance increased substantially. In fact, the variance nearly doubled from that for the population of cylinders. The total length of line required for the 10-percent margin of error was approximately 9,500 feet (2895.61 m). This points out an extreme case where the required sample size would double simply due to a change in geometric shape of the pieces.

It is interesting to note that the method of estimation using the length and mid-length diameter produces the same variance for both the cylinder and cone populations. Thus, the effect of taper can be eliminated by estimating with midpoint diameter and length.

Frustum. The same number of pieces were created for the third population, but in this instance we generated pieces that we felt were similar to the configuration of top logs in trees. These pieces were frustums with 3-inch (7.62 cm) and 8-inch (20.32 cm) end diameters, and 20 feet (6.10 m) long. The average volume of residue for this population of 1,000 pieces was reduced to 800 cubic feet per acre (55.98 m^3/ha).

This change reduced the population variance approximately 27 times over the cone population. The frustum population contained only one-fourth the volume of the cone population. Where the average residue volume per acre was reduced, the length of line required to meet the specified 10-percent margin of error was increased. These two combined effects produced a net increase in the sampling effort for the 10-percent margin of error of approximately 40-percent over the cylinder population. In other words, it required approximately 7,000 feet (2133.60 m) of line on the frustum population to produce the same precision that was accomplished with 5,500 feet (1676.40 m) of line in the cylinder population.

To get a better feeling for the effect of density alone on sampling effort, we increased the density of this frustum population from 1,000 to almost 4,000 pieces. The effect on the population variance and sample size was entirely predictable. The variance increased proportionally to the increase in density, but the sample size for a specified level of precision decreased to approximately one-fourth that for the 1,000 piece population.

The explanation is that the number of pieces intersecting a given length of line is a random process, and thus are Poisson-distributed with equal mean and variance. The volume per line is proportional to the number of pieces, with the constant of proportionality being:

$$\frac{\pi^2 \frac{\Sigma d^2}{n}}{8L}$$

where $\Sigma d^2/n$ is the mean of the intersected diameters squared and L is the line length.

The final set of simulations in this portion of the study involved a population of 1,000 pieces, but this time with varying piece diameter and length distributions. To get realistic values for diameters and length, 12 clearcuts were measured in western Oregon and a frequency count of pieces was made by diameter and length classes. A new frustum population was then created using these empirical frequency distributions. A taper of 1 inch of diameter in 4 feet (1.22 m) of length was assumed. The mean residue volume was 2,341 cubic feet per acre (163.80 m^3/ha). This produced a realistic population except for the spatial distribution and orientation of the pieces in it.

The key result of sampling this population was that the population variance increased 8.8 times over the population of uniform frustums and the total length of line required to satisfy a specified level of precision increased 16 times. The increase in sample size is primarily a result of the variability introduced by varying the piece diameter and length distributions.

Figure 1 illustrates the relationship between number of sampling units and margin of error. Until such time as the characteristics of real populations are mapped into the computer and used for these simulations, this figure provides our best guide to the effort required in sampling forest residue in the Pacific Northwest.

3. CONCLUSION

Our simulations with randomly distributed cylinders, cones, and frustums have shown the following:

1. The line intersect method requires a large sampling effort even under ideal conditions; however, it still may be more efficient than area sampling.

2. Introduction of taper into the population substantially increases the sampling effort; yet, taper does not introduce bias, confirming Van Wagner's (1968) unproven statement to that effect.

3. Sampling becomes more efficient as the density of pieces in the population increases. Although predictable from sampling theory, this result had not previously been explored empirically.

4. Estimates of population variance change proportionally to changes in mean volume per acre, and sample size decreased proportionally to increasing line length for a specified margin of error. Again, this is predictable from sampling theory but has not been emphasized in the literature.

5. Sampling a random population of tapered pieces whose length and diameter distribution resembles actual populations required five times the sampling effort required for the population of cylinders for the same precision. The sampling effort required for this population is close to the sampling effort discussed by Howard and Ward (1972) for

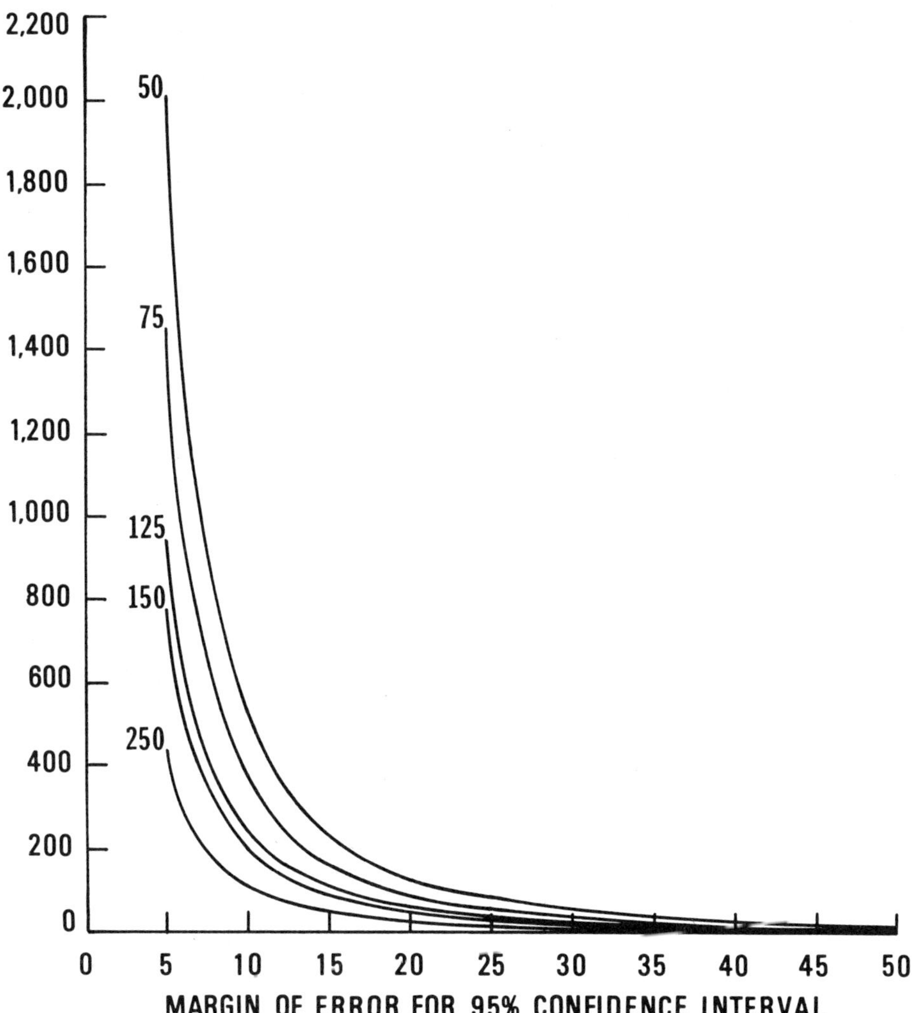

FIG. 1: Number of lines and lengths of line in feet (1 *foot* = .3048 *m*) *required to meet specified levels of precision. The population is composed of* 1,000 *pieces averaging* 2,341 *cubic feet per acre* (163.80 *cu. m/ha*). *Pieces are frustums of variable length and diameter possession* 1 *inch of taper in* 4 *feet of length* (2.54 *cm in* 1.22 *m*).

actual residue populations.

6. The total length of line is the important consideration in establishing expected precision. In random populations, the product of number of lines times individual line length is approximately constant for a specified level of precision.

We have not studied the performance of the line intersect method on strongly non-uniform populations such as cableyarded residue, nor have we studied the performance of the sampling method using common systematic sampling patterns. Also, we have not compared the line intersect method with existing area sampling methods. These studies are underway and will be reported in the future.

REFERENCES

Bailey, G. F. (1969). Evaluation of the line-intersect method of logging residue. Canadian Department of Fisheries and Forestry, Forest Products Laboratory, VP-X-23.

Bailey, G. F. (1970). A simplified method of sampling logging residue. *Forestry Chronicle*, 46, 288-294.

Brown, J. K. (1970). A planar intersect technique for sampling fuel volume and surface area. *Forest Science*, 17, 96-102.

Brown, J. K. (1974). Handbook for inventorying downed woody material. U.S. Department of Agriculture Forest Service General Technical Report INT-16. Intermountain Forest and Range Experiment Station, Ogden, Utah.

Brown, J. K. and Roussopoulos, P. J. (1974). Eliminating biases in the planar intersect method for estimating volumes of small fuels. *Forest Science*, 20, 350-356.

Dell, J. D. and Ward, F. R. (1971). Logging residues on Douglas-fir region clearcuts--weights and volumes. U. S. Department of Agriculture Forest Service Research Paper PNW-115. Pacific Northwest Forest and Range Experiment Station, Portland, Oregon.

De Vries, P. G. (1973). A general theory on line intersect sampling with application to logging residue inventory. Medelingen Landbouw Hogeschool No. 73-11. Wageningen, The Netherlands.

De Vries, P. G. (1974). Multistage line intersect sampling. *Forest Science*, 20, 129-134.

Howard, J. O. (1971). Volume of logging residues in Oregon, Washington and California--initial result from a 1969-70 study. U.S. Department of Agriculture Forest Service Research Note PNW-163. Pacific Northwest Forest and Range Experiment Station, Portland, Oregon.

Howard, O. and Ward, F. R. (1972). Measurement of logging residue--alternative applications of the line intersect method. U.S. Department of Agriculture Forest Service Research Note PNW-183. Pacific Northwest Forest and Range Experiment Station, Portland, Oregon.

Van Wagner, C. E. (1968). The line intersect method in forest fuel sampling. *Forest Science*, 14, 20-256.

Van Wagner, C. E. (1976). Diameter measurement in the line intersect method. *Forest Science*, 22, 230-232.

Warren, W. G. and Olsen, P. E. (1964). A line intersect technique for assessing logging waste. *Forest Science*, 10, 267-276.

[*Received December* 1977. *Revised June* 1978]

SECTION V

APPLIED STATISTICAL ECOLOGY

G. P. Patil and M. Rosenzweig, (eds.),
Contemporary Quantitative Ecology and Related Ecometrics, pp. 507-532.
International Co-operative Publishing House, Fairland, Maryland.

MAN AS PREDATOR

RICHARD C. HENNEMUTH

United States Department of Commerce
National Oceanic and Atmospheric Administration
National Marine Fisheries Service
Northeast Fisheries Center
Woods Hole, Massachusetts 02543 USA

SUMMARY. The ecological basis of fisheries is discussed in relation to the validity of existing concepts and the needs of managers. The failure to maintain the very intensive fisheries achieved in the 1970's requires more appropriate feedback between harvesters and their impacts on the fish stocks.

KEY WORDS. fisheries, models, management, recruitment, multi-species.

1. INTRDOUCTION

It is probably valid to assume that the natural ecosystem represents the optimum in terms of balance and survival of its living resources. Man has not co-evolved in the marine ecosystem, and, thus, intrudes as a totally foreign element.

That this intrusion has been forceful is well demonstrated by reviewing the fishery statistics over the last 25 years. World catch of marine fish[1] increased at a rate of about 21% per year

---

[1]FAO classification: includes the flounders, cods, redfishes, jacks, herrings, tunas, mackerels, sharks, smelts.

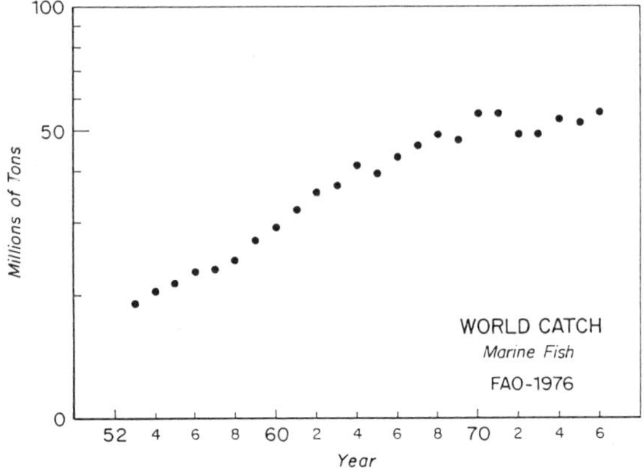

FIG. 1: World catch of marine fish. FAO, 1976.

until 1970, increasing from 2 million metric tons (mmt) to 50 mmt (Figure 1). During this period, most of the fishable stocks (within economic constraints) have been heavily exploited, and, in spite of increased fishing-vessel capacity, the catch over the last 6 years has not increased. It is doubtful that any significant sustained increase in catch of this category can be made. The more relevant question is whether or not the yield can be maintained.

The answer to this question depends on the ability to correctly determine and model resource productivity. Marine fishery resources are common property. Prior to extended national jurisdictions, they were globally common. Because of this, national governments (as opposed to internal political divisions or private owners) had the responsibility for management and development of fisheries. This brought to bear a great amount of effort on the study of fisheries resources and modeling of potential yields. These models have played a powerful and important role in formulating policy and management measures.

Because of the pervasive force of fishing and marked declines of many traditional and highly valued fisheries, the countries of the world have extended national jurisdictions to include most of the productive coastal ecosystems. It is expected that this will lead to more and more extensive efforts to manage fisheries in order to optimize national benefits. Successful management, the fulfillment of expectations, will depend to a large extent on adequate advice based on good models.

The simpler models include only one cause-effect relationship. There are no interactions and multiple effects are ignored. For example, yield-per-recruit models, which lead to control of size or age-specific mortality (e.g., mesh regulation), assume that changing the proportion of ages and, concomitantly in many cases, sexes in the population has no effect on growth of the regulated population or any others. Regulation of a fishing mortality on a single species assumes no interactions with any other components of the system. Abiotic effects are also ignored.

In a sense, the first attempts at regulation based on the 'simple' models assume a knowledge that is quite advanced. Attempting to regulate *en masse* would require less knowledge but is considered to be highly advanced management. We seem to have the cart before the horse.

We are really only capable of prevention of deleterious effects of man's activities. We are dealing largely with unknown causes of ecosystem change. We cannot cure the disease nor eliminate the causes. Most often we are faced with radical end-point events. It is, therefore, necessary to adjust management objectives to the reality of our knowledge of cause and effect.

Optimal benefits are based on societal values. The management models will thus include man's interventions primarily in terms of economics. Most, if not all, of the bioeconomic models include man as an exogenous variable, eliminating feedback mechanisms. The effects of fishing and the control of it thus depends on external perceptions which are very much muted and extended by economic efficiencies which generally bear no relationship to ecosystem efficiencies.

This paper will deal mostly with a definition of the problem. Some general proposals for solution will be discussed. Mostly, we hope to stimulate thought about development of useful models.

2. BIOLOGICAL FISHERY MODELS

Most management has been based on single-species population models. Within the last few years, multispecies modeling has received more attention, but has had very little impact on management. However formulated, the basis of sustained exploitation involves assumptions of some general concepts which have had limited verification.

The primary assumption is that there is a limited capacity of the environment to support a given population of fish. The limiting factors, which are often not explicitly included in the models, tend either to increase mortality or to suppress the physiological growth potential. The central thesis of sustainable fisheries is that exogenous mortality through fishing either replaces the natural mortality or increases the intrinsic net natural rate of growth by reducing the standing stock. Both of these are hypothesized to have finite limits, which limits the potential yield.

These are not unreasonable concepts, and have been confirmed by experiment for some animals and empirical observation of long-term fisheries. As will be pointed out later, however, the stability of fisheries has decreased markedly as fishing activity has increased. There have also been many demonstrations that the total mortality rate increases in proportion to removals, and that natural mortality, when estimated before heavy fishing takes place, is relatively small. There is, therefore, some doubt that man can be a prudent predator (Slobodkin, 1962).

The models, to serve as useful guidance, are usually carried to point equilibriums. This implies a unique maximum yield at some fixed fishery mortality rate termed MSY. Although generally accepted as an objective of management, the estimation and application of MSY has not provided satisfactory results and has, in fact, led to many failures in maintaining yields (Hennemuth, 1977; Edwards and Hennemuth, 1975; Holt and Talbot, 1978). Theoretical concern about the adequacies of these models has been discussed also by May (1975).

The major implications of these models are that fishing mortality is a significant controllable force in changing population biomass, and that productivity of the population will react in a predictable manner to changes in population biomass.

The major fluctuations in biomass are caused by changes in production of eggs and survival in the egg, larvae, and juvenile stages--the whole process generally called recruitment. Managing fisheries yields by regulating fishing mortality to produce changes in biomass implies that there does exist a dependency between stock biomass and recruitment.

Explicit stock-recruitment models have been developed over the last 30 years or so, and their applicability has been treated in a great deal of literature (Ricker, 1954; Beverton and Holt, 1957; Larkin *et al.*, 1964). With the exception, perhaps, of some anadromous species, they have never directly formed the basis of regulatory measures. Rather they have led to the generalized hypothesis that reduction of stock biomass to the point where the models indicate a sharply descending left-hand limb begins will reduce long-run average recruitment.

Three typical S/R models are those developed by Beverton-Holt, Ricker, and Larkin *et al.* (a 'depensatory' type) (Figure 2). The

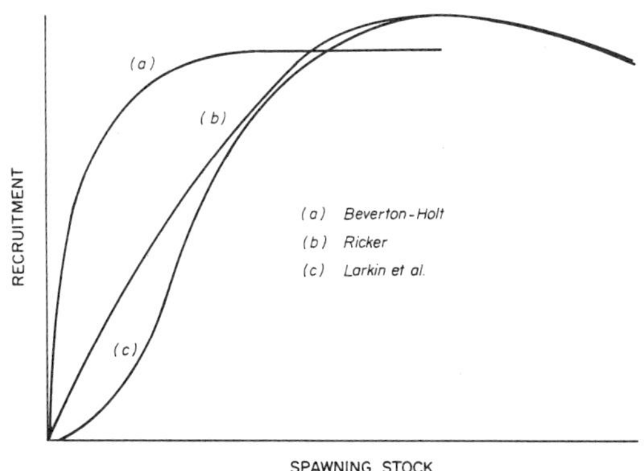

FIG. 2: Three typical stock-recruitment curves.

models were developed to accommodate the traditional set of available fishery data on stock biomass and year-class strength. There are few, if any, cases where empirical confirmation of the effects of stock biomass on recruitment has been obtained by fitting these models to observed data. A selected set of case histories illustrating this point are given in Figure 3 to 8. These were not selected because they did not fit the S/R models, but are rather typical of pelagic and demersal species.

The models do not, of course, account for effects of environment, nor for any other effects independent of stock density. There may well be an effect of biomass on recruitment which is masked by equal or greater effects of other factors. Very probably this is the case, but considering stock recruitment in the above manner leads one to a rather loose and unconvincing argument for controlling recruitment by manipulating stock biomass. Only single stocks are considered, which may be a very significant factor, but more of that aspect later.

There is another way of looking at recruitment which is more integrative, but which may not, in fact, cause a loss of real (valid) information. One may consider the sum total of all effects on recruitment over time leading to a joint probability density function (Figure 9). Since it will require much study and a long time series of data to identify and delineate the marginal distributions, one of which is stock dependency, the more general treatment is necessary and probably a more practical and useful approach for management advice.

Definition of the joint p.d.f. is not initially as important as recognizing its existence and general shape. One of the

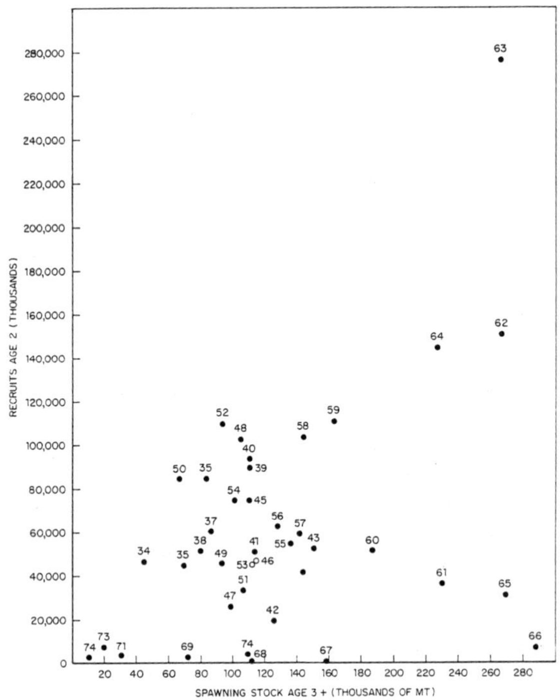

FIG 3: Stock-recruitment observations--haddock, Georges Bank. Clark and Overholtz, 1979.

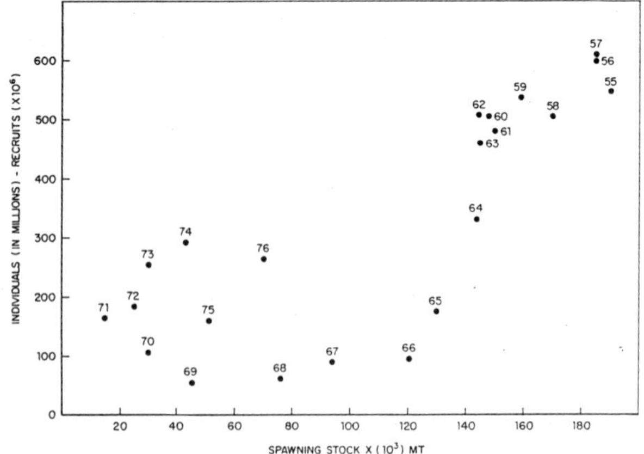

FIG 4: Stock-recruitment observations--silver hake, Gulf of Maine. Almeida and Anderson, 1978b.

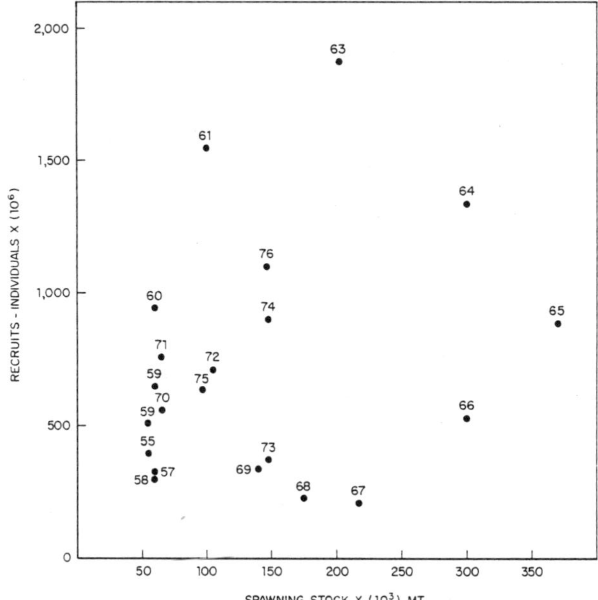

FIG 5: Stock-recruitment observations--silver hake, Southern New England. Almeida and Anderson, **1978a.**

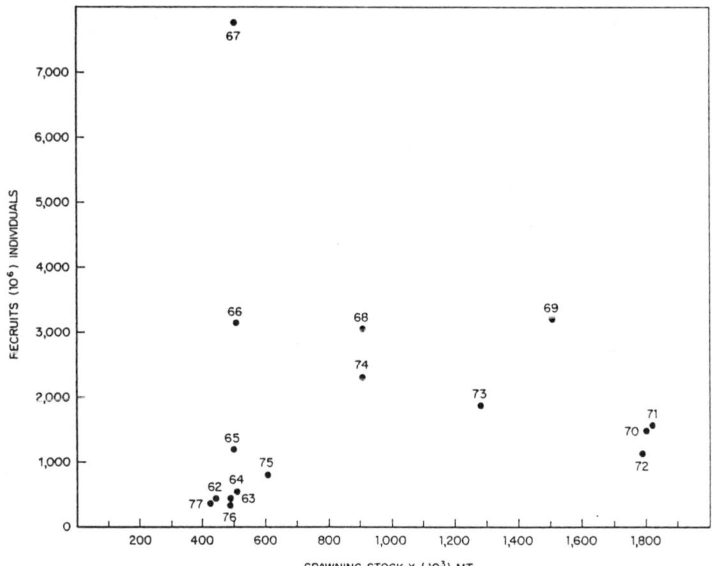

FIG 6: Stock-recruitment observations--mackeral, Northwest Atlantic. Anderson and Paciorkowski, in press.

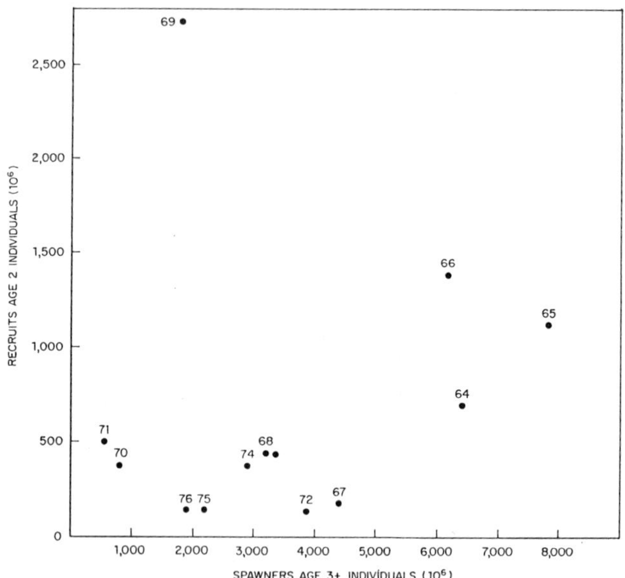

FIG. 7: Stock-recruitment observations--herring, Arcto-Norwegian. Dragesund et al., in press; Schumacher, in press.

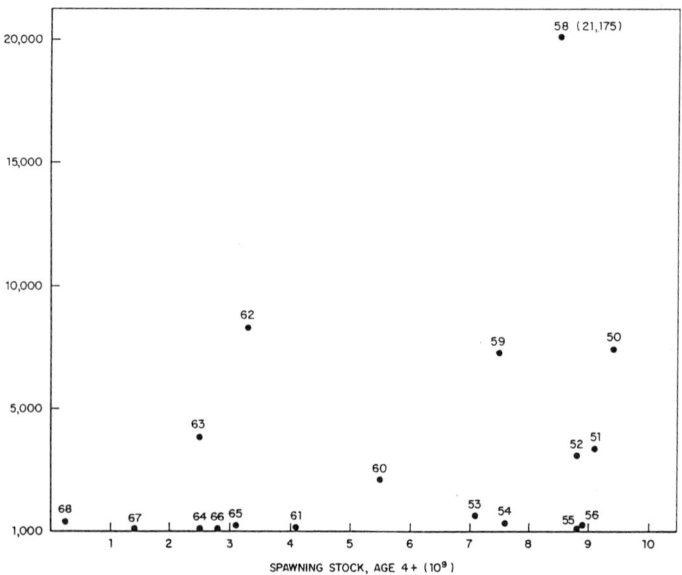

FIG 8: Stock-recruitment observations--mackerel, North Sea. Hamre, 1978, *in press.*

important considerations it bears on is the oft-used phrase 'recruitment failure.' It would be convenient to consider that this reflects a finite probability of no recruitment. There is not, to my knowledge, any evidence for such an event, and without this one must consider the continuum from very small to very large recruitment. (In most fish populations, reproduction is an annual event, and hence recruitment is defined in terms of year classes, the survivors of an annual, not necessarily single, spawning.)

In such a situation, recruitment failure is a somewhat arbitrary concept. At the most, it is based on practical fishery considerations, that is, the level of recruitment needed to maintain the fishery, but this is relative and related to past experience. Without some probability of extinction, consideration of stock biomass in management is a social option. That does not mean, of course, that there are not significant deleterious effects with lower recruitment. It is a matter of whether or not it can be controlled by manipulating stock biomass through control of fisheries

The observed frequencies of year-class size categories for the same stocks used in the S/R plots are shown in Figure 10. The number of years involved varies from 10 to 30, depending on the stock. One general characteristic is the high degree of skewness with long right-hand tails due to annual variations greater than a factor of 10. There are, in fact, many more 'poor' year classes than 'good' ones, if the average size is the dividing line. The bimodal frequency (silver hake, 5Y) would be of some special interest if it reflects a real phenomenon. This leads to discontinuities in catastrophic theory (Stewart, 1975), and would be related to the concept of recruitment failure.

More often than not, the beginning or expansion of a fishery is the consequence of the occurrence of a good year class. This is borne out by the time sequence of the same major Atlantic fisheries mentioned above (Figures 11-16). The subsequent 'collapse' of these fisheries is often because of the assumption that such good recruitment is not a rare event. Fishing mortality can be kept high, because of efficiencies in locating aggregations, until the stock biomass becomes very low indeed.

Management based on the use of such p.d.f.'s might be much more successful than assuming a stock-recruitment relation which leads to controversy over minimum stock size, i.e., that below which 'recruitment failure' occurs, or promises greater or more stable long-term recruitment than probably will be the case.

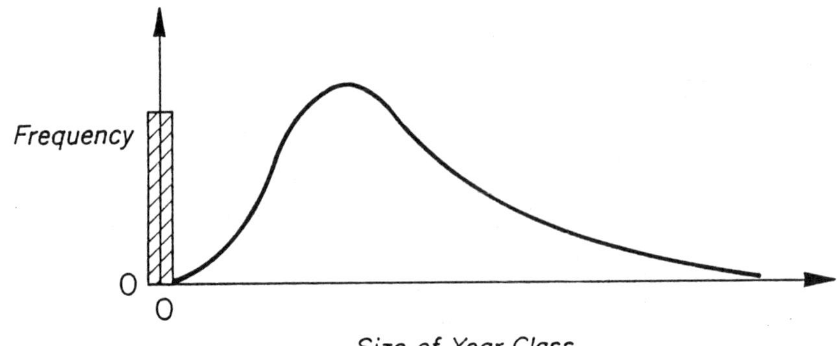

FIG. 9: Generalized recruitment function.

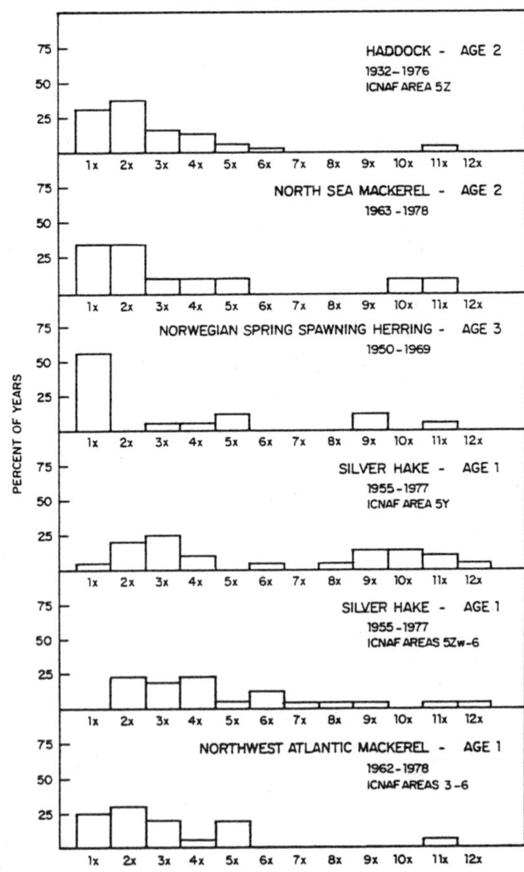

FIG. 10: Observed size-frequencies of recruitment.

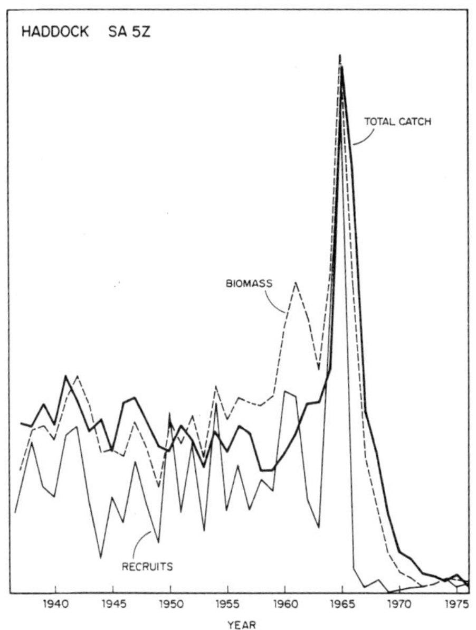

FIG. 11: Time trends in catch, population size, and recruitment for haddock, Georges Bank.

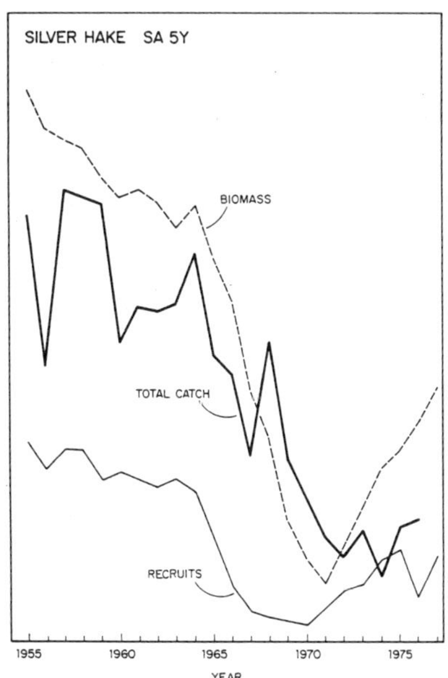

FIG. 12: Time trends in catch, population size, and recruitment for silver hake, Gulf of Maine.

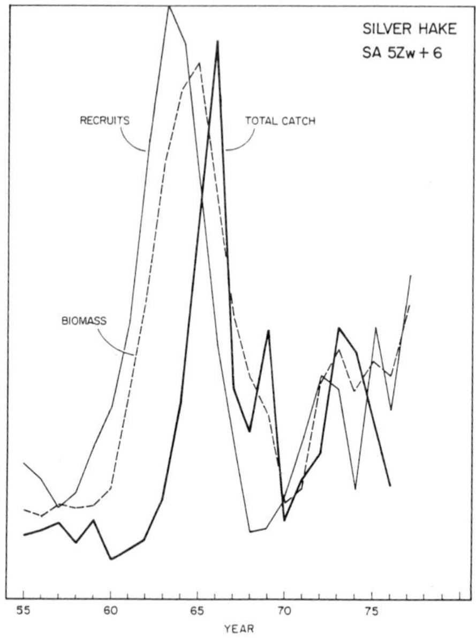

FIG. 13: Time trends in catch, population size, and recruitment for silver hake, Southern New England.

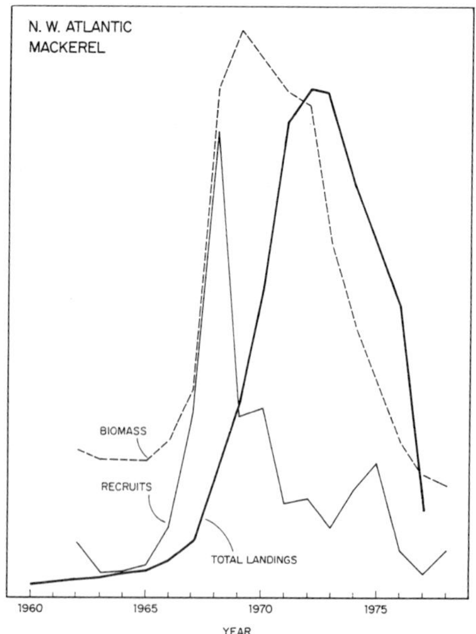

FIG. 14: Time trends in catch, population size, and recruitment for mackerel, Northwest Atlantic.

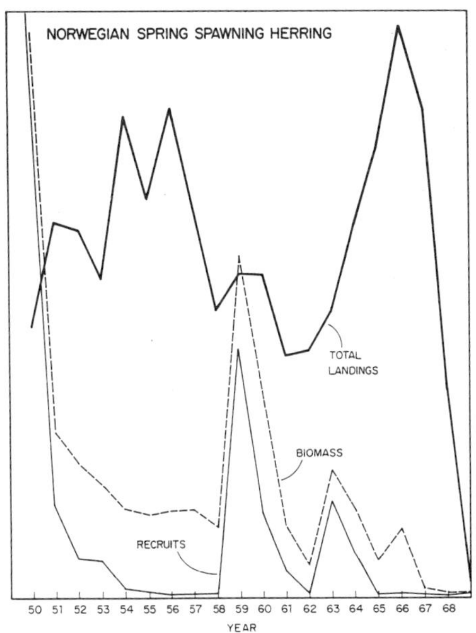

FIG. 15: Time trends in catch, population size, and recruitment for mackerel, Arcto-Norwegian.

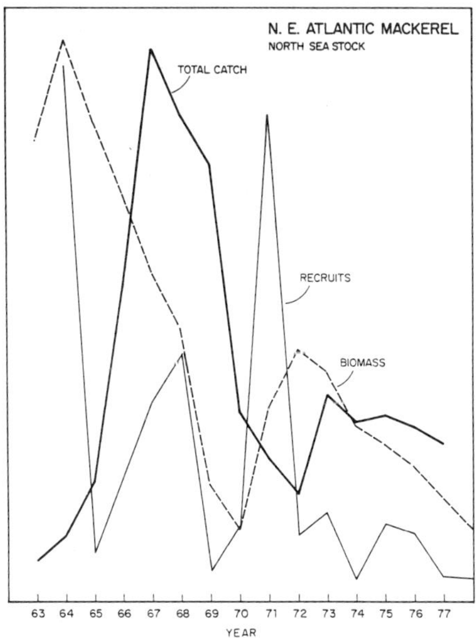

FIG. 16: Time trends in catch, population size, and recruitment for mackerel, North Sea.

3. MULTISPECIES MODELS

Very few marine fish, if any, live in complete isolation. Most, if not all, have a significant degree of interdependence based either on direct competition or predation. There is, of course, a wide range of degree of interdependence, and the dependence may be several steps removed through intermediaries. Despite the axiomatic nature of this aspect, there have been few cases where this fact has entered into management consideration.

It is no accident that the most prolific ecosystems are characterized by a great number of species. This must provide the most efficient utilization of energy. In spite of the preference for certain species in the marketplace and selective fishing gear, the major fisheries are based on a mixed-species catch.

It is very difficult to account for the total mortality exerted on ecosystem resources because most studies deal with the primary, species-directed fisheries. The incidental species taken in these fisheries can be quite significant. It has been estimated, for example, that in the Northwest Atlantic 38% of the total mortality was generated as by-catch, and generally ignored in assessing the effects of fishing (ICNAF, 1973).

There are two primary problems in multispecies management. One is the balance of species biomasses, and the rather severe perturbation of this that directed fisheries can produce. The other is the effect of species interrelations on productivity and potential yields to man.

The desires of man are seldom compatible with the constraints of natural production. The allocations of energy expenditure within the ecosystem and within the fishery as desired and directed by man have always been at odds (Figure 17).

The point is often made that management must be based on models of fisheries rather than models of the ecosystem. However, failure to achieve objectives is almost guaranteed if they are not set within the constraints of natural productivity. The initial development of fisheries has been to seek out a few of the species. This produces changes in the system productivity. At the same time natural changes are occurring; these are not predictable and cause great confusion since it is usually only the fishery effects that are monitored. Over the same time span, new fisheries interests are developed or forced by declines in the originally desired species or encouraged by the occurrence of a good year class of some species which creates a high density and profitable opportunity. These events are characterized in Figure 17, and are

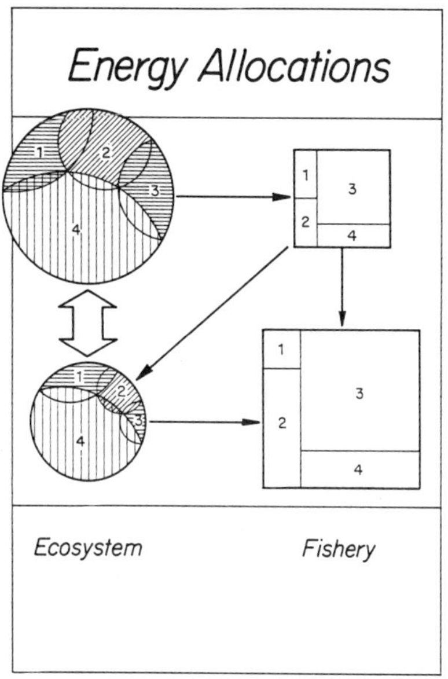

FIG. 17: Energy allocations in population and fisheries.

clearly traced in the North Atlantic fisheries (Figures 18 and 19). The Gulf of Thailand illustrates a different trend in that all species seem to be decreasing at the same rate (Figure 20). The fishery there may be less selective, but catch statistics are not available to ascertain the selectivity of gear or discarding in this fishery. The actual catches (as opposed to landings) would show more variety than these illustrations because many minor species are not shown or are not landed.[2]

The effect of the synergism in the system is a rather important aspect most often neglected. The problem may be stated in the simple question of whether the estimated overall productivity of fish biomass is greater than, equal to, or less than the sum of the estimated individual population's productivity. The question arises, of course, because of the use of traditional single-species models applied to directed-species fisheries to estimated productivity.

[2]Note. Data for the North Sea were obtained from *Bulletin Statistique, Cons. int. Explor. Mer*, and various working group reports published in ICES Cooperative Research Report Series. Data for the Gulf of Thailand were obtained by personal correspondence from John Pope, Min. Ag., Fish, and Food, Fish. Lab., Lowestoft, UK.

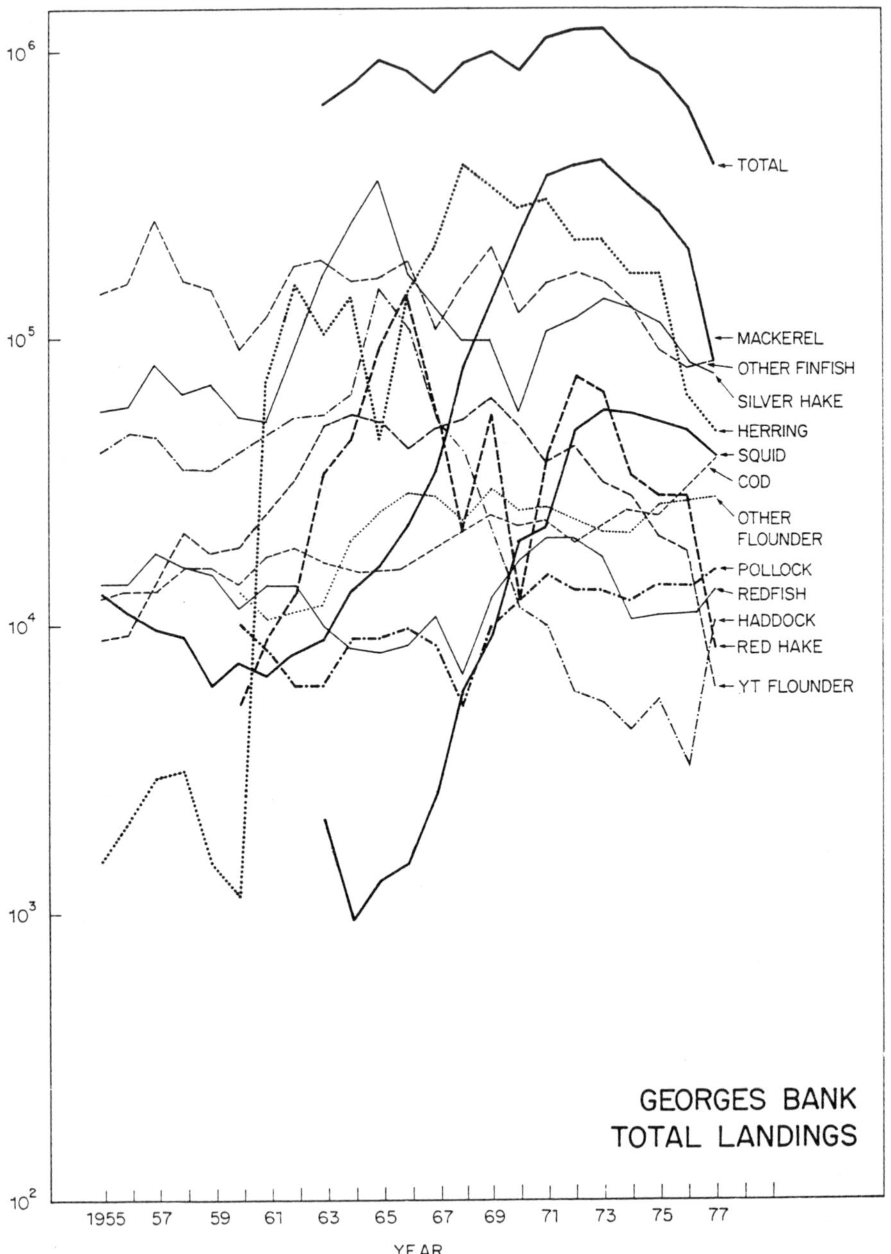

FIG. 18: Trends in landings of major species caught on Georges Bank. Brown et al., **1976.**

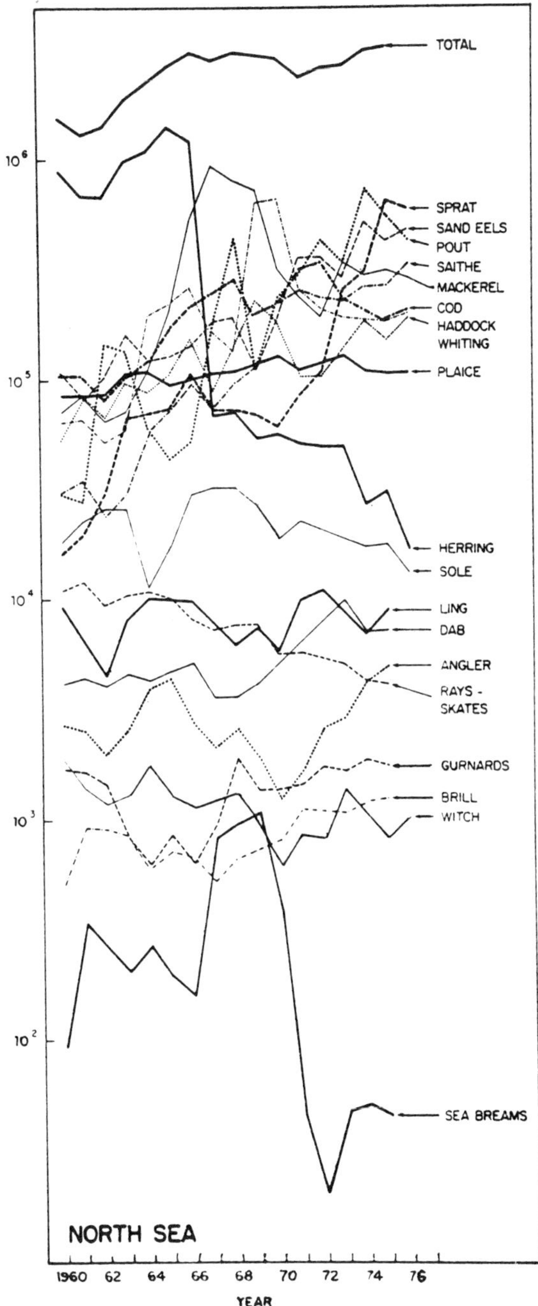

FIG. 19: Trends in landings of major species caught in North Sea.

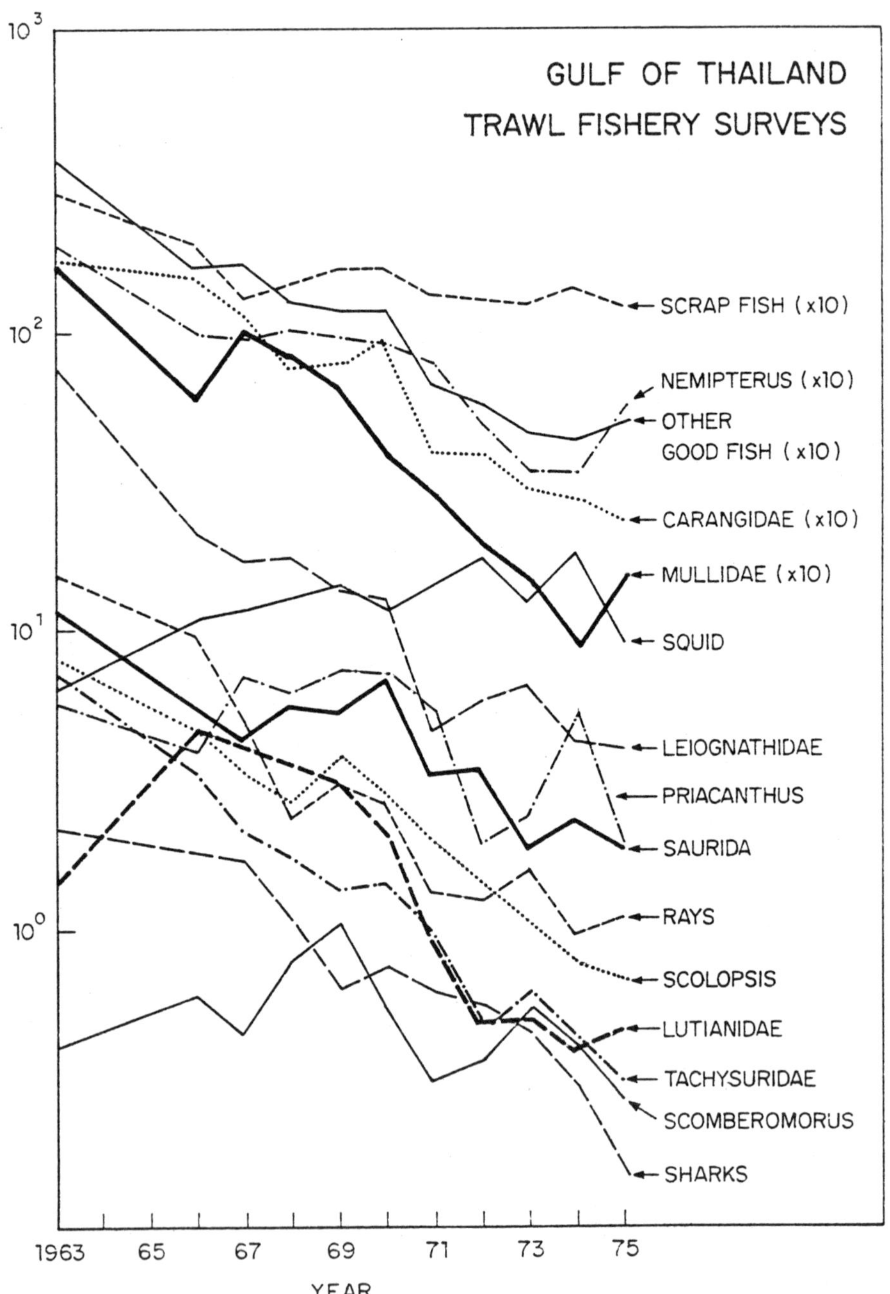

FIG. 20: Population trends, Gulf of Thailand.

FINFISH+SQUID

SUBAREA 5 + 6 · 1976 TAC's

SPECIES (FIRST TIER)	TAC
COD	43
HADDOCK	6
SILVER HAKE	103
RED HAKE	42
HERRING	(69)
MACKEREL	254
SQUID	74
POLLOCK	17
YELLOWTAIL FLOUNDER	20
REDFISH	17
OTHER FLOUNDER	20
OTHER FISH	150
SPECIES TOTAL	816
ALL SPECIES (SECOND TIER)	650

FIG. 21: Two tier system of total allowable catches implemented by ICNAF (thousands of metric tons).

In practical terms, a unit of effort is directed at a given species (or a combination of species much less than the total available), and the fishing mortality is assessed only against this species. By-catch mortality is neglected. In this case, the total fishing mortality the system can tolerate is less than the sum of the directed fisheries mortalities over all the species in the system (Brown *et al.*, 1973; Hackney and Minns, 1974). The studies of scientists for ICNAF (1973) demonstrated that the answer to the question above was 'less than,' and ICNAF in 1974 set an all-species total allowable catch (TAC) which was less than the sum of the directed-fisheries TAC's (Figure 21).

Many multispecies models are constructed in the same practical manner. There are individual species components, and the interactions are variously added within limitations of mathematical convenience.

The Lotka-Voltair-based models do indicate that, while there is greater efficiency in the species competing state, i.e., the total productivity is greater in a mixed-species situation, the total is still less than one would obtain by adding up the yields estimated by applying the single-species analogues (Hackney and Minns, 1974). Silliman (1975) has confirmed this experimentally. Brown *et al.* (1976) have demonstrated this empirically for the Northwest Atlantic Fisheries (Figure 22) by applying the logistic yield model to the total finfish catch. Lett and Kohler (1976) have modified interspecific effects in the Gulf of St. Lawrence. They concluded that there were trade-offs through predation, at the larval stage primarily, between production of herring and mackeral.

The more complex models have some surprises with respect to this general behavior. De Angelis *et al.* (1975) have shown, for example, that under some conditions an increased mortality rate on a consumer species may lead to an increased biomass. But it is at least intuitively acceptable to hypothesize that total yield in a multispecies ecosystem will be less than that obtained by summing up individual-species estimations.

The more sophisticated multispecies models (Smith, 1975; Goh, 1975) are not meant to be operational models. These models are valuable in learning about behavior of complex systems with respect to stability, persistence, and vulnerability to various perturbations.

Man is almost always considered an external perturbating force. His effect on the system, through fishing or habitat modification, may easily change the state of the system even to the extreme of eliminating certain components. Models like those of Smith and Goh tell us that the system will or will not stabilize or return to initial state after a finite perturbation or with a continuing perturbation. There is, however, no feedback within the models to this external force exerted by man so that modification of it before deleterious alterations occur is not directly simulated or determinable from analysis.

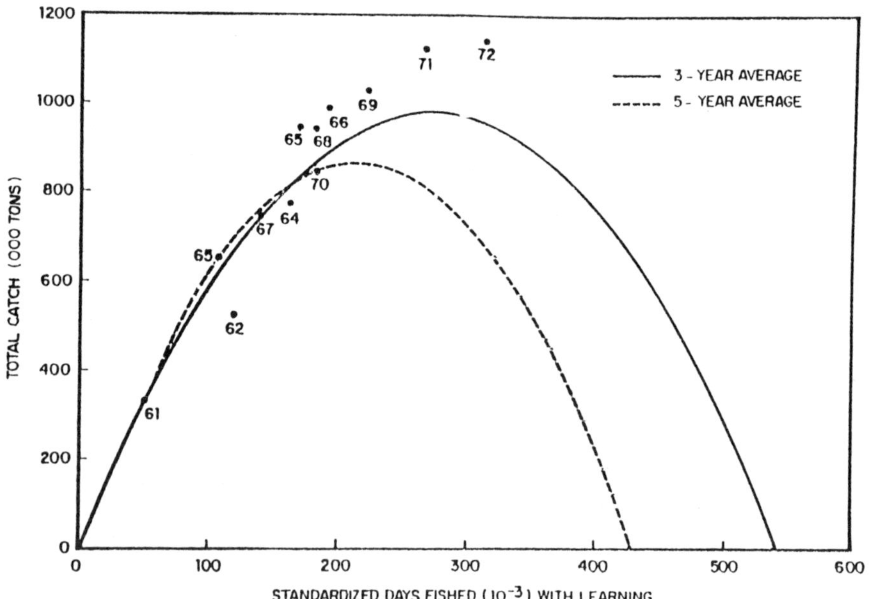

FIG. 22: The effect of fishing on the marine finfish biomass off New England. Brown et al., 1976.

4. BIOECONOMIC CONSIDERATIONS

There are direct feedbacks that affect man, e.g., a lack of fish to eat, but the same technology that produced the effect on the resource can be used to make it inconsequential, except in the very long run or in local situations. In general, then, we are dependent on observing the effect of man's activities and providing scientific proof that the observed changes are indeed his fault. Even given this, the remedial acts often demand corrective procedures, usually short of cessation, which demand even more observation and proof of cause-effect.

It goes quite beyond even this, however, in that the marine ecosystem and its resources are considered to exist only for changing man's condition; that is, they have no intrinsic merit. It is thus primarily a social (economic) mechanism that provides the feedback link, not survival of man within the ecosystem. This brings in bioeconomic models which typically include man as an exogenous variable. The objectives, e.g., maximizing profit, and their attainment through fishing do not have useful built-in feedbacks.

Devanney *et al.* (1977) has shown that the fishermen's surplus dollar yield has two maxima, one that occurs at high population density and one at low population density (Figure 23). Under the conditions of this model, it is obvious that there is no strong economic feedback to restrict effort and maintain high population levels.

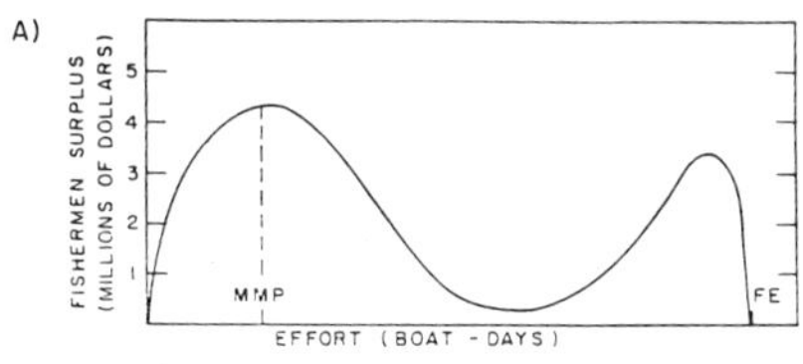

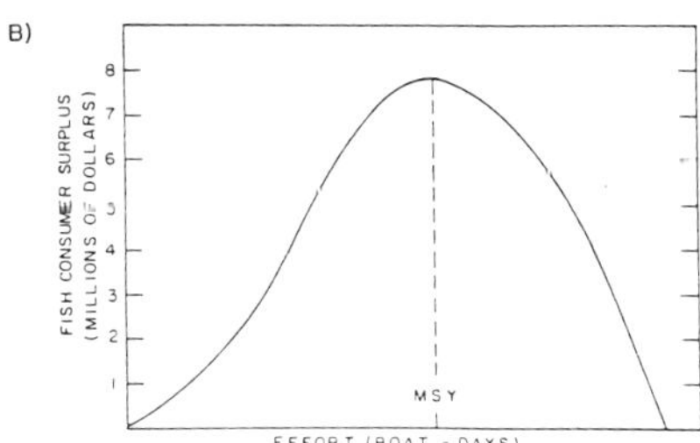

FIG. 23: Results of a traditional bioeconomic-yield model:
A. Fisherman suprlus vs. fishing effort. (Figure 2.2.4 of Devanney et al., 1977.*)*
B. Consumer surplus vs. fishing effort (Figure 2.3.1 of Devanney et al., 1977.*)*

Clark (1974) and Gulland (1975) have pointed out that changes in 'catchability' may often mask even severe changes (decreases) in population magnitude, so that neither monitoring of catch-and-effort statistics nor fisheries perceptions provide useful feedback.

It seems desirable and necessary to construct bioeconomic models based on multispecies effects and including both man and fish as endogenous variables. This will require some new and innovative study, but perhaps as a first approach some of the current models might be suitably modified. For example, the input-output analysis (see for example Miernyk, 1965) because of its matrix structure may be modified to include the ecosystem production as one of its sectors. It is limited because production is described by a set of linear equations, but this does permit more empirical use than more real, but abstract, nonlinear equations. Meanwhile, the management plans and development of regulation measures should and can include some general qualitative considerations.

5. MANAGEMENT IMPLICATIONS

A perusal of the exploratory multispecies models provide some clues as to how the necessary feedback might be included. At the very least, regulatory measures should be robust with respect to errors in assumptions about effects.

1. Permit fishing mortality only as a passive force. This could be achieved by the use of passive gear such as set lines, traps, set nets, etc. The prime criterion is that a constant exploitation rate be generated so that as density (biomass) decreases, the catch also decreases, and inversely. This same effect can be achieved for active gear by regulating fishing effort.

2. Minimize perturbation by removing species in proportion to their biomass. This requires either the use of nonselective gear, or the control of gear by selective factors. A less efficient but similar result could be obtained by prohibition of discarding. This would tend to constrain severe selective removals (fish caught in the net have low survival, thus discards are a loss to the system) by tending to prevent filling up boats with a single species. This is related to exerting a 'white noise' perturbation which, for some models of the ecosystem, creates less extreme perturbation. The alternative of specific removals is a good strategy, in terms of minimizing effects, only if the fishing mortality is placed on species which have a small interaction with other species. If this is known, it could be used to minimize

vulnerability of the system to the fishery by permitting directed fisheries only on such species. Fishing mortality should not, however, be overly directed to species with long time-reactions, i.e., late maturation, slow growth.

3. Direct fishing mortality more on ecosystems with higher spatial heterogeneity. This may be a matter of geographical, or even seasonal, definition of management units.

4. Regulate fishing mortality to the smallest possible unit (area or subpopulations).

5. Reduce the number of differently regulated fishing units in order to deal with a single homogeneous disturbance. In fishing, multiple regulated fisheries may be perverse enough to combine and drive the system more rapidly to an undesirable state. In combination with Measure 2 above, a regulation of the total of all species catch would tend to allow this.

REFERENCES

Almeida, F. P. and Anderson, E. D. (1978a). Status of the Southern New England - Middle Atlantic silver hake stock--1978. Woods Hole Laboratory Reference No. 78-55. Northeast Fisheries Center, Woods Hole, Massachusetts. (mimeo.)

Almeida, F. P. and Anderson, E. D. (1978b). Status of the Gulf of Maine silver hake stock--1978. Woods Hole Laboratory Reference No. 78-56. Northeast Fisheries Center, Woods Hole, Massachusetts. (mimeo.)

Anderson, E. D. and Paciorkowski, A. L. (in press). A review of the Northwest Atlantic mackerel fishery. Proc. Symp. on the Biological Basis of Pelagic Fish Stock Managenment. *Rapports et Procès-verbaux des Reunions. Conseil Permanent International Pour L'Exploration de la Mer.*

Beverton, R. H. and Holt, S. (1975). *On the Dynamics of Exploited Fish Populations.* Ministry of Agriculture, Fisheries, and Food, London. Fisheries Investigations, Series 2, Vol. 19.

Brown, B. E., Brennan, J. A., Grosslein, M. D., Heyerdahl, E. G. and Hennemuth, R. C. (1976). The effect of fishing on the marine finfish biomass in the Northwest Atlantic. *International Commission for the Northwest Atlantic Fisheries, Research Bulletin,* 12, 49-68. Dartmouth, Nova Scotia.

Brown, B. E., Brennan, J. A., Heyerdahl, E. G., and Hennemuth, R. C. (1973). Effects of by-catch on management of mixed-species fisheries in Subarea 5 and Statistical Area 6. International Commission for the Northwest Atlantic Fisheries, *Redbook*, 1973, Part III, 217-231.

Clark, C. W. (1974). Possible effects of schooling on the dynamics of exploited fish populations. *Journal Conseil Permanent International Pour L'Exploration de la Mer*, 36, 7-14.

Clark, S. H. and Overholtz, W. J. (1979). Review and assessment of the Georges Bank and Gulf of Main haddock fishery. Woods Hole Laboratory Reference No. 79-05. Northeast Fisheries Center, Woods Hole, Massachusetts. (mimeo.).

De Angeles, D. L., Goldstein, R. A., and O'Neill, R. V. (1975). A model for trophic interaction. *Ecology*, 56, 881-892.

Devanney, J. W., Simpson, H., and Geisler, Y. (1977). The MIT simple species simulator: application to the Georges Bank yellowtail. *Report No. MITSG 77-21*, Massachusetts Institute of Technology, Cambridge, Massachusetts.

Dragesund, O., Hamre, J., and Ultang, O. (in press). Biology and population dynamics of the Norwegian spring spawning herring. Proc. Symp. on the Biological Basis of Pelagic Fish Stock Management. *Rapports et Procès-verbaux des Reunions. Conseil Permanent International Pour L'Exploration de la Mer.*

Edwards, R. L. and Hennemuth, R. C. (1975). Maximum yield: assessment and attainment. *Oceanus*, Winter 1975, 3-9.

Goh, B. S. (1975). Stability, vulnerability, and persistence of complex ecosystems. *Ecological Modelling*, 1, 105-116.

Gulland, J. A. (1975). The stability of fish stocks. *Journal Conseil Permanent International Pour L'Exploration de la Mer*, 37(3), 199-204.

Hackney, P. A. and Minns, C. K. (1974). A computer model of biomass dynamics and food competition with implications for its use in fishery management. *Transactions of the American Fisheries Society*, 103(2), 215-225.

Hamre, J. (1978). The effect of recent changes in the North Sea mackerel fishery on stock and yield. Proc. Symp. on the Biological Basis of Pelagic Fish Stock Management. *Rapports et Procès-verbaux des Reunions. Conseil Permanent International Pour L'Exploration de la Mer.*

Hamre, J. (in press). Biology, exploitation, and management of the Northeast Atlantic mackerel. Proc. Symp. on the Biological Basis of Pelagic Fish Stock Management. *Rapports et Procès-verbaux des Reunions. Conseil Permanent International Pour L'Exploration de la Mer.*

Hennemuth, R. C. (1977). Some biological aspects of Optimum Yield. In *Marine Recreational Fisheries* 2, Henry Clepper, ed. Proceedings of the 2nd Annual Marine Recreational Fisheries Symposium, Sport Fishing Institute, Washington, DC. 17-27.

Holt, S. J. and Talbot, L. M. (1978). New principles for the conservation of wild living resources. *Wildlife Monograph* 59, The Wildlife Society, Inc., Washington, DC.

Int. Comm. Northw. Atlant. Fish. (1973). Proc. of the Statistics and Research Committee. International Commission for the Northwest Atlantic Fisheries, *Redbook*, 1973, Dartmouth, Nova Scotia.

Larken, P. A., Raleigh, R. F., and Wiliamonsky, N. J. (1964). Some alternative premises for constructing theoretical reproduction curves. *Journal of the Fisheries Research Board of Canada*, 21(3), 477-484.

Lett, P. F. and Kohler, H. C. (1976). Recruitment: a problem of multispecies interaction and environmental perturbation, with special reference to Gulf of St. Lawrence Herring. *Journal of the Fisheries Research Board of Canada*, 33, 1353-1371.

May, Robert M. (1975). Stability in ecosystems, some comments. In *Unifying Concepts in Ecology*, W. H. van Doblein and R. H. Lowe-McConnel, eds. Report of Plenary Session, 1st International Congress on Ecology. W. Junk, Publishers, The Hague. 161-168.

Miernyk, W. H. (1965). *The Elements of Input-Output Analysis*. Random House, New York.

Ricker, W. E. (1954). Stock and recruitment. *Journal of the Fisheries Research Board of Canada*, 11(5), 559-623.

Schumacher, A. (in press). Review of North Atlantic catch statistics. Proc. Symp. on the Biological Basis of Pelagic Fish Stock Management. *Rapports et Procès-verbaux des Reunions. Conseil Permanent International Pour L'Exploration de la Mer.*

Silliman, R. (1975). Experimental exploitation of competing fish populations. *Fishery Bulletin*, Vol. 73, No. 4, 875-888. National Marine Fisheries Service, Washington, DC.

Slobodkin, L. B. (1962). *Growth and Regulation of Animal Populations*. Holt, Rinehart, and Winston, New York.

Smith, D. F. (1975). Quantitative analysis of the functional relationships existing between ecosystem components. III. Analysis of ecosystem stability. *Oecologia*, 21, 17-29.

Stewart, I. (1975). The seven elementary catastrophies. *New Scientist*, 20 Nov. 1975, 447-454.

[*Received July* 1978. *Revised May* 1979]

G. P. Patil and M. Rosenzweig, (eds.),
Contemporary Quantitative Ecology and Related Ecometrics, pp. 533-544.

ON ASSESSING POPULATION CHARACTERISTICS OF MIGRATORY MARINE ANIMALS

VINCENT F. GALLUCCI

College of Fisheries
University of Washington
Seattle, Washington 98195 USA

SUMMARY. Selected models of population dynamics are critically reviewed with a focus on problems for the application to migratory animals. Selected models of population assessment (mark and recapture and line transect) are reviewed with the same focus. A Markov model of the inter-mixture of fur seal stocks is constructed. A formalism and field technique to estimate school-level intermixture of porpoise is presented.

KEY WORDS. Population models, migratory, marine, Markov models, Pribilof fur seals, porpoise assessment, mark and recapture, line transect.

1. INTRODUCTION

The population characteristics of migratory marine animals are often very difficult to determine. Among the many obvious reasons is the simple elusiveness of some animals over vast areas. Equally important, however, is the sketchy state of our knowledge about many relevant supporting food webs.

For purposes of discussion, only animals that migrate over extensive distances during the stages from juvenile to adult are considered. Excluded are animals with larval stages that drift from spawning area to nursery area, and so on. Although the focus is upon marine mammal populations, similar, if not as dramatic, difficulties affect studies of many other organisms, e.g., Alaskan king crab and salmonid populations.

An unfortunate practical limitation is a focus upon animals that are, or were recently, harvested by man. Many marine mammal populations have been subject to intensive study because they have been harvested (directly or incidentally): e.g., whales, harp seals from the North Atlantic, Pribilof seals from the North Pacific and dolphins from the Eastern Tropical Pacific Ocean (ETP). We simply do not know a great deal about manatees and dugongs, which are not usually harvested.

A variety of mathematical models are used in the determination of population characteristics. In one sense, models may contribute insights into, and predictions of, the energy flow within a population and with its environment, as is the case for age structure or surplus yield models. Other mathematical formalisms are the actual tools by which the properties or parameters of a model are determined. Such formalisms (e.g., mark-recapture and line transect estimation) are known as population assessment methods and often have theoretical probabilistic characteristics. One important way models and assessment methods differ is that the model is defined with respect to the population and should be discarded if the parameters and qualitative predictions are unrealistic (Gallucci and Quinn, 1979). The formalism for assessment is defined independently of the population and has perhaps a 'ball-and-urn' (Quinn and Gallucci, 1979) origin.

Both models and assessment methods are discussed, quickly reviewing where necessary; the focus is on those aspects which contribute to the difficulties in the context of migratory populations.

2. A BRIEF DISCOURSE ON MODELS

Mathematical modeling has an increasingly major roll in the study of population dynamics--a fact some would argue is misplaced faith in a new mythology successful only when evaluated in its own terms of reference, and not those of biological reality. The accusation is by and large well aimed. There is reason to think, however, that today's models do relate better to reality, but a large dose of faith must still accompany most models. Everyone should recognize that a model is simply an abstraction of reality constructed because reality is entirely too confusing with so many details. The modeler seeks an abstraction which is far simpler than the observed dynamics of the population (but not so far as to be useless) but which is consistent in hypotheses and assumptions with reality. As a result of such efforts, the model or abstraction can lead to projections and insights that are potentially useful.

One of the controversies in the modeling of animal population dynamics concerns the use of single-species vs. multi-species models. Single-species models are the traditional approach for the prognostication of the state of the population under various harvest strategies. These models have the advantage of providing straightforward, expected yield estimates, subject to an array of assumptions not the least of which is that the population functions independently from that of other populations' activities.

The difficulty with multi-species models is not so much in the methodology as in the philosophy and the data requirements. Multi-species models abandon simplicity and try to replace it with reality (or a closer version) but we usually do not know enough about reality to formulate the correct functions and to collect the appropriate data to model. Although classical approaches to multi-species models have been disappointing, contemporary approaches based on control theory principles hold some promise.

Maximum sustainable yield, often cited as one of the major strengths of single-species models of exploited populations, is itself a concept under hot debate these days (Roedel, 1975). The usual formulation of maximum sustainable yield (MSY) is based upon a 'logistic type' model; for example, the Schaefer exploitation model (surplus yield model) in which the dynamics are expressed as a differential equation. MSY occurs at the population level corresponding to the maximum rate of increase of the population. At that point, the maximum number of animals are added to the population and hence a maximum number may be subtracted, and yet remain in equilibrium. One problem is that harvests are not strictly biological processes. Harvests are economic enterprises that exist for an economic profit and not necessarily to preserve the population. Therefore, some argue (Clark, 1976) that maximum sustainable yield should have a definition which involves economics rather than being a purely biological definition. The economic statement involves the use of discount and interest rates and the economic 'rent' for a particular resource. While few decisions are made today on purely economic or biological criteria, a rational analysis of the situation demands both an economic and a biological interpretation. My prognostication of where modeling will be ten years from now is that MSY will survive but with both economic and biological interpretations, the manager taking a composite into account. At present, management typically selects harvesting rates in such a way that a percentage of MSY is harvested, thereby providing a safety margin. (The International Whaling Commission frequently sets allowed catch at 90% MSY.) This rule of thumb is likely to be satisfactory in simple situations but more complex scenarios including those of highly coupled species such as the Antarctic baleen whale competing for krill with the crabeater seal and a commercial krill operation require more insight into what constitutes a safety margin.

It is likely that renewed interest in age-structure, single-species models (e.g., Vaughan and Saila, 1976; Horst, 1977) will be the basis for rapid progress across a broad front in modeling populations. This is implicitly noted by Anderson and Ursin (1977) when they say that their coupled Beverton-Holt type stock-recruitment models have limited success because predation and competition vary with the relative ages of predators and prey. Despite the advantages of age structure models, a problem with their use for fish should be noted. The models are generally formulated to allow for the average number of offspring per animal in each age class. Spawning marine fish carry egg masses whose density could be estimated and incorporated in some form, but it continues to be difficult to relate numbers of eggs to the probability of a juvenile occurring. Until this difficulty is resolved such models will be difficult to apply.

Note that questions concerning the use of multi-species or single species models, biological or economic MSY evaluations, or age structure vs. biomass models have all been asked in the course of studying such well-known populations as North Sea plaice. Models for highly migratory species, about which only localized data are generally available, are more difficult because parameter estimates may vary greatly from region to region. Mammalian populations often have well-developed behavioral characteristics which greatly influence biological responses to the environment of a region. Until these responses are better understood, accurate prognostication in a non-steady state will be guesswork.

3. THEORETICAL CONSIDERATIONS ABOUT ASSESSMENT METHODS

The first question usually asked in the beginning of an assessment is how many animals are in the population, followed by questions about rate parameters and so on. Regardless of the methodology of assessing population characteristics, they involve the statistical evaluation of data. At the present time, the operational and supporting theory for mark-recapture methods are well known (e.g., see Seber, 1973). More recently, transect methods have undergone intensive development in one brief burst of activity (e.g., see Eberhardt, 1978; Quinn and Gallucci, 1979). It is likely that estimation problems associated with acoustic or sonar sampling will soon receive new attention--based on a methodology that exists around the detection of submarines. Finally, catch statistics for harvested populations are often the only data available. These data are usually reported as catch per unit of effort. Certain characteristics of migratory marine animals frequently strain the theoretical frameworks which underlie the methodologies.

Mark-recapture methods are based upon the validity of the equality of the ratios: the total number in the population (N)/ the total number of animals recaptured (C) : the number marked (M) of N/the number of marks (R) in C. That is,

$$N/C = M/R \quad . \tag{1}$$

The theory is frequently developed using the ball-and-urn probability model, with marks being M discolored balls. If each of the C balls is replaced after examination, thus not changing the probability of recovering a mark, a binomial model is appropriate; otherwise, a hypergeometric model is used. In either case, the sampling properties of the ratio are well known.

Usually, $\hat{N}$ is estimated to obtain knowledge of the size of the population to within a certain confidence interval and with a certain precision (Robson and Regier, 1964). It should be noted that in the course of doing a mark-recapture experiment a great deal of important population data is often collected. If the marked animals have a variety of sizes and are sampled frequently enough over a long enough period of time, information about age structure, growth rates, mortality and birth rates, and migration paths may be expected to follow. Additional information obtained by sacrificing (or regurgitating) animals and examining stomach contents, gonads, etc. is also potentially available.

However, the formation of schools, herds, or pods, stratification by age and/or sex, the 'loss' of marks, additions and losses to the population, inaccessibility for recapture, are factors which must be considered and corrected if statistical accuracy is to be attained. Although many animals do have a general migration path and thus may be labeled as distinct 'stocks,' stocks may mingle during some or most of the migrational phase. Animals such as seals, porpoise, and salmon present such difficulties. In addition, all animals, but especially mammals, may exhibit a learning curve which would affect recaptures and violate the common statistical assumption that the experiment was done under constant conditions.

The standard approach is to compute the estimate $\hat{N}$ from (1) as best as possible and then to incorporate corrections to the estimate for tag loss, serial correlation of recaptures because of clumping, mortality and dispersal. In all, the animals seem to seek ways to contradict the fundamental property underlying an urn model: randomness.

The operational aspects of the *transect study* should involve a careful experimental design based on preliminary knowledge of the population. The sighting data are used to define a sighting model $g(y)$ where y is the distance on either side of the transect. If $g(y)$ is fixed, the experiment is a strip census. If y is measured from the time of first sighting to the animal it will often be at other than perpendicular to the transect and thus an angle of the sighting must also be noted. Assume, as is commonly done, that y is measured perpendicular to the transect. A parameter c, the effective width, is estimated from

$$c = \int_{y=0}^{\infty} g(s)ds \quad , \quad g(0) \equiv 1$$

to quantify the decreased likelihood of sighting animals further from the transect. The abundance estimate is

$$\hat{N} = \frac{n}{2}\frac{A}{L}\hat{c}^{-1} \quad ,$$

where A is the area of the region (km^2), L is the length of the total transect (km), and n is the number of sightings and a random variable. If $g(y)$ is not fixed, it is guessed from the empirical distribution of the y_i sighting values.

Many of the animal's behavioral characteristics, such as clumping or schooling, present statistical difficulties in the estimation of N. The effects of such characteristics on line transect estimates have been considered in a 1977 thesis by Mr. T. Quinn, a student in the College of Fisheries at the University of Washington. (Also see Quinn, 1977; Quinn and Gallucci, 1979; and Quinn, 1979). While the estimate of N from a transect study sometimes may be the more desirable procedure (based on criteria such as cost, accuracy, time, etc.), generally little about the population is learned besides $\hat{N}$.

When catch data are available, they are commonly utilized in the form of catch-per-unit of effort (cpue). Despite its long use in fisheries management, effort remains difficult to quantify. The reasons for the difficulty stem from the diverse natures of individual vessels in a fleet, the biological characteristics of the prey, etc. The underlying assumption is that searching for animals (and thus their capture) is random and thus catch-per-effort is directly proportional to density. However, for any animals which school, a measure of abundance based on effort introduces a statistical bias (Paloheimo and Dickie, 1964).

Some characteristics of animals which cause difficulties for quantitative studies of the population have been presented, accompanied by a rapid review of the mark-recapture and line transect theories. These are among the best and most frequently used methods.

Obviously the reliability of an estimate is highly dependent upon what is initially known about the population. School behavior, dispersal of animals to unknown locations, and the possible intermixture between different stocks are important factors to be considered in the following examples. The first example involves Pribilof Island seals and 'stock-intermixture' arising from the yearly harvest, and attempts to simultaneously maximize both the harvest and the well-being of the stocks. This limited migration is modeled by the use of a discrete state Markov chain model. The model is well adapted to the migration problem but as will be seen, a semi-Markov model would be even better adapted.

The second involves porpoise in the Eastern Tropical Pacific Ocean (ETP) and their aggregation or schooling nature and the possible intermixture between schools. The problems arose from the large numbers killed incidental to the tuna harvests in the ETP. This schooling problem is quantified by an experimental design using resighting of marked porpoise data with unorthodox estimators invented for this particular problem.

4. EXAMPLES

The Pribilof Islands consist of two islands: St. Paul and St. George, with seals on each. Harvests used to occur on both of these islands but St. George has been a sanctuary since 1973. The total population size is about 1.3 million with 300 to 400 thousand pups born each year. On St. Paul, 90% of the 1977 harvest contained 16,500 three-year olds and 9,400 four year olds (i.e., relatively few two and five-year olds). The predicted kill for 1978 is 27,500 on St. Paul and a potential subsistence harvest of 4,000 on St. George. Basically, 3 and 4 year old (subadult non-breeding male or SAM) animals are harvested, in contrast to harp seal harvests which focus on pups.

The problem of stock-intermixture between St. Paul and St. George arises. In particular, are SAMs from St. Paul, perhaps recognizing impending catastrophe, or by luck, migrating to St. George (the SAM density is increasing) or do St. George SAMs react to the crowding and unwittingly move into the St. Paul harvest? A SAM is considered to be either on one of the two islands or somewhere in the water.

The analytical approach used involves a Markov model in which State 1 (a_1) is St. George, State 2 (a_2) is St. Paul, and State 3 (a_3) is neither St. Paul nor St. George, e.g., in between, or dead or joined the Kommandorski Island stock. Thus the state space is (a_1, a_2, a_3). A corresponding probability matrix of p_{ij} $(i,j=1,2,3)$ terms called the transition matrix is thus defined to indicate the probability that a SAM will be found in any one state from one time interval to another.

$$\underset{\sim}{P} = \begin{bmatrix} p_{11} & p_{12} & p_{13} \\ p_{21} & p_{22} & p_{23} \\ p_{31} & p_{32} & p_{33} \end{bmatrix} \begin{matrix} \text{St. George} \\ \text{St. Paul} \\ \text{unknown} \end{matrix}$$

The probability estimation could be done over weekly time intervals (for biological reasons) thus possibly redefining p_{ij} each week. If $\underset{\sim}{P}$ is not made a function of time, but is an average over the field season, the standard Markov model applies. It is clear that τ units in the future

$$\underset{\sim}{n}(t+\tau) = \underset{\sim}{P}^{\tau}\underset{\sim}{n}(t)$$

where $\underset{\sim}{n}(t) = [n_1(t)\ n_2(t)\ n_3(t)]^T$ and $n_i(t)$ is the proportion of SAMs in state a_i at t. The total number in state a_i at time $t+1$ is

$$n_i(t+1) = \sum_{j=1}^{3} n_i(t) p_{ij} \quad .$$

The operational procedure would be to mark the seals so they can be recognized as individuals and to count 're-observations' on both islands (states) to estimate the proportion that 'intermix.'

The advantages of this formulation are that the entire problem is embedded in a single model encompassing all possibilities. There are obvious ways in which the model may be fallacious. The Markovian assumption of independence of position in state space at $(t+1)$ with that at time (t) may be a major weakness. This field and modeling study is now underway by Mr. Michel Griben, a graduate student in the College of Fisheries at the University of Washington.

The porpoise of the Eastern Tropical Pacific Ocean (ETP) are frequently associated with yellowfin tuna. The commercial fishermen use the presence of porpoise as one of several cues to the likelihood of tuna being below the surface. Therefore, they set their nets around the porpoise, unfortunately inducing high levels of mortality on the porpoise. However, there has been a negative trend in the yearly mortality of porpoise: in 1960 about 853,000 porpoise were killed and in 1976, about 106,000 were killed. Part of this decline may be due to the U.S. Marine Mammal Protection Act of 1972. A measure of the success of efforts to lower incidental mortality is that from a 51,995 permissible kill for 1978, 4,029 porpoises were killed from 1 January-18 June, a period which encompasses the bulk of the fishing effort. Based on the need to monitor the populations to obtain mortality rates from fishing and the need to collect the necessary data for optimal sustainable population computations, it might appear that a detailed knowledge of the populations is available. However, little population-level data are actually known. The range of the several stocks of dolphins covers about 90° to 150°W longitude and 0° to 20°N latitude (roughly from Central America three-fourths of the way to Hawaii and from Peru to Acapulco). There are several species and possibly many stocks (*Stenella* spp. mainly) moving through these areas in patterns about which only vague information is available.

An earlier section discussed the problems associated with estimation when animals school or migrate or when only a fraction are available to be seen above the surface of the water at any given time. A gedunken experiment (by V. Gallucci and T. Quinn) which had as its objective the development of estimators *to correct* the Petersen estimate for school structure (since mark-recovery estimates assume random dispersal of individuals with marks or of recapture effort) is now discussed. Two aspects are discussed here: cohesion or school structure when schools are being actively-fished and cohesion when schools are undisturbed. It is thought likely that in both cases, groups of individuals frequently 'change schools.'

Essentially, the experiment is to track a school which has at least one radio-tagged and many marked animals. Successive net sets are made on the school, noting each time the recaptures and the fraction of the school recaptured (thus, the size of the school before attempted encirclement). This fraction is followed over time and the identification numbers of those caught in each net set are recorded. Ultimately, an estimate of the proportion of marked members from the original school remaining at the time of the *tth* set is computed.

Of the several parameters and their estimators that were defined, the two of central interest here are:

1. The proportion of the school captured, e_t, is

$$e_t = \frac{I_t}{N_t} = \frac{\text{the true number in the net}}{\text{the true number before attempted encirclement}} ,$$

where $\hat{e}_t$ would be the ratio of an actual count of the number in the net to the observer's estimate of the number in the school before the encirclement operation.

2. The estimate of the proportion of the marked members from the original school first encircled, remaining at time t

$$\hat{p}_t = \frac{\hat{M}'_t}{M_t} = \frac{\text{computed estimate of number of marks in school at t}}{\text{number of marks applied}} ,$$

where M_t is known and need not be estimated.

If $\hat{p}_t$ is not 'small' and if the set of $\hat{p}_t$ has consistent properties, it is possible to correct an estimate $\hat{N}$ computed from a major mark-recapture open population study in which neither randomization of marks nor randomization of recapture effort could be assumed. In particular, due to schooling, the probability of finding the $(i+1)th$ mark is greater once the ith mark is found. The correction could be found under a variety of values of the relevant 'school parameters' by computing a series of estimates $\hat{N}^*_i$ from a computer simulated mark-recovery experiment when school structure is not present. The correction follows from the ratio of $\hat{N}^*_i$ and $\hat{N}_i$ for $i=1,2,\cdots,T$, and properties of the ratio.

To obtain estimates of cohesiveness under fishing and non-fishing conditions the time interval is adjusted between successive sets of the net. For example, to simulate fishing, one set is made per day and to simulate the non-fishing conditions one set every five days is made. Radio-tagged animals (Evans, 1974) allow the boat to track at a far distance to minimize disturbance.

It is well to recognize that the execution of such an experiment is just barely possible under today's technology. It would all occur in the open ocean, involve a seine that could encircle the 700 animals in an average-sized porpoise school, permit animals to be shunted into holding pens (without drowning) and allow men to work inside the net, marking, counting, etc.

A variety of marks are possible to apply. A tag must be rapidly applied and highly visible but tag longevity can be sacrificed. Such an experiment should lead to a great deal of additional data on school structure (e.g., relative ages, sex, sizes, etc.). In the event a correction such as the above is not applied to the estimate, $\hat{N}$, there is certain to be a bias due to the aggregated occurrence of marks.

The above examples represent only two of many possible situations where migration, schooling and intermixture over time will cause the analyst to adopt alternatives to the standard population models and to do specific experiments to develop corrections to the standard population assessment methods.

REFERENCES

Anderson, K. and Ursin, E. (1977). A multispecies extension to the Beverton and Holt theory of fishing with accounts of phosphorous circulation and primary production. *Meddelelser Fra Danmarks Fiskeriog Havundersogelser*, N.S. 7, 319-435.

Clark, C. (1976). *Mathematical Bioeconomics*. Wiley-Interscience, New York.

Eberhardt, L. L. (1978). Transect methods for population studies. *Journal of Wildlife Management*, 42(1), 1-31.

Evans, W. M. (1974). Radio-telmetric studies of two species of small odontocete cetaceans. In *The Whale Problem*, W. E. Scheville, ed. Harvard University Press, Cambridge. 385-394.

Gallucci, V. F. and Quinn, T. II. (1979). Reparameterizing, fitting and testing a simple growth model. *Transactions of the American Fisheries Society*, 108(1), 14-25.

Horst, T. (1977). Use of the Leslie matrix for assessing environmental impact with an example for a fish population. *Transactions of the American Fisheries Society*, 16, 253-257.

Paloheimo, J. E. and Dickie, L. M. (1964). Abundance and fishing success. *International Council for the Exploration of the Sea*, 155, 152-163.

Quinn, T. J., II. (1977). The effects of aggregation and the choice of an observer sighting response function upon line transect estimates of abundance. Abstract. Second Conference on Biology of Marine Mammals.

Quinn, T. J., II. (1979). The effects of school structure on line transect estimators of abundance. In *Contemporary Quantitative Ecology and Related Ecometrics*, G. P. Patil and M. L. Rosenzweig, eds. Satellite Program in Statistical Ecology, International Co-operative Publishing House, Fairland, Maryland.

Quinn, T. J., II and Gallucci, V. F. (1979). Parametric models for line transect estimators of abundance. *Ecology* (in press).

Robson, D. and Regier, H. A. (1964). Sample size in Peterson mark-recapture experiments. *Transactions of the American Fisheries Society*, 93(3), 215-226.

Roedel, P., ed. (1975). *Optimum Sustainable Yield as a Concept in Fisheries Management*. American Fisheries Society, Washington, D. C.

Seber, G. A. F. (1973). *The Estimation of Animal Abundance*. Hafner Press, New York.

Vaughan, D. and Saila, S. (1976). A method for determining mortality rates using the Leslie matrix. *Transactions of the American Fisheries Society*, 105, 130-183.

[*Received September* 1978. *Revised June* 1979]

G. P. Patil and M. Rosenzweig, (eds.),
Contemporary Quantitative Ecology and Related Ecometrics, pp. 545-567.
International Co-operative Publishing House, Fairland, Maryland.

MESA CONTRIBUTIONS TO SAMPLING IN MARINE ENVIRONMENTS

PAUL A. EISEN

JOEL S. O'CONNOR

NOAA Environmental Research Laboratories
MESA New York Bight Project
Old Biology Building, SUNY
Stony Brook, New York 11794 USA

SUMMARY. Four MESA-sponsored studies are described that use statistical methods to address improved sampling in the New York Bight. The first study finds that metals in sediments can be sampled more efficiently if their observed log normal spatial distribution is considered. The proposed strategy concentrates sampling in regions of highest, nonuniform contamination. The second contribution outlines a method to monitor the influence of waste disposal on benthic faunal assemblages. If sediment strata are first identified by physical/chemical factors (associated with waste inputs), statistically significant differences in benthic species abundances among strata then can be used to monitor waste impacts. The third study compares standard and alternate methods for estimating tidal datum planes at locations where only a short series of observations is available. A t-test is used to determine if the estimated datum calculated by each method is acceptable. Although the standard method of calculating datums is generally acceptable, the alternative method can be used to advantage on the West Coast of the United States. Finally, weather data from the East Coast have been subjected to cross-correlation analyses to study the feasibility of interpolating and hind-casting meteorological forcing in data-sparse marine areas. Sufficient correlation is shown to exist in space and time to proceed with experiments in hindcasting.

KEY WORDS. marine sediment contamination, benthic organisms, tidal datum planes, atmospheric forcing, New York Bight, ecosystem sampling.

1. INTRODUCTION

Since 1973, the Marine EcoSystems Analysis (MESA) New York Bight Project has been examining regional problems arising from people's uses of marine and estuarine resources. The New York Bight was selected as a prototype for several studies conducted by the U. S. National Oceanic and Atmospheric Administration because of the significance and urgency of the region's environmental problems. The New York Bight extends seaward from Long Island and New Jersey to the edge of the continental shelf and encompasses more than 39,000 km^2 (Figure 1). It is adjacent to one of the most populated and industrialized regions of the world, which supports almost 10 percent of the United States population. It also is the repository for wastes from more than twenty million people and many major industries, and is the recipient of the largest ocean dumping operation in the United States.

Domestic and industrial wastes are released into the rivers and harbors adjoining the Bight and to the Bight itself by harbor flushing, direct ocean dumping, and atmospheric fallout. These quantities, which have been summarized by Mueller *et al.* (1976), have average daily loadings of 264 tons (metric) of seven measured heavy metals, 520 tons of nitrogen, 2,600 tons of organic carbon, and 24,000 tons of suspended solids. In recent years, ocean dumping operations have contributed more than 35,000 tons per day of chemical wastes, sewage sludge, dredged materials, and construction debris.

Although contaminant loadings are relatively easy to estimate, reliable measures of contaminant impacts and their geographic extents are more difficult to obtain, and some MESA studies have been designed specifically to improve sampling strategies. These studies have helped define existing contaminant impacts and indicate the types of monitoring activities that will detect significant changes in the Bight environment and ecosystems.

In this paper we discuss the results of several MESA-sponsored studies of field sampling questions and assess their utility in sampling coastal environments and ecosystems. A technique is described and used to sample marine sediment. This sampling scheme conforms to the expected, small-scale heterogeneity of contaminant concentrations in sediments. Another study presents a novel and useful method for measuring small-scale variability in benthic invertebrate community structure. A sequential sampling scheme is outlined for use after sampling strata have been defined on physical/chemical bases. The sampling plan is designed to detect shifts in species abundance over time.

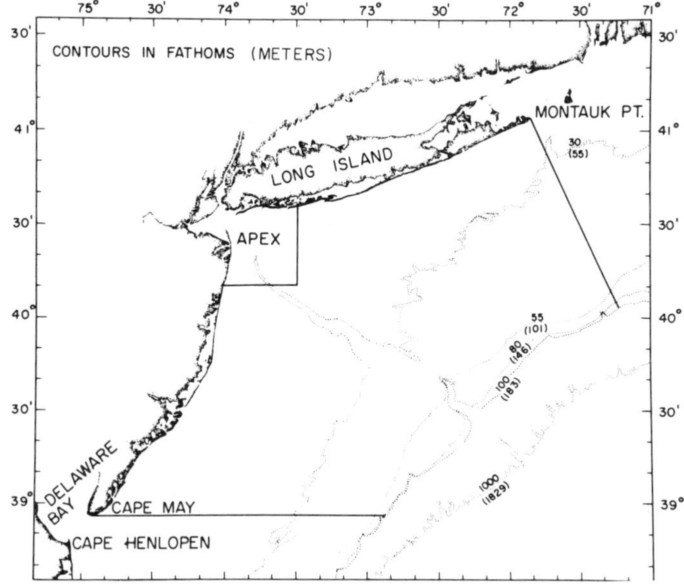

FIG. 1: Map of area included in the New York Bight.

Two methods are presented for determining tidal datums from shorter than usual series of observations. A statistical analysis provides a basis for comparing the two methods.

Meteorological statistical studies have helped show that important weather features such as surface pressure and wind stress are highly correlated in space and time. Thus when a sufficiently dense network of coastal sampling stations is present, hindcast experiments on weather features may succeed in specifying atmospheric forcing over data-sparse marine areas.

2. PROPOSED SAMPLING DESIGN FOR METALS IN SEDIMENTS

MESA has used a grid or systematic sampling plan (Pearce *et al.*, 1977) to sample sediments, benthic organisms, and trace metals simultaneously in the Bight Apex. Data from this plan and from samples outside the Apex are an adequate basis from which to draw preliminary correlations between sediment contaminant levels and benthic community structure. Below, we describe how these data can also be used in developing more efficient sampling designs.

Saila *et al.* (1978) have suggested a sampling design for the spatial distribution of contaminants that have been dispersed from a central point. This is the type of spatial distribution of sediment contaminants in the Bight. Contaminants initially are dumped or settle out of contaminant-laden estuarine waters to sediments of the relatively deep Christiaensen Basin near the center of the Bight Apex. From there they are resuspended and dispersed to less contaminated regions. Saila *et al.* (1978) demonstrate that the decreasing concentration of several heavy metals with distance from the source is approximately log normal (e.g., see plot of copper concentrations in Figure 2). Similar plots were obtained for other metals (chromium, nickel, lead, and zinc) and along other transects. Although contaminant concentrations do not decrease as smoothly along all radii from the source (Figure 3), the log normal distribution model seems to be the best simple generalization.

The proposed sampling design attempts to distribute efficiently the sampling effort to contour the contaminant concentrations in space. The data in Figure 2 are from samples evenly spaced at approximately 5-km intervals. In the model to distribute sampling effort evenly over the range of contaminant concentrations, we fit a least squares regression line to values of log contaminant concentration plotted against distance from the contaminant source. We then substitute the desired number of equally spaced concentration values into the regression equation to get appropriate sampling locations.

This proposed strategy concentrates sampling effort in the regions of highest, nonuniform contaminant levels (Figure 4). The proposed strategy is more efficient for contouring because the distance between adjacent measurements conforms to the actual distribution of contaminant concentrations. The new sampling points (open circles of Figure 4) more precisely define the geographical extent of high copper concentration; however, sampling also shows where the copper concentrations are relatively low and slowly declining to background levels.

This sampling strategy may require adaptation to accommodate uneven dispersal or multiple contaminant sources. For instance, in the New York Bight this sampling design cannot be applied strictly because three major sources of contaminants (sewage sludge, dredged material, and wastewater) have left a complex configuration of contaminant isopleths. Figure 3 illustrates this by showing average concentrations of lead in sediments of the Bight Apex. The data suggest two point sources (values of 140 and 137 ppm near the dredge spoil and sewage sludge dump sites, respectively) and a tendency for the metal to accumulate in topographic lows (e.g., the Hudson Shelf Valley). Thus several sampling transects

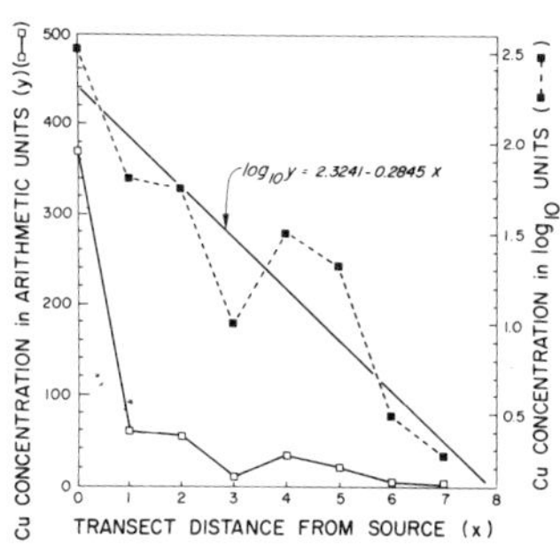

FIG. 2: Plot of copper concentrations (ppm, dry weight) along a transect in the New York Bight and the transformed data fitted by linear regression. The start of the transect is the center of the sewage sludge dump area.

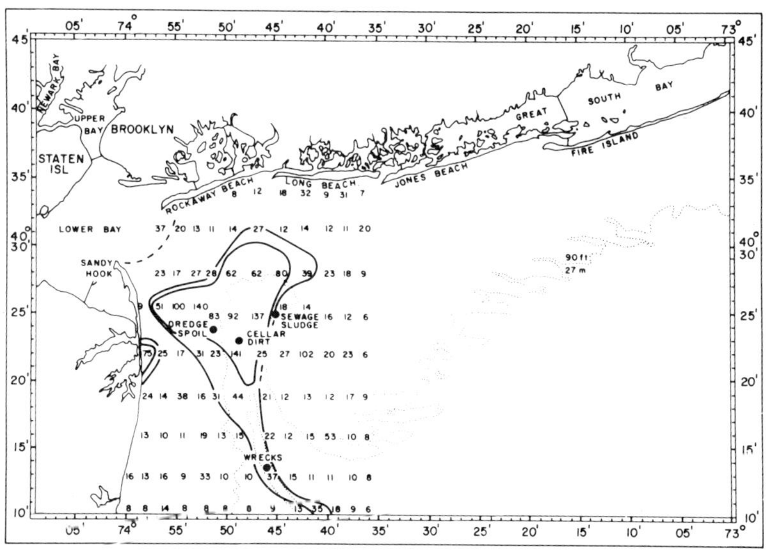

FIG. 3: Average concentrations of lead in sediments (as ppm of dry sediment) of the New York Bight Apex. Values are average concentrations measured over four seasons (August and October 1973 *and January-February and March-April* 1974*). The outer contour is* 25 *ppm (dashed in areas of uncertainty); the inner contour is* 50 *ppm. The numbers are values of lead concentration in sediment samples taken at those points. Outside the contours, the greatest concentrations observed are in the high* 30's *(with two isolated exceptions).*

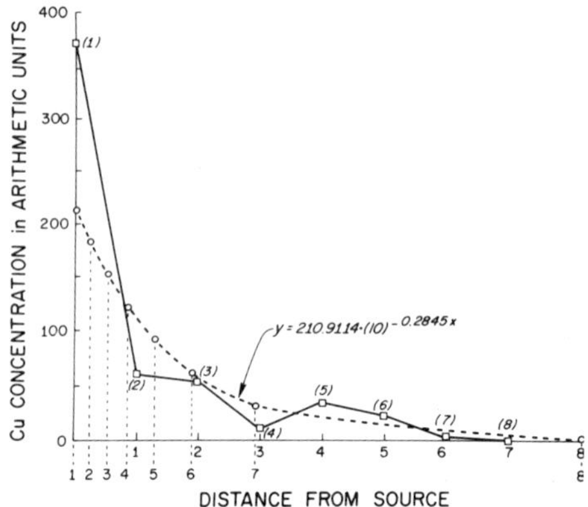

FIG. 4: New allocation of samples based on the initial survey and an assumed decaying exponential distribution of copper along the transect.

that take into account the major sources of contaminants and their observed spatial distributions may be appropriate for the Bight Apex.

For instance, based on observed spatial distribution, we could select a west-to-east transect through both the sewage sludge and dredge spoil dump site areas. Although the data from such a transect would not be expected to fit a simple log normal distribution, they probably would fit a bimodal curve that is the sum of two independent log normal distributions. Other transects that focus on individual sources of contaminants do fit simple log normal distributions.

Transect data, which are fit to a statistical model, can provide a basis for quantitatively evaluating temporal changes in metal concentrations. For instance, if baseline data are used to create a curve as shown in Figure 4, then the area under the curve (the integral of the regression equation) is a quantitative measure of total sediment metal content along the transect. For the curve in Figure 4, the following calculation would apply:
$y = 210.9114 \cdot (10)^{-.2845x}$ is of the form b^{ax}, where $b = 10$ and $a = -.2845$. Then

$$A = 210.9114 \cdot \int_{x_1}^{x_2} b^{ax}\, dx = 210.9114 \cdot \{b^{ax}/(a \log b)\}\Big]_{x_1}^{x_2} ,$$

where A = area under the curve. If we integrate from $x = 0$ to $x = 8$, $A = 736.08$ square units.

Future sampling data could be used to create new curves. The percent change in metal content between two cruises would be: $\Delta P = \{(A_2-A_1)/A_1\} \cdot 100$.

Changes in the shape of the curve over time could provide a rational basis for directing sampling efforts. Monitoring could be focused on regions with maximal distributional changes but reduced to a minimum where the curve did not change significantly over time.

3. BENTHIC INVERTEBRATE SAMPLING STRATEGY

Benthic invertebrates are exposed to a wide variety of contaminants that accumulate in bottom sediments of the Bight. These organisms perform several significant ecosystem functions, and their communities have been modified substantially over large areas by high contaminant concentrations (Pearce *et al.*, 1976; O'Connor, 1976). It is important to understand the degree to which benthic invertebrate communities have been modified by contaminant exposure and long-term temporal changes in the degree of impact. In the New York Bight these impacts must be distinguished from natural influences, such as sediment grain size, water column depth, water temperature, reproductive success, and predation. The well-known aggregated or clumped distributions of most benthic invertebrate species increase the cost of precise abundance estimates. All these factors underline the significance of improving benthic invertebrate sampling strategies.

Walker *et al.* (1979) present a sampling strategy that uses data from the physical/chemical environment more effectively than heretofore. They search for patterns among the physical variates and then search for related patterns among the biological variates. Their objective is to examine the influence of waste disposal on the benthic faunal assemblages of the Bight Apex area and propose a sampling plan to monitor these influences. The data set used (Pearce *et al.*, 1977) is composed of sediment mean grain size, percent organic matter, trace metal concentrations, and benthic species abundance, each measured from the same sediment grab. The grabs were collected on four cruises to the Bight Apex stations shown in Figure 5.

Investigators (Johnson, 1971; Sanders, 1958) have established that the spatial distribution of benthic animals is determined largely by sediment characteristics. In this study the authors examine the assumption that the distribution of benthic animals is affected by waste inputs. Sediment characteristics associated with waste inputs

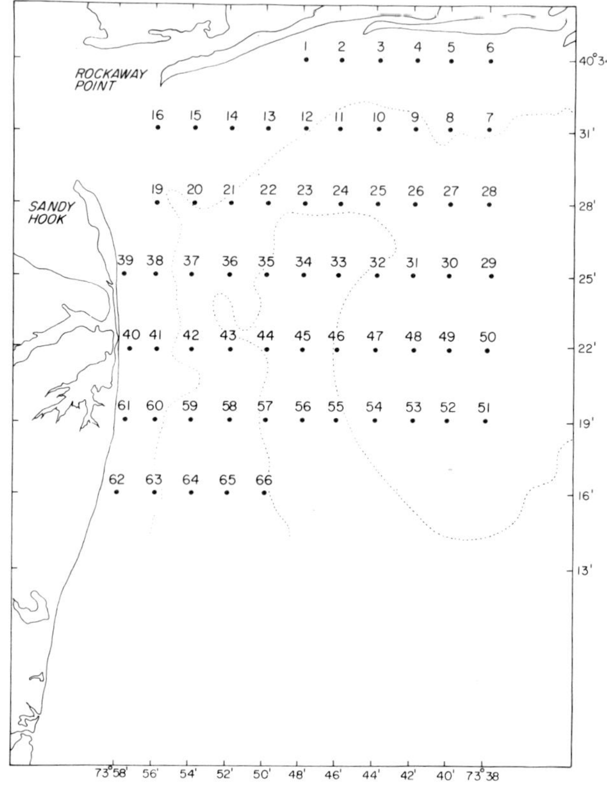

FIG. 5: New York Bight Apex area showing the station numbers and locations utilized for the cruise data analysis. This station grid was established by the MESA NY Bight Project.

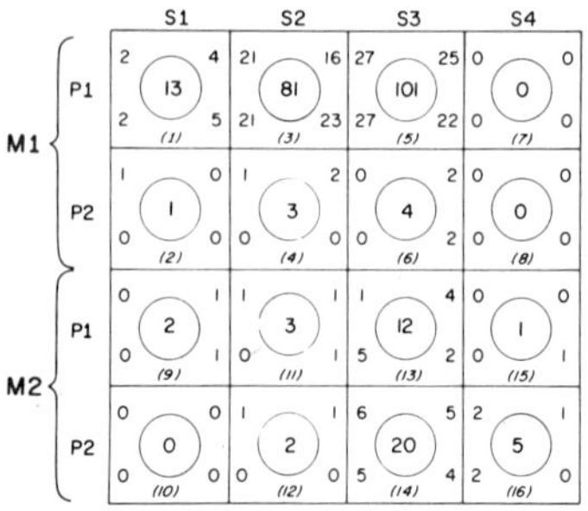

		S1	S2	S3	S4
M1	P1	2 4 / **13** / 2 5 (1)	21 16 / **81** / 21 23 (3)	27 25 / **101** / 27 22 (5)	0 0 / **0** / 0 0 (7)
	P2	1 0 / **1** / 0 0 (2)	1 2 / **3** / 0 0 (4)	0 2 / **4** / 0 2 (6)	0 0 / **0** / 0 0 (8)
M2	P1	0 1 / **2** / 0 1 (9)	1 1 / **3** / 0 1 (11)	1 4 / **12** / 5 2 (13)	0 0 / **1** / 0 1 (15)
	P2	0 0 / **0** / 0 0 (10)	1 1 / **2** / 0 0 (12)	6 5 / **20** / 5 4 (14)	2 1 / **5** / 2 0 (16)

FIG. 6: Breakdown of physical/chemical strata used in this study.

are used to designate sediment strata. Once strata are defined, the effect of waste inputs on the distribution of benthic species within and among strata can be evaluated.

A sediment stratum is defined in this study as independent of geographic location. This approach avoids the complexities associated with microenvironmental variability. For example, small-scale spatial variability of sediment type within the Bight Apex is high. There is no assurance that repeated grab sampling at one location (even at approximately the same time) will produce grabs with very similar sediment characteristics.

Sixteen physical strata (Figure 6) are identified based on all possible combinations of three variables broken down into eight distinct groups. Sediment grain size values are divided into four groups, and trace metals and percent organic matter values each are divided into two groups. The cut points for these groups are given in Table 1. These divisions were not chosen arbitrarily. For example, inspection of plots of value vs. occurrence for percent organic matter and trace metals revealed bimodal distributions. Minima between the high and low modes were used as cut points.

The strata are depicted as having low or high metals concentrations (M1 or M2, respectively), low or high percent organic matter (P1 or P2, respectively), and mean sediment grain sizes ranging from coarse (S1) to fine (S4). The strata are numbered in parentheses at the bottom, center of each cell.

The total number of times that each stratum was sampled in all four cruises is shown in the center circle of each stratum cell of Figure 6. The numbers at the corners of each cell represent the stratum sampling frequencies for each cruise as follows: top left = cruise 1, top right = cruise 2, lower left = cruise 3, and lower right = cruise 5. Some stratum cells are empty or virtually empty as expected. For example, high levels of organic matter or metals would not be found in very coarse sediment regimes (strata 2, 9, and 10).

Strata 1, 3, 5, 13, and 14 occur fairly frequently. Strata 1, 3, and 5 each represent grabs with low metal and low percent organic matter values. They are distinguished by their mean grain size, which varies from coarsest (stratum 1) to fine but not claylike (stratum 5). Strata 13 and 14 have low and high percent organic matter values, respectively, and appear to represent grabs from two types of waste impacted regions. Their trace metal (high) and mean grain size (fine, but not claylike) categories are identical.

TABLE 1: Cut points of sediment physical/chemical characteristics used for stratum definitions in the New York Bight (Walker et al., 1979).

Metals (parts per million)		Sediment grain size (Phi units)				Organic matter (percent)	
M1	M2	S1	S2	S3	S4	P1	P2
150		0.0	2.0	5.0		5.0	

Walker *et al.* (1979) developed and applied a technique for determining the relative influence of natural factors and contaminants on the abundance of selected benthic species. Transformed species counts were used {ln (x+1), where x = count} to bring the distribution of species counts closer to normal and reduce the dependence of the variance on the mean. To identify those species whose mean abundances clearly were influenced by the physical/chemical variables used as stratification criteria, a one-way analysis of variance over strata was performed, after rarely-occurring species were eliminated. Those species having a between-stratum 'F' statistic greater than 2.0 (corresponding to 95% confidence that significant between-stratum differences in abundance are identified) in any cruise are shown in Table 2.

Since the authors wished to examine the effects of waste disposal on benthic faunal assemblages, the biological responses to contaminants were compared graphically in two strata with similar sediment grain size. This comparison is between relatively uncontaminated sediments with low percentages of organic material and low heavy metal concentrations (stratum 5) vs. waste impacted sediments with high percentages of organic material and high heavy metal levels (stratum 14). Species were selected from Table 2, and 90% confidence limits for within stratum mean densities were computed and plotted as shown in Figure 7. Stratum 5 mean abundances are connected by dashes; stratum 14 mean abundances are connected by solid line. The five species in Figure 7 have mean abundances that are significantly different between the two strata for all cruises. These species may be significant indicators of stressed vs. unstressed environments of similar sediment characteristics in the New York Bight Apex. For each species a significantly reduced abundance is evident in the stressed environment. The undesirable length of many of the 90% confidence bands in Figure 7 results from the small number of observations within strata.

These results may be used to help design future monitoring efforts. Walker *et al.* (1979) propose a two-step sequential random

Table 2. Species with significant differences ($P<.05$) among strata in the New York Bight.

Edwardsia sp.
Edwardsia elegans
Edwardsia sipunculdides
Cerianthus americanus
Rhynchocoela
Protodrilus symbioticus
Harmothoe extenuata
Harmothoe imbricata
Lepidonotus squamatus
Hartmania moorei
Sthenelais limicola
Pholoe minuta
Eteone longa
Autolytus cornutus
Nereis grayi
Nereis succinea
Aglaophamus circinata
Nephtys bucera
Nephtys incisa
Glycera americana
Glycera dibranchiata
Hemipodus sp.
Hemipodus armatus
Goniadella gracilis
Ophioglycera gigantea
Pisione remota
Ophelia denticulata
Mediomastus ambisetae
Aricidea jeffreysii
Paraonis gracilis
Prionospio malmgreni
Prionospio steenstrupi
Spiophanes bombyx
Lumbrineris tenuis
Lumbrineris fragilis
Lumbrineris acuta
Ninoe nigripes
Drilonepeis longa
Stauronereis caecus
Haploscoloplos robu
Cossura longocirrat
Tharyx annulosus
Tharyx acutus
Owenia fusiformis
Asabellides oculata
Polycirrus sp.
Pherusa affinis
Potamilla neglecta
Euchone rubrocincta
Hydrobia minuta
Crepidula plana
Unid bivalve #2
Nucula proxima
Nucula delphinodont
Yoldia limatula
Mytilus edulis
Astarte castanea
Cerastoderma pinnul
Pitar morrhjana
Tellina agilis
Mulinia lateralis
Copepoda
Heteromysis formosa
Leptocuma minor
Cirolana polita
Unciola inermis
Unciola irrorata
Phoxocephalus holbc
Cancer borealis
Cancer irroratus
Phoronis architecta
Asterias forbesii

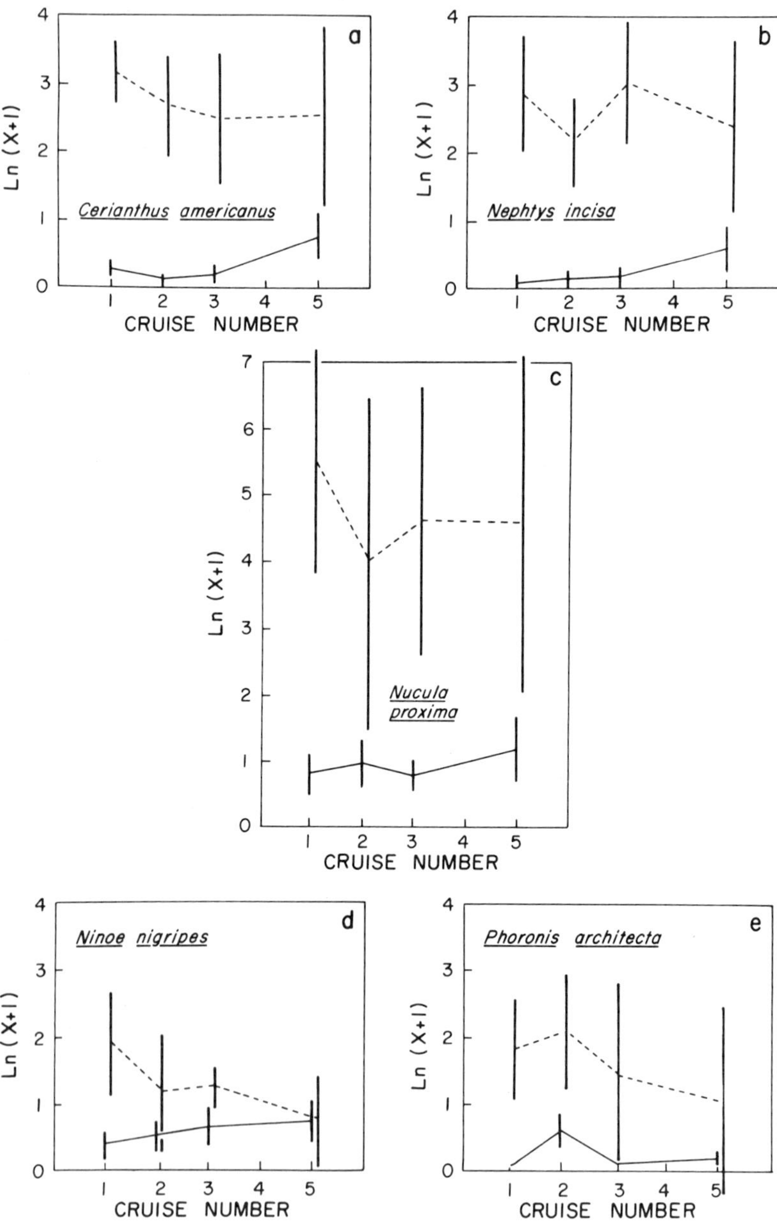

FIG. 7: Changes in mean abundance of selected invertebrate species over the four sampling periods (cruises) in the New York Bight Apex area, 1973-74.

sampling plan where replicate grabs would be obtained from a sampling grid of concentric circle pattern centered on the sludge dump site. The distances between sampling circles would increase exponentially with distance from the dump site, following the findings of Saila *et al.* (1978). Each grab first would be classified as to stratum based on the sediment physical/chemical characteristics. Then more extensive species counts would be made within each stratum for only the number of grabs necessary to achieve the desired precision in estimates of mean density.

This approach avoids large microhabitat variability in the density of benthic invertebrates that necessitates larger sample sizes than often are practical to estimate mean densities with desired precision. The variability of density is much less within the strata as defined by Walker *et al.* (1979). Additional advantages derive from cumulative records of sampling intensity within each stratum. Not only is it possible to insure that adequate numbers of samples are analyzed for each combination of physical/chemical variables desired, but the proposed geographic sampling allocation insures that data points are distributed appropriately for contouring purposes (Saila *et al.*, 1978).

Since the Apex of the Bight has an irregular distribution of contaminants, the proposal by Walker *et al.* (1979) to sample at the intersection of radii and concentric circles centered at a point requires minor adaptations. The sample location strategy would be more straightforward in many other environments.

4. ANALYSIS OF METHODS FOR DETERMINING TIDAL DATUMS

Tidal datum planes are used to determine the positions of boundaries, to serve as planes of reference for maps and charts, and to delineate the extent of land uses in coastal areas. The planes of mean higher high water, mean high water, mean sea level, mean tide level, mean low water, and mean lower low water commonly are used in the United States. These tidal datums are based on tidal records over 19-yr periods, the time period used because it is the closest full year to the 18.6-yr cycle of changing inclination of the Moon's orbit relative to the plane of the Earth's Equator. This motion creates an 18.6-yr periodic fluctuation of the low and high water diurnal inequalities of the tides. Swanson (1974) has statistically examined two methods for determining tidal datums from shorter series of observations. Such methods are necessary because complete tidal records (over a 19-yr period) with close geographical spacing over the entire United States coastline are impractical and prohibitively expensive.

Each method uses a control station in determining the tidal datums for a subordinate station. The control station has observations for a number of years. The subordinate station has a short series of observations (e.g., 1 mo, 3 mo, 6 mo, 1 yr) that are compared to simultaneous observations at the control station and reduced to mean values representative of a datum derived from 19 years of observations.

The relationship in the fluctuation of monthly mean values of a reference datum at two stations is illustrated in Figure 8. Here we have monthly mean tide level values at Sandy Hook, NJ, and Atlantic City, NJ, plotted over the 19-yr period from 1941 to 1959. The accepted value for the datum of mean tide level (MTL) for each station would be the mean of the station's values over the entire period. As shown in this figure, if the accepted value is known for one station, a reliable estimate of the accepted value for the datum at the other station is possible (even if a considerable number of data were missing).

The methods for estimating tidal datums at a subordinate station with a short series of observations can be presented as follows:

For both methods: $\Delta TL = TL_1 - TL_2$, $CTL_1 = \Delta TL + MTL_2$,

where 1 refers to the subordinate station,
2 refers to the control station,
TL = observed monthly mean tide level,
MTL = 19-yr accepted value of mean tide level,
CTL = estimate of 19-yr accepted value of mean tide level.

Standard Method	*Alternate Method*
$F = R_1/R_2$	$\Delta LW = LW_1 - LW_2$
$CR_1 = F \cdot MR_2$	$CLW_1 = \Delta LW + MLW_2$
$CLW_1 = CTL_1 - \frac{1}{2}CR_1$	$\Delta HW = HW_1 - HW_2$
$CHW_1 = CLW_1 + CR_1$	$CHW_1 = \Delta HW + MHW_2$

where R = observed monthly mean range,
MR = the 19-yr accepted value of mean range,
CR = estimate of MR,
LW = observed monthly mean low water,
MLW = the 19-yr accepted value of mean low water,
HW = observed monthly mean high water,
MHW = the 19-yr accepted value of mean high water,
CLW = estimate of MLW,
CHW = estimate of MHW.

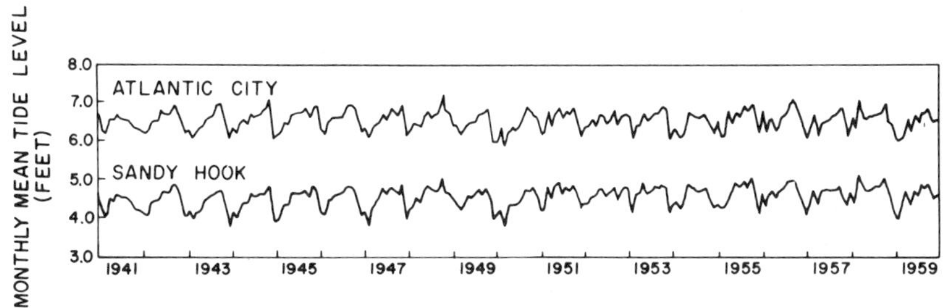

FIG. 8: Time series of monthly mean tide level at Sandy Hook and Atlantic City, NJ.

The alternate method provides flexibility in computation. For example, if data are lacking on LW (e.g., because of partial instrument failure), then CHW still can be calculated.

Swanson (1974) statistically analyzed the error associated with each method. He compared data from pairs of control stations having 19 years of simultaneous observations. The data from one station (B), assumed to be the control station, were used to adjust a short data series from the other station (A), representing a subordinate station. This was done for monthly mean values and for running means of monthly mean values over 3, 6, and 12 mo. A set of residuals was generated by subtracting the accepted 19-yr mean (datum) of station A from each value estimated from short series of observations at station A adjusted through control station B. The mean and variance of each set of residuals (assuming normal distribution) were computed for each datum plane for numerous station pairings around the United States.

A t-test was used for each station pairing to test the hypothesis that the estimated datum from a short series of observations accurately estimates the 19-yr accepted value. The hypothesis is not rejected when

$$-t_{0.025} < t < t_{0.025}$$

where $t_{0.025}$ is the 2.5% point with $n-1$ degrees of freedom of Student's t-distribution and

$$t = (\bar{y} - \mu_0)/(s^2/n)^{\frac{1}{2}} \quad .$$

TABLE 3: Percentage of station pairings for which no significant differences (P<.05) are found between estimates from short term observations and datums from 19 years of observations.

Period*	Percentage of values accepted			
	Standard Method		Alternate Method	
	**MLW	MHW	MLW	MHW
East Coast				
1---------------	73	77	87	90
3---------------	73	67	73	80
6---------------	67	47	73	70
12---------------	43	43	60	53
Gulf Coast				
1---------------	100	88	100	100
3---------------	100	88	100	100
6---------------	100	75	100	100
12---------------	100	62	88	100
West Coast				
1---------------	36	36	100	73
3---------------	36	36	100	73
6---------------	36	36	82	73
12---------------	27	27	82	64

*Length of record in months.
**MLW, mean low water; MHW, mean high water.

In the computation $\bar{y}$ is the difference between the value of the datum computed from the short series of observations and the accepted value of the datum. The term μ_0 is the difference between the measured datum and the actual datum, always assumed to be zero. Both computational methods were treated in this manner. Some paired tests on different coasts of the United States are shown in Table 3.

The percentage of acceptance generally decreases with an increase in the time period over which the datum is computed. This results because the mean difference between the computed and accepted values does not improve with increasing time; however, the standard deviation does decrease considerably with time because the measurements are averaged over longer time intervals. The hypothesis is rejected most frequently when the standard method of calculation is used for the respective datums on the West Coast. This condition is partly a result of fewer, more widely spaced, control stations on the West Coast. In most instances the alternate method of estimating MLW and MHW is accepted more often than the standard method.

The largest difference between estimated and datum values on the East Coast was 0.049 ft (15 mm). This mean difference is large enough so that it does not statistically represent a true datum. From a practical viewpoint, the difference is small because we are concerned with errors on the order of tenths of feet. On the West Coast, however, there is a clear advantage in using the alternate method. This judgment is made using the percentage acceptance of the hypothesis as a criteria.

5. STATISTICAL ANALYSIS OF METEOROLOGICAL FORCING DATA

The forcing of surface water currents by weather features such as wind stress is a significant phenomenon in the New York Bight and other coastal regions. Movements of large water masses, forced by winds, cause substantial alterations in the physical/ chemical environment, move oil and debris onto beaches, and result in upwelling of nutrient-rich waters which stimulate plant growth.

Beardsley and Butman (1974) have described an example of large-scale meteorological forcing in the middle Atlantic region. In what they call a scale-matching phenomenon, the entire middle Atlantic continental shelf region, from Maine to North Carolina, tends to interact with winter storms of appropriate size and motion so that very intensive southwestward flows result.

Because of the large surface area of the New York Bight (39,000 km^2) actual measurements of weather features must be sparse in both space and time. Mooers *et al.* (1976) examined coastal data from 1975 to determine how reliably these sparse measurements can be extrapolated to develop better models of surface circulation. We summarize some of their findings in this section.

One part of the analysis involved a cross-correlation of weather features derived from National Weather Service coastal stations in the middle Atlantic region. Cross-correlations of surface atmospheric pressure (P) and the two components of surface wind stress (τ^x) and (τ^y) were performed. The east-west wind stress is τ^x and the north-south wind stress is τ^y. Previous analyses had shown little spatial coherence for time scales less than 2 days or greater than 2 weeks except for the diurnal and annual components. Therefore, it was decided to focus on a 1.67-40-day, band passed data series. The cross-correlations for all combinations of coastal station pairs were plotted as a function of the magnitude of the station separation distance. Cross-correlations in time were not performed at this point. The results for P, τ^x and τ^y are shown in Figure 9. The zero separation correlation

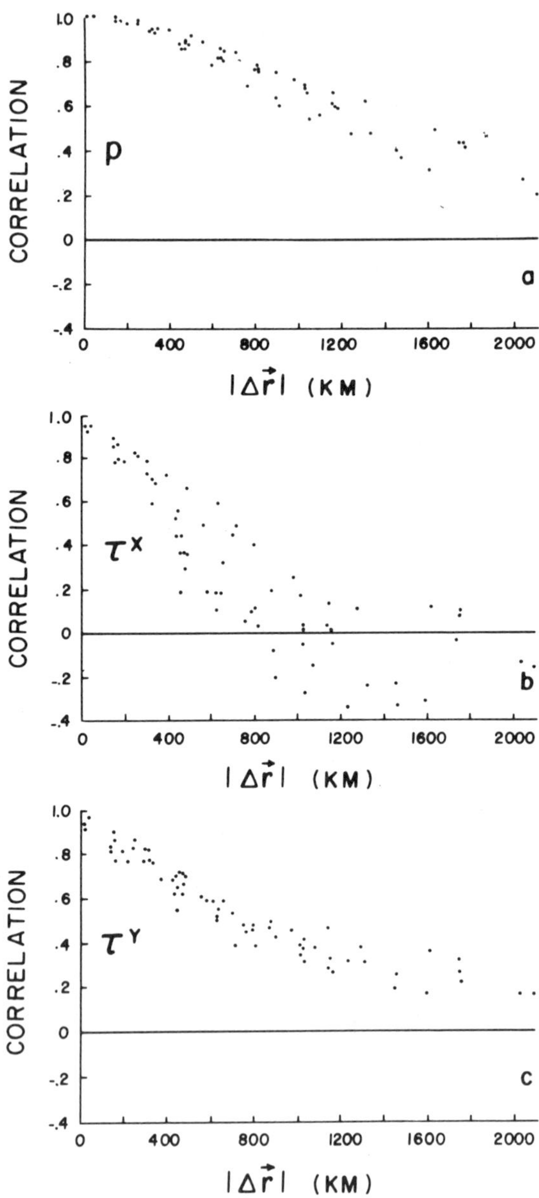

FIG. 9: Cross-correlation of P(A), τ^x(B), *and* τ^y(C) *at all National Weather Service coastal station pairs as function of radial separation distance* (km) *between stations.*

(a measure of observational noise) is approximately 0.99 for P and 0.95 for the wind stress components. These values also indicate the best possible results to be expected from a statistical hindcasting scheme.

If we arbitrarily assume that correlations less than 0.4 are not significant, we can estimate the furthest distance that a coastal station observation can be used for extrapolation in space. For P , τ^x , and τ^y (in Figure 9) these distances are about 1,600 km, 650 km, and 850 km, respectively, in the alongshore direction. Also, if P and τ^y have a zero crossing, it occurs for a separation distance of more than 2,000 km; in contrast, τ^x tends to have a zero crossing at about 1,000 km, a distance commensurate with the size of a subtropical cyclone.

Mooers *et al.* (1976) extended the cross-correlation analysis to include time as a factor. They calculated cross-correlation between paired coastal observations for different times as well as at different separation distances. In each calculation an observation from John F. Kennedy International Airport (JFK) was one half of the station pair. The result for P is shown in Figure 10; each part (A and B) is a two-dimensional presentation and considerably more complex than Figure 9. The time axis is the same for each part. The space axis for part A represents east-west (longitudinal) separation distance between JFK and the other observation point (east is positive). The space axis for part B represents north-south (latitudinal) separation distance between JFK and the other observation point (south is positive).

To estimate the limit in time that a coastal station observation can be used for extrapolation we apply the same assumption used previously (i.e., correlations less than 0.4 are not significant). From Figure 10, we see that a time limit of ± 24 h is a representative estimate for P .

We also have drawn an axis of symmetry (dashed line) on correlation contours of each part of Figure 10. Points along each axis of symmetry represent maximum correlations given certain time and space separations. These lines allow for a useful interpretation. When the JFK observation is being correlated with observations taken at a prior time (negative Δt), highest correlations occur with stations to the west and south. However, when the JFK observation is being correlated with observations taken at some future time (positive Δt), highest correlations occur with stations to the east and north. These findings indicate that pressure disturbances propagate to the north and east as they pass through the middle Atlantic region. Their speed of propagation (found from the slope of the symmetry axis) is 15 m s^{-1}.

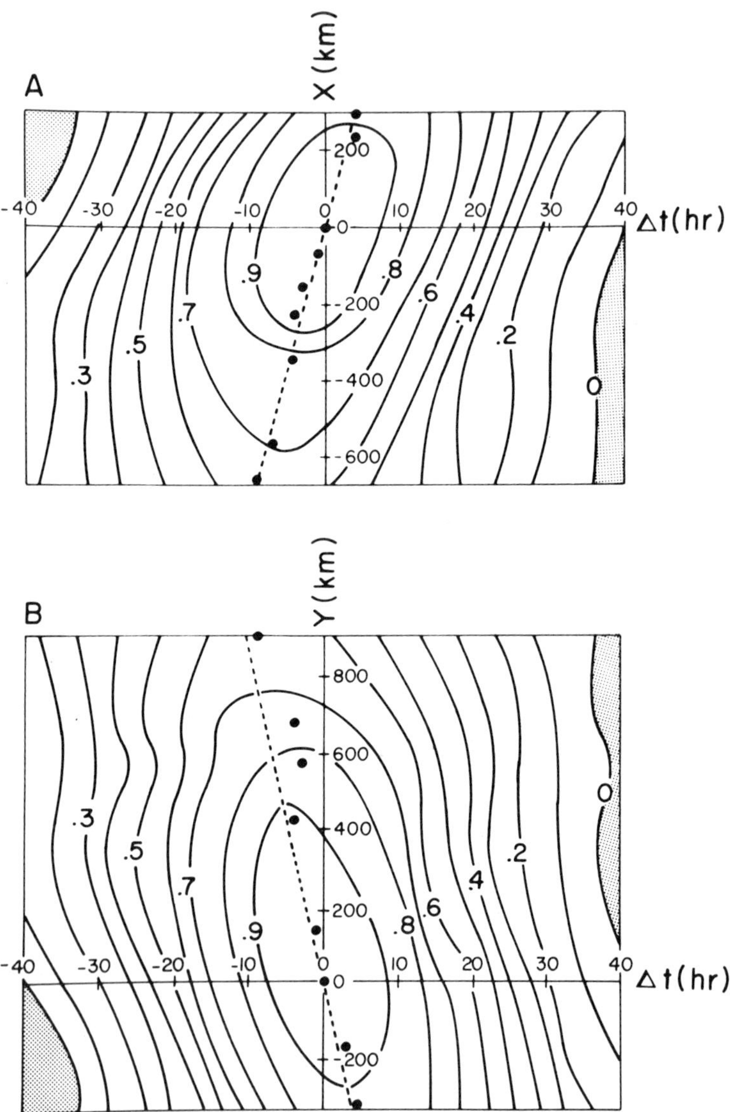

FIG. 10: Space and time lagged cross-correlation field of pressure. JFK *is the base station, and the space coordinate is longitudinal* (A) *and latitudinal* (B) *separation (km) between* JFK *and* NWS *coastal station. Data were band pass filtered. Stippled areas are cross-correlation less than zero.*

Although these results are not surprising (we know that storms move through the region in this manner), the development of an accurate objective statistical model of this process indicates that we also can develop statistical interpretations of other processes.

Similar analyses were performed for the stress components τ^x and τ^y. The propagation again was found to be to the north and east, but speeds were somewhat slower (τ^x = 5 m s^{-1}, τ^y = 10 m s^{-1}). The estimated time limit for useful correlation for the stress components was ± 18 h.

Mooers *et al.* (1976) suggest that a hindcast scheme is possible because of the high degree of space-time correlation demonstrated. It might benefit from including values for up to ±18 h centered on the time of interest or, equivalently, from stations located along a northeastward track at distances of ± 700 km for P, ±200 km for τ^x, and +300 km for τ^y.

6. DISCUSSION

The four studies herein contribute to the solutions of New York Bight ecosystem sampling questions through use of statistical methods. These methods will aid MESA as more sophisticated experimental designs and data analysis techniques are developed for monitoring and synthesis.

Saila *et al.* (1978) and Walker *et al.* (1979) have performed complementary analyses of baseline data in the Bight Apex. They have proposed methods for monitoring trace metal content in sediments and for separating naturally occurring changes in benthic invertebrate community structure from those changes induced by contaminants. Some minor adaptation will be required because the Bight Apex does not have a well defined center of contaminant concentration.

Swanson (1974) and Mooers *et al.* (1976) have shown that tidal and meteorological data exhibit enough correlation in space and/or time for successful application of data interpolation and extrapolation techniques. Short series of tidal observations yield statistically acceptable estimates of tidal datum planes when adjusted through a nearby control station where observations are available for a number of years. The use of coastal weather data in experiments to specify atmospheric forcing over data-sparse marine areas has been shown statistically to be promising.

ACKNOWLEDGEMENTS

Special appreciation is expressed to Dolores Toscano who typed the original draft and all revisions to this paper and to Karen Henrickson for drafting the figures. We thank the MESA investigators for providing the New York Bight Project with research results that are helping the Project define future sampling strategies.

REFERENCES

Beardsley, R. C. and Butman, B. (1974). Circulation on the New England continental shelf: response to strong winter storms. *Geophysical Research Letters*, 1, 181-184.

Johnson, R. C. (1971). Animal-sediment relations in shallow water benthic communities. *Marine Geology*, 11, 93-104.

Mooers, C. N. K., Fernandez-Partagas, J. and Price, J. F. (1976). Meteorological forcing fields of the New York Bight (first year's progress report). University of Miami, Rosenstiel School of Marine and Atmospheric Science Technical Report 76-8.

Mueller, J. A., Jeris, J. S., Anderson, A. R., and Hughes, C. F. (1976). Contaminant inputs to the New York Bight. NOAA Technical Memorandum ERL MESA-6.

O'Connor, J. S. (1976). Contaminant effects on Biota of the New York Bight. *Proceedings 28th Annual Session Gulf and Caribbean Fisheries Institute*. 50-63.

Pearce, J. B., Caracciolo, J. V., Halsey, M. B., and Rogers, L. H. (1976). Temporal and spatial distributions of benthic macroinvertebrates in the New York Bight. *American Society of Limnology and Oceanography Special Symposium*, 2, 394-403.

Pearce, J., Rogers, L., Caracciolo, J., and Halsey, M. (1977). Distribution and abundance of benthic organisms in the New York Bight Apex, five seasonal cruises, August 1973 - September 1974. NOAA Data Report ERL MESA-32.

Saila, S. B., Anderson, E. L., and Walker, H. A. (1978). Sampling design for some trace elemental distributions in New York Bight Sediments. In *Biological Data in Water Pollution Assessments: Quantitative and Statistical Analyses*, K. L. Dickson, J. Cairns, Jr., and R. J. Livingston, eds. American Society for Testing and Materials, Philadelphia. 166-177.

Sanders, H. L. (1958). Benthic studies in Buzzards Bay. I. Animal sediment relationships. *Liminology and Oceanography*, 3, 245-258.

Swanson, R. L. (1975). Variability of tidal datums and accuracy in determining datums from short series of observations. NOAA Technical Report NOS 64.

Walker, H. A., Saila, S. B., and Anderson, E. L. (1979). Analysis of benthic data collected from the New York Bight Apex area. *Journal of Estuarine and Coastal Marine Science* (in press).

[*Received September* 1978. *Revised March* 1979]

G. P. Patil and M. Rosenzweig, (eds.),
Contemporary Quantitative Ecology and Related Ecometrics, pp. 569-617.

ECOLOGICAL AND STATISTICAL FEATURES OF SAMPLING INSECT POPULATIONS IN FOREST AND AQUATIC ENVIRONMENTS

WILLIAM E. WATERS AND VINCENT H. RESH

Department of Entomological Sciences
University of California
Berkeley, CA 94720 USA

SUMMARY. Forest and aquatic insects are important to man in both adverse and beneficial contexts, and sound quantitative methods are necessary to obtain reliable data for specific applications. They share some ecological and statistical features which point to the development of sampling techniques and methods of analysis useful to each. Relevant work is further advanced in sampling forest insect populations, and many of the approaches developed thus far have application to aquatic insects. Some basic features, requisites, and constraints of sampling forest and aquatic insect populations are described, with specific examples of the problems encountered and suggestions for their resolution.

KEY WORDS. forest insects, aquatic insects, sampling distributions, aggregation, population dynamics, biomonitoring.

1. INTRODUCTION

Insects affect our lives in many ways and require corresponding attention. Some of their activities threaten our very existence, some are merely a nuisance, others are beneficial in one way or another. Sound quantitative methods are necessary to obtain reliable data for analysis and interpretation of the important relationships among insects and between these highly evolved creatures and other organisms, including man. It is essential that we have the capability to predict changes in the abundance and distribution of those that affect our health, food, and natural resources, and that we have the knowledge and technology to manipulate the beneficial and harmful ones alike to our advantage. For the basic biological information and for the development of appropriate methodologies, we are dependent on data obtained by sampling.

Sampling studies, approached as an investigation of the nature and sources of variation in numbers of organisms over space and time, provide a direct link to the analysis and modeling of single- and multi-species population dynamics. They also provide, of course, the basis for biologically meaningful and statistically efficient sampling schemes for estimation of population parameters of interest. This includes the development of techniques for biomonitoring man-induced changes in environmental quality.

Small organisms, such as insects, characterized by variable life stages or forms, mobility, and high reproductive rate present special problems in sampling. These are compounded by the heterogeneity of microenvironments in which the organisms operate and by the diversity of biotic and abiotic factors affecting their numbers and distribution. In addition, there are often physical and logistic difficulties in sampling such populations effectively. All of these considerations apply most stringently to forest and aquatic insects.

The term *forest insects*, as used here, denotes only those insects occurring on or in living trees. Many of the species cited are very destructive and are considered to be major forest pests (Figure 1). By *aquatic insects* we mean those species which spend at least one life stage in an aquatic environment. Most of those cited have been studied because of their importance as vectors of disease, as food for fish and other larger organisms, and more recently, as biological indicators of water quality.

2. FOREST AND AQUATIC ENVIRONMENTS AS HABITAT FOR INSECTS

Natural and man-made forests, rivers and streams, and lakes and ponds are uniquely structured ecosystems within which insects reproduce, grow, disperse, and die, and in so doing co-evolve with other biotic components of the environment. The variation in numbers of an insect species among its habitats reflects the interplay of different adaptive strategies by the insect and the many intrinsic and extrinsic processes and agents that regulate and limit its density and spatial distribution. The nature and consistency of the relationships between insect and habitat characteristics are of critical interest to the ecologist. Quantification of these relationships in the form of descriptive and predictive models is often required.

FIG. 1: Upper Panel: Defoliation by the spruce budworm (Choristoneura fumiferana) *in a spruce-fir forest managed for pulpwood production (New Brunswick, Canada). Lower Panel: Whole-tree and top-killing of white fir by the fir engraver* (Scolytus ventralis) *in a Northern California forest recreation area.*

Despite what might appear to be an extremely large number of variables affecting the space-time dynamics of insect populations in forest and aquatic environments, there are in fact definite patterns of occurrence and change. For any given insect species, the number of determining factors may be limited. The basic purpose of sampling studies is to ferret out the really significant factors and to characterize the spatial and temporal patterns in ways that will be useful for specific applications (e.g., biomonitoring, life table studies, and modeling).

The factors or processes determining the numbers and distribution patterns of insects in both forest and aquatic environments may be categorized as follows: (1) mode and rate of reproduction, (2) number and kind of life forms, (3) behavioral responses to physical factors (i.e., light, temperature, humidity, pressure, chemicals), (4) behavioral responses to host plant(s) and other substrates, (5) behavioral responses to insects of the same kind and to other organisms, and (6) differential mortality. Which of these are operative and the specific nature of the interactions involved will vary between insect species and between stages of the same species. They will vary between habitats and between vertical and horizontal strata within a given habitat. They will vary over time in a given habitat as insect numbers and behavior change with density. Their relative importance differs generally between forest and aquatic environments. Examples of their influence on sampling with respect to specific forest and aquatic insects will be described subsequently.

For purposes of this discussion, variation deriving from factors or processes operating at the regional or geographic level will not be covered. We will focus on forest and aquatic environments at the level of the natural ecological units that logically comprise the sampling universe for an insect population: namely, the individual forest stand and the stream or lake. Differences between and within these entities make up the physical and biological template for insect population estimation and prediction.

2.1 Forest Habitats. Forest and forest-related environments are ecologically diverse, complex, and dynamic. For most survey and research purposes, they are considered in a 2-dimensional context with sampling units distributed in the horizontal plane. For statistical measurement and analysis of insect populations, forests must be viewed in three dimensions. For a variety of reasons, most forest insects do not distribute themselves in a random manner on or in host trees. The distributional pattern may change in successive life stages, and it may be altered by competition for food or space and by differential mortality. Overall, vertical stratification of insect numbers is a major source of variation and must be accounted for statistically. It may also have considerable ecological significance.

Dispersion of Eggs. The distributional pattern of an insect is initiated by the dispersion of eggs, which strictly is a function of the dispersal capabilities and behavior of the adult female. In some cases, the insects fly directly to their host trees in response to visual and/or olfactory stimuli. Certain species of pine sawflies (*Neodiprion* sp.) and engraver beetles (*Ips* sp.) operate this way. Others apparently fly or move about more or less at random and either remain on a tree or leave it depending on chemotactile stimuli arising from the tree surface (foliage or bark). Studies of the landing rates of the mountain pine beetle (*Dendroctonus ponderosae*) and the smaller European elm bark beetle (*Scolytus multistriatus*) on host and non-host trees, for example, indicate this pattern. Still others, such as bark aphids (e.g., *Adelges* sp.), are even less directed, being carried more or less passively by air movements, in which case those that land on non-host trees or other unsuitable substrates die. In this phase of insect distribution, stand factors such as location with respect to elevation and topography, directional exposure, slope, moisture conditions, and the density and condition of host trees determine the probability of a particular insect species occurring there. Individual tree factors such as color, form, odor, surface characteristics (foliage and bark), relative size and position in the stand largely determine the probabilities of eggs being laid on them given that the insect is present.

Where the eggs are laid within a tree is largely a matter of the female's response to light and substrate conditions. Geotaxis and simple orthokinesis in response to a variety of other stimuli may further explain the vertical movement of the females and why they stop where they do. Light intensity and quality is greatly influenced by the density and spectral quality of the foliar canopy. The character of tree substrates depends on the tree species involved, the age and vigor of the tree, and in the case of foliage, whether it is exposed to the sun or shaded. Some insects tend to fly at about the general forest canopy level, remaining above the darkened interior. The spruce budworm moth (*Choristoneura fumiferana*) acts this way, with the result that its egg masses are laid generally in the upper portion of the tree crowns, and a disproportionate number may be laid on larger trees whose crowns project above the general canopy level. When populations increase and the foliage in the tree tops is depleted, more egg masses are laid in lower sections of the tree crowns.

Other insects tend to remain below the canopy level, or at least attack and lay eggs in or on lower portions of the tree. The western pine beetle (*Dendroctonus brevicomis*) and the southern pine beetle (*D. frontalis*) attack and construct egg galleries in the boles of host trees from near the ground to well within the live crown. There is a definite stratification, however, with the greatest proportion of attacks occurring in the mid-bole area. A close relative, the red turpentine beetle (*D. valens*) attacks only near the ground, rarely above two meters. A less definite egg laying pattern is exemplified by the hemlock loopers (*Lambdina* sp.) which lay eggs in shaded places on the ground, on the tree boles, and on branches and twigs. Some stratification may be evident in a given habitat, but it is not generally consistent and the behavioral basis for the seeming indiscriminate pattern is not known.

The distribution of eggs of some species whose adult females are sessile or nearly so is essentially determined by where they pupate (or transform to adults in the case of hemimetabolous insects). The gypsy moth (*Lymantria dispar*), for example, pupates on trees, on shrubs and herbaceous vegetation, and on rocks and other litter on the ground. The female moth is winged but cannot fly and moves only a limited distance, if at all. It mates and normally lays a single egg mass in place.

Larval Behavior and Distribution. The greatest diversity of behavior occurs generally in the immature, larval stages. Much of this is inherent, or fixed, but there also is considerable flexibility, or adaptability, with respect to habitat conditions.

Among the most complex are those species in which the behavior of the larvae changes markedly in successive stages. The spruce budworm, for example, on hatching from an egg mass (on a needle) crawls to the needle base or to a suitable twig scale or flower bract nearby and settles for the winter. Since these first-stage larvae come from masses of 20-25 eggs, there is a tendency to clumping in the overwintering population. In the spring, the emerging larvae move outward toward the terminals of the branches mining a few needles on the way. Many spin fine threads and dangle briefly from the ends of branches, at which time they are subject to dispersal by the wind. In any case, the larvae safely ensconsed on a suitable host then proceed to bore into the current year's buds, at which time a very large proportion of the budworms on a particular tree will be located on the outside periphery of the crown. As the buds open (i.e., those not attacked and injured), the developing larvae feed on the expanding new foliage and move about as needed to find fresh, tender needles. At this time (4th-6th instars), when disturbed or when the foliage in their immediate surrounds is depleted, they will spin down on threads to foliage at a lower level. This behavior results in a more random distribution of larvae within the tree. Differences in the timing of movements and in the related proportional distributions of the budworm occur between trees of different sizes and different species, due largely to different light conditions and the phenologies of the host trees. Budworm emergence and movement on a large, dominant tree exposed to full sunlight may be 2-3 weeks ahead of that on a smaller shaded tree. Similarly, budworm activity and distribution on a balsam fir, synchronized with foliage development, will be 1-3 weeks in advance of that on red spruce. Some differences in lateral movement and distribution of the budworm occurs at different crown levels within a tree, especially in stands where tree density is high.

Certain forest insects have daily larval activity patterns which cause major shifts in spatial distribution. Young larvae (1st-3rd instars) of the gypsy moth generally stay in place within the crowns of their host trees, but in the later stages (4th-6th) they move downward at daylight to the underside of large branches, to the tree trunk, and sometimes to the ground. They remain in these sheltered niches throughout the day, returning to the tree crown at dusk for feeding during the nighttime hours. Where they stop to rest during the day depends on the availability of large branch stubs, bark flaps, and other hiding places on the trees. In stands of young trees with smooth, clear boles, a large proportion of the older larvae will move to and remain on the ground during the day. Interestingly, these larvae utilize silk trails in their daily travels. This results in characteristic larval aggregations. Other lepidopterous species, such as the fall webworm (*Hyphantria cunea*) and the forest tent caterpillar (*Malacosoma disstria*), behave similarly.

The movement and spatial distribution of many insects infesting forest trees are more limited, and they remain more or less in place throughout the larval period. Their distributions are determined largely by where and how the eggs are deposited. Habitat features have relatively little influence on subsequent larval behavior. The gregarious feeding habit of certain pine sawflies of the genus *Neodiprion*, for example, arises from the fact that the eggs are laid in the needles in clusters, and the larvae tend to remain grouped as they feed. They disperse only when the foliage is essentially depleted. Larvae of species of *Rhyacionia*, generally termed shoot and tip moths, bore directly into the buds and shoots of host trees, which is facilitated by the eggs being deposited at the bases of needles in the outer periphery of the tree. The larval period is spent entirely in this portion of the tree. Tip-infesting weevils of the genus *Pissodes* are limited entirely to leading terminals of host trees. The eggs are laid in groups within 6 inches of the terminal bud cluster, and the larvae bore downward for up to 36 inches or more (killing up to 3-4 years' growth in the process). This, then, is the population universe for a critical period in their life history.

The most limited larval distribution patterns are exemplified by the numerous bark- and wood-boring beetles that infest living, dying, or recently killed trees. The movement of these insects is generally limited to a few centimeters. Some species, particularly of the family Scolytidae, occur in characteristically patterned groups.

Distribution of Pupae. The distribution of the pupae of most insects is determined by where the last stage larvae cease feeding. With some, however, there is a complete shift of the population to another stratum of the habitat. The mature larvae of some pine sawflies (*Neodiprion* sp.) and the larch sawfly (*Pristiphora erichsonii*), for example, drop to the ground and crawl into the litter or humus to pupate. Larvae of the fall cankerworm (*Alsophila pometaria*) and many other Geometridae crawl to the ground and pupate under leaves, rocks, and other debris. Obviously, the physical characters of the surface of the forest floor influences the spatial distribution of these insects in the pupal stage.

Adult Emergence. To return full cycle, we spoke of the dispersal behavior and capabilities of the adult insects. It remains to point out an additional and very important variable of adult activity: namely, the timing of emergence. This is essentially under genetic physiological control. There are some ecologically significant differences, however. Predictive accuracy is very high with insects that overwinter as eggs or larvae, and then develop into pupae in the milder seasons of the year. With such insects, temperature and other physical factors normally are not limiting, and the timing of adult emergence in the late spring or summer can be predicted within a narrow range of days from the initiation of pupation. With insects that overwinter as pupae or adults, cumulative temperatures and occasional frosts in the early spring are critical. These vary tremendously in forest environments: in gross ways according to latitude, elevation, directional exposure, and other geographic and topographic features and to a finer degree according to tree canopy density, tree height levels, and other within-stand variables.

Differential Mortality. Another major source of variation in forest insect numbers is the differential mortality occurring in all stages. Natural mortality ordinarily is very high, the reproductive rate of most insects is such that equilibrium is maintained with approximately 2% generation survival. Pest outbreaks are characterized by a survivorship of 3%-10%.

The mortality factors operating on insects in forest environments include the direct effects of physical factors, primarily temperature, host tree resistance, intraspecific competition, and the myriad of natural enemies, including disease, that attack all stages. The action of these factors or agents vary greatly with habitat conditions. Host resistance is determined in part at least by habitat or site conditions that affect tree vigor. The rate and spatial variations in mortality from parasites, predators, and disease are determined in part by the behavioral responses of the individual agents to habitat factors which bring them into proximity with the target insect and by the numerical and functional responses of each agent to changes in density of the insect. Though complex, here too there are patterns of mortality to be discerned and quantified. This is the essence of forest insect population dynamics.

2.2 Aquatic Habitats. Insects occupy a broad array of aquatic habitats, ranging from rainwater accumulations in animal tracks and plants (e.g., bromeliads) to the violent intertidal zone of the world's oceans. Approximately 20,000 species in thirteen insect orders are obligate inhabitants of aquatic systems during some stage of their development. (The insect orders Odonata, Ephemeroptera, Plecoptera, Trichoptera, and Megaloptera are entirely composed of species that are aquatic in one or all of their immature stages; the Coleoptera, Diptera, and Hemiptera contain large numbers of aquatic species, but also have a major terrestrial fauna; the Collembola, Hymenoptera, Lepidoptera, and Neuroptera are mainly terrestrial, each having a few aquatic representatives.) Although insects utilize both still (lentic; e.g., lakes and ponds) and running water (lotic; e.g., rivers and streams) systems, far more species occur in the latter with many groups limited to such environments and others reaching their greatest diversity there.

Just as tree and stand characteristics may influence the distribution of insects in forests, certain habitat features operate similarly in streams and lakes. The size of the habitat, its permanence, the vegetation type and distribution, temperature range, current velocity, substrate type (and underlying soil and geological features), food availability, and chemical factors all influence species composition and diversity, population density, and spatial distribution patterns. The relative importance of these factors to aquatic life differs, however, in lotic and lentic environments. For example, oxygen concentrations are often at, or even greater than, saturation levels in streams and rivers, whereas they may be greatly reduced in lake habitats. Substrate composition is generally much more diverse in streams. In lakes, any sorting of mineral particles generally occurs only on shores exposed to wave action. In this littoral zone the substrate may be quite heterogeneous and, interestingly, such lentic habitats are often occupied by typical lotic species. In localized segments of streams, temperature, water chemistry, and dissolved oxygen may be relatively homogeneous, but current velocity, substrate composition, and food source distribution usually differ within very short distances (Rabeni and Minshall, 1977). In localized areas of lakes, homogeneity in all of these features often occurs.

For many years, insects in aquatic environments like their counterparts in forests were sampled and analyzed in a two-dimensional context. Insects were described as benthic inhabitants, but were generally considered to occur only on the surface and in the uppermost layer of the substrate. However, insects in lakes very often occupy a third dimension, layers of the open water. This is in addition to the backswimmers, water striders, and other obvious surface dwellers. Many benthic species (i.e., those traditionally considered as occupying the substrate areas) also occur in the open water because of 1) a response to environmental disturbances (e.g., midge larvae), 2) a diel behavioral pattern (e.g., phantom midge larvae), and 3) an obligate part of their life cycle (e.g., planktonic first-instar midge larvae). In streams, this third dimension may be viewed as either 1) the water column in which insects enter and become part of the 'stream drift' (described below) or 2) the vertical distribution through the substrate area (the hyporheic zone). In fact, insects may be found deep within the substrate (in some cases, up to 1 m in depth) down to the level of bedrock.

In the following discussion, microenvironmental variation and behavioral factors influencing spatial distributions of insects in aquatic habitats will be examined. It may be appropriate to consider a comment made by the Greek philosopher Heraclitus almost 2500 years ago: "No man steps into the same stream twice." Although this was intended to be taken in the same fashion as his oft-quoted "There is nothing permanent except change," it may serve as a useful prologue to considering the dynamic processes that occur in aquatic habitats, and their implications to sampling design. In fact, comparisons of depth, current, and substrate features for the same segments of stream less than a season apart (Ulfstrand, 1967, Figures 1-6) suggest that Heraclitus' statement might be taken literally!

Oviposition and Egg Distribution. As in forest environments, the initial distribution of any aquatic population will depend on the deposition of eggs by the adult insects. As with insect populations generally, a species may be absent from an otherwise suitable habitat because conditions are not attractive to the ovipositing female. Selective oviposition occurs in a number of species of aquatic insects. For example, certain backswimmers insert their eggs directly into plant tissues. Ovipositing flood water mosquitoes lay their eggs in sites that will later be inundated by tidal water, seepage, overflow, or rainwater.

The oviposition behavior of insects in aquatic environments can be summarized as follows: 1) some species lay their eggs terrestrially, with first instar larvae either hatching after inundation with water or hatching after dropping to the water surface from eggs laid on overhanging vegetation; 2) many species scatter their eggs or egg masses directly on the water surface, often with some selectivity in terms of distance from the shoreline; and 3) other species may actually enter the water and select specific vegetation or other objects for egg attachment.

The selection of a particular oviposition site in terms of future larval survival and population success is probably less important in aquatic than in forest environments. Adult midges, for example, generally exhibit poor powers of flight and oviposition site selection, and the first instar larvae must disperse through the system and select substrates on which to live (Davies, 1976). This may be typical for many aquatic insects.

Aquatic insect eggs are deposited singly or in masses, with clutch size ranging from as low as 2-5 (Spongilla flies) to over 5000 (some mayflies) eggs/individual female. Generation mortality normally is very high. It has been estimated that as many as 1,000 mayfly eggs may be required to produce one adult of the next generation (Clifford and Boerger, 1974). Most aquatic insects have been described as r-strategists.

As with forest insects, the timing of egg hatch is very important. A large number of newly hatching larvae all appearing at the same time could result in a food shortage through overgrazing. Eggs of floodwater mosquitoes, for example, do not hatch at one time. Most of the eggs may hatch after the first flooding, but some remain for a second and subsequent floodings. Stonefly eggs undergo a diapause broken by a few at a time, with the result that larvae may emerge from a single batch of eggs for several months. Thus, in both these situations individuals of different larval ages may result from the same egg mass. This has a survival advantage, but it makes individual cohort distinctions, and consequently studies of population and production dynamics, very difficult.

Habitat Diversity and Larval Responses. The potential array and interactions of biotic and abiotic factors that can influence the distribution of newly-hatched insect larvae in an aquatic environment are immense. In lotic systems, species often select an area with a particular range of current velocities. This selection is largely a physiological response to the availability of dissolved oxygen. Habitat selection may also be related to the effect of current on feeding habits. Net-spinning, filter-feeding caddisfly larvae appear to select a particular current velocity based on the size of particles carried in suspension. Species that graze on detritus and periphyton located on rock surfaces also appear to select habitats according to current velocity (Mackay and Wiggins, 1979).

Current velocity and substrate composition also are interactive in regard to larval behavior and distribution. Species associated with coarse substrates may be reacting more to fast current than substrate size (Rabeni and Minshall, 1977), whereas species associated with finer substrates may either be adapted for burrowing, seeking deposited food materials, or merely responding to slower currents. Aquatic insects that feed on dead organic matter are usually found in slow current areas, but a few species also occupy leaf packs that are trapped in fast current areas (Mackay and Wiggins, 1979).

In lentic environments, heterogeneity in habitat features and the related spatial patterns of insect inhabitants is largely restricted to the littoral zone. In this zone, some species are restricted to microhabitats within the rooted aquatic vegetation, whereas others appear to distribute themselves on top or within the substrate. Although relatively few species occupy the deeper areas of lakes, the profundal zone, the uniform substrate composition of these areas often supports benthic populations with non-aggregated spatial distribution patterns.

With the exception of certain pond hemipterans and riffle beetles, immature stages and adults of the same species rarely occupy the same habitat and rarely can they be quantitatively collected with a single sampling technique. Unfortunately, population shifts also occur throughout the larval period of many species. For example, first instar larvae of many lentic midges are planktonic, with dispersal largely resulting from wind-induced water currents. They exhibit adaptations to a planktonic existence by way of positive phototaxis and an ability to feed on certain true plankton. Subsequent larval instars are generally confined to the benthic substrates (Davies, 1976).

Insects in lotic habitats may disperse as stream drift, the downstream transport of individuals by current. Drifting is a temporary event in the life of many stream insects, but it occurs very frequently in the younger life stages. Several species exhibit a diel periodicity in drift pattern, usually with a sharp increase in drifting activity at about the time of full darkness, some pattern of change during the night, and a sharp return to daylight levels at dawn (Waters, 1972). The tremendous quantities of drifting organisms have brought into question the productive capacity of streams to withstand such a high rate of attrition and the possibility of an upstream return by adults.

In addition to differential drift rates with age, many aquatic insects change habitats as they develop. Early instar larvae may occupy the hyporheic zone within the stream substrate, moving to the surface substrate layers as they get older. Many stonefly nymphs begin moving shoreward as they develop, some moving from gravel to moss or dead leaf areas of the stream as they grow, others moving from leaves or moss to stone. In the extreme case of certain southern hemisphere stoneflies, they actually leave the water and become semi-terrestrial! Seasonal changes in distribution are also common. Some caddisfly larvae move periodically to gravel or sand from other substrate areas when seeking particles of a certain size for casemaking or other activities.

Many aquatic insects have highly restricted distributions as a result of their feeding requirements. Filter-feeders (some caddisflies, blackflies, mosquitoes) are an example of this, as are *Ceraclea* caddisflies and spongilla flies that are predators of freshwater sponge and midge larvae that live in obligatory associations with aquatic vegetation, blue-green algae, and bryozoans. The distribution of these populations must reflect that of their food sources.

Pre-pupation and pre-emergence population movements are common in aquatic insects. For example, *Pycnopsyche* caddisflies burrow in specific grades of gravel for pre-pupation diapause, although at earlier stages in larval development, no positive response to particle size occurs (Mackay and Wiggins, 1979). Some pond caddisflies exhibit pre-pupal migrations from deep to shallow water. In a study of four coexisting populations of stream mayflies, one species occupied the same habitat during the entire portion of its life cycle, two species moved to shallower water prior to emergence, and one species moved to deeper water (Harker, 1953). The caddisfly, *Gumaga nigricula*, migrates from riffle and pool habitats to stream margins where hundreds of individuals may clump together for pupation and emergence (Figure 2). Aquatic beetles, megalopterans, and neuropterans actually leave the water and pupate terrestrially.

Pupation and Adult Emergence. Pupation in holometabolous aquatic insects may occur within a sealed cocoon (e.g., some Diptera and all Trichoptera) or the pupa may be active and free-winged (e.g., mosquitoes). The length of emergence is a function of the population structure. Many aquatic insects have multiple cohort populations, with adults emerging from early spring to late autumn. Single cohort populations usually emerge over a relatively narrow time span in temperate regions. Year-round emergence occurs in

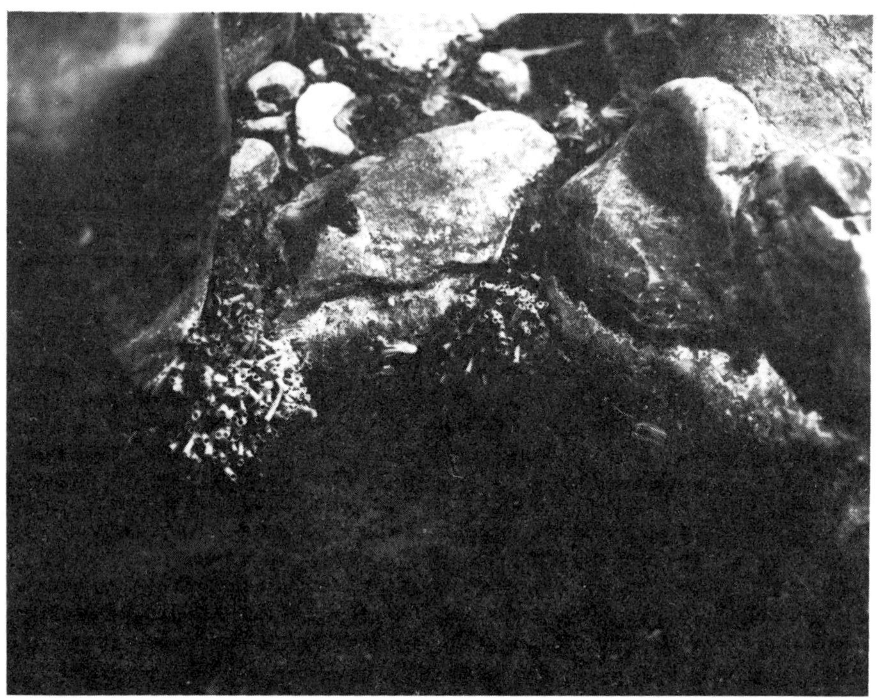

FIG. 2: Pre-emergence aggregations formed by final instar larvae of the caddisfly, Gumaga nigricula *(Trichoptera: Sericostomatidae) in a California stream (Big Sulphur Creek, Sonoma County, California, USA).*

some tropical populations. Diurnal periodicity in emergence patterns has been reported in several aquatic insect populations, also. Coexisting, systematically related species of insects with overlapping emergence periods have been shown to have relatively distinct flight activity periods, thus reducing interference between them in their search for mates. Since the adults of many types of aquatic insects are non-feeding (e.g., mayflies, stoneflies, caddisflies), adult life may be quite short and confined to mating and oviposition.

Differential mortality of insects in stream and lake environments may result from size-specific predation by fish and certain invertebrates, changes in the physical environment (e.g., spates, intermittent flow), and man-made alterations such as the addition of effluents and siltation. In regard to this latter feature, smaller individuals may be affected more severely than larger ones, e.g., from mechanical abrasion by silt.

5. SAMPLING -- PROCESS AND PROBLEMS IN FOREST AND AQUATIC ENVIRONMENTS

The complexities of insect biology, behavior, and habitat relations present distinct problems in the development of efficient sampling designs. As with the sampling of all biological populations, there are two basic requirements: (1) unequivocal specification of the objective(s) of the sampling, and (2) definition of the sampling universe, and recognition of its limitations relative to the population universe of the insect (or more particularly, of the stage of the insect in question). The second requirement logically relates to the first. In research studies, specification of the universe is usually selective and precise and this is generally recognized in the analysis and interpretation of the data obtained. In survey and monitoring operations, the sampling universe often is restricted in unspecified or inadequately specified ways, or it may not be really defined at all. Such data result in invalid, or questionable, extrapolations.

When successive samples of a population are required -- for either research or survey purposes -- and there is a population shift in whole or part from one segment of the habitat to another (as when the larvae of a forest defoliator are found on the trees and the pupae occupy the soil or litter, or when the larvae of an aquatic insect occupy riffles and pools and the pupae only shoreline areas), the numerical and statistical relationships of the population in these two sampling universes must be determined. This difficulty is apparently resolved when an absolute measure of population density, i.e., number per unit area, is used. But then, when such units are sufficiently large to be not limiting with respect to the two sampling universes (e.g., a hectare or acre), differences in critical features such as number of trees or range of aquatic substrate types become major sources of variation in insect numbers and spatial distribution.

For this and other reasons, two other aspects of sampling design which also relate to objectives are especially important in forest and aquatic environments: (1) selection of the sampling unit, and the number and distribution of units within the specified universe, and (2) selection of the unit of measurement, i.e., a direct count per unit of area or habitat feature (e.g., leaf, tree, stone); or an indirect count or measurement of frass drop, webs, nests, or feeding injury to host plants.

The timing of sampling is another basic element to be decided upon, as it too may constrain or bias estimates.

The relative importance of these elements of sampling design, or at least the attention given them, has differed somewhat in studies of forest and aquatic insects.

With forest insects, measures both of absolute population and population intensity are used. Absolute population data (number of organisms per unit area of land) are necessary for life table studies and other approaches whose purposes are to describe and analyze mortality and survival and to assess the individual and joint effects of mortality-causing factors in different habitats and over successive years on a common basis (Morris, 1960; Southwood, 1966). Population intensity (e.g., number of organisms per leaf, tree, stone) reflects insect response to food supply and living space in terms of habitat units actually utilized. Such data are most applicable when population levels are to be related to tree or stand damage (e.g., Carolin and Coulter, 1975), when comparisons of the numbers on different tree species or in different parts of a tree are desired (e.g., Stark, 1952), when spatial patterns are to be analyzed in terms of different tree or habitat units (e.g., Waters and Henson, 1959), or when density-related interactions of an insect with its natural enemies are of interest (e.g., Amman, 1969). In general, population intensity measures are more commonly used in forest insect studies and surveys, with emphasis on utilizing natural units of the habitat to enhance ecological interpretation (Waters, 1962; Cole, 1970).

Also, with forest insects, the great variety of life forms and behavior and the diversity of habitat conditions (e.g., tree sizes and foliage characteristics) have generated a corresponding variety of collection, extraction, and counting techniques and equipment. In fact, nearly every forest insect pest for which intensive sampling methods have been developed has its own more or less unique set of devices and procedures for sampling the stages of interest.

In contrast, with aquatic insects measurements of absolute population (i.e., numbers per quadrat sample area) are in general use (e.g., Waters and Crawford, 1973; Resh, 1977). Population intensity measures have been used in specific ecological studies, including insects on individual stones (e.g., Stout and Vandermeer, 1975; Resh, 1979), on aquatic plant shoots, and in specific substrate segments. Such measures have been used also in environmental monitoring programs utilizing both natural and artificial substrates.

Much time and effort has been given to the development of standardized equipment and procedures for sampling aquatic insect populations. Despite this, considerable diversity still exists (Cummins, 1962). Several hundred sampling devices have been described for collecting insects in running water environments alone (Elliott and Tullett, 1978). A comprehensive schematic listing of sampling devices appropriate for particular aquatic situations has been developed by Merritt, *et al.* (1978).

Aquatic habitats may change drastically within and between years due to factors such as changing water levels, within-stream plant growth, intermittent flow, and ice cover. The method and mechanics of sampling during one period may not be feasible for another, and the logistics of the overall procedure may necessarily have to be changed. Logistic problems arise similarly in forest environments at different times of the year.

A few examples will suffice to indicate the state of the art (or science?) in sampling insects in forest and aquatic environments.

3.1 Sampling Forest Insect Populations. The most fundamental problem in sampling forest insect populations is in developing a complete and unambiguous statement of objectives and related specifications and adhering to these rigorously in the design of the sampling plan. As generally recognized (by statisticians, at least), the following must be stated explicitly: 1) the purpose of the sampling -- what the data are to be used for, 2) the population and/or sampling universe concerned, 3) the precision of estimates desired, 4) the risk or probability levels for the errors of estimate, and 5) the particular cost criteria or limitations to be complied with.

Objectives and Design. In only a few cases have the sampling plans for forest insects -- usually species of major economic importance -- been based on specifications of all of the foregoing. The long-term study of the population dynamics of the larch bud moth (*Zeirphera diniana*) in the Engadin Valley of Switzerland (Baltensweiler, *et al.*, 1977), for example, is based on a hierarchical stratified random sampling scheme which provides detailed data on the spatial and temporal dynamics of the insect and its principal host, European larch, in a discrete geographic area (Kaelin and Auer, 1954). Adjustments have been made in the amount and distribution of sampling as population densities have fluctuated and as requirements for additional information and greater representativeness have been met. However, the effects of these changes on the precision of estimates have been ascertained, and precision has been maintained at acceptable levels (Auer, 1971). The study of the dynamics of epidemic spruce budworm populations in the Green River Watershed of New Brunswick evolved from intensive analysis of the sampling characteristics of budworm populations at high densities and related dendrometric studies of host trees according to specified objectives (Morris and Miller, 1954; Morris, 1955). In this instance, the desired precision of estimate per basic population unit was relaxed considerably to allow more representative sampling of budworm populations and of tree and stand conditions in the Watershed with the resources available.

Most often, the purpose of the sampling and the universe of interest are specified, along with the cost functions and limitations that pertain. However, the statistical specifications (items 3 and 4 above), rather than being related to the purpose at hand, are taken from prior or ancillary studies which have shown the sampling errors to be expected from the procedure to be used. At best, adjustments are made in the amount of sampling, without any significant change in the method itself. This is particularly true for forest insect surveys and population monitoring. Omission of the desired precision and risk levels in research studies can have serious consequences. For example, in numerous life table studies of forest insects, the sampling errors involved in the estimation of densities in successive stages have been ignored, and their effects are unknown (e.g., Campbell, 1967). The seriousness of such errors when multiple regression and related techniques are used to discern the 'key factors' affecting an insect population has been shown graphically by Kuno (1971).

The quintessence of inattention to objectives, perhaps, is in surveys utilizing laborious and time-consuming direct counting methods to arrive at a mean density estimate, which then is converted to some *class* of infestation, rather than utilizing a sequential sampling technique to classify the infestation directly (e.g., Waters, 1955; Shepherd and Brown, 1971), or a relative population index requiring less sampling effort (Morris, 1960).

The Universe -- What is it? What constitutes the universe of interest is usually arbitrary. It may be delimited spatially (i.e., geographically or by within-habitat strata) or temporally (i.e., relative to times of the year or to particular insect stages). Or it may not be delimited at all.

In the larch bud moth study in Switzerland, the population universe was specified to be that portion of the upper Engadin Valley containing natural stands of larch. This was essentially an elevational band encompassing 62 km^2 of the 120 km^2 total area of the Valley. Only the larval stage was censused each year. As the larvae occupy and feed on the foliage, the foliated portion of the trees' branches constituted the sampling universe for purposes of the study. Neither the egg stage, which is found under bark scales and lichens on twigs, branches, and the tree trunk, nor the pupal stage, which occurs on the ground, was sampled. Both of these stages would have required a different sampling universe and would have been subject to considerable error in the mechanics of sampling. To obtain wider representation of the larch host type, the study was eventually extended to four other subalpine outbreak areas, and supplementary sample plots of restricted size were established at 20 additional locations. Data from these places provide a basis for generalization of the findings in the Engadin Valley. However, the annual estimates of larval density and distribution permit analyses only of year-to-year changes in population levels and associated tree damage. This precludes the analysis of within-generation mortality and determination of the relative importance of the within-generation factors influencing the annual fluctuations.

For the spruce budworm study in New Brunswick, the population universe was not precisely defined. Rather, a series of plots approximately 10 ha in size were distributed throughout the Green River Watershed, which covers about 1050 km^2. The plots were located in stands with homogeneous conditions, all differing, however, in characteristics related to budworm activity. They, in effect, represented different population universes of the insect; they did not comprise in any way a random sample of the Watershed. This was consistent with the primary objective of the study, which was to gain insight and understanding of the dynamics of budworm populations under different forest conditions that could be adequately described and compared quantitatively (Morris, 1955). The sampling universe, then, was the foliated portion of host trees (primarily balsam fir) in which all stages of the insect, except the winged adults, were found.

For other research purposes, the population universe of interest may be even more limited, e.g., a particular forest stand or plantation (such as in the study of the pine looper, *Bupalus piniarius*, by Klomp (1966)), or a small group of trees (for the study of the winter moth, *Operophtera brumata*, and associated insects by Varley and Gradwell (1968), for example). Interpretations of analyses of data from such studies must be limited accordingly.

In most forest insect surveys, the population universe is not defined, or only broadly so. Survey observations and/or sample collections are made at various points, sometimes systematically, and the data obtained are used to prepare maps of the infestation or to make control decisions. Each sample contributes a discrete bit of information, but it is not expanded in any sense to a 'population.' In some cases, the universe of interest is defined arbitrarily by civil units (e.g., province, state, county, town) or by ownerships for administrative purposes.

With surveys directed at pest insects, the sampling universe may comprise all habitat components utilized by the stage(s) for which a population fix is desired. More often sampling is restricted in some way: to certain stands or areas of high value or high hazard, to particular tree species (those of commercial interest), to certain age-size classes of direct concern to forest managers, or even to particular strata within trees (for ease of sampling). Restriction of the sampling universe, just as limitation of the population universe, may have a practical advantage and the statistical justification that it reduces variability. But there are risks and uncertainties involved. How representative is the restricted sample? Does it comprise a consistent proportion of the population in different forest habitats, in different stages of the insect, in successive years? Will some useful information on the spatial distribution of the pest be missed? The utility of restricted sampling may seem evident. But its reliability and interpretation rests on a thorough understanding of the biology and behavior of the insect in question.

Sampling Unit Selection and Allocation. The choice of the sampling unit often defines the sampling universe.

For foliage feeding insects, the basic sampling unit may be either a specified biomass (weight) or a surface area measure of foliage. The larch bud moth study, for example, uses a weight measure. The complete sampling frame for this study comprises five levels (Auer, 1971).

(1) The TREE -- approximately 1 kg of branches with foliage is collected at random from each sample tree, and the number of larvae found is recorded as number/kg foliage and twigs. Different sets of sample trees are selected each year.

(2) The CLUSTER -- a group of sample trees in a small area with homogeneous conditions, selected at random each year.

(3) The STRATUM -- a geographic area of a given elevation and directional exposure, containing a number of clusters.

(4) The SECTION -- a transection of the entire valley, encompassing 2-4 strata.

(5) The VALLEY -- the entire area of the study, i.e., the larch stands in the upper Engadin Valley.

For purposes of analysis, both numbers of larvae per kg of foliage and twigs (population density) and numbers per hectare (absolute population) are used. The latter are derived by conversion factors of average foliage biomass per unit tree crown volume and average tree crown volume per hectare. The sampling plan for this study is probably the most complete yet devised for a forest insect research study.

The basic observational unit in the spruce budworm study was the foliated portion of a branch, with the numbers of insects present expressed as number per 10 ft^2 of foliage. This unit was used for eggs, larvae, and pupae, since all these stages occupy the foliage area of host trees. Because vertical distribution of the budworm varied significantly between and within stages as well as between plots, sample branches were taken at random from four crown levels, two each from the upper two quarters and one each from the lower two quarters. This distribution of sample branches approximated the actual distribution of foliage surface area in the live crowns of normal balsam fir trees, as determined by adjunct dendrometric studies. Thus, the sampling intensity, i.e., the amount of branch surface area examined, was approximately the same for all crown levels. This provided a representative sample, whatever the distribution of the insect in the tree, and a weighting of the counts by the method of collection rather than by calculation later on. For purposes of analysis, the group of six branches from each sample tree was considered the sampling unit, with the individual branch termed simply the collection unit (Morris, 1955). Statistically, it was found necessary only to count the insects on one side (longitudinally) of each sample branch. The side to be examined was selected by a random procedure.

Initially, the sample trees in each plot were selected in clusters of 10 trees each, with at least two clusters per plot. The variance in counts between clusters proved to be not significant -- as might be expected since the plots were in stands of homogeneous character -- and the data for all trees in each plot were then pooled. From calculations using both original and transformed data, it was determined that optimum sampling in terms of the number of sampling units required for a given precision was attained by taking just one sample (i.e., six half-branches) per tree and a variable number of trees per plot, depending on population density and stage of the insect. For the range of population densities in the study plots over the epidemic period of the study, it was found that 50 trees or less per plot generally would provide estimates of mean densities within ± 10 percent (odds 2:1) for all stages sampled. Sampling errors in excess of this undoubtedly occurred when egg masses (the most variable stage) were sampled at low densities.

With this study, also, regressions were calculated from the dendrometric data to convert the basic population intensity data to absolute population terms for analysis. Overall, the purposes of the Green River study were well served by its sampling plan, as evidenced by the many publications that came forth on the dynamics of spruce budworm populations, insect population theory and modeling, and methods of sampling this and other defoliators of forest trees (e.g., Morris, 1955, 1963, 1965; Morris and Miller, 1954; Miller, 1955; Watt, 1959, 1961). Studies of the spruce budworm under endemic conditions in New Brunswick were continued, but unfortunately the results have not been published as yet.

Sampling and Spatial Distribution. Sampling studies approached strictly from the standpoint of spatial distribution of major life stages can provide much basic information for the development of efficient methods of sampling. Lyons (1964), for example, made detailed counts and analyses of the densities and frequency distributions of egg clusters and cocoons of the Swaine jack pine sawfly (*Neodiprion swainei*) on its native host tree and of the European pine sawfly (*N. sertifer*) on red pine in eastern Canada, using various tree and ground area units and different means of stratification to characterize spatial distributions and obtain density estimates of desired precision at minimal cost. He found with both species that the density of egg clusters, typically laid on the needles of their host trees, varied directly with height in the trees, branch length, and tree size, evidently reflecting the adult females' preference for the more illuminated, or more exposed, parts of the tree for egg laying. The frequency distributions of egg clusters per tree also were influenced by population density, tree density, and size of sampling unit (in the case of ground area units). The precision of mean estimates was increased and the tendency to overdispersion, i.e., non-randomness, was reduced greatly by stratification by tree size and density and by reduced quadrat size. Sampling time (and cost) was related directly to population density, tree density, and tree size. Optimal sampling schemes for estimation of egg cluster densities and the interpretation of observed frequency distributions were proposed, taking these factors into account.

Cocoon densities were largely a function of the size of tree and density of the larval population above the ground sampling units. This was especially pronounced when the trees were sparsely distributed, or at least not overlapping in crown projection area. In a dense stand of trees, however, dispersal of the larvae on the ground prior to the forming of cocoons tended to make cocoon densities more uniform on an area basis. No benefit of stratification was indicated generally, but the cost of sampling could be minimized by adjusting unit size to stand conditions. Cocoon density, degree of aggregation, soil type, and method of extracting cocoons from soil material also affected optimal sampling unit size.

The basic information on the spatial distributions and sampling statistics of these two pine sawflies provided by Lyons have been applied to studies of their population dynamics, and have proved extremely useful in monitoring the long-term effects of insecticide applications on Swaine jack pine sawfly populations in Quebec (McLeod, 1972).

Considerable experience has been gained from sampling studies of forest insects in distinguishing the biological and statistical components of spatial distribution (Waters, 1962) and in avoiding some artifacts of sampling that affect frequency distributions (e.g., Waters and Henson, 1959; Lyons, 1964). This information has been important in both the design of sampling schemes and in the analysis of sample data.

Bark Beetles and Borers -- A Special Problem. Bark beetles (Scolytidae) and other insects infesting the main boles of trees generally present a simpler situation for sampling. The tree bole is their universe; the only departure from it occurs when the new adult beetles emerge and fly to other trees. Detailed studies have been made of the patterns of attack and emergence and of within-tree populations of numerous bark beetles, mostly species of economic importance in the genera *Ips*, *Dendroctonus*, and *Scolytus*. The process of sampling generally has been one of taking samples of bark of a specified size (surface area) and shape either at a fixed level or at several levels over the infested length of the bole.

Indirect estimates are made by counting the entrance and/or emergence holes using some device to delimit the sample unit area. While this is a non-destructive procedure and allows sampling to be conducted in the field, it invariably is inaccurate since each hole can represent more than one insect. With entrance hole counts, the relationship to the true number of individuals depends on the attack density and mating behavior of the insect. Typical complications with the western pine beetle, for example, include (1) more than one female (the sex initiating attack on a tree) utilizing the same hole, with one or more males following, and (2) individual females attacking more than one tree, with or without male companions (Miller and Keen, 1960). The density and patterns of attack by this species are strongly influenced by attractant chemicals produced by the insect and its host tree (Wood and Bedard, 1977). With species of *Ips*, each male (which initiates attack and bores into the inner bark) normally is accompanied by 2-6 females. Attractant, aggregating chemicals are involved in this behavior, also.

For exit holes, the number of insects represented by each hole is largely a function of the density of surviving insects and the proximity of pupal sites within the tree. Also, the activity of woodpeckers and other predators often obscures the holes.

'Correction factors' have been developed to obtain estimates from hole counts of the 'true' numbers of bark beetles entering and leaving host trees (e.g., Sartwell, 1971). These, however, must be considered to have limited applicability and reliability because of the statistical, biological, and mechanical factors involved, and the hole counts themselves are best used as *indices* of population levels and trends.

Direct estimates of within-tree densities of bark beetles require the removal of bark sections from the tree and utilization of dissection, radiographic, or rearing techniques to obtain counts. For emergence counts, traps placed on sample trees also have been used (McClelland, *et al.*, 1978). All of these methods have some technical and mechanical problems. Their accuracy and reliability depend on the insect involved and the physical nature of the bark substrate. Dissection, for example, is slow and laborious, and extreme care must be taken to find all of the beetles in each bark section. With radiography, it often is difficult to distinguish species of similar size and shape and to separate the stages of a particular insect (when more than one stage is present, as is usually the case). Also you cannot be sure whether an insect is alive or dead. In rearing and trapping, some insects may die in the process before emergence. With these methods, dissection of a subsample of bark sections generally is recommended to determine how much, if any, mortality has occurred and to correct the rearing or trapping counts accordingly.

The vertical distribution of bark beetles within trees is not uniform (e.g., Berryman, 1968; DeMars, 1970; Safranyik, 1971; Mayyasi, *et al.*, 1976). Attack patterns differ, but the intensity of attack generally is highest in the central portion of the infested area of the bole, tapering off toward the ground and upward in the tree. The infested length may be quite limited (i.e., the attacks are concentrated) or they may extend from 1 m or less from the ground well up into the tree crown. Generally, the higher the population level in a forest area, the greater will be the infested length on individual trees. This relationship is not a close one -- not sufficiently so for estimation purposes at least -- since tree size and condition, predation, and other factors affect within-tree distributions of bark beetles.

A variety of sampling unit sizes, shapes, and configurations have been tested for sampling populations of the major bark beetle species (e.g., Carlson and Cole, 1965; Berryman, 1968; Stephen and Taha, 1976). Final selection of unit size and shape generally has been a compromise between statistical efficiency and practical expediency. For example, the basic unit for estimating attack densities and brood numbers of the western pine beetle is an 88 cm^2 circular bark section, cut and removed from the tree with a portable circular-cut saw (Dudley, 1971). In recent studies of the southern pine beetle, Coulson, *et al.* (1975) have used a 100 cm^2 circular section of bark, cut with a similar instrument. The method of Carlson and Cole (1965) for the mountain pine beetle utilizes a square unit 15.24 cm × 15.24 cm.

Because of the within-tree gradients in insect densities, either stratified random sampling or systematic sampling at specified height levels generally is recommended. The most detailed study and mathematical analysis of within-tree populations has been of the southern pine beetle, using a tree geometry model to estimate the infested surface area of individual trees (Foltz, *et al.*, 1976) and a probability density function to estimate average insect density (Mayyasi, *et al.*, 1976). The entire procedure is described, with practical examples, by Pulley, *et al.* (1977).

The TG-PDF method has been extended to the estimation of populations of the southern pine beetle in infested groups of trees (Foltz, *et al.*, 1977). It first requires stratifying the group into subgroups of trees containing approximately the same life stage(s) and then selecting at random a given number of trees within each such stratum to obtain estimates of the desired precision. With this approach it was determined that acceptable estimates could be obtained by taking four 100 cm^2 bark disks at one height level, either 5 m or 2 m above the ground depending on the life stage predominating, from a reasonable number of trees. In operation, the method requires preliminary information on the total number of infested trees in the group, the diameter sizes (measured 2 m above the ground) of the infested trees, and the predominant life stage of the beetle within each tree.

Less detailed methods for estimating the average densities and/or total numbers of bark beetles in infested stands have been proposed (e.g., Carlson and Cole, 1965; Stephen and Taha, 1976). All of these procedures, including the one described above for the southern pine beetle, assume that all trees (living and dead) containing insects can be identified. Since this is not always the case, the sampling universe, i.e., the set of trees from which the trees to be sampled will be drawn, may represent a large, but unknown proportion of the population universe of the insects in that location. How representative they are of the population with respect to density and other characteristics being measured, then, is questionable. More work is needed in general to better quantify the between-tree variation in bark beetle numbers in discrete universes, and to apply this information to practicable field survey methods on a larger scale.

Improvement of Sampling Efficiency. Two approaches have been developed to improve the efficiency of sampling forest insects on an extensive scale -- for both research and survey purposes. First is the use of aerial reconnaissance and remote sensing techniques to identify areas of infestation, to classify them in terms of intensity and extent of visible damage, and to provide the basis for the first level of multistage sampling of the causal organisms (e.g., Langley, 1971). Most recently, color aerial photography has been incorporated into the monitoring aspects of the pest management systems for the southern pine beetle (Schreuder, *et al.*,

1978), the mountain pine beetle (Klein, 1973), and the Douglas-fir tussock moth, *Orgyia pseudotsugata* (Heller, *et al.*, 1977), in the United States. The procedures developed provide the basis for (1) stratification of forested areas according to stand conditions and other habitat features related to the probability of outbreaks, and (2) designation of high hazard stands in which more intensive sampling of the pest populations by ground methods should be conducted.

Second is the application of sequential sampling for direct classification of insect densities and other population characteristics. The rationale, utility, and methodology of this approach to forest insect surveys have been described by Waters (1955, 1974). Sequential sampling plans have been developed for a wide variety of forest insect pests in the United States and Canada, including active, feeding stages causing damage, e.g., winter moth (Reeks, 1956), Douglas-fir tussock moth (Mason, 1969), and roundheaded borers (Safranyik and Raske, 1970), and egg populations -- for prediction of potential population levels and damage -- e.g., forest tent caterpillar (Shepherd and Brown, 1971), larch sawfly (Ives and Prentice, 1958), and Nantucket pine tip moth (Waters, 1974). To be reliable, sequential sampling requires detailed knowledge of the frequency distribution of counts or observations with the sampling unit used and of the relationship of insect numbers or infestation level to tree and stand damage and other habitat conditions. It is directly applicable where threshold values can be specified for assessment, prediction, or decision. It is most helpful where the examination and counting of sample material is slow and tedious relative to the time and effort needed to acquire the samples.

It should be evident from this discussion that behind the biological and statistical elements of sampling insects in forest environments lie many, seemingly intractable, problems of logistics and simple mechanics. The requisites of representativeness and randomness inevitably are compromised by time and cost considerations. In research studies, a sampling plan may be developed from the standpoint of achieving acceptable precision at minimum cost. In forest insect surveys, it invariably is a matter of maximizing precision with the resources available. In both cases, a great deal of reliance is placed on *replication* -- in time and space -- to reveal patterns of occurrence and change in insect numbers.

3.2 Sampling Aquatic Insect Populations. In aquatic insect studies and surveys, more so than in most other fields of ecology, the difficult four 'S's' -- *sampling, sorting, species,* and *statistics* -- all must be taken into account. In addition to the basic considerations of where, when, by what method, and how often to sample, there are special problems of removing the specimens from the samples and then identifying them accurately. Each of these features, together with appropriate data analysis, must be integrally related to specific objectives. The success of any study ultimately depends on how well this integration is achieved.

Unfortunately, because of the time and cost generally involved in sampling aquatic insects, the tendency has often been to follow a predetermined, 'cookbook' sampling design regardless of the study objectives. Benthic sampling procedures are too often prescribed as a set number of samples taken at regular intervals over an annual cycle, with sampling limited to restricted habitats. Extrapolations then are made to other, often dissimilar, areas. As Cummins (1975) has noted, the "procedures are merely stated in the methods section of publications with no consideration of the implication for, or constraints on, the conclusions presented later -- usually calculations of diversity indices or production (turnover) rates."

For example, secondary production estimates of aquatic insects and other benthic species are of prime interest to both ecologists and resource managers. Single-species estimates of annual production of aquatic insects vary greatly between species, populations of the same species, and often within subdivisions of a single population. These studies require repeated sampling of a population over time. This in itself involves logistic difficulties since seasonal changes in depth and discharge (e.g., Figures 3 and 4) may hinder sampling accessibility, and the calculation of production estimates based on a small number of samples per interval may lead to compounded errors and gross misinterpretations of the energetics of aquatic insect populations (Resh, 1979).

Equally important in the design and analysis of such studies is the sampling frequency. In the majority of aquatic insect secondary population analyses, the experimental design has been developed for benthic samples to be collected at regular intervals (e.g., monthly, bimonthly) throughout an annual cycle. This raises a basic question regarding study objectives: is this regular spacing of sampling intervals related in any way to the life history of the insect -- will it provide the information desired? Cummins (1975) and Waters (1969) have stressed that when the objective of a sampling program is secondary production, the use of long, equally spaced sampling intervals provides maximum information about the least frequently encountered type of survivorship, linear decrease, while it provides minimal information on nonlinear survivorship, the pattern that is typical for most benthic species (Resh, 1979). Long sampling intervals may, in fact, mask the true form of survivorship. Moreover, in production studies, the mean individual weight (MIW) of the larval insect

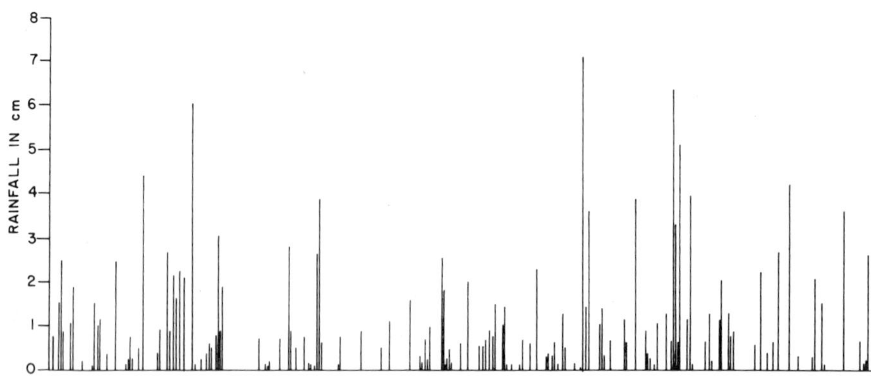

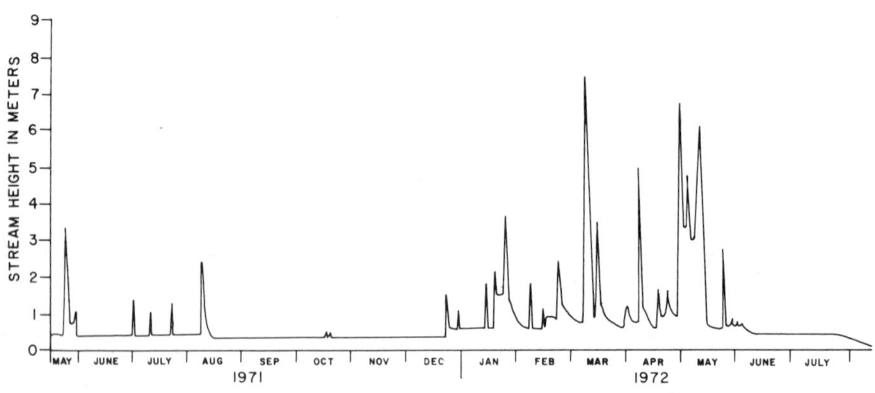

FIG. 3: Changes in precipitation and stream height over an annual cycle (Salt River at Taylorsville, Spencer County, Kentucky, USA).

FIG. 4: Seasonal changes in Brashears Creek (at Rivals, Spencer County, Kentucky, USA): summer (1971), *winter and spring* (1972), *(Resh,* 1977, *Plate 1).*

population must be accurately determined, especially for full grown, final instar individuals (Resh, 1977). A sampling regime with long, regularly spaced sampling intervals may fail to accurately measure the maximum MIW and consequently result in an underestimation of secondary production (Resh, 1979).

Faunal inventories and surveys used in the preparation of site selection and pre-impoundment reports and other types of environmental impact analysis also may have sampling designs that are not in keeping with study objectives. For example, aquatic insect surveys in these types of studies are generally based almost exclusively on benthic collections, i.e., those that concentrate on the aquatic, immature stages of insects. In addition to the taxonomic difficulties involved (Resh, 1979a), almost all sampling devices used in such studies do not capture insects occurring in the hyporheic zone. This may result in gross underestimations of numbers and biomass, as well as incomplete faunal listings. For analysis of possible environmental deterioration, data on the hyporheic fauna may be of prime importance, since siltation and other perturbations may drastically reduce this segment of the fauna, and consequently, the aquatic habitat's productivity.

An alternative to benthic sampling in faunal studies is adult collections using ultraviolet light-trap techniques (Resh, *et al.*, 1975). Unfortunately, these are selective for dusk or night flying species, and generally provide no information about habitat preference. A combined benthic and adult sampling approach, along with an aquatic rearing program, may be the best way to proceed in such inventories. However, the value of the inventory approach in environmental assessment has itself been questioned (Andrews, 1973; Patil and Taillie, 1979).

The use of community diversity indices in environmental impact analyses also illustrates the problems that may arise when sampling regimes do not follow study objectives. For example, if a measure of faunal diversity in a particular aquatic system is desired for baseline information, transect sampling which traverses the range of habitats in the system will yield far more information than confining samples to a single stratum, or habitat. However, in impact studies of aquatic environments, benthic diversity analyses are generally confined to single habitats (e.g., riffle areas) because of apparent sampling device limitations or the cost and time constraints in sampling several habitats. Consequently, they do not provide a valid data base for subsequent comparisons.

Community analyses are currently in vogue as a means of quantifying the dynamics of benthic systems. However, when the objectives of a study involve environmental monitoring, long-term evaluations of biotic patterns, or other such features of aquatic insect dynamics, financial and logistic considerations often require some design limitations. Undoubtedly, the object of any sampling design should be to maximize the relevant information obtained per unit effort. Cummins (1975) has reported that from previous studies of stream insects it appears that the compromises made when limited manpower is involved in a faunal study often render the data of little value and that under most conditions, it is much more profitable to focus on several key populations than to attempt to monitor the entire fauna.

The Sampling Universe. The above discussion of sampling problems in production and diversity analysis leads logically into a consideration of the importance of carefully defining the sampling universe and the recognition of its limitations relative to the population universe of the insect. Since the majority of freshwater species have aquatic immature stages and terrestrial adult stages, the same sampling approach will rarely collect both. Because of difficulties in sampling adults with both attractant and non-attractant traps, and relating catch to population density and distribution, quantitative sampling is generally confined to the aquatic stages.

Thus, the distinction between the population universe and the sampling universe is very important in developing sampling designs for quantitative aquatic insect studies. The population universe of aquatic insects includes a three dimensional distribution aspect (i.e., hyporheic and planktonic as well as substrate surface occurrence). The sampling universe, in older studies particularly, has ignored this third dimension. This has drastic implications in terms of the reliability of many past studies. For example, estimates of benthic secondary production in the Speed River, Ontario, Canada (Hynes and Coleman, 1968) are almost an order of magnitude greater than other benthic community production estimates (listed by Waters, 1977, Table IV) due to the inclusion of large numbers of invertebrates from the hyporheic region of the stream bottom, a population component generally not considered in the other studies.

Examples of differences between the population universe and the traditional sampling universe of aquatic insects in lakes can also be cited. For instance, early instar, planktonic midge larvae in lakes (Diptera: Chironomidae) require special sampling devices quite different from those used in sampling benthic portions of the population. Likewise, the phantom midge (Diptera: Chaoboridae) which dominates the benthic fauna of the profundal zone of many lakes has early larval instars that are always limnetic and positively phototactic. In the third instar, larvae are mostly limnetic but also occur in the benthic sediments. The fourth (and final) instar of many *Chaoborus* species exhibit a diurnal rhythm including movement from the sediments (or their daytime depth) to the surface strata at sunset; they stay there until about sunrise before descending (Wetzel, 1975). In this case, both age and specific behavioral activities must be considered in defining the sample universe and designing sampling programs.

In stream environments, many insects (including some that appear to occupy the hyporheic zone during part of their lives) also enter the stream drift. Like planktonic lake insects, this part of their population universe must be sampled separately from the benthic portion. Some studies of benthic production have been based on rates of insect drift with the hypothesis that the difference between the amount of drift entering and leaving a given section of stream is equivalent to the secondary production of that section (Waters, 1972).

After delineating the sampling universe, the choice of sampling approach in assessing aquatic insect dynamics is of major importance. Simple random sampling, although unbiased, is usually the least effective in terms of understanding general biotic associations (Cummins, 1975). Stratified random sampling, in which subdivisions of the main habitat (e.g., riffles, pools) are sampled by a random design (Cummins, 1962), is much more efficient (Elliott, 1977). Transect sampling, which takes maximum advantage of previous information about the ecology of the environment (e.g., flow and substrate distribution) may be most suitable for many studies since the chances of missing any of the general biotic associations are extremely small (Cummins, 1975).

The Sampling Unit. In contrast to terrestrial habitats such as forests, major difficulties in making absolute population estimates of aquatic insects lie in actually taking a sample of a known unit size (Southwood, 1966). For measures of larval population intensity, the sampling unit may be a vascular plant shoot or leaf, an algal mass (e.g., *Nostoc*, Brock, 1960), individual stones (e.g., Stout and Vandermeer, 1975; Resh, 1979), or other sessile organisms (e.g., sponges, Resh, 1976). These are extremely useful as units of measure since they may be considered as the 'effective habitat' where all inter- and intra-specific interactions occur. However, in most aquatic insect population studies the sampling unit has generally been a standard quadrat measurement (e.g., ft^2, m^2).

Studies of sampling unit size and sampling efficiency have suggested that a small unit is more efficient than a larger one when the dispersion of a population is contagious or aggregated (Beall, 1939; Finney, 1946; Elliott, 1977). However, if the dispersion of a population is truly random, all quadrat sizes are equally efficient in the estimation of population parameters (Elliott, 1977). Unfortunately, the vast majority of aquatic insect studies have relied on sampling devices that have quadrats of a fixed dimension. In streams, this is generally the Surber sampler (or some modification of it). The base of this sampler is 1 ft^2; however, depending on the depth to which the substrate is removed with this sampler, the total unit volume sampled will vary. For example, if substrate is removed to a depth of 3 inches, the unit volume would be 0.25 ft^3; if substrate was only removed 1 inch, the unit volume would only be 0.083 ft^3. In lakes where various dredges and grabs are used, different substrate characteristics may also result in differential penetration and similar variability in sample volume.

Catch per unit effort measures are widely used in dip sampling of immature mosquito populations and in time-interval sampling, particularly in faunal surveys. Adult mosquitoes and other biting insects whose immature stages are aquatic are sampled in a similar catch/unit effort manner using human, animal, or other attractants as bait. By their nature, these techniques are highly dependent on the sampling attributes of the individual collector.

Sampling units have been developed for comparative analysis of aquatic insect drift by measuring the total quantity of organisms drifting past a given point per 24-hr period, divided by the total discharge of the stream as a measure of stream size (Waters, 1972). Unfortunately, drift rates and species composition vary greatly in response to water temperature and light intensity and such factors compound difficulties in quantifying and comparing these estimates both temporally and spatially.

The size of the sampling unit in aquatic insect population studies should relate better to study objectives and the population universe than normally (or by chance) occurs with a fixed quadrat size. When the number of individuals collected in replicate samples with a fixed dimension quadrat is to be used to relate environmental features with aquatic insect numbers and distributions, the scale of these features has to be large in relation to the sampler dimension to enable significant differences to be shown (Allen, 1959). The lack of flexibility in modifying sampler dimensions is a major shortcoming of the majority of quantitative aquatic insect studies.

Sample Size Requirements. Presumably, given specific study objectives, precision of estimates needed, and appropriate methods for making observations to obtain this information, the necessary number of samples in a field study could be calculated and then this number of samples collected. However, in aquatic insect studies, as in many other ecological analyses, this procedure is rarely carried out. In fact, Eberhardt (1978) has commented that in the application of sample size considerations, the procedure is usually done in the opposite direction. That is, sample sizes are fixed largely by the money and manpower available, with calculations being carried out mainly to see what degree of precision is achieved.

The number of samples necessary in any ecological study is a function of three factors: the density of the population under examination, its degree of aggregation, and the desired level of precision. Rare species are extremely difficult to sample for both mathematical and biological reasons, i.e., in addition to requiring large numbers of samples because of low densities, rare populations are often highly aggregated.

In lotic environments, there is probably no more complete a study of benthic sampling variability and spatial heterogeneity than that of Needham and Usinger (1956). This study did exhibit various 'state of the art' limitations in that some physical measurements (e.g., current, substrate composition) were incomplete or lacking entirely, taxa were not identified (or distinguished) below the generic or family level, and as indicated by Chutter (1972), there are both sampler use limitations and a need for statistical corrections in their data analysis. Despite this, the publication of Needham and Usinger's raw data from 100 quantitative samples taken in a uniform riffle provides a basis for detailed examination of both sampling variability and the spatial relationships of the aquatic insect species in that benthic community.

In terms of actual numbers of samples required for a specified degree of precision, Needham and Usinger's (1956) results indicate that sample sizes (for the Surber square foot sampler) ranging from 12 to 2500 are required for estimating mean density $\pm$ 40% at odds 19:1 for the 44 populations occurring in that riffle (Resh, 1979). This range of sample sizes and especially the minimum number indicated (even at this relatively low precision) to quantitatively sample populations in this single habitat is very significant when compared to the number of samples actually collected in the vast majority of quantitative studies of stream insect populations. In experimental designs of these projects, predetermined sample sizes of as few as 2 quantitative samples per sampling interval (e.g., monthly) are extremely common and rarely are more than 5-10 samples collected per interval! These low sample sizes are in part due to the difficulty and expense in sorting, identifying, and enumerating organisms collected with fixed dimension quadrat samplers. Without question, too few samples is another major shortcoming in the design of many aquatic insect studies.

Spatial Distribution Pattern. The scale of the sampling unit must also be considered in regard to population distribution. Aggregated or contagious distributions can result from an organism's response to variation in environmental features. Depending on the size of unit used, however, the frequency distribution observed may range from random to aggregated. Similarly, if the sampling universe contains a significant area that is not part of the population universe, the distribution may appear nonrandom because of the preponderance of zero counts.

Some aquatic insect populations will show broader ranges than others in their ability to occupy areas that cover a variety of environmental conditions (e.g., current velocities, substrate characteristics, water chemistry), but all populations generally exhibit density differences in response to these (and other) environmental irregularities. Obviously, a judicious choice of the sampling universe, sampling unit size, and the distribution of samples will enable the spatial arrangement of individuals in the population to be more accurately determined and not merely to appear to be random or aggregated as the result of sampling artifacts.

In terms of spatial distribution patterns of component populations, a reexamination of Needham and Usinger's (1956) data for a uniform riffle indicates that with the exception of certain extremely rare species (mean densities = 0.01 - 0.2 organisms/ft^2) all populations in their study riffle exhibited aggregated spatial distribution patterns (Figure 5). This characteristic typically holds true when the data from similar studies are examined closely (Resh, 1979). In fact, because of the range of microhabitat variability in stream environments, a truly random quadrat sampling scheme will almost invariably show indications of aggregated spatial distributions among component aquatic insect populations. Habitat stratification, however, may more accurately indicate organism-habitat interactions. Lamberti and Resh (1979) analyzed substrate composition and sampling variability of a caddisfly population in a pool segment of a California stream. In areas of near uniform substrate size, populations exhibited non-aggregated distribution patterns, whereas in mixed substrate areas aggregated distributions occurred. This was, in part, related to the scale of environmental irregularities and the body size of the study organisms. However, it was clearly shown that stratification of the entire pool area would result in less sampling effort required for a given precision. Other studies in running water have also suggested that even within a well-differentiated, relatively uniform habitat (e.g., a stream riffle), variability may be even further reduced by more narrowly defining the sampling area with regard to current (Chutter and Noble, 1966), depth (Chutter, 1972), or a combination of physical factors (Allen, 1959; Mason, 1976). Thus, it would appear that in many aquatic insect studies, habitat

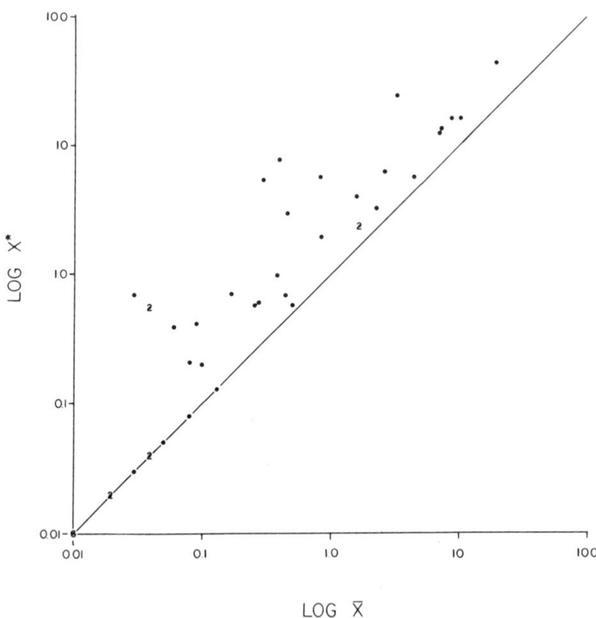

FIG. 5: 'Mean crowding' (x^*) *as a function of mean density* ($\bar{x}$) *for the data reported by Needham and Usinger* (1956) *On* 44 *macroinvertebrate taxa from* 100 *Surber samples collected in a uniform riffle in Prosser Creek, California. Each point represents a single population; where several points coincided, an Arabic numeral and point is plotted. Points along the center line represent species whose distribution was indistinguishable from random.*

stratification to reduce sampling variability and improve sampling efficiency should be incorporated into the sampling scheme.

In contrast to lotic systems, both aggregated and non-aggregated patterns have been reported in lentic insect populations. In fact, early and late instar midge larvae may change from aggregated to non-aggregated patterns temporally (Thut, 1969; Shiozawa and Barnes, 1977). Unfortunately, many lentic studies have considered spatial distributions at the family or generic level and possibly different patterns would have emerged had they been examined at the species level.

The profundal zone of many lakes is relatively homogeneous and requires far less sampling effort than heterogeneous stream systems. In fact, Jonasson's (1972) excellent study on benthic population dynamics in the profundal zone of Lake Esrom reflects the sampling attributes of this habitat type.

Ceraclea ancylus -- An Example of Within-Stream Spatial Pattern Analysis. In both lake and stream environments, aggregated distribution patterns of aquatic insects may result from the initial deposition of an egg mass by an ovipositing female. Some populations may subsequently approach randomness due to dispersal and competition, whereas others may not. For example, a population of *Ceraclea ancylus* (a stream caddisfly) in Brashears Creek, Kentucky, exhibits a highly clumped spatial pattern conforming to a logarithmic distribution throughout larval life (Resh, 1977). Significant differences can be seen, however, when the entire population (i.e., all habitats considered together) and the subpopulations occupying riffle and pool habitats are analyzed separately using Lloyd's (1967) 'mean crowding' index and Iwao's (1968, 1970) regression method for calculating 'the index of basic contagion', α, and 'the density contagious coefficient', β. These parameters indicate whether the basic component of the distribution is a single individual ($\alpha = 0$) or a group of individuals ($\alpha > 0$) and how these components are distributed in time and space ($0 < \beta < \infty$). When all habitats are considered together, sample counts of *C. ancylus* are characteristic of an overdispersed (*sensu* Iwao) population that can be described by a negative binomial series ($\alpha = 0.04$; $\beta = 5.16$). However, when the riffle population is considered separately the values of $\alpha = 10.35$ and $\beta = 1.26 \pm 0.49$ (95% confidence limits) indicate that riffle populations of *C. ancylus* exhibit an overdispersed distribution that follows a model of *randomly* distributed colonies with the mean colony size fixed. The pool also exhibits a high α ($= 8.81$) but a $\beta > 1$ ($= 2.61 \pm 0.80$), which indicates an overdispersed distribution that follows a model of *contagiously* (*sensu* Iwao) distributed colonies with mean colony size fixed.

The different patterns exhibited by the entire *C. ancylus* population, and the subpopulations in the riffle and in the pool, relate both to biological attributes of the species and physical features of the environment. *C. ancylus* adults oviposit a single egg mass directly on the water surface which sinks and then adheres to submerged stones (Resh, 1977). The tendency for aggregation noted in the entire population and in the selected habitats is explained by the deposition of the eggs in masses and the sedentary habits of the larvae. The values of α and β for the entire population undoubtedly reflect the heterogeneous nature of the habitats sampled and the combining of the counts, whereas the differences in the distributional characteristics of the riffle and pool habitat subpopulations may be due to differences in the abundance and location of the substrate particles to which egg masses can adhere in each habitat. For example, in the riffle there are substrate particles for attachment located throughout almost the entire habitat, with less than 7% of the riffle being bedrock to which egg masses cannot adhere. The pool habitat is characterized by a highly patchy distribution of particles, with over 76% of the surface area being bedrock. Thus, distribution of egg masses in the riffle can be close to random because of the wide availability of substrate attachment sites, whereas distribution in the pool is aggregated, due to the nature of its substrate.

When those areas in the pool in which only bedrock occur are eliminated from consideration because they cannot serve as attachment sites (i.e., the sampling universe is made to conform more closely to the population universe) and the data are analyzed in the same way, the *C. ancylus* distribution in the pool more closely agrees with a model of randomly distributed colonies with mean colony size fixed (where β is not significantly different from 1) than with a model of contageously distributed colonies (where $\beta>1$). The above example illustrates the useful combination of spatial analysis and an understanding of the biological and behavioral characteristics of an organism to provide basic information for the design of a sampling regime.

Improvement of Sampling Efficiency. It is obvious that the three dimensional aspect of distribution, microhabitat preferences based on food and other physiological requirements, and changes in habitat preference with age, all potentially compounded by the presence of multiple cohort populations, will give rise to spatial and temporal variability in aquatic insect populations. This clearly can result in complications affecting the quantitative analysis of their population dynamics. Yet, insufficient sample sizes, fixed dimension quadrat sampling units, and numerous other sampling design foibles are still characteristics of far too many 'quantitative' benthic studies. Analyses of insect populations in forest environments have considered these features and made appropriate adjustments in sampling design to a far greater extent than has been done in aquatic insect studies. There is a touch of irony in this fact since studies of forest insect population dynamics have generally dealt with organisms chosen for their pest status (rather than for their usefulness as ecological study organisms *per se*, e.g., see Coulson, 1979); whereas, with aquatic insects, there is essentially no restriction in the choice of a study organism; it can usually be made from any of the dozens or literally hundreds of available populations in most aquatic environments. With this in mind, two approaches should be considered in the sampling design of future population studies of aquatic insects.

First, the sampling designs in current benthic studies have an advantage over those of previous studies in that major technological improvements for the rapid and highly efficient extraction of organisms from samples have recently been developed (e.g., Stewart, 1975; Resh, 1979, Table 2). The use of these techniques should go hand in hand with a judicious choice of study organism to concentrate attention on those species that can provide maximum information with a reduced sampling effort. Cummins (1975) and Resh (1977, 1979) have given examples of how 'sampling attributes' of certain organisms can be effectively incorporated into sample designs for quantitative analysis of population dynamics.

A second approach that should be considered in quantitative aquatic insect studies, particularly those involved in biomonitoring, is sequential sampling. As mentioned previously, sequential sampling is most useful in situations where the examination and counting of sample material is more time consuming than the collection of samples. This is certainly true in benthic studies where field collection of samples is a mere fraction of the required sorting and identification time. In environmental monitoring programs, baseline data can be used in developing a sequential sampling procedure. The practical requisites of threshold or class limits and desired reliability are easily incorporated into this approach. In addition to monitoring specific populations, community analysis using standard diversity indices, or modifications of these such as the Sequential Comparison Index (Cairns, *et al.*, 1968) may also be included in a sequential sampling design. Although not yet used in traditional lake and stream benthic studies, this approach has been used in quantitative studies of mosquito (e.g., Mackey and Hoy, 1978; Service, 1976), planktonic (Möller and Bernhard, 1974), and marine benthic invertebrate (e.g., Saila, *et al.*, 1965; Loesch, 1974) populations.

4. FINAL COMMENTARY

As a final summation, we would like to emphasize several basic points.

First, in coping with both the ecological complexities and the logistic and mechanical difficulties that are inherent in sampling insect populations in forest and aquatic environments, it is essential that *all* of the basic requisites for the design of a sampling program be met. That is, the objectives of the sampling regime, the limitations of the population/sampling universe involved, the sampling unit and the basis of measurement to be used, the timing and frequency of observations or collections, and cost and time restraints must all be considered and clearly specified *before* beginning the actual field sampling program. Failure to do this -- equivocation in any aspect -- inevitably leads to indecision and inefficiency in field operations and, ultimately, to potential difficulties and necessary compromises in the analysis and interpretation of data.

Second, it is particularly difficult in field studies of forest and aquatic insects to define the precision of the estimates needed. Unfortunately, the impact of sampling error and bias on the results of population dynamics studies has not been fully assessed. Researchers rely mainly on the replication of sample estimates over time and space to reveal significant relationships and patterns. Whatever the form of analysis, failure to account for sampling error or to distinguish artifacts of sampling from biological reality makes interpretations subject to question. This is an area of statistical research greatly in need of further attention.

Third, in monitoring both forest and aquatic insect populations, experience with specific methods and procedures has provided some background on the range of sampling errors that may be expected. Combined with cost criteria and other constraints, attempts have been made in some cases to specify the desired precision and risk levels in advance, e.g., sequential sampling plans for major forest insect pests. Unfortunately, these are point or one-time-only estimates. However, more and more, these types of data are intended to provide input to predictive models. The reliability of such models depends on the validity and precision of the estimated parameter values *and* on the accuracy and precision of the estimates when applied in field situations. When predictions are too far in error, is it the fault of the model or of the monitoring procedure? This too is an area that requires attention if the use of predictive models in monitoring forest and aquatic insect populations -- in fact, in resource management generally -- is to be developed and used in a realistic manner.

ACKNOWLEDGMENTS

The research leading to part of this report (Resh) was supported by the Office of Water Research and Technology, U. S. Department of the Interior, under the allotment program of Public Law 88-379, as amended, and by the University of California Water Resources Center, as part of Office of Water Research and Technology Project No. A-063-CAL and Water Resources Center Project UCAL-WRC-W-519.

REFERENCES

Allen, K. R. (1959). The distribution of stream bottom fauna. *Proceedings of the New Zealand Ecology Society*, 6, 5-8.

Amman, G. D. (1969). A method of sampling the balsam woolly aphid on Fraser fir in North Carolina. *The Canadian Entomologist*, 101, 883-889.

Andrews, R. N. L. (1973). A philosophy of environmental impact assessment. *Journal of Soil and Water Conservation*, 28, 197-203.

Auer, C. (1971). Some analyses of the quantitative structure in populations and dynamics of larch bud moth 1949-1968. In *Statistical Ecology, Vol. 2*, G. P. Patil, E. C. Pielou, and W. E. Waters, eds. The Pennsylvania State University Press, University Park. 151-173.

Baltensweiler, W., Benz, G., Bovey, P., and Delucchi, V. (1977). Dynamics of larch bud moth populations. *Annual Review of Entomology*, 22, 79-100.

Beall, G. (1939). Methods of estimating the population of insects in a field. *Biometrika*, 30, 422-439.

Berryman, A. A. (1968). Development of sampling techniques and life tables for the fir engraver *Scolytus ventralis* (Coleoptera: Scolytidae). *The Canadian Entomologist*, 100, 1138-1147.

Brock, E. M. (1960). Mutualism between the midge *Cricotopus* and the alga *Nostoc*. *Ecology*, 41, 474-483.

Cairns, J., Jr., Albaugh, D. W., Busey, F., and Chaney, M. D. (1968). The sequential comparison index -- a simplified method for non-biologists to estimate relative difference in biological diversity in stream pollution studies. *Journal of the Water Pollution Control Federation*, 40, 1067-1613.

Campbell, R. W. (1967). The analysis of numerical change in gypsy moth populations. *Forest Science Monograph* 15.

Carlson, R. W. and Cole, W. E. (1965). A technique for sampling populations of the mountain pine beetle. *U. S. Forest Service Research Paper INT-20*. U. S. Department of Agriculture, Forest Service, Intermountain Forest and Range Experiment Station, Ogden, Utah.

Carolin, V. M. and Coulter, W. K. (1975). Comparison of western spruce budworm populations and damage on grand fir and Douglas-fir trees. *USDA Forest Service Research Paper PNW-195*. Pacific Northwest Forest and Range Experiment Station, Forest Service, Portland, Oregon.

Chutter, F. M. (1972). A reappraisal of Needham and Usinger's data on the variability of a stream fauna when sampled with a Surber sampler. *Limnology and Oceanography*, 17, 139-141.

Chutter, F. M. and Noble, R. G. (1966). The reliability of a method of sampling stream invertebrates. *Archiv für Hydrobiologie*, 62, 95-103.

Clifford, H. F. and Boerger, H. (1974). Fecundity of mayflies (Ephemeroptera), with special reference to mayflies of a brown-water stream of Alberta, Canada. *The Canadian Entomologist*, 106, 1111-1119.

Cole, W. E. (1970). The statistical and biological implications of sampling units for mountain pine beetle populations in lodgepole pine. *Researches on Population Ecology*, 12, 243-248.

Coulson, R. N. (1979). Population dynamics of bark beetles. *Annual Review of Entomology*, 24, 417-447.

Coulson, R. N., Hain, F. P., Foltz, J. L., and Mayyasi, A. M. (1975). Techniques for sampling the dynamics of southern pine beetle populations. *Texas Agricultural Experiment Station MP-1185*. Texas Agricultural Experiment Station, College Station, Texas.

Cummins, K. W. (1962). An evaluation of some techniques for the collection and analysis of benthic samples with special emphasis on lotic waters. *American Midlands Naturalist*, 67, 477-504.

Cummins, K. W. (1975). Macroinvertebrates. In *River Ecology*, B. Whitton, ed. Blackwell, London. 170-198.

Davies, B. R. (1976). The dispersal of Chironomidae larvae: A review. *Journal of the Entomological Society of South Africa*, 39, 39-62.

DeMars, C. J. (1970). Frequency distributions, data transformations, and analysis of variations used in determination of optimum sample size and effort for broods of the western pine beetle. In *Studies on the Population Dynamics of the Western Pine Beetle Dendroctonus brevicomis LeConte (Coleoptera: Scolytidae)*, R. W. Stark and D. L. Dahlsten, eds. University of California, Division of Agricultural Sciences, Berkeley. 42-65.

Dudley, C. O. (1971). A sampling design for the egg and first instar larval populations of the western pine beetle, *Dendroctonus brevicomis* (Coleoptera: Scolytidae). *The Canadian Entomologist*, 103, 1291-1313.

Eberhardt, L. L. (1978). Appraising variability in population studies. *Journal of Wildlife Management*, 412, 207-238.

Elliott, J. M. (1977). *Some methods for the statistical analysis of samples of benthic invertebrates.* 2nd ed. Freshwater Biological Association Scientific Publication No. 25.

Elliott, J. M. and Tullett, P. A. (1978). *A bibliography of samplers for benthic invertebrates.* Freshwater Biological Association Occasional Publication No. 4.

Finney, D. J. (1946). Field sampling for the estimation of wireworm populations. *Biometrics*, 2, 1-7.

Foltz, J. L., Mayyasi, A. M., Pulley, P. E., Coulson, R. N., and Martin, W. C. (1976). Host-tree geometry models for use in southern pine beetle population studies. *Environmental Entomology*, 5, 714-719.

Foltz, J. L., Pulley, P. E., Coulson, R. N., and Martin, W. C. (1977). Procedural guide for estimating within-spot populations of *Dendroctonus frontalis*. *Texas Agricultural Experiment Station MP-1316*. Texas Agricultural Experiment Station, College Station, Texas.

Harker, J. E. (1953). An investigation of the distribution of the mayfly fauna of a Lancashire stream. *Journal of Animal Ecology*, 22, 1-13.

Heller, R. C., Sader, S. A., and Miller, W. A. (1977). Identification of preferred Douglas-fir tussock moth sites by photo interpretation of stand, site, and defoliation conditions. Final report to the USDA Douglas-fir Tussock Moth Expanded R & D Program, University of Idaho, College of Forestry, Wildlife, and Range Sciences, Moscow, Idaho.

Hynes, H. B. N. and Coleman, M. J. (1968). A simple method of assessing the annual production of stream benthos. *Limnology and Oceanography*, 13, 569-573.

Ives, W. G. H. and Prentice, R. M. (1958). A sequential sampling technique for surveys of the larch sawfly. *The Canadian Entomologist*, 90, 331-338.

Iwao, S. (1968). A new regression method for analyzing the aggregation pattern of animal populations. *Researches on Population Ecology*, 10, 1-20.

Iwao, S. (1970). Analysis of spatial patterns in animal populations: Progress of research in Japan. *Review of Plant Protection Research*, 3, 41-54.

Jonasson, P. M. (1972). Ecology and production of the profundal benthos in relation to phytoplankton in Lake Esrom. *Oikos Supplement*, 14, 1-148.

Kaelin, A. and Auer, C. (1954). Statistiche Methoden zur Untersuchung von Insektenpopulationen des Grauen Lärchenwicklers (*Eucosma griseana* Hb. = *Semasia diniana* Gn.). *Zeitschrift für Angewandte Entomologie*, 36, 241-282; 423-461.

Klein, W. H. (1973). Beetle-killed pine estimates. *Photogrammetric Engineering*, 39, 385-388.

Klomp, H. (1966). The dynamics of a field population of the pine looper, *Bupalus piniarius* L. *Advances in Ecological Research*, 3, 207-305.

Kuno, E. (1971). Sampling error as a misleading artifact in "key factor analysis." *Researches on Population Ecology*, 13, 28-45.

Lamberti, G. A. and Resh, V. H. (1979). Substrate relationships, spatial distribution patterns, and sampling variability in a stream caddisfly population. *Environmental Entomology*, 8, 561-567.

Langley, P. G. (1971). The benefits of multi-stage variable probability sampling using space and aircraft imagery. In *Application of Remote Sensors in Forestry*, G. Hildebrandt, ed. Druckhans Rombach, Freiburg. 119-126.

Lloyd, M. (1967). 'Mean crowding'. *Journal of Animal Ecology*, 36, 1-30.

Loesch, J. G. (1974). A sequential sampling plan for hard clams in lower Chesapeake Bay. *Chesapeake Science*, 15, 134-139.

Lyons, L. A. (1964). The spatial distribution of two pine sawflies and methods of sampling for the study of population dynamics. *The Canadian Entomologist*, 96, 1373-1407.

Mackay, R. J. and Wiggins, G. B. (1979). Ecological diversity in Trichoptera. *Annual Review of Entomology*, 24, 185-208.

Mackey, B. E. and Hoy, J. B. (1978). *Culex tarsalis*: sequential sampling as a means of estimating populations in California rice fields. *Journal of Economic Entomology*, 71, 329-334.

Mason, J. C. (1976). Evaluating a substrate tray for sampling the invertebrate fauna of small streams, with comments on general sampling problems. *Archiv für Hydrobiologie*, 78, 51-70.

Mason, R. R. (1969). Sequential sampling of Douglas-fir tussock moth populations. *U. S. Forest Service Research Note PNW-102*. U. S. Department of Agriculture, Forest Service, Pacific Northwest Forest and Range Experiment Station, Portland, Oregon.

Mayyasi, A. M., Pulley, P. E., Coulson, R. N., DeMichele, D. W., and Foltz, J. L. (1976). Mathematical description of within-tree distributions of the various developmental stages of *Dendroctonus frontalis* Zimm. (Coleoptera: Scolytidae). *Researches on Population Ecology*, 18, 45-55.

McClelland, W. T., Hain, F. P., DeMars, C. J., Fargo, W. S., Coulson, R. N., and Nebeker, T. E. (1978). Sampling bark beetle emergence: a review of methodologies, a proposal for standardization, and a new trap design. *Bulletin of the Entomological Society of America*, 24, 137-140.

McLeod, J. M. (1972). The Swaine jack pine sawfly, *Neodiprion swainei*, life system: evaluating the long-term effects of insecticide applications in Quebec. *Environmental Entomology* 1, 371-381.

Merritt, R. W., Cummins, K. W., and Resh, V. H. (1978). Collecting, sampling, and rearing methods for aquatic insects. In *An Introduction to the Aquatic Insects of North America*, R. W. Merritt and K. W. Cummins, eds. Kendall-Hunt Publishing Company, Iowa. 13-28.

Miller, C. A. (1955). A technique for assessing spruce budworm larval mortality caused by parasites. *Canadian Journal of Zoology*, 33, 5-17.

Miller, J. M. and Keen, F. P. (1960). Biology and control of the western pine beetle. *U. S. Department of Agriculture, Forest Service Miscellaneous Publication 800*. U. S. Department of Agriculture, Washington, D. C.

Möller, F. and Bernhard, M. (1974). A sequential approach to the counting of plankton organisms. *Journal of Experimental Marine Biology and Ecology*, 15, 49-68.

Morris, R. F. (1955). The development of sampling techniques for forest insect defoliators, with particular reference to the spruce budworm. *Canadian Journal of Zoology*, 33, 225-294.

Morris, R. F. (1960). Sampling insect populations. *Annual Review of Entomology*, 5, 243-264.

Morris, R. F., ed. (1963). The dynamics of epidemic spruce budworm populations. *Memoirs of the Entomological Society of Canada No. 31.*

Morris, R. F. (1965). Contemporaneous mortality factors in population dynamics. *The Canadian Entomologist*, 97, 1173-1184.

Morris, R. F. and Miller, C. A. (1954). The development of life tables for the spruce budworm. *Canadian Journal of Zoology*, 32, 283-301.

Needham, P. R. and Usinger, R. L. (1956). Variability in the macrofauna of a single riffle in Prosser Creek, California, as indicated by the Surber sampler. *Hilgardia*, 24, 383-409.

Patil, G. P. and Taillie, C. (1979). An overview of diversity. In *Ecological Diversity in Theory and Practice*, J. F. Grassle, G. P. Patil, W. Smith and C. Taillie, eds. Satellite Program in Statistical Ecology, International Co-operative Publishing House, Fairland, Maryland.

Pulley, P. E., Foltz, J. L., Coulson, R. N., Mayyasi, A. M., and Martin, W. C. (1977). Sampling procedures for within-tree attacking adult populations of the southern pine beetle *Dendroctonus frontalis* Zimmerman (Coleoptera: Scolytidae). *The Canadian Entomologist*, 109, 39-48.

Rabeni, C. F. and Minshall, G. W. (1977). Factors affecting microdistribution of stream benthic insects. *Oikos*, 29, 33-43.

Reeks, W. A. (1956). Sequential sampling for larvae of the winter moth, *Operophtera brumata* (Linn.) (Lepidoptera: Geometridae). *The Canadian Entomologist*, 88, 241-246.

Resh, V. H. (1976). Life cycles of invertebrate predators of freshwater sponge. In *Aspects of Sponge Biology*, F. W. Harrison and R. R. Cowden, eds. Academic Press, New York. 299-314.

Resh, V. H. (1977). Habitat and substrate influences on population and production dynamics of a stream caddisfly, *Ceraclea ancylus* (Leptoceridae). *Freshwater Biology*, 7, 261-277.

Resh, V. H. (1979). Sampling variability and life history features: basic considerations in the design of aquatic insect studies. *Journal of the Fisheries Borad of Canada*, 36, 290-311.

Resh, V. H. (1979a). Biomonitoring, species diversity indices, and taxonomy. In *Ecological Diversity in Theory and Practice*, J. F. Grassle, G. P. Patil, W. Smith, and C. Taillie, eds. Satellite Program in Statistical Ecology, International Co-operative Publishing House, Fairland, Maryland.

Resh, V. H., Haag, K. H., and Neff, S. E. (1975). Community structure and diversity of caddisfly adults from the Salt River, Kentucky. *Environmental Entomology*, 4, 241-253.

Safranyik, L. (1971). Some characteristics of the spatial arrangement of attacks by the mountain pine beetle, *Dendroctonus ponderosae* (Coleoptera: Scolytidae), on lodgepole pine. *The Canadian Entomologist*, 103, 1608-1625.

Safranyik, L. and Raske, A. G. (1970). Sequential sampling plan for larvae of *Monochamus* in lodgepole pine logs. *Journal of Economic Entomology*, 63, 1903-1906.

Saila, S. B., Flowers, J. M., and Campbell, R. (1965). Applications of sequential sampling to marine resource surveys. *Ocean Sciences and Ocean Engineering*, 2, 782-802.

Sartwell, C. (1971). *Ips pini* (Coleoptera: Scolytidae) emergence per exit hole in ponderosa pine thinning slash. *Annals of the Entomological Society of America*, 64, 1473-1474.

Schreuder, H. T., Clarke, W. H., and Barry, P. J. (1978). A flexible three-stage sampling design with double sampling at the second stage in estimating southern pine beetle spot characteristics. *Final Report -- Expanded Southern Pine Beetle Program*. U. S. Department of Agriculture, Washington, D. C.

Service, M. W. (1976). *Mosquito Ecology*. Applied Science Publishers, London.

Shepherd, R. F. and Brown, C. E. (1971). Sequential egg-band sampling and probability methods of predicting defoliation by *Malacosoma disstria* (Lasiocampidae: Lepidoptera). *The Canadian Entomologist*, 103, 1371-1378.

Shiozawa, D. K. and Barnes, J. R. (1977). The microdistribution and population trends of larval *Tanypus stellatus* Coquillett and *Chironomus frommeri* Atchley and Martin (Diptera: Chironomidae) in Utah Lake, Utah. *Ecology*, 58, 610-618.

Southwood, T. R. E. (1966). *Ecological Methods, with Particular Reference to the Study of Insect Populations*. Methuen, London.

Stark, R. W. (1952). Analysis of a population sampling method for the lodgepole needle miner in Canadian Rocky Mountain parks. *The Canadian Entomologist*, 84, 316-321.

Stephen, F. M. and Taha, H. A. (1976). Optimization of sampling effort for within-tree populations of southern pine beetle and its natural enemies. *Environmental Entomology*, 5, 1001-1007.

Stewart, K. W. (1975). An improved elutriator for separating stream insects from stony substrates. *Transactions of the American Fisheries Society*, 104, 821-823.

Stout, J. and Vandermeer, J. (1975). Comparisons of species richness for stream-inhabiting species in tropical and mid-latitude streams. *The American Naturalist*, 109, 263-280.

Thut, R. N. (1969). A study of the profundal bottom fauna of Lake Washington. *Ecological Monographs*, 39, 79-100.

Ulfstrand, S. (1967). Microdistribution of benthic species (Ephemeroptera, Plecoptera, Trichoptera, Diptera: Simuliidae) in Lapland streams. *Oikos*, 18, 293-310.

Varley, G. C. and Gradwell, G. R. (1968). Population models for the winter moth. In *Insect Abundance*, T. R. E. Southwood, ed. Blackwell, Oxford and Edinburgh. 132-142.

Waters, T. F. (1969). The turnover ratio in production ecology of freshwater invertebrates. *The American Naturalist*, 103, 173-185.

Waters, T. F. (1972). The drift of stream insects. *Annual Review of Entomology*, 17, 253-272.

Waters, T. F. (1977). Secondary production in inland waters. *Advances in Ecological Research*, 10, 91-164.

Waters, T. F. and Crawford, G. W. (1973). Annual production of a stream mayfly population: a comparison of methods. *Limnology and Oceanography*, 18, 286-296.

Waters, W. E. (1955). Sequential sampling in forest insect surveys. *Forest Science*, 1, 68-79.

Waters, W. E. (1962). The ecological significance of aggregation in forest insects. *Proceedings 11th International Congress of Entomology (Vienna, 1960)*, 2, 205-210.

Waters, W. E. (1974). Sequential sampling applied to forest insect surveys. In *Monitoring Forest Environment Through Successive Sampling*, T. Cunia, ed. State University of New York College of Environmental Science and Forestry, Syracuse. 290-311.

Waters, W. E. and Henson, W. R. (1959). Some sampling attributes of the negative binomial distribution with special reference to forest insects. *Forest Science*, 5, 397-412.

Watt, K. E. F. (1959). A mathematical model for the effect of densities of attacked and attacking species on the number attacked. *The Canadian Entomologist*, 91, 129-144.

Watt, K. E. F. (1961). Mathematical models for use in insect pest control. *The Canadian Entomologist Supplement*, 19.

Wetzel, R. G. (1975). *Limnology*. Saunders, Philadelphia.

Wood, D. L. and Bedard, W. D. (1977). The role of pheromones in the population dynamics of the western pine beetle. In *Proceedings of XV International Congress of Entomology (Washington, D. C., August 19-27, 1976)*. Entomological Society of America, College Park, Maryland. 643-652.

[*Received November* 1978. *Revised April* 1979]

G. P. Patil and M. Rosenzweig, (eds.),
Contemporary Quantitative Ecology and Related Ecometrics, pp. 619-634.

STATISTICAL PALEOECOLOGY: PROBLEMS AND PERSPECTIVES

ROGER L. KAESLER

Department of Geology
The University of Kansas
Lawrence, Kansas 66045 USA

SUMMARY. Paleoecology has remained largely nonquantitative and nonstatistical, primarily because it is a historical science. Most kinds of experimentation are not possible in paleoecology, and retrodiction rather than prediction is the characteristic inference. Special difficulties arise because the taphonomic overprint and time averaging often make the ecological meaning of paleoecological samples unclear. Rate of growth of a statistical methodology in paleoecology will increase, but careful attention must be paid to the correct use of statistical methods and the consequences of limitations of the fossil record.

KEY WORDS. paleoecology, paleontology, statistics, retrodiction, taphonomy, time averaging.

1. INTRODUCTION

The science of ecology has become increasingly quantitative during the past two decades, so much so, in fact, that it has been described as an essentially mathematical subject because the variates it deals with, such as numbers of individuals, biomass, and measurements of physical and chemical properties of the environment, are inherently quantitative (Pielou, 1977, p. v). Moreover, much of the application of quantitative techniques to ecology has involved the use of statistical analysis. From the univariate and bivariate statistics used in the initial stages, our science has moved toward application of multivariate methods, in which the means of computation remain decades ahead of interpretation.

Development of a statistical underpinning, of course, has not been equal throughout ecology. Of all its branches, none except for classical natural history has been slower than paleoecology to adopt the modern, quantitative posture. This has resulted in spite of the availability of a useful introductory textbook on the subject (Reyment, 1971). The goals of my paper are to examine the reasons for the nonstatistical basis of most paleoecology and to attempt to predict the future of statistical analysis in this field. The paper is in no sense intended as a comprehensive review of the subject, and lack of reference to much of the fundamental literature should not be misconstrued.

Some of the apparent reticence of paleoecologists with respect to quantitative and statistical work no doubt stems from a general lack of familiarity with current ecological literature. Jackson (1978), for example, in a review of *Patterns of Evolution* (Hallam, ed., 1977), a volume with strong paleoecological overtones, found "...only 15 references to the basic English language ecological literature." Moreover, most of the activities of the Paleoecology Section of the Ecological Society of America are devoted to the study of postglacial lacustrine paleoecology rather than to paleoecology of ancient organisms from the remote geological past (Solomon, 1977, 1978). Evidently, these activities reflect the interests of the members of the section, and few people from the main stream of geological paleoecology are involved.

The introversion of paleoecology is by no means universal, however, and it does not explain fully the nonquantitative approaches of paleoecologists to the problems in their field. Neither can general lack of training in mathematics provide a suitable explanation. In fact, most paleoecologists have entered the field through geology, a science in which coursework in mathematics, physics, chemistry, and engineering has long been a part of the standard curriculum. The emphasis, however, has been on the calculus, which is likely to lead to a deterministic philosophy of science rather than to a probabilistic one (Griffiths, 1966). A curriculum with more than one semester of statistics is a rarity, so that most geologists who use statistics are self taught, a poor procedure in this multivariate world. Nevertheless, the mathematical foundation is usually present, even if the statistical structure remains to be built. Moreover, geology is a *user science*, drawing heavily on fundamental development in other fields. Part of the tradition of paleoecologists as geologists, then, is to involve themselves in the affairs and especially in the literature of other sciences. Thus, although Jackson's (1978) observation about the lack of familiarity of paleoecologists with current ecological literature points up an important shortcoming of much of paleoecology, it does not explain the question at hand.

2. PALEOECOLOGY AND GEOLOGY

Paleoecology is the study of the ecology of ancient or fossil organisms, and it is as much a branch of geology as it is of ecology. While most of the goals of paleoecology are dictated by its affinity to ecology, the methods of research and the material for study are controlled by geological factors. Consider, for example, the study of predation. The ecological literature abounds with reports of research on the interactions of predators with their prey. Curio's (1976) monograph on the subject lists 687 references that he regarded as having "...contributed 'key' discoveries or ideas" (p.3). Study of predator-prey interactions, then, might safely be regarded as a fundamental aspect of all ecology, paleoecology as well as *neoecology*. Yet in a recent, careful paleoecological study of Devonian brachiopods that had been attacked by a shell-boring predator, the nature of the preserved material in the fossil community prevented the authors from determining which species of gastropod had bored the shells or, indeed, if the predator was a gastropod at all (Sheehan and Lesperance, 1978). Few ecologists would be content in their work if they faced the degree of indeterminancy that has been imposed on this important ecological problem by the nature of the fossil record. To avoid some of the uncertainty, they have come to rely increasingly on experimentation. Yet in paleoecology, where experimentation is impossible, we must often settle for indications rather than conclusions because the information needed to solve the problems at hand simply no longer exists.

2.1 Paleoecology as an Historical Science. Geology is fundamentally an historical science, and the geological-historical imprint on paleoecology has contributed substantially to its enduring qualitative methodology. Simpson (1963, p. 25) has defined historical science as "...the determination of configurational sequences, their explanation, and the testing of such sequences and explanations." Of course, geologists also concern themselves with originating new generalizations, but as Kitts (1977, p. 5) has pointed out, they regard such generalizations "...as means to an end rather than as ends in themselves, the end being the construction of a chronicle of specific events occurring at specific times." Elsewhere, Kitts (1977, p. xvi) has made this point again, equally emphatically:

> The difference between the theoretical sciences and the historical sciences does not lie in the theories which are invoked or in the inferential use to which these theories are put. It lies rather in what those engaged in the two kinds of sciences see as their goal. For historical scientists, singular descriptive statements are the end and theories are a means to that end. For theoretical scientists, theories are the end and singular descriptive statements are a means to that end.

The reason for divergence of the methodologies of paleoecology and modern ecology, then, is just that paleoecology is an historical science and modern ecology is increasingly a theoretical one. The fact that modern ecology has its roots in natural history and historical biology and might, thus, have developed a similar methodology has no bearing here. We are considering the tree, not its roots, and ecology as a whole has long since outgrown the state Ager (1963, p. 4) described admiringly from the outside as "...the refuge of those [biologists] who were interested in animals and plants as living things rather than in ...mathematics." Perusal of some of the milestones of modern theoretical ecology reveals that if refuge was sought it was not refuge from mathematics!

The implications of paleoecology's historical nature for the development of its methodology and, for our purposes specifically, for development of a methodology that routinely includes statistical analysis are overwhelmingly important.

2.1.1 Experimentation in Paleoecology. Much of the statistical analysis in ecology has stemmed from the rapid growth of the experimental approach with its attendant testing of statistical hypotheses. No such change has overcome paleoecology. Although Reyment (1971, p. 5-6) has listed several experimental studies in paleoecology and more have been done since his compilation, for most of the discipline the experiments were run by Nature millions of years ago. The task of the paleoecologist is to decide, if possible, what those experiments were. Unfortunately, we can never be sure of our ability to infer correctly from the record Nature has been kind enough to leave us. Thus, although Valentine (1973, p. 9) has assured us that "The fossil record deserves to be taken very seriously" and Hallam (1977, p. v) has urged us to "...emphasize the more positive aspects..." of the fossil record, nevertheless all paleoecologists are faced with analyzing only a portion of the results of the original experiment. Of course, such a situation may mean that the problem being investigated is insoluble because the answer is lost with the missing results. At best, it is a situation that does not lend itself conveniently to statistical analysis.

2.1.2 Retrodiction in Paleoecology. Fretwell (1975, p. 3) has suggested that one of the changes wrought by Robert MacArthur and his use of the hypothetico-deductive method in ecology was a ten-fold increase in prediction (at least in papers published in *Ecology*). A characteristic of historical scientists, on the other hand, is a concern not for prediction but for *retrodiction*, which Kitts (1977, p. 39) has called "...the most characteristic geologic inference." This is, of course, a consequence of our having had our experiments run for us. Kitts (1977, p. 43-45) has discussed prediction and retrodiction. He has pointed out that retrodictive uncertainty results when a consequence or effect can have more than one antecedent or cause. Geologists, however, "...are more inclined to suppose that the same antecedents have different consequences than to suppose the contrary." That is, faced with seemingly indistinguishable end products (consequences) of different geological or paleoecological processes (antecedents), the geologist or paleoecologist will return to the field or laboratory in the belief that he can learn how to discriminate between them. Again methodology results that is not conducive to statistical analysis.

2.1.3 Test of Significance in Paleoecology. Pielou (1977, p. 299), in a discussion of ecological diversity, has made a point about sampling that applies particularly well to paleo-ecology:

> Observe that it is rarely legitimate to treat a censused collection as a sample from a larger (conceptually infinite) parent community. This is permissible only if the boundaries of the postulated parent community can be precisely specified and the collection at hand is a truly random sample from it, which is seldom the case.

Griffiths (1967, p. 12-30) has discussed in detail the problems of sampling in geology, and he, too, has emphasized the necessity of carefully defining the population to be investigated. The likelihood of defining in a paleoecologically interesting way the boundaries of a population or community of fossils is not great. The opportunities for truly random sampling are even more limited, especially if one takes into account the nature of the fossil record, which I shall discuss in more detail in Section 3.

The paleoecologist may overcome the problem of sampling (Griffiths, 1967, p. 12-30) or, as is more frequently done in ecology, overlook it. In formulating interesting statistical hypotheses to test, however, he faces a further difficulty that is not usually recognized. This difficulty stems from the historical perspective of paleoecology, its concern with retrodiction, and the fact that unknown experiments have been run by Nature.

The null hypothesis that two populations do not differ with respect to some property that can be measured must always be rejected in paleoecology. Moreover, it can be rejected *a priori*, before taking a single measurement. The basis for rejection lies in the fact that the questions of interest in paleoecology always involve comparisons between different populations--populations of fossils of different age or from different environments. If a paleoecologist finds no basis for rejecting the null hypothesis, the meaning of his result is that he has not tried hard enough to find the differences that must be there. He needs to measure more specimens or to remeasure his collection with greater precision.

Another way of expressing this is to say that paleoecological studies are always Model II (see Sokal and Rohlf, 1969, p. 192, 201-202). Clearly definable *fixed treatment effects* do not exist as they do in the Model I studies of experimental ecology. Only the random, added variance components of Model II can occur, and only the search for patterns of variation and partitioning of variances is appropriate. Of course, in any statistical testing it is at least theoretically possible to devise a plan of sampling that will reduce the standard error and allow small differences to be detected. The problem in paleoecology is conceptual rather than methodological since one must always deal with samples that cannot be demonstrated to have formed contemporaneously, although the most interesting paleoecological problems may require contemporaneity. Given this constraint, the gross scale on which environments of the past must be measured, and the penchant of geologists for making maps to express their conclusions, it is not surprising that the application of statistical analysis has been held back.

2.2 The Role of Paleoecology. In addition to its implcations for statistical methodology, the geological-historical nature of paleoecology determines to a considerable extent the place of paleoecology in the sciences. The question "What is paleoecology for?" has been answered in many ways. The days when paleoecology was equated with interpreting ancient environments of deposition are fortunately behind us. Ladd's (1957) treatise approached the subject in this way, and as late as 1964 McKee's paper "Inorganic Sedimentary Structures" was included in a paleoecological symposium volume. It began (McKee, 1964, p. 275): "Ancient environments may be interpreted, at least in part, from primary [inorganic] structures in sedimentary rocks." Such work is interesting and important, but it has little to do with paleoecology as we practice it today.

Ager (1963, p. 4) and Imbrie and Newell (1964, p. 1) considered ecology to be a subset of paleoecology, which was necessarily broader in scope than ecology because of the geological factors to be considered. More recently, several paleoecologists have expressed the hope that paleoecology will contribute to the development of the principles and theories of ecology. Paraphrasing Archibald Geikie's maxim, Valentine (1973, p. 14) declared "The past is a key to the present" and expressed his view (p. 16) that the ecological configurations of the past, being so unlike those of the present, "...provide configurational tests which ecological and evolutionary process models must satisfy, regardless of their sources." Schopf (1975), p. 130) expressed a similar view: "Using communities over geologic time, paleontologists may be in the position to clarify the relative importance of these different ways to approach ecology....It remains to be seen whether paleoecology will contribute to the theory of ecosystem regulation, over geologic time." Such a contribution to theory seems doubtful in light of Kitts' (1977, p. xvi) arguments about the ends and means of historical science and his assertion (p. 159):

> To 'flesh out' a fossil organism we almost always have to presuppose the properties of greatest theoretical interest. For this reason extant plants and animals will always take precedence over fossil organisms in the testing of biological theories.

3. PALEOECOLOGY, PROBABILITY, AND PROBLEMS

This somewhat gloomy description of the state of statistical analysis in paleoecology is not intended to prescribe that paleoecologists continue to address research problems without the benefit of statistics. No doubt can remain that use of computer-based methods and statistical tests of hypotheses will enable paleoecologists to manipulate ever larger sets of data in ever more intricate and, we hope, meaningful ways. The development of a statistical methodology in paleoecology, however, will require that paleoecologists take special care in the uses to which they put statistical methods. Two pitfalls, especially, must be avoided: lack of understanding of the statistical methods to be used and lack of appreciation of ways in which the nature of the fossil record limits the usefulness of data derived from it. It is these pitfalls that I shall address in this section.

3.1 Methodological Limitations. Every statistically oriented ecologist or paleoecologist has been approached by a less quantitative colleague who wants "to do some statistics on" a set of data he has on hand, representing perhaps three years' research. Most of us have agreed reluctantly to help. Too seldom do our colleagues realize that statistical hypothesis testing presupposes data collected with a specific hypothesis and a specific test in mind. Moreover, sets of data are, of course, not always appropriate for some kinds of statistical analysis. Unfortunately one can rarely be sure from published reports that authors have tested the assumptions required of the data by the methods they have used (e.g., normal distribution, homoscedasticity, and homogeneous variance-covariance matrix).

As the computation of statistical analysis becomes increasingly involved mathematically, as in the use of some multivariate methods such as discriminant function analysis and factor analysis, a further complication may arise. Wright and Van Dyne (1971, p. 92) have pointed out that "...most biologists are now having their data analyzed by others...". Successful completion of research may require effective communication between the ecologist or paleoecologist, a biometrician, and a computer programmer. In some instances the specific computer program and the kind of computer used may have an important effect on the accuracy of the results. These views apply to paleoecology as well as to ecology.

As the use of statistical analysis in paleoecology grows, these methodological difficulties will arise repeatedly and will need to be addressed. Overcoming them is possible but will require astute editors, careful reviewers, compassionate colleagues, and a commitment to success in the science as a whole. More serious, however, are the substantive problems that arise because of the nature of the fossil record, making it uninterpretable for some purposes and, in some instances and from some points of view, not suitable for statistical analysis.

3.2 Substantive Limitations. Table 1 shows four areas of paleoecology that are now the subject of extensive research: functional morphology, organism-sediment interactions, population structure and dynamics, and community structure and

TABLE 1: Some effects of the taphonomic overprint and time averaging on interpretation of samples in four areas of paleoecological research.

	Factors Affecting Interpretation of Samples	
	Taphonomic Overprint	Time Averaging
Functional Morphology	a. loss of soft parts b. breaking of skeletons c. loss of symbionts, predators, etc. d. solution of skeletons	a. mixing of faunas b. mixing of ecophenotypes
Organism-Sediment Interactions	a. transportation b. loss of soft-bodied organisms	a. environmental change masked by bioturbation b. burrowing into sediment having no relationship to feeding environment
Population Structure and Dynamics	a. loss of immature stages b. mixing by transportation	a. mixing of successive populations by bioturbation
Community Structure and Diversity	a. loss of some aragonite-shelled species b. loss of species with no skeletons	a. mixing of successive communities b. effects of differential longevity

diversity. Across the top of the table are the two major obstacles to understanding paleoecology--the taphonomic overprint and time averaging. Both of these obstacles affect primarily the meaning of paleoecological samples. A brief discussion of their effects on the four major areas of paleoecological research follows.

Taphonomy has been discussed in detail by Lawrence (1968, 1971). It is concerned with the series of events that occur from the time of death of an organism and the time of its discovery and collection by a paleontologist. The effects of these taphonomic events are called the taphonomic overprint. Depending on the nature of the organisms, the nature and intensity of the taphonomic factors, and the length of geologic time involved, the taphonomic overprint may range from mere decay of soft parts to relocation by transportation to complete destruction of any trace of the organism. In almost every instance, soft parts are lost, and this loss naturally hampers the study of functional morphology. Transportation from one kind of substrate to another, if it has occurred, makes study of organism-sediment interactions impossible. Both transportation and destruction of some organisms make study of population structure difficult. The small, immature stages are usually most likely to be destroyed by abrasion and crushing during taphonomy, giving a distorted picture of the age- or size-structure of a population. Moreover, small individuals may be washed out of the sediment before lithification and deposited in another area altogether.

Study of community structure and diversity are made particularly difficult by the taphonomic overprint. What does species diversity mean if computed for only that subset of the macrobenthic community that has durable shells and was not destroyed? What does species diversity mean in a community where the top-level carnivores are rarely if ever fossilized with the organisms upon which they preyed? A common approach has been to compute diversity of only a small subset of the community comprising fossils of organisms that presumably had about equal impact on the environment and equal likelihood of fossilization, such as the Ostracoda (Brondos and Kaesler, 1976), Foraminiferida (Gibson and Buzas, 1973), or Brachiopoda and Bivalvia. Differences in diversity are then compared from area to area or from one geological time to another. This approach is interesting, but it is not really community paleoecology as it is sometimes labeled. Of course, ecologists have been forced to adopt a somewhat similar procedure, analyzing only a subset of the organisms comprising the communities they study because of the impossibility of quantifying an entire community. A fundamental difference exists, however, because it is at least theoretically possible for the ecologist to know what he is omitting from study.

Time averaging is the process whereby remains of organisms that lived at different times in an area are mixed together. Mixing may result from wave and current action, bioturbation, or simply faunal change that occurred on a time scale finer than the vertical scale at which samples were collected. Time averaging is not always a disadvantage to paleoecologists because it enables them to get a composite sample that includes organisms from all seasons and local microhabitats. It does, however, affect the meanings of samples and has implications for the use of statistical methods. For example, the study of functional morphology can become terribly complicated if ecophenotypes from successive environments are mixed. Study of organism-sediment interactions presupposes that we know what substrate an organism occupied. This is not necessarily possible to know if the evidence from successive, rapidly changing environments has been homogenized by bioturbation or if the organisms burrowed into a substrate deposited under different environmental conditions from those in which it fed. Successive populations and successive communities may be mixed by bioturbation, greatly distorting the measures of survivorship and diversity. Moreover, the diversity of a collection of fossils can be a poor measure of the community structure if species had dramatically different longevities. Other things being equal, individuals of a species in which life spans are short will have many more individuals in the fossil record than a species in which individuals have long life spans. The former, however, may have had much less impact on the community and played a much smaller role in the ecosystem.

3.3 Statistical Methods. Statistical methods that are useful in paleoecology are, for the most part, the same as those that are useful in ecology. The greatest difference is in the difficulty the paleoecologist has in interpreting the meaning of his samples and, hence, his results. To what extent is a sample representative of the ancient population or community? The question is unanswerable and dictates that extreme caution be used in drawing conclusions from paleoecological research--be it statistical or otherwise.

Krumbein (1965) has discussed sampling in paleontology, and Chang (1967) and Dennison and Hay (1967) have discussed some specific applications of sampling theory. In addition, all paleoecologists should familiarize themselves with the thinking that has led to Sokal and Rohlf's (1969, p. 246-249) discussion of what constitutes an adequate sample size. Several books are available on sampling and experimental design (see e.g. Winer, 1962; Cochran, 1963).

Reyment (1971) has discussed the use of many quantitative methods in paleoecology, including several methods of univariate statistics. His book is particularly useful because of the many examples it contains. Davis (1973) and Poole (1974) have also provided introductions to applied statistics that are useful in paleoecology.

It is interesting that the statistical methods used by

plant ecologists are also among the most useful to paleoecologists. Fossils hold very still, even more so than plants; and many of the important faunal elements of ancient communities were either sessile or very sluggish. For the purposes of sampling, they can be considered immobile. Greig-Smith (1964) and Kershaw (1964) have compiled and discussed many useful methods, especially univariate and bivariate methods and means of measuring fidelity of species to specific habitats and communities. In addition, a number of excellent introductory textbooks deal with nonparametric statistical methods (Siegel, 1956; Bradley, 1969; Conover, 1971). In many instances, nonparametric statistics are more applicable to paleoecological problems than the more rigorous parametric statistics.

Multivariate statistical methods have occupied the forefront of statistical ecology and paleoecology in recent years, and they seem likely to continue to do so. These methods, however, are often used for fishing expeditions or as aids to graphic display of results rather than for testing hypotheses. Elsewhere in these volumes, Reyment (1979) has discussed the use of some multivariate methods in paleoecology. Several textbooks are also available on the subject of multivariate statistics, some of which deal with ecological and paleoecological problems (Cooley and Lohnes, 1971; Blackith and Reyment, 1971; Davis, 1973; Sneath and Sokal, 1973; Poole, 1974).

4. THE FUTURE

One can confidently predict that the use of statistical analysis in paleoecology will increase, especially as the science moves farther away from the descriptive phase that now involves the efforts of so many paleoecologists. It seems doubtful, however, that the potential of univariate and bivariate statistics will ever be fully realized. The availability of multivariate methods fosters the tendency to collect all the data possible and "run them through a computer." As was mentioned previously, results of such analyses are more commonly used in constructing graphic aids than in testing hypotheses. It seems likely, also, that the use of Markov models will increase since these are ideally suited to studying stratigraphic sequences (Doveton, 1971; Davis, 1973, p. 280-288; Krumbein, 1975).

Not too many years ago I attended a conference in which perhaps fifty papers were presented on many aspects of paleoecology and biostratigraphy. The statistical papers were grouped into a session entitled "Statistical Methods," and the afternoon set aside for their presentation gave at least three-fourths of the conferees a much needed rest from the busy schedule. Part of the lack of interest resulted from peoples' apprehension that they would not understand the papers. The other part stemmed from the preoccupation of the speakers with methods rather than results, a preoccupation that has plagued statistical paleoecology from the first. As we look to the future of statistical paleoecology, we can expect to see less emphasis on methods and more emphasis on results, with the use of statistical methodology taken in stride where appropriate.

ACKNOWLEDGMENTS

I have benefitted immeasurably from discussions with my colleague A. J. Rowell on the subjects of paleoecology, statistical methodology, and their intersection. John C. Griffiths and two anonymous reviewers provided constructive critiques that led to improvement of the manuscript.

REFERENCES

Ager, D. V. (1963). *Principles of Paleoecology.* McGraw-Hill, New York.

Blackith, R. E. and Reyment, R. A. (1971). *Multivariate Morphometrics.* Academic Press, New York.

Bradley, J. V. (1968). *Distribution-Free Statistical Tests.* Prentice-Hall, Englewood Cliffs, New Jersey.

Brondos, M. D. and Kaesler, R. L. (1976). Diversity of assemblages of late Paleozoic Ostracoda. In *Structure and Classification of Paleocommunities*, R. W. Scott and R. R. West, eds. Dowden, Hutchinson, & Ross, Stroudsburg, Pennsylvania.

Cochran, W. G. (1963). *Sampling Techniques*, 2*nd* ed. Wiley, New York.

Conover, W. J. (1971). *Practical Nonparametric Statistics.* Wiley, New York.

Cooley, W. W. and Lohnes, P. R. (1971). *Multivariate Data Analysis.* Wiley, New York.

Curio, E. (1976). *The Ethology of Predation.* Springer-Verlag, New York.

Davis, J. C. (1973). *Statistical and Data Analysis in Geology.* Wiley, New York.

Doveton, J. H. (1971). An application of Markov chain analysis to the Ayrshire Coal Measures succession. *Scottish Journal of Geology*, 7, 11-27.

Fretwell, S. D. (1975). The impact of Robert MacArthur on ecology. *Annual Review of Ecology and Systematics*, 6, 1-13.

Gibson, T. G. and Buzas, M. A. (1973). Species diversity: patterns in modern and Miocene foraminifera of the eastern margin of North America. *Geological Society of America Bulletin*, 84, 217-238.

Greig-Smith, P. (1964). *Quantitative Plant Ecology, 2nd* ed. Butterworths, London.

Griffiths, J. C. (1966). Future trends in geomathematics. *Mineral Industries*, 35, 1-8.

Griffiths, J. C. (1967). *Scientific Method in Analysis of Sediments.* McGraw-Hill, New York.

Hallam, A., ed. (1977). *Patterns of Evolution as Illustrated by the Fossil Record.* Elsevier, New York.

Imbrie, J. and Newell, N. D. (1964). Introduction: the viewpoint of paleoecology. In *Approaches to Paleoecology*, J. Imbrie and N. D. Newell, eds. Wiley, New York.

Jackson, J. C. B. (1978). Reviews. Patterns of evolution as illustrated by the fossil record. *Journal of Paleontology*, 52, 1164-1166.

Kershaw, K. A. (1964). *Quantitative and Dynamic Ecology.* Edward Arnold, London.

Kitts, D. B. (1977). *The Structure of Geology.* SMU Press, Dallas, Texas.

Krumbein, W. C. (1965). Sampling in paleoecology. In *Handbook of Paleontological Techniques*, B. Kummel and D. M. Raup, eds. W. H. Freeman and Company, San Francisco.

Krumbein, W. C. (1975). Markov models in the earth sciences. In *Concepts in Geostatistics*, R. B. McCammon, ed. Springer-Verlag, New York.

Ladd, H. S., ed. (1957). *Treatise on Marine Ecology and Paleoecology, Vol. 2.* Geological Society of America, Boulder, Colorado.

Lawrence, D. R. (1968). Taphonomy and information losses in fossil communities. *Geological Society of America Bulletin*, 75, 1315-1330.

Lawrence, D. R. (1971). The nature and structure of paleoecology. *Journal of Paleontology*, 45, 593-607.

McKee, E. G. (1964). Inorganic sedimentary structures. In *Approaches to Paleoecology*, J. Imbrie and N. D. Newell, eds. Wiley, New York.

Pielou, E. C. (1977). *Mathematical Ecology.* Wiley-Interscience, New York.

Poole, R. W. (1974). *An Introduction to Quantitative Ecology.* McGraw-Hill, New York.

Reyment, R. A. (1971). *Introduction to Quantitative Paleoecology.* Elsevier, New York.

Reyment, R. A. (1979). Multivariate analysis in statistical paleoecology. In *Multivariate Methods in Ecological Work*, L. Orloci, C. R. Rao, and W. M. Stiteler, eds. Satellite Program in Statistical Ecology, International Co-operative Publishing House, Fairland, Maryland.

Schopf, T. J. M. (1975). Theory in paleoecology. *Paleobiology*, 1, 129-131.

Sheehan, P. M. and Lesperance, P. J. (1978). Effect of predation on the population dynamics of a Devonian brachiopod. *Journal of Paleontology*, 52, 812-817.

Siegel, S. (1956). *Nonparametric Statistics for the Behavioral Sciences.* McGraw-Hill, New York.

Simpson, G. G. (1963). Historical science. In *The Fabric of Geology*, C. C. Albritton, Jr., ed. Freeman, Cooper, & Company, Stanford, California.

Sneath, P. H. A. and Sokal, R. R. (1973). *Numerical Taxonomy.* W. H. Freeman and Company, San Francisco.

Sokal, R. R. and Rohlf, F. J. (1969). *Biometry.* W. H. Freeman and Company, San Francisco.

Solomon, A. M. (1977). Paleoecology society. *Bulletin of the Ecological Society of America*, 58, 18.

Solomon, A. M. (1978). Paleoecology section. *Bulletin of the Ecological Society of America*, 58, 17.

Valentine, J. W. (1973). *Evolutionary Paleoecology of the Marine Biosphere*. Prentice-Hall, Englewood Cliffs, New Jersey.

Winer, B. J. (1962). *Statistical Principles in Experimental Design*. McGraw-Hill, New York.

Wright, R. G. and Van Dyne, G. M. (1971). Comparative analytical studies of site factor analysis. In *Statistical Ecology, Vol. 3*, G. P. Patil, E. C. Pielou, and W. E. Waters, eds. The Pennsylvania State University Press, University Park. 59-95.

[*Received December* 1978. *Revised March* 1979]

G. P. Patil and M. Rosenzweig, (eds.),
Contemporary Quantitative Ecology and Related Ecometrics, pp. 635-647.

THE DESCRIPTION OF A NON-LINEAR RELATIONSHIP BETWEEN SOME CARABID BEETLES AND ENVIRONMENTAL FACTORS

S. A. L. M. KOOIJMAN

Department of Biology
Division of Technology for Society
P. O. Box 217
2600 AE Delft, The Netherlands

R. HENGEVELD

Department of Geobotany
Catholic University
Toernooiveld
Nijmegen, The Netherlands

SUMMARY. A model is presented which helps to clarify the choice of the environmental variable with the strongest influence on the number of small terrestrial animals caught. In the model a non-linear relationship between these numbers and the hypothetical environmental variable is assumed. This is illustrated by applying the model to catches of Carabid beetles in a large area of more or less homogeneous grassland. A Poisson distribution was assumed for the catches and a Gaussian distribution was taken to describe the non-linear relationship. To evaluate the improvement provided by the new model, a multiple correlation analysis was performed between the actual catches of the nine species in question as well as the estimated factor values and two environmental variables: the lutum content of the soil and the elevation of the trap sites.

KEY WORDS. spatial distribution, Carabidae, non-linear relationship, Poisson distribution, gradient analysis.

1. INTRODUCTION

In planning a study on factors determining the spatial distribution of organisms in the field, we are faced with the problem of the choice of method. The experimental method of investigation, although the most conclusive, is not the appropriate one to start with, because it gives little information about relevance of the conditions prevailing in the field. When we lack prior information, we need a technique that will give rise to a limited number of hypotheses directly from the observations made in the field.

This paper describes such a technique and its application to data collected in an investigation on the spatial distribution of Carabid beetles caught in pitfall traps in an experimental area.

A procedure which leads to a hypothesis concerning determinative factors in the spatial distribution of Carabid beetles, should yield the density of a particular species as a function of one or more hypothetical variables. The usual way of selecting the most relevant variable(s) from a large number of possible variables is to determine their correlation with the observed species densities. The more this correlation deviates from zero, the more relevant we expect this variable to be in determining the species density. If such correlation analyses are to be meaningful, it must be assumed that a monotonous relationship exists between the environmental variables and the densities. However, this assumption is not applicable in many biological cases, where some optimum relationship is more often expected. The interpretation of a principal components analysis, which is frequently used in this connection, may be hampered by this drawback, since the procedure is based on the same assumption of monotony as underlies almost all other multivariate procedures.

To avoid this kind of difficulty, a model that does not involve this unrealistic assumption is required. Such a model is presented here.

2. MODEL SPECIFICATION

Application of the present model to the collection of Carabid beetles from the field raises the question of which of the assumptions made in such a model seem reasonable. The first of these assumptions is that the catches per pitfall trap are mutually independent. This is the case if care is taken to place the traps sufficiently far apart.

The second assumption is that interference between individual beetles of either the same or a different species is not very strong. This may seem to be a rather stringent assumption, but statistical analysis is hardly feasible without it.

The third assumption is that the collecting period should be short compared with the duration of changes in the environmental variable, but long enough to assure catches large enough for statistical analysis. This assumption is necessary for the detection of a meaningful correlation between the estimated factor value of each trap and the measured environmental variable(s).

The fourth and more stringent assumption, which is implicit in the functional form of the expected densities, is that the numerical response of each species follows the same function with respect to the environmental variable. This has the consequence that the present model can be expected to be successful only when it is applied to ecologically related species. Our findings indicate that the ground-beetle species investigated fulfil this requirement.

The first two assumptions imply that the number of individuals of each species collected per trap, represents a trial from a Poisson distribution. The parameter λ_{ts} of this distribution represents the expected cumulative catch over the sampling period for species s in trap t.

The relationship between the expected densities, λ_{ts}, and environmental variables can now be formulated as an optimum one. The simplest non-negative function is the Gaussian, given by the formula $\lambda_{ts} = \alpha_s \exp[-(f_t-\mu_s)^2/\beta_s]$, in which:

- f_t is the intensity of the environmental variable in the vicinity of trap t;
- μ_s is the optimum of species s for the particular variable;
- β_s is the tolerance of species s for the particular variable;
- α_s is the density of species s when the variable has a value which is optimal for that particular species.

For the estimation of the species and factor parameters, the procedure of Kooijman (1977) was used.

The most useful feature of this model is that correlations can be calculated between estimated factor values and measured environmental factors. When the factor is identified with some (function of) environmental variables, the species parameters can be used for the characterization of the ecological behavior of the species under study.

For less general deterministic models related to the present one, the reader is referred to Gauch and Chase (1974) and Ihm and Van Groenewoud (1975. The present model differs from these in that it represents the sampling process and provides a valid testing procedure as to the goodness-of-fit.

3. MATERIAL

The sampling area, called The Ellerslenk Ecological Reserve, is situated in The Netherlands at the border of Oostelijk Flevoland, one of the recently reclaimed IJsselmeer polders. The reserve measures 800 × 300 m.

The populations of Carabid species occurring in this area were sampled with pitfall traps, which were arranged in a rectangular grid system of 168 points, each of which consisted of five traps. The catches of each set of five traps were combined, because of the 5-meter distance between the traps was considered too small to treat the samples from each individual trap as independent. The catches per sampling-point were considered to be independent because the shortest distance between these points was 40 meters.

The sampling device (see Figure 1) consisted of two cups one suspended inside the other. This arrangement speeded up collection of the specimens and left the surrounding soil undisturbed. Because water flowing over the soil surface after a heavy rainfall tended to accumulate in the outer cup, the concentration of formalin in the solution (used as a killing and conservation fluid) remained almost constant in the inner cup. It is evident that this arrangement could not work perfectly: rainwater may have reached the inner cup at times and the rim of this inner cup would rise when too much water accumulated in the outer cup, thus preventing the trapping of the beetles. However, this did not occur often enough to seriously affect the results.

The killing and conservation fluid consisted of a 4% formalin solution to which a little detergent had been added to lower the surface tension of the water. As a result, the beetles immediately sank to the bottom of the trap and could not swim to the side and climb out of the cup before the fluid took effect.

Collections were made twice, and covered two successive periods of one month each at the end of the summer of 1974: August 1st to September 2nd (period A) and September 2nd to October 1st (period B). Some 60 species were collected although only 9 of these were in sufficiently large numbers for reliable statistical analysis. (Table 1). Initially, a tenth species, *Calathus melanocephalus*, was also included, because it was reasonably abundant, but the maximization procedure required for the estimation of the parameters of the model presented in this paper, failed to converge, which indicated that this species is not ecologically related to the other nine. The inclusion of this species would therefore require a model for the analysis of at least two factors.

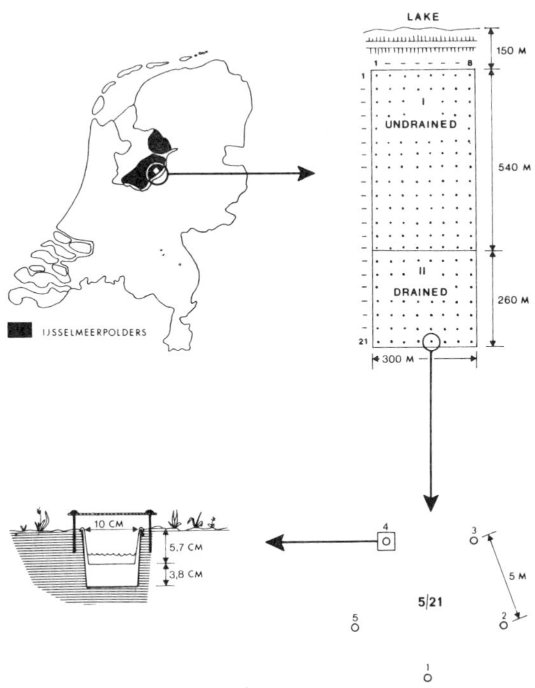

FIG. 1: Location of the sampling area in The Netherlands, and diagram of the grid system of 8 × 21 *sampling sites, each consisting of* 5 *sampling devices. The area has an undrained* (I) *and a drained part* (II), *and is separated from a large inland lake by a dike and a hedgerow.*

TABLE 1: Total number of specimens per species collected during two one-month periods in 1974 *in the Ellerslenk Ecological Reserve. Period A: August* 1 *to September* 2; *Period B: September* 2 *to October* 1.

Species	Period A	Period B
Amara communis	70	71
Amara familiaris	141	32
Bembidion properans	100	253
Clivina fossor	194	23
Clivina collaris	70	21
Dyschirius globosus	2282	1517
Pterostichus coerulescens	6576	6832
Pterostichus cupreus	649	1115
Trechus obtusus	1449	2020

We decided to sample for two periods of one month for a number of reasons: (1) Collection over one month would provide enough data to give statistically reliable results. (2) The reproductibility of the first month's results could be checked by comparison with the results of the second collecting period. (3) To determine the relevant time-scale on which a possible dynamic process operates, since a process of this kind might govern the spatial patterns of the species. If so, the catches from the same sites would show a systematic, though small, difference over the area. The entire area has a clay soil which is partly covered by a layer of sand varying in thickness. Consequently, there are differences in elevation as well as in the chemical and physical soil conditions over the area.

The height of each pitfall trap was measured with a theodolite and related to height of the lowest trap. The height data of only one trap per sampling point were used for this paper. Measurements were made to an accuracy of 1 millimeter to obtain an accuracy in the calculations of 1 centimeter.

The lutum content at each grid point was estimated qualitatively in the field by an experienced pedologist and expressed in percentages.

Apart from its dependence on soil type, the moisture condition over the area is influenced by: (1) the influx of ooze from the nearby lake and (2) the drainage of part of the area by an underground system emptying into two ditches running along the longer sides of the field (Figure 1). Remains of trenches from the original reclamation period (before 1968) are still present, especially in the wet parts, running straight across the area without reaching these two ditches. A system of locks keeps the water in different parts of the ditches at the same level relative to the slope of the field.

The vegetation consists mainly of grasses, its composition varying over the field. The soil has not been fertilized since 1968; the reserve is mowed twice a year, i.e., at the beginning of July and October. The vegetation is mainly dense in the drained parts with *Festuca rubra* predominating and sparse in the sandy parts.

4. RESULTS

The factor and species parameters were estimated according to the maximum-likelihood criterion. The parameter estimates of both sets of figures for the two periods (A and B) are shown in Figure 2, where the species parameters are depicted by the shape of the Gaussian curves. The value of the factor parameter per sampling point, i.e., the point on the x axis, determines the expected number of individuals on the y axis.

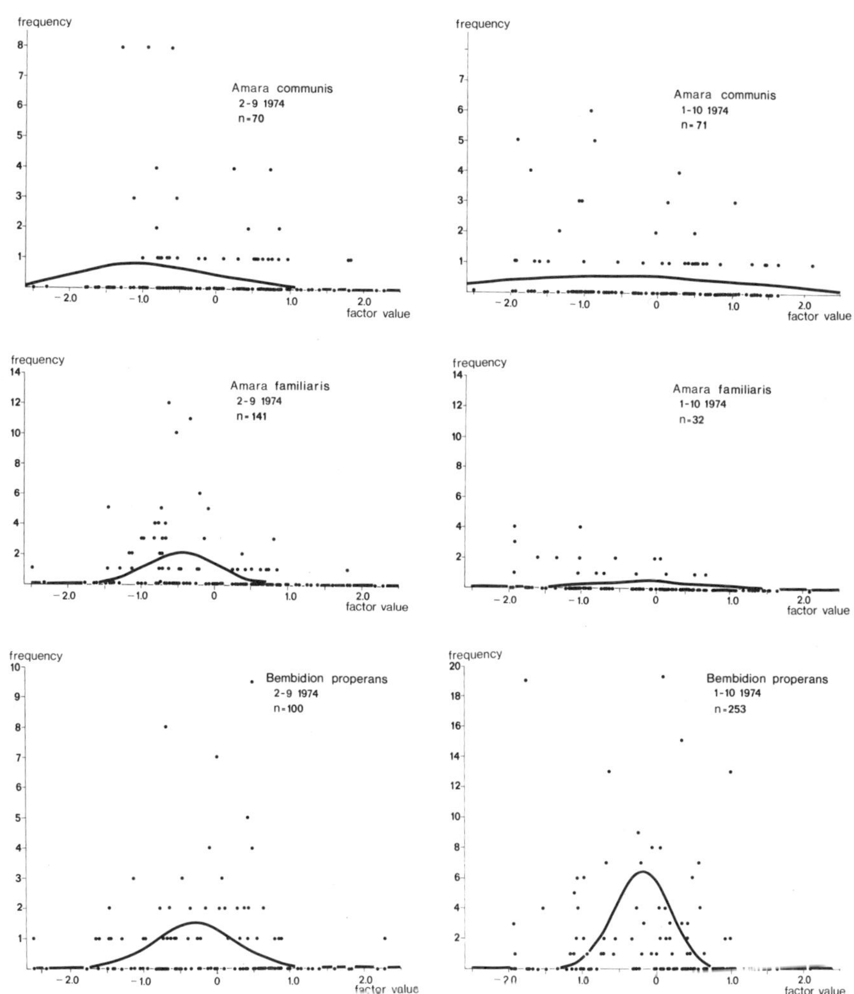

FIG. 2: Graphs for each of the nine species in the two sampling periods, showing curves with a Gaussian distribution for the expected values of the number of a given species at each sampling point representing the actual number of individuals collected. x *axis: estimated factor value of the sampling sites;* y *axis: expected (or actual) number of individuals.*

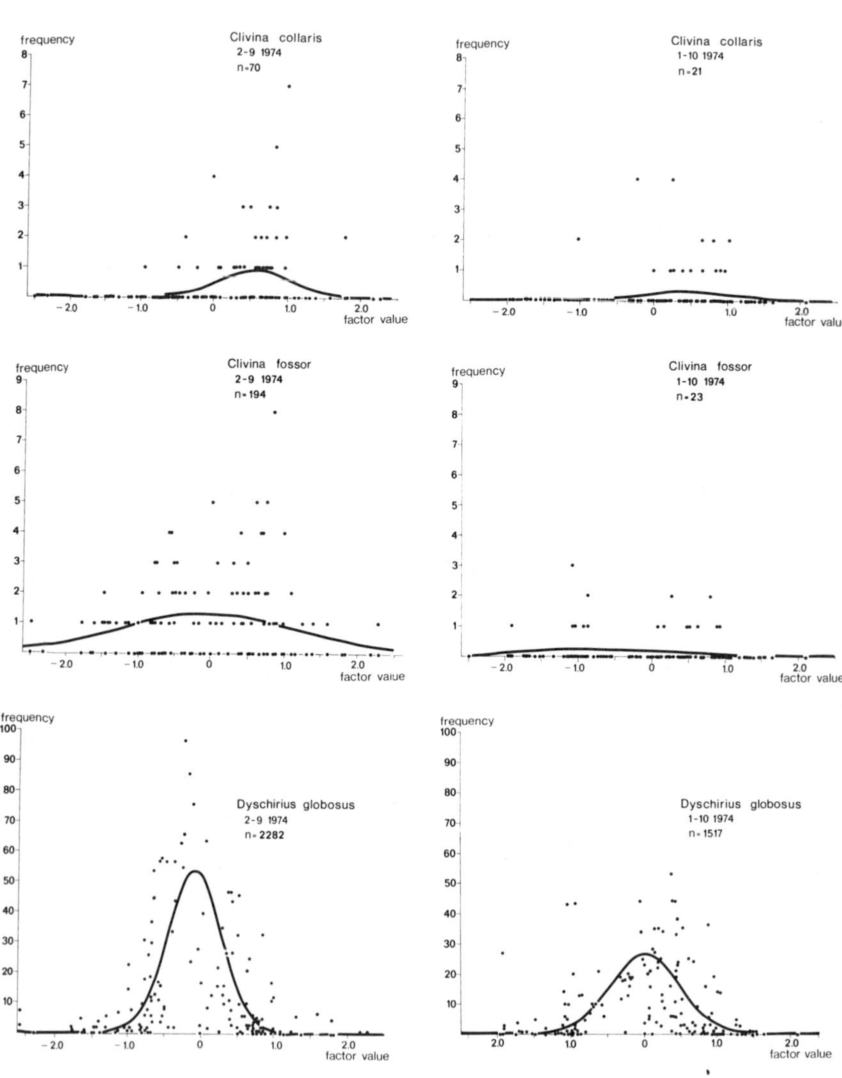

FIG. 2: (Continued)

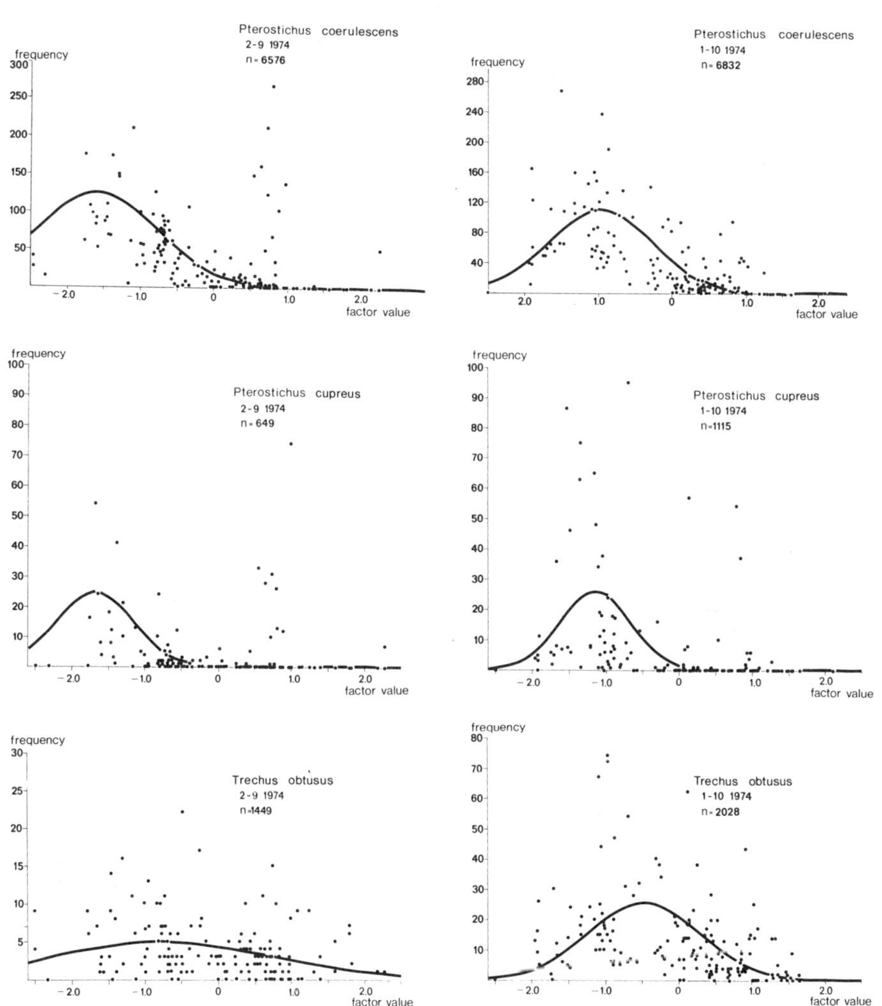

FIG. 2: (Continued)

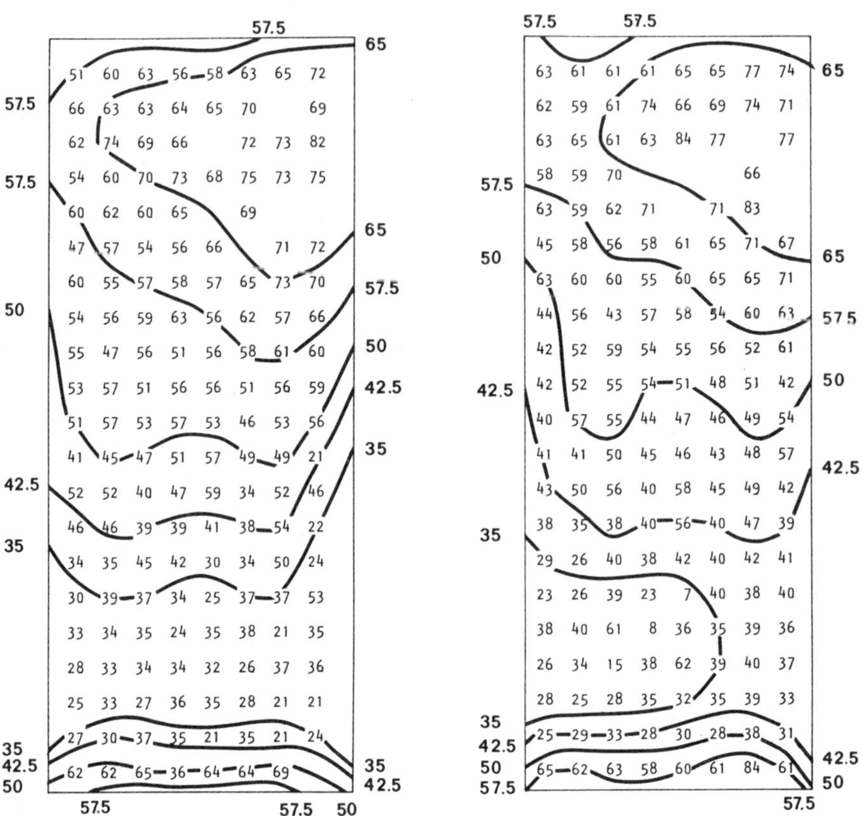

FIG. 3: Factor values estimated per sampling point, with the two-dimensional 5th-degree polynomial fitted to these estimates. Seven points were omitted because none of the species had been collected there. Sampling dates: September 2, 1974 (left hand side) and October 1, 1974 (right hand side).

Figure 3 shows the spatial distribution in the field of the estimated factor values for sampling periods A and B, respectively. In each collecting period there were seven sampling sites at which no representative of any of the nine species was caught. These sites differed in each period. Because the maximum-likelihood estimates for these factor values would be $\pm \infty$, these zero catches were deleted before the data were processed. A two-dimensional 5*th*-degree polynomial was fitted to describe the spatial variation of the postulated factor. The polynomial coefficients were estimated according to the algorithm of Whitten (1970) for polynomials which are orthogonal with respect to the coordinates of the sampling sites.

The general features of the spatial variation of the factor estimates correspond very well with the lutum content as well as with the variation in the elevation of the sites. Although of course elevation itself cannot be a determinant of the distribution of individuals, several other variables such as lutum and water content of the soil can be expected to be dependent on it. The correlation coefficients are given in Table 2. The correlation between the height and the lutum content over the 168 sampling sites is -0.91.

TABLE 2: Correlation between factor estimates and lutum content or elevation of the traps for the two collecting periods, estimated over the 161 *sampling sites.*

Variable	Period A	Period B
lutum content	-0.67	-0.75
elevation	-0.72	-0.78

The goodness-of-fit was found to be poor; the likelihood-ratio statistic of each period is roughly 2×10^5, and has approximately a Chi-square distribution with 1,261 degrees of freedom under the null hypothesis. It is doubtful, however, that the Chi-square approximation is satisfactory in this case, since the two period 51% and 56% of the 9 species × 161 sites catches were zero.

5. DISCUSSION

Although the goodness-of-fit test indicated that the model was not entirely satisfactory, the value of this test should not be overestimated. The model used in the present work serves primarily as a hypothesis-generating tool. As such, it is more satisfactory than the linear models commonly used. This can be illustrated by a straightforward correlation between the species numbers and the lutum content of the soil and the variation in elevation of the sites across the study area. Table 3 shows that the linear models would not have pointed so clearly to the part played by these two variables as the model presented here has done.

TABLE 3: Multiple correlation, called R, *between the lutum content or the variation in elevation of the* 161 *sampling sites and either the factor values or the numbers of individuals of the nine Carabid beetle species caught. The values given are* $N-q/q-1 \cdot R^2/1-R^2$ *when* $N = 161$ *and* $q = 2$ *for the factor values* 10 *for the beetles. Under the hypothesis of non-correlation, the entries are* F-*distributed with* $q - 1$ *and* $N - q$ *degrees of freedom.*

	Period A		Period B	
	Factor values	No. of beetles	Factor values	No. of beetles
lutum content	130.8	13.2	201.2	9.6
elevation	169.6	11.5	289.7	9.3

It is not difficult to explain a poor goodness-of-fit. The choice of this model assumes, as already stated, that the environmental conditions are static, i.e., that the fluctuation of the relevant environmental variables is negligible with respect to the duration of the collecting period. The difference in the parameter estimates of the postulated factor values for the two sampling periods indicates that such an assumption is only partially justified. The non-systematic differences can probably be explained by the stochastic nature of the actual number of individuals caught per species.

Systematic differences may, however, indicate that neither elevation nor lutum content can be the determining factor. Therefore, these differences may provide some suggestions as to the nature of the environmental factor(s) mainly responsible for the abundance distributions of the species concerned. In this case the water content of the soil could be such a factor, because it can be expected to change slowly, depending on the lutum content and elevation. The time scale on which the change in water content occurs, justifies the assumption that it did not vary during the collecting periods. Later observations, made in 1976, support this assumption.

Of course, a high correlation between factor values estimated from the number found of a particular species and one or more environmental variables does not necessarily imply a causal relationship. Such a relationship can only be investigated by an experimental approach.

ACKNOWLEDGEMENTS

This research has partly been supported by a grant from the Netherlands Organization of the Advancement of Pure Research (Z.W.O.).

REFERENCES

Gauch, H. G. and Chase, G. B. (1974). Fitting the Gaussian curve to ecological data. *Ecology,* 55, 1377-1381.

Ihm, P. and van Groenewoud, H. (1975). A multivariate ordering of vegetation data based on Gaussian type gradient response curves. *Journal of Ecology,* 63, 767-777.

Kooijman, S. A. L. M. (1977). Species abundance with optimum relations to environmental variables. *Annals of Systematic Research,* 6.

Whitten, E. H. T. (1970). Orthogonal trend surfaces for irregular spaced data. *Mathematical Geology,* 2, 141-152.

[*Received December* 1978. *Revised July* 1979]

G. P. Patil and M. Rosenzweig, (eds.),
Contemporary Quantitative Ecology and Related Ecometrics, pp. 649-662.

SAMPLING BARK BEETLE POPULATIONS FOR ABUNDANCE

PAUL E. PULLEY

Data Processing Center
Texas A&M University
College Station, Texas 77843 USA

ROBERT N. COULSON

Department of Entomology
Texas A&M University
College Station, Texas 77843 USA

JOHN L. FOLTZ

Department of Entomology & Nematology
University of Florida
Gainesville, Florida 32611 USA

SUMMARY. Procedures for quantiatively sampling within-tree and within-infestation populations of bark beetles (Coleoptera: Scolytidae) are discussed. Several different options are available for use. The general approach taken was one of purposeful selection of samples relative to defined distributions of the various life stages of the insect.

KEY WORDS. sampling, bark beetles, Coleoptera: Scolytidae, *Dendroctonus*.

1. INTRODUCTION

Bark beetles (Coleoptera: Scolytidae) are major mortality agents in mature pine forests throughout North and Central America and Europe. This family of insects has attained notoriety primarily as a result of their economic significance and have been studied extensively with forest management goals guiding the course of much of the research (Coulson, *et al.*, 1979). Fortunately, bark beetles are also marvelous subjects for the study of many aspects of population and community ecology and a large volume of published literature is available for many species.

There are generally three basic circumstances or reasons why one might wish to sample bark beetle populations for abundance: (1) to study some aspect of population or community ecology; (2) to evaluate the results of preventive, suppressive, or regulatory tactics; and (3) to survey for the presence and numbers of the insect over a large geographical area. Different estimation procedures are required for each of the sampling goals as the requirements for precision are different.

Bark beetle populations have several attributes which on the one hand complicate and on the other simplify the development of quantitative estimation procedures. Among the complicating variables are the following: (1) there are six life stages that may be of sampling interest [attacking adults, reemerging adults, eggs, larvae, pupae-callow adults, and emerging adults]; (2) where investigated, these life stages have been found to have different spatial distributions within the tree; (3) bark beetles often colonize single trees, trees in clumps [infestations], and clumps of trees over large forested areas; (4) the life cycle may vary from six weeks to one year depending on the species and season of the year; and (5) beetle development within infestations is often asynchronous with multiple overlapping generations. Among the simplifying variables are the following: (1) about 97% of the life cycle of the insect is spent within the inner-bark region of the tree; (2) the sampling universe has discrete boundaries which can be defined by inspection, i.e., the length of the infested bole and the number of trees infested; and (3) the habitat boundaries and beetles within the habitat are amenable to rather precise measurement.

In this discussion we are limiting our comments to the within-tree population. Very little is known about flight and dispersal of bark beetles and the sampling considerations for flying populations are clearly different from within-tree populations.

Our objectives in the following discussion are to present a framework for the development of sampling plans for within-tree populations of bark beetles in general. We will consider two levels of complexity: (1) estimation of beetles in single trees and (2) estimation of beetles within trees occurring in clumps, i.e., infestation spots. The basic framework has been constructed from our research on the southern pine beetle, *Dendroctonus frontalis* Zimm., and in terms of complexity this insect probably represents a 'worst case' experimental animal. The basic methodologies presented should have application to other species as well.

2. GENERAL BACKGROUND

2.1 Life History of Dendroctonus frontalis. Before examining the development of specific sampling plans for bark beetles it is necessary to have an understanding of the general life history of the species of concern. Although there is considerable variation between the different species, we are including a brief description of the life cycle of *D. frontalis* to provide a general framework for the estimation procedures.

The general life history of the *D. frontalis* begins with the selection of suitable host trees by adults through either random and/or directed behavior (the exact process is unknown). Colonization of trees is regulated by a blend of both insect-produced pheromones and host-produced attractants. Females initiate the construction of 'egg galleries' by boring into the inner bark region where they are joined by males. Mating takes place within the galleries and eggs are oviposited in niches at intervals along the lateral walls. Both the males and females reemerge and are capable of attacking and colonizing new host trees. Eggs hatch shortly after oviposition and the ensuing larvae begin to excavate 'larval galleries.' The first two instars remain in the phloem region and the last two migrate into the outer corky bark where pupation and adult emergence occur. Development can be completed in as few as 35 days and six to eight generations per year occur in some regions of the South. Numerous parasites, predators, and associates have been identified to occur with *D. frontalis*, but their effects on within-tree populations have not been quantitatively defined. Infestation spots often enlarge dramatically in a short period of time, are characterized by an active front or head, and are comprised of multiple asynchronous generations of the insect developing concurrently. Both reemerging and parent adults are involved in the colonization of trees (Coulson *et al.*, 1979).

2.2 Data Base Accumulation. There are numerous possible ways to collect data on within-tree populations of bark beetles, and several of these procedures have been reviewed by Stark and Dahlsten (1970) and Coulson *et al.* (1975). The format we selected for the accumulation of a data base on *D. frontalis* consisted of extracting four 100-cm^2 bark disks at the NE, NW, SE, and SW starting aspects at 1.5 m interval at 2.0 m and continuing to the top of the infested bole. Emerging beetles were collected in cages deployed at the same heights and aspects. The sampling intervals, and sample unit size, shape and intensity were based on previous studies of the western pine beetle, *D. brevicomis* LeConte (DeMars *et al.*, 1970) and the mountain pine beetle, *D. ponderosae* Hopkins (Safranyik, 1968).

Bark disks containing the desired life stages were removed with hole-cutting saws and radiographs were prepared. The insect inclusions on the radiographs were then counted.

In addition to the insect data, various parameters of the host trees were measured: tree height, height of the infestation, and diameter and bark thickness at each sample height.

Data were collected on attacking adults, eggs, larvae, pupae-callow adults, and emerging adults over a three year period. In all, 134 trees were sampled. The data were then structured into a machine-readable format.

We elected to use the procedure of sub-sampling many infested trees because bark beetle populations are characterized by a tremendous inherent variation. We judged that consideration of this variation was of paramount importance in the development of applicable estimation procedures. An indication of the range in variation of the data collected over the three years of the study is contained in Table 1.

TABLE 1: Characteristics of tree data used to simulate infestation spots of D. frontalis *attacking adults.*

Year	No. trees	Beetles per tree	Diam. (cm)	Beetles per cm	Infested area dm^2	Beetles per dm^2
1972						
Mean	54	14023	36.1	388	1381	10.16
Range	-	1497-32499	24-50	286-664	702-2464	2.53-15.69
1973						
Mean	44	6333	31.9	199	1009	6.28
Range	-	1213-15255	22-45	53-382	369-1700	1.66-11.78
1974						
Mean	34	15317	39.3	390	1375	11.14
Range	-	6748-25018	30-49	222-554	749-1934	6.48-17.16
Overall						
Mean	132	11793	35.3	332	1255	9.40

An alternative approach would have been to collect all of the bark from a few trees and sub-sample this data. Indeed, this practice has been followed by Stephen and Taha (1976) and Nebeker and Hocking (personal communication). The main disadvantage to this approach is that it is necessary to assume that the intensive sample from a few trees represents the true general condition in nature.

With the data base on 134 trees we are now in a position to consider the development of various types of estimation procedures for within-tree and within-infestation populations. In the following discussion we will consider *what* was done in the development of estimation procedures. The details and *how* the procedures were developed have been described in a series of publications: Coulson *et al.* (1976), Foltz *et al.* (1976, 1977), Mayyasi *et al.* (1976a,b), Pulley *et al.* (1976, 1977a,b,c). The general procedure followed in developing sampling plans was to select a life stage (e.g. attacking adults), test various potential plans, make a decision on the suitability of the plans for application purposes, treat the remaining life stages, and finally describe the procedures in a manner useable by the practitioner, i.e. the entomologist or forester.

3. WITHIN-TREE POPULATION ESTIMATION

The development of within-tree population estimation procedures was accomplished in a series of steps progressing from the simple to the more complex.

The first step was to investigate beetle distribution within the sample unit (100-cm^2 bark disks), the density of beetles in the samples at the four directions, and density along the infested portion of the tree bole. Analyses of the distributions within sample disks provided no reason to reject the hypotheses that the beetles were uniformly distributed within the average disks. There was no consistent directional bias associated with the four aspects. The distribution along the infested bole varied with life stages. The general pattern was for beetle density to be greatest towards the center part of the infested bole and taper towards the extremes. The distribution of eggs was more trapezoidal in shape. The functional distributions of the various life stages in relation to the normalized infested bole were described using two and three parameter nonlinear models: attacking adults (Coulson *et al.*, 1976), eggs (Foltz *et al.*, 1976b), larvae, pupae-callow adults, emerging adults (Mayyasi *et al.* 1976b). Mayyasi *et al.* (1976a) described the statistical distributions and developed probability density functions (proportional density functions) for the various life stages.

The second step was to develop a methodology for computing the total number of beetles on a tree. This methodology had to (1) be consistent with the previously defined information on beetle distributions, (2) utilize systematically collected data, and (3) minimize errors resulting from smoothing or extrapolation. The procedure developed was the topological mapping routine (Pulley *et al.*, 1976a). This procedure may be understood by visualizing the insects as a film covering the tree and the thickness of this film proportional to the insect density. If the investigator knows the volume of this film, he knows the number of insects on the tree. It is difficult if not impossible to integrate over the surface to get the volume of count because of the undulations of the thickness of this configuration. However, by use of topological transformations we can get an equivalent result and do so in a tractable fashion. The topological mapping technique assumes a gradual change in insect density between observation points, which is completely consistent with the defined within-tree distributions. This procedure was used as the basis or target value for calculating the bias and variability of other estimation procedures.

The third step was to select a series of candidate sampling plans from a large number of possibilities and test their suitability for estimating within-tree populations. At this stage a single life stage (attacking adults) was used. Later the promising procedures were tested with the other life stages as well.

A computerized (brute force) methodology was used to ascertain the accuracy and precision of various estimation techniques. In all cases, as indicated above, the values estimated by the topological procedure were used as the population of inference and served as target values for the estimates, since this technique used all available information from each tree and minimized the errors of smoothing, averaging, and extrapolation. Each time an estimate was made by any technique, it was compared to the topological mapping estimate and a percentage error term was computed, a negative value indicating an underestimate and a positive value an overestimate of the target number. Computing the mean and standard deviation of the percent error for the total trees in the data base showed the bias and variability of the procedure being tested and provided information for securing unbiased estimates and confidence intervals for those estimates.

All within-tree population sampling procedures must contain two separate estimates: one of the surface area or volume of the habitat and the other of insect density within this habitat. We selected five procedures each for calculating surface area and beetle density. For surface area the procedures included the following: (1) the stacked cylinder technique, (2) truncated cone technique, (3) topological mapping technique, (4) long cylinder technique, and (5) the tree geometry model technique. For beetle density the procedures included: (1) the truncated cone technique, (2) the stacked cylinder technique, (3) the topological estimation technique, (4) the extrapolated disk technique, and (5) the probability density function technique. Each of these techniques is described in detail by Pulley *et al.* (1977a.)

For convenience these procedures for estimating tree surface area and beetle density were combined and classed into 'large' and 'small' sampling plans based on the amount of data required. The first three procedures for estimating surface area and beetle density are 'large' sample techniques and the remaining two in each category are 'small' sample. Our primary interest is in the small sample procedures as they are likely to be of value in operational sampling programs. Again, there are several possible combinations of 'small' sample surface area and 'small' sample beetle density techniques.

The tree geometry model (TG) probability density function (PDF) combination is a good illustration of a 'small' sampling plan. In this procedure surface area of the tree is obtained from the tree geometry model described by Foltz *et al.* (1976a). This model utilizes as input parameters the tree diameter and bark thickness at 2.0 m and the infested bole height. Integrating over the infested bole of the tree yields the surface area of the infestation. Tables of surface area, based on this model, have been provided for loblolly pine (*Pinus taeda* L.) of 25-40 cm diameter outside bark (DOB) for three bark thickness classes (Coulson *et al.*, 1976a).

Beetle density is obtained from the PDF procedure (Pulley *et al.*, 1976a). The mathematical derivations used in the PDF technique are provided by Mayyasi *et al.* (1976a). In this application the ordinate value of the PDF, which is the ratio between the insect density at a given height and a uniform insect density along the infested bole, is used as a weight for adjusting the observed density to estimate average density. Ordinate values of the PDF for *D. frontalis* life stages are provided in Coulson *et al.* (1976a). The abscissa value X is the normalized infested bole height which is the ratio of the sample height to the infested bole height.

To estimate the number of beetles on a tree using the TG-PDF procedure, the observed density per sample unit is divided by the ordinate value of the PDF and multiplied by the number of sample units in the infested bole. This procedure requires only one sample or set of samples at a given level from the tree and utilizes the defined within-tree distributions of the *D. frontalis* life stages. Procedures have also been developed for combining multiple sampling levels as well.

The methodology for evaluating the 'small' sample procedures consisted of taking all possible sample combinations of 1, 2, 3, and 4 disks at a sample level, calculating the estimated within-tree population, and comparing each estimate with the value estimated by the topological procedure. Since all possible combinations of four disks taken at a given height were used in calculating bias, it follows that the biases for the various sample sizes for that height are the same. As expected the standard deviation decreased with increasing sample size.

If the bias of the sampling procedure is known, it is possible to calculate a correction weight and secure an unbiased estimate. For the 'small' sample techniques the appropriate correction weight (W) was calculated as

$$W = 100/(B_h + 100),$$

where B_h = percentage bias at height h.

To use the correction weight a count is estimated by one of the techniques and then multiplied by the corresponding W to obtain the corrected estimate. The corrected standard deviation for calculating the confidence interval is obtained by multiplying the tabular standard deviation (Pulley *et al.*, 1977b; Foltz *et al.*, 1977) by W.

In some instances, the precision of the estimate may be increased by sampling more than one level. To do this type of sampling it is necessary to have a table of biases and the variance-covariance matrix for the estimation procedure under consideration. Procedures and the necessary tables for multi-level sampling using the TG-PDF procedure and the long cylinder-PDF procedure are provided in Coulson *et al.* (1976a).

When utilizing the 'small' sample procedures, the location on the tree where the samples are taken influences greatly the precision of the estimate obtained. Pulley *et al.* (1977c) described in detail the results of various sample level locations using the TG-PDF procedure for sampling intensities ranging from one to seven levels (four samples/level). In general, gain in information became marginal after three levels were sampled, and dispersal of sample levels over the infested bole, while avoiding the extremes, produced the best results.

The fourth and final step in the development of within-tree population sampling procedures was to present the material in a form suitable for use by practitioners, i.e., entomologists and foresters. In the case of *D. frontalis* a user's guide was prepared (Coulson *et al.*, 1976a) which (1) described the basic procedures of the sampling plans, (2) included all needed statistical tables, and (3) outlined step-by-step instructions, with examples, for using the procedures. The guide focused on the small sample procedures as they are economical and provide suitable precision for most applications. Where greater precision is required, the topological estimation procedure was recommended.

4. WITHIN-INFESTATION SPOT ESTIMATION

The problem of estimating total numbers of a given life stage of a bark beetle in an infestation is rather complex. Clearly it is completely impractical to massively sample each and every tree. However, as discussed above, procedures do exist for estimating within-tree numbers (Pulley *et al.*, 1976a; Coulson *et al.*, 1976a).

For sampling purposes it is logical to consider an infestation spot as a population of infested trees stratified according to the predominant life stage within the tree. After stratification, sample trees are selected, estimates made of the within-tree populations, and this information is then extrapolated to include all trees in the stratum. Among the factors which need to be considered in sampling infestations are (1) the precision of the within-tree estimates, (2) the total number of infested trees, (3) the number of trees sampled, and (4) the methods of selecting sample trees and scaling the observed counts.

In our studies of within-infestation spot sampling three different procedures were used to estimate within-tree populations: (1) the topological mapping technique, (2) the single level (5.0 m) TG-PDF procedure, and (3) the two level (3.5, 6.5 m) TG-PDF procedure. These particular sample height arrangements are among the most precise of the single and two-level estimating procedures.

As it is impractical to actually sample many infestations and compare the results of different procedures, infestation spots of varying sizes were simulated. From the 134 trees sampled during the three year study, trees were selected randomly and without replacement to form spots of K trees $(1 \leq K \leq 50)$. Within each spot k trees $(1 \leq k \leq 10)$ were sampled without replacement and the estimated population within the k trees E_k was then extrapolated by one of five scalers to obtain an estimate the population in the K trees, E_K. This E_K value was then compared with T_K, the sum of the topological mapping estimates for the K trees, to determine the error in estimating the true population. The proportional error $\phi = (E_K - T_K)/T_K$ was also calculated. The mean $\bar{\phi}_{K,k}$ and standard deviations $\sigma_{K,k}$ of ϕ from 50,000 trials showed the bias and variability of a particular estimating procedure for a sample of k trees in a spot of size K. Where there was evidence of bias, the trials were repeated and each E_K was weighted by $W_{K,k} = 1/(1 + \bar{\phi}_{K,k})$ and the resultant $\sigma_{K,k}$ for the unbiased estimates were used for comparisons of precision between procedures.

Ten procedures were evaluated as estimates of within-spot populations of attacking adults (Table 2). The effects of within tree precision were studied by comparing the single level within tree procedure with the two level procedure. Methods of

TABLE 2: Procedures for estimating within-spot populations of attacking adult D frontalis.

Procedure No.	Sample heights, m	Tree selection	Scaling factor
1	5	Random	S_k
2	3.5 + 6.5	Random	S_k
3	5	Random	S_d
4	3.5 + 6.5	Random	S_d
5	5	Random	S_a
6	3.5 + 6.5	Random	S_a
7	5	Largest DBH	S'_d
8	3.5 + 6.5	Largest DBH	S'_d
9	5	Largest area	S'_a
10	3.5 + 6.5	Largest area	S'_a

selecting the k trees were evaluated by comparing random selection with the purposeful selection of the k largest trees, largest being determined by measurements of the diameter at breast height (DBH) and the infested phloem area. Three methods of extrapolating the E_k to calculate E_K were also tested and consisted of scaling the E_k according to the proportion of the tree numbers, tree diameters, and infested phloem areas included in the sample (Pulley *et al.*, 1976b).

There are numerous other combinations of within-tree precision, tree selection, and scaling arrangements which could be tested. However, the ten that were tested have a sufficient range to suggest the relative merits of other possible arrangements.

The results of this study revealed several interesting features of the within-spot estimation problem. The procedures in which trees were randomly selected provided unbiased estimates of within-spot populations, selecting the largest trees tended to overestimate the true number with the bias diminishing to zero as $k \to K$.

When $k=K=1$, the precision of all within-spot estimators was equivalent to the precision of the within-tree estimate. For $k=K$, the precision improved approximately as the square root of K. For each procedure, precision improved as $k \to K$. Sampling the k trees at two sample heights (3.5 and 6.5 m, four 100 cm^2 disks/height) was more precise than single level sampling (four disks at 5.0 m), but equally precise estimates could be obtained by single level sampling of just one or two additional trees in the spot. Random selection of the k trees with scaling by the number of infested trees was the least precise of the estimating procedure; scaling by diameter and by infested surface area increased precision. Best precision was obtained by selecting the k trees of greatest infested phloem area, but selecting the largest diameter trees was nearly as precise. The least costly procedure for obtaining a desired level of precision consists of selecting the k trees of largest diameter and extracting 4 disks/tree at 5.0 m.

As with the within-tree sampling procedures the final step in the development of within-spot sampling plans was to select those procedures which would be amenable to application and prepare a user's guide explaining the basic procedures, step-by-step instructions for the procedures, and append the necessary statistical tables. The following aspects were treated in the user's guide: within-tree population estimation, within-spot population estimation by random sampling, and within-spot estimation by sampling the largest trees (Foltz *et al.*, 1977).

5. CONCLUSIONS

1. There are many possible options for sampling within-tree and within-spot populations of *D. frontalis* and other bark beetles species in addition to those discussed herein. The procedures presented, and described in greater detail in the references cited, should be helpful in judging other potentially fruitful approaches as well as identify those of less potential value.

2. The general approach taken has been one of purposeful selection of samples relative to the defined within-tree distributions of the various life stages of the insect. This approach was necessary because the random sampling alternatives required far too many samples to be practical. In the case of stratified random sampling, the sampling was concentrated to areas of high variability.

3. The general format followed in the development of within-tree and within-spot population estimating procedures for *D. frontalis* should be applicable to other bark beetle species. Indeed, many of the bark beetle species which have been studied in detail have been 'sampled' using a systematic procedure similar to the one used for *D. frontalis*.

ACKNOWLEDGMENTS

Texas Agricultural Experiment Station Paper No. TA 13606. The work reported herein was funded in part by the National Science Foundation project entitled "The Principles, Strategies, and Tactics of Pest Population Regulations and Control in Major Crop Ecosystems," NSF GB 34718 and in part by the U.S. Dept. of Agriculture program entitled "The Expanded Southern Pine Beetle Research and Applications Program," grant numbers CSRS G6163 and CSRS 516-15-58. The findings, opinions, and recommendations expressed herein are those of the authors and not necessarily those of sponsoring agencies.

REFERENCES

Coulson, R. N., Hain, F. P., Foltz, J. L. and Mayyasi, A. M. (1975). Techniques for sampling the dynamics of southern pine beetle populations. Texas Agricultural Experiment Station, Miscellaneous Publication 1185.

Coulson, R. N., Pulley, P. E., Foltz, J. L., and Martin, W. C. (1976a). Procedural guide for quantitatively sampling within-tree populations of *Dendroctonus frontalis*. Texas Agricultural Experiment Station, Miscellaneous Publication 1267.

Coulson, R. N., Mayyasi, A. M., Foltz, J. L., and Hain, F. P. (1976b). Resource utilization by the southern pine beetle. *Canadian Entomologist*, 108, 353-62.

Coulson, R. N., Leuschner, W. A., Foltz, J. L., Pulley, P. E., Hain, F. P., and Payne, T. L. (1979). Approach to research and forest management for southern pine beetle control. In *New Technology of Pest Control*, C. B. Huffaker, ed. Wiley, New York (in press).

DeMars, C. J. (1970). Frequency distribution, data transformations, and analysis of variations used in determination of optimum sample size and effort for broods of the western pine beetle. In *Studies on the Population Dynamics of the Western Pine Beetle*, R. W. Stark and D. L. Dahlsten, eds. University of California Press, Berkeley, 42-65.

Foltz, J. L., Mayyasi, A. M., Pulley, P. E., Coulson, R. N., and Martin, W. C. (1976a). Host-tree geometric models for use in southern pine beetle population studies. *Environmental Entomology*, 5, 640-43.

Foltz, J. L., Mayyasi, A. M., Hain, F. P., Coulson, R. N., and Martin, W. C. (1976b). Egg-gallery length relationship and within-tree anslyses for the southern pine beetle, *Dendroctonus frontalis* Zimm. (Coleoptera: Scolytidae). *Canadian Entomologist*, 108, 341-52.

Foltz, J. L., Pulley, P. E., Coulson, R. N., and Martin, W. C. (1977). Procedural guide for estimating within-spot populations of *Dendroctonus frontalis*. Texas Agricultural Experiment Station, Miscellaneous Publication 1316.

Mayyasi, A. M., Pulley, P. E., Coulson, R. N., DeMichele, D. W., and Foltz, J. L. (1976a). Mathematical descriptions of within-tree distribution of the various developmental stages of *Dendroctonus frontalis* Zimm. Coleoptera: Scolytidae. *Researches on Population Ecology*, 18, 135-45.

Mayyasi, A. M., Coulson, R. N., Foltz, J. L. and Hain, F. P. (1976b). The functional distribution of within-tree larval and progeny adult populations of *Dendroctonus frontalis* (Coleoptera: Scolytidae). *Canadian Entomologist*, 108, 363-72.

Pulley, P. E., Mayyasi, A. M., Foltz, J. L., Coulson, R. N., and Martin, W. C. (1976). Topological mapping to estimate numbers of bark-inhabiting insects. *Environmental Entomology*, 5, 714-19.

Pulley, P. E., Foltz, J. L., Mayyasi, A. M., Coulson, R. N., and Martin, W. C. (1977a). Sampling procedures for within-tree attacking adult populations of the southern pine beetle, *Dendroctonus frontalis* Zimm. *Canadian Entomologist*, 109, 39-48.

Pulley, P. E., Foltz, J. L., Coulson, R. N., and Martin, W. C. (1977b). Evaluation of procedures for estimating within-spot populations of attacking adult *Dendroctonus frontalis*. *Canadian Entomologist*, 109, 1325-34.

Pulley, P. E., Coulson, R. N., Foltz, J. L., and Martin, W. C. (1977c). An evaluation of the relationship between sampling intensity, informational content of samples, and precision in estimating within-tree populations of *Dendroctonus frontalis*. *Environmental Entomology*, 6, 607-15.

Safranyik, L. (1968). *Development of a technique for sampling mountain pine beetle populations in lodgepole pine*. Ph.D. dissertation, University of British Columbia.

Stark, R. W. and Dahlsten, D. L. (1970). Studies on the population dynamics of the western pine beetle, *Dendroctonus brevicomis* LeConte (Coleoptera: Scolytidae). University of California Press, Berkeley.

Stephen, F. M. and Taha, H. A. (1976). Optimization of sampling effort for within-tree populations of southern pine beetle and its natural enemies. *Environmental Entomology*, 5, 1001-1007.

[*Received August* 1977. *Revised September* 1977]

SECTION VI

A BIBLIOGRAPHY

G. P. Patil and M. Rosenzweig, (eds.),
Contemporary Quantitative Ecology and Related Ecometrics, pp. 665-678.

BIBLIOGRAPHY OF BOOKS ON QUANTITATIVE ECOLOGY AND ECOMETRICS

B. Dennis, G. P. Patil, M. V. Ratnaparkhi, and S. Stehman
The Pennsylvania State University, University Park, PA 16802 USA

INTRODUCTION

Suppose 100+ statistical ecologists were to gather together for several weeks of intensive workshops, seminars, and research activities. What books should be on hand for them to use?

That such a question needed serious attention was one of the many unique aspects of the Satellite Program in Statistical Ecology. The list of books presented here is an outgrowth of that question.

We have included books in this list on the basis of their interdisciplinary flavor. These books discuss mathematical or statistical matters, yet they also contain significant books on such topics as measure theory or birds of Italy, of obvious interest to many Satellite Program participants, were excluded. However, certain basic ecology texts appear (surprisingly few such texts have been written); we feel that these ecology texts identify many quantitative issues and may be of help to statisticans who are involved with ecological data. In addition, certain books on probability or statistical methods (surprising multitudes of which are in existence) are to be found on this list, though with these we have been strict, and have selected only those with explicit ecological applications or examples.

Investigators should also consult the excellent bibliography by Schultz, Eberhardt, Thomas, and Cochran (1976) which contains journal articles and books (see also an earlier work by Schultz in Patil, Pielou, and Waters 1971, vol. 3).

We hope this present list becomes antiquated quickly by the publication of many new books on quantitative ecology!

BOOKS ON QUANTITATIVE ECOLOGY AND ECOMETRICS

Adams, L. (ed.). (1970). *Population Ecology*. Dickenson Publishing, Belmont, California. 1

Allee, W. C.; Emerson, A. E.; Park, O.; Park, T.; and Schmidt, K. P. (1949). *Principles of Animal Ecology*. W. B. Saunders, Philadelphia. 2

Allee, W. C. *See:* Heese, R.; Allee, W. C.; and Schmidt, K. P. (1951).

Anderson, D. R. *See:* Brownie, C.; Anderson, D. R.; Burnham, K. P.; and Robson, D. S. (1978).

Andrewartha, H. G. (1971). *Introduction to the Study of Animal Populations*. The University of Chicago Press, Chicago. 3

Andrewartha, H. G.; and Birch, L. C. (1954). *The Distribution and Abundance of Animals*. The University of Chicago Press, Chicago. 4

Anonymous. (1972). *3rd Conference Advisory Group of Forest Statisticians. Institut National de la Recherche Agronomique, Paris, France*. I.N.R.A. Publ. 72-3. 5

Armitage, P. (1971). *Statistical Methods in Medical Research*. Blackwell Scientific, Oxford. 6

Arnason, A. N.; and Baniuk, L. (1978). *POPAN-2. A Data Maintenance and Analysis System for Mark-Recapture Data*. Charles Babbage Research Centre, Manitoba. 7

Bailey, N. T. J. (1957). *The Mathematical Theory of Epidemics*. Charles Griffin, London. 8

Bailey, N. T. J. (1967). *The Mathematical Approach to Biology and Medicine*. John Wiley & Sons, New York. 9

Bancroft, T. A. *See:* Kempthorne, O.; Bancroft, T. A.; Gowen, J. W.; and Lush, J. L. (eds.). (1964).

Bang, F. B. (eds.). *See:* Sladen, B. K.; and Bang, F. B. (eds.). (1969).

Baniuk, L. *See:* Arnason, A. N.; and Baniuk, L. (1978).

Bartlett, M. S. (1960). *Stochastic Population Models in Ecology and Epidemiology*. Methuen, London. 10

Bartlett, M. S. (1975). *The Statistical Analysis of Spatial Pattern*. John Wiley & Sons, New York. 11

Bartlett, M. S.; and Hiorns, R. W. (eds.). (1973). *The Mathematical Theory of the Dynamics of Biological Populations*. Academic Press, New York. 12

Batschelet, E. (1970). *Introduction to Mathematics for Life Scientists*. Springer-Verlag, New York. 13

Behnke, J. A. (ed.). (1972). *Challenging Biological Problems - Directions Toward Their Solution*. Oxford University Press, London. 14

Bendix, S.; and Graham, H. R. (eds.). (1978). *Environmental Assessment: Approaching Maturity*. Ann Arbor Science Publishers, Ann Arbor, Michigan. 15

Berry, B. J. L.; and Marble, D. F. (1968). *Spatial Analysis*. Prentice Hall, Englewood Cliffs, New Jersey. 16

Beverton, R. J. H.; and Holt, S. J. (1957). *On the Dynamics of Exploited Fish Populations*. Fish, Invest. Minist. Agric. Fish., Food (Great Britain) Ser. 2, Vol. 19. 17

Beyer, J. E. (1976). *Ecosystems (An Operational Research Approach)*. The Institute of Mathematical Statistics and Operation Research, The Technical University of Denmark. 18

Beyer, J. E. (1976). *Fish I (Survival of Fish Larvae - A Single Server Queue Approach)*. The Institute of Mathematical Statistics and Operation Research, The Technical University of Denmark. 19

Beyer, J. E. (1976). *Fish II (Survival of Fish Larvae - A Single Server Queue Approach)*. The Institute of Mathematical Statistics and Operation Research, The Technical University of Denmark. 20

Bibbero, R. J.; and Young, I. G. (1974). *Systems Approach to Air Pollution Control*. John Wiley & Sons, New York. 21

Birch, L. C. *See:* Andrewartha, H. G.; and Birch, L. C. (1954).

Blackith, R. E.; and Reyment, R. A. (1971). *Multivariate Morphometrics*. Academic Press, New York. 22

Bodenheimer, F. S. (1938). *Problems of Animal Ecology*. Oxford University Press, London. 23

Bookstein, F. L. (1978). *The Measurement of Biological Shape and Shape Change*. Springer-Verlag, New York. 24

Bossert, W. H. *See:* Wilson, E. O.; and Bossert, W. H. (1971).

Bromley, D. W.; McMillan, M.; Robertson, M.; and Schoeder, A. (1977). *Institutional Design for Improved Environmental Quality*. University of Wisconsin Sea Grant Program, Technical Report No. 232. 25

Brown, G. M. Jr. *See:* Hammack, J.; and Brown, G. M. Jr. (1974).

Brownie, C.; Anderson, D. R.; Burnham, K. P.; and Robson, D. S. (1978). *Statistical Inference from Band Recovery Data - A Hand Book*. Fish and Wildlife Service Resource Publication No. 131, Washington, D. C. 26

Brunig, E. F. (ed.). (1977). *Transactions of the International MAB-IUFRO Workshop on Tropical Rainforest Ecosystems Research*. Chair of World Forestry, Hamburg-Reinbek. 27

Brussard, P. F. (1977). *Ecological Genetics: The Interface*. Springer-Verlag, New York. 28

Burnham, K. P. *See:* Brownie, C.; Anderson, D. R.; Burnham, K. P.; and Robson, D. S. (1978).

Cain, S. A. (1944). *Foundations of Plant Geography*. Harper and Row, New York. 29

Cain, S. A.; and de Oliveira Castro, G. M. (1959). *Manual of Vegetation Analysis*. Harper & Row, New York. 30

Cairns, J. Jr. (ed.). (1977). *Aquatic Microbial Communities*. Garland Publishing, New York. 31

Cairns, J. Jr.; and Dickson, K. L. (eds.). (1973). *Biological Methods for the Assessment of Water Quality*. American Society for Testing and Materials, Philadelphia. 32

Cairns, J. Jr.; and Dickson, K. L. (1974). *The Environment: Costs, Conflicts, Action*. Marcel Dekker, New York. 33

Cairns, J. Jr.; Dickson, K. L.; and Maki, A. W. (1978). *Estimating the Hazard of Chemical Substances to Aquatic Life - STP 657*. American Society for Testing and Materials, Philadelphia. 34

Cairns, J. Jr.; Dickson, K. L.; and Westlake, G. G. (eds.). (1975). *Biological Monitoring of Water and Effluent Quality*. American Society for Testing and Materials, Philadelphia. 35

Cairns, J. Jr.; Patil, G. P.; and Waters, W. E. (eds.). (1979). *Environmental Biomonitoring, Assessment, Prediction, and Management - Certain Case Studies and Related Quantitative Issues*. Satellite Program in Statistical Ecology, International Co-operative Publishing House, Fairland, Maryland. 36

Calabrese, E. J. (1978). *Pollutants and High-Risk Groups*. John Wiley & Sons, New York. 37

Calhoun, A. (ed.). (1966). *Inland Fisheries Management*. California Department of Fish and Game, San Francisco. 38

Canale, R. P. (ed.). (1976). *Modeling Biochemical Processes in Aquatic Ecosystems*. Ann Arbor Science Publishers, Ann Arbor, Michigan. 39

Carle, F. L. (1976). *An Evaluation of the Removal Method for Estimating Benthic Populations and Diversity*. M. S. Thesis, Virginia Polytechnic Institute and State University. 40

Caughley, G. (1977). *Analysis of Vertebrate Populations*. John Wiley & Sons, New York. 41

Chapas, L. C. *See:* Wadsworth, R. M.; Chapas, L. C.; Rutter, A. J.; Solomon, M. E.; and Wilson, J. W. (eds.). (1967).

Chapman, D. G.; and Gallucci, V. (eds.). (in press). *Quantitative Population Dynamics*. Satellite Program in Statistical Ecology, International Co-operative Publishing House, Fairland, Maryland. 42

Chapman, S. B. (1977). *Methods in Plant Ecology*. Halsted Press, New York. 43

Charnes, A.; and Lynn, W. R. (eds.). (1975). *Mathematical Analysis of Decision Problems in Ecology*. Springer-Verlag, New York. 44

Chaston, I. (1971). *Mathematics for Ecologists*. Butterworths, London. 45

Cheremisinoff, P. N.; and Morresi, A. C. (eds.). (1978). *Air Pollution Sampling and Analysis Deskbook*. Ann Arbor Science Publishers, Ann Arbor, Michigan. 46

Cheremisinoff, P. N.; and Morresi, A. C. (1977). *Environmental Assessment and Impact Statement Handbook*. Ann Arbor Science Publishers, Ann Arbor, Michigan. 47

Chiang, C. L. (1978). *Life Table and Mortality Analysis*. World Health Organisation, Geneva, Switzerland. 48

Christiansen, F. B.; and Fenchel. T. M. (1977). *Ecological Studies - Volume 20, Theories of Populations in Biological Communities*. Springer-Verlag, New York. 49

Clark, C. W. (1976). *Mathematical Bioeconomics*. John Wiley & Sons, New York. 50

Clark, L. R.; Geier, P. W.; Hughes, R. D.; and Morris, R. F. (1967). *The Ecology of Insect Populations in Theory and Practice*. Methuen, London. 51

Clarke, G. L. (1965). *Elements of Ecology*. John Wiley & Sons, New York. 52

Clepper, H. (ed.). (1977). *Marine Recreational Fisheries*. Sport Fishing Institute, Washington, D. C. 53

Cliff, A. D.; and Ord, J. K. (1973). *Spatial Autocorrelation*. Pion Limited, London. 54

Cochran, M. I. *See:* Schultz, V.; Eberhardt, L. L.; Thomas, J. M.; and Cochran, M. I. (1976).

Cody, M. L.; and Diamond, J. M. (eds.). (1975). *Ecology and Evolution of Communities*. Harvard University Press, Cambridge. 55

Cohen, J. (1977). *Statistical Power Analysis for the Behavioral Sciences*. Academic Press, New York. 56

Cohen, J. E. (1966). *A Model of Simple Competition*. Harvard University Press, Cambridge. 57

Cohen, J. E. (1971). *Casual Groups of Monkeys and Men: Stochastic Models of Elemental Social Systems*. Harvard University Press, Cambridge. 58

Cohen, J. E. (1978). *Food Webs and Niche Space*. Princeton University Press, Princeton. 59

Cole, A. J. (ed.). (1969). *Numerical Taxonomy*. Academic Press, New York. 60

Colgan, P. (ed.). (1978). *Quantitative Ethology*. John Wiley & Sons, New York. 61

Colinvaux, P. (1973). *Introduction to Ecology*. John Wiley & Sons, New York. 62

Cooke, K. L. *See:* Ludwig, D.; and Cooke, K. L. (1975).

Cormack, R. M.; and Ord, J. K. (eds.). (1979). *Spatial and Temporal Analysis in Ecology*. Satellite Program in Statistical Ecology, International Co-operative Publishing House, Fairland, Maryland. 63

Cormack, R. M.; Patil, G. P.; and Robson, D. S. (eds.). (1979). *Sampling Biological Populations*. Satellite Program in Statistical Ecology, International Co-operative Publishing House, Fairland, Maryland. 64

Costlow, J. D. (ed.). (1971). *Fertility of the Sea*. Gordon and Breach, New York. 65

Cragg, J. B.; and Pirie, N. W. (eds.). (1955). *The Numbers of Man and Other Animals*. Oliver and Boyd, London. 66

Croft, B. A. (eds.). *See:* Tummala, R. L.; Haynes, D. L.; and Croft, B. A. (eds.). (1976).

Crow, J. F.; and Kimura, M. (1970). *Introduction to Population Genetics Theory*. Harper & Row, New York. 67

Cunia, T. (ed.). (1974). *Monitoring Forest Environment Through Successive Sampling*. State University of New York, College of Environmental Science and Forestry, Syracuse, New York. 68

Cushing, D. H. (1975). *Marine Ecology and Fisheries*. Cambridge University Press, Cambridge. 69

Cushing, D. H.; and Walsh, J. J. (eds.). (1976). *The Ecology of the Seas*. W. B. Saunders, Philadelphia; and Blackwell Scientific, Oxford. 70

Cushing, J. M. (1977). *Integro-* 71
differential Equations and Delay Models in Population Dynamics. Springer-Verlag, New York.

D'Ancona, U. (1954). *The Struggle for* 72
Existence. E. J. Brill, Leiden, Netherlands.

Dajoz, R. (1971). *Precis d'Ecologie.* 73
Dunod, Paris.

Dajoz, R. (1974). *Dynamique des Popu-* 74
lations. Masson, Paris.

Dansereau, P. (1957). *Biogeography:* 75
An Ecological Perspective. Ronald Press, New York.

Darlington, P. J. (1957). *Zoogeog-* 76
raphy: the Geographical Distribution of Animals. John Wiley & Sons, New York.

Davies, R. G. (1971). *Computer* 77
Programming in Quantitative Biology. Academic Press, New York.

Day, J. W. Jr. (eds.). *See:* Hall, C. A. S.; and Day, J. W. Jr. (eds.). (1977).

Deininger, R. A. (ed.). (1974). *Models* 78
for Environmental Pollution Control. Ann Arbor Science Publishers, Ann Arbor, Michigan.

DeWit, C. T.; and Goudriaan, J. 79
(1974). *Simulation of Ecological Processes*. Centre for Agricultural Publishing and Documentation, Wageningen, Netherlands.

Diamond, J. M. (eds.). *See:* Cody, M. L.; and Diamond, J. M. (eds.). (1975).

Dickson, K. L. (eds.). *See:* Cairns, J. Jr.; and Dickson, K. L. (eds.). (1973).

Dickson, K. L. *See:* Cairns, J. Jr.; and Dickson, K. L. (1974).

Dickson, K. L. *See:* Cairns, J. Jr.; Dickson, K. L.; and Maki, A. W. (1978).

Dickson, K. L. *See:* Cairns, J. Jr.; Dickson, K. L.; and Westlake, G. G. (eds.). (1975).

Douglas, J. B. (1979). *Analysis with* 80
Standard Contagious Distributions. Satellite Program in Statistical Ecology, International Co-operative Publishing House, Fairland, Maryland.

Drake, E. T. (ed.). (1968). *Evolution* 81
and Environment. Yale University Press, New Haven, Connecticut.

Eberhardt, L. L. *See:* Schultz, V.; Eberhardt, L. L.; Thomas, J. M.; and Cochran, M. I. (1976).

Eckman, S. (1953). *Zoogeography of* 82
the Sea. Sidgwick and Jackson, London.

Edington, J. M.; and Edington, M. A. 83
(1978). *Ecology and Environmental Planning*. Halsted Press, New Jersey.

Edington, M. A. *See:* Edington, J. M.; and Edington, M. A. (1978).

Elandt-Johnson, R. C. (1971). 84
Probability Models and Statistical Methods in Genetics. John Wiley & Sons, New York.

Ellenberg, H. (ed.). (1971). *Integrated* 85
Experimental Ecology - Methods and Results of Ecosystem Research in the German Solling Project. (Ecological Studies - Springer-Verlag, New York.

Ellenberg, H. *See:* Mueller-Dombois, D.; and Ellenberg, H. (1974).

Elliot, J. M. (1971). *Statistical Analysis.* 86
Freshwater Biological Association.

Elliot, J. M. (1977). *Some Methods for* 87
the Statistical Analysis of Benthic Invertebrates. Freshwater Biol. Assoc. Sci. Publ. No. 25, West Moreland, United Kingdom.

Elton, C. S. (1968). *Animal Ecology.* 88
Methuen, London.

Elton, C. S. (1969). *The Ecology of In-* 89
vasions by Animals and Plants. Methuen, London.

Emerson, A. E. *See:* Allee, W. C.; Emerson, A. E.; Park, O.; Park, T.; and Schmidt, K. P. (1949).

Emlen, J. M. (1973). *Ecology: An Evo-* 90
lutionary Approach. Addison-Wesley, Reading, Massachusetts.

Endler, J. A. (1977). *Geographic Varia-* 91
tion, Speciation, and Clines. Princeton University Press, Princeton.

Engen, S. (1978). 92
Stochastic Abundance Models. Chapman and Hall, London.

Evans, G. C. (1972). *The Quantitative* 93
Analysis of Plant Growth. University of California Press, Berkeley.

Everhart, W. H. *See:* Rounsfell, G. A.; and Everhart, W. H. (1953).

Falkenborg, D. H. *See:* Middlebrooks, E. J.; Falkenborg, D. H.; and Maloney, T. E. (eds.). (1977).

Fenchel. T. M. *See:* Christiansen, F. B.; and Fenchel. T. M. (1977).

Fife, P. C. (1979). *Mathematical As-* 94
pects of Reacting and Diffusing Systems. Springer-Verlag, New York.

Fisher, R. A. (1930). *The Genetical* 95
Theory of Natural Selection. Clarendon Press, Oxford.

Fletcher, W. W. *See:* Lenihan, J.; and Fletcher, W. W. (1978).

Flieger, W. *See:* Keyfitz, N.; and Flieger, W. (1971).

Fransz, H. G. (1974). *The Functional* 96
Response to Prey Density in an Acarine System. Centre for Agricultural Publishing and Documentation, Wageningen, Netherlands.

Freese, F. (1967). *Elementary Statistical* 97
Methods for Foresters. U. S. Forest Service, Agricultural Handbook 317, Washington, D. C.

Fretwell, S. D. (1972). *Populations in* 98
Seasonal Environment. Monographs in Population Biology 5, Princeton University Press, Princeton, New Jersey.

Gallucci, V. (eds.). (in *See:* Chapman, D. G.; and Gallucci, V. (eds.). (in press).

Gates, D. M.; and Schmerl, R. B. 99
(1975). *Perspectives of Biophysical Ecology*. Springer-Verlag, New York.

Gause, G. F. (1935). *Verifications Ex-* 100
perimentales de la Theorie Mathematique de la Lutte Pour la Vie. Hermann et Cie, Paris.

Gause, G. F. (1969). *The Struggle for* 101
Existence. Hafner Publishing, New York.

Geier, P. W. *See:* Clark, L. R.; Geier, P. W.; Hughes, R. D.; and Morris, R. F. (1967).

Gerking, S. D. (ed.). (1978). *Ecology of* 102
Freshwater Fish Production. John Wiley & Sons, New York.

Ghetti, P. F. (1974). *L'acqua* 103
nell'ambiente umano di val Parma. Editrice, Studium Parmense, Parma, Italy.

Giles, R. H. (ed.). (1969). *Wildlife Man-* 104
agement Techniques. The Wildlife Society, Washington, D. C.

Giles, R. H. Jr. (1977). *A Watershed* 105
Planning and Management System: Design and Synthesis. Virginia Water Resources Research Center, Bull. 102, Blacksburg, Virginia.

Gilpin, M. E. (1975). *Group Selection in* 106
Predator Prey Communities. Princeton University Press, Princeton, New Jersey.

Ginzburg, L. R. (1978). *Species Inter-* 107
actions in Ecosystem Modelling. Springer-Verlag, New York.

Goel, N. S.; and Richter-Dyn, N. 108
(1974). *Stochastic Models in Biology*. Academic Press, New York.

Goel, N. S.; Maitra, S. C.; and 109
Montroll, E. W. (1971). *On the Volterra and Other Nonlinear Models of Interacting Populations*. Academic Press, New York.

Gold, H. J. (1977). *Mathematical Mod-* 110
eling of Biological Systems: An Introductory Guidebook. John Wiley & Sons, New York.

Goldberg, E. D. (1976). *Strategies for* 111
Marine Pollution Monitoring. John Wiley & Sons, New York.

Goldberg, E. D.; McDave, I. N.; 112
O'Brien, J. J.; and Steele, J. H. (1977). *The Sea - Volume 6, Marine Modeling*. John Wiley & Sons, New York.

Goldstein, G. *See:* Thomas, W. A.; Goldstein, G.; and Wilcox, W. H. (1976).

Gopal, B. (eds.). *See:* Singh, J. S.; and Gopal, B. (eds.). (1978).

Goudriaan, J. *See:* DeWit, C. T.; and Goudriaan, J. (1974).

Gounot, M. (1969). *Methodes D'Etude* 113
Quantitative de la Vegetation. Masson, Paris.

Gowen, J. W. *See:* Kempthorne, O.; Bancroft, T. A.; Gowen, J. W.; and Lush, J. L. (eds.). (1964).

Gradwell, G. R. *See:* Varley, G. C.; Gradwell, G. R.; and Hassell, M. P. (1973).

Graham, H. R. (eds.). *See:* Bendix, S.; and Graham, H. R. (eds.). (1978).

Grassle, F.; Patil, G. P.; Smith, W.; 114
and Taillie, C. (eds.). (1979). *Ecological Diversity in Theory and Practice*. Satellite Program in Statistical Ecology, International Co-operative Publishing House, Fairland, Maryland.

Gray, T. R. G. *See:* Parkinson, D.; Gray, T. R. G.; and Williams, S. T. (eds.). (1971).

Gregg, J. R. (1954). 115
The Language of Taxonomy. Columbia University Press, New York.

Greig-Smith, P. (1964). 116
Quantitative Plant Ecology. Butterworths, London.

Griffith, A. L.; and Prasad, J. (1949). 117
The Silviculture Research Code, Vol. 3, The Tree and Crop Measurement Manual. The Manager of Publications, Delhi, India.

Griffith, A. L.; and Ra, B. S. (1947). 118
The Silviculture Research Code, Vol. 2, The Statistical Manual. The Manager of Publications, Delhi, India.

Gulland, J. A. (ed.). (1977). 119
Fish Population Dynamics. John Wiley & Sons, New York.

Gulland, J. A. (1971). 120
The Fish Resources of the Ocean. Fishing News, Ltd., West Byfleet, Surrey.

Hall, C. A. S.; and Day, J. W. Jr. 121
(eds.). (1977). *Ecosystem Modeling in Theory & Practice: An Introduction With Case Histories*. John Wiley & Sons, New York.

Hallam, A. (1977). 122
Patterns of Evolution as Illustrated by the Fossil Record. American Elsevier, New York.

Hammack, J.; and Brown, G. M. Jr. 123
(1974). *Waterfowl and Wetlands: Towards Bioeconomic Analysis*. Johns Hopkins University Press, Baltimore, Maryland.

Hammen, C. S. (1972). 124
Elementary Quantitative Biology. John Wiley & Sons, New York.

Harbaugh, J. W.; and Merriam, D. F. 125
(1968). *Computer Applications in Stratigraphic Analysis*. John Wiley & Sons, New York.

Harden-Jones, F. R. (ed.). (1974). 126
Sea Fisheries Research. Logos Press, London.

Hardy, J. T. (1975). 127
Science Technology and the Environment. W. B. Saunders, Philadelphia.

Hargrave, B. *See:* Parsons, T. R.; Takahashi, M.; and Hargrave, B. (1973).

Harper, J. L. (1977). 128
Population Biology of Plants. Academic Press, New York.

Hassell, M. P. (1978). 129
The Dynamics of Arthropod Predator-Prey Systems. Princeton University Press, Princeton, New Jersey.

Hassell, M. P. *See:* Varley, G. C.; Gradwell, G. R.; and Hassell, M. P. (1973).

Hauck, B.; and Keenan, P. C. (1977). 130
Abundance Effects in Classification. Reidel, Boston.

Hawley, A. H. (1950). 131
Human Ecology: A Theory of Community Structure. Ronald Press, New York.

Haynes, D. L. *See:* Tummala, R. L.; Haynes, D. L.; and Croft, B. A. (eds.). (1976).

Hazen, W. E. (1964). 132
Readings in Population and Community Ecology. W. B. Saunders, Philadelphia.

Hazlett, B. A. (ed.). (1977). 133
Quantitative Methods in the Study of Animal Behavior. Academic Press, New York.

Heaslip, G. B. (1975). 134
Environmental Data Handling. John Wiley & Sons, New York.

Hedgepeth, J. W. (ed.). (1957). 135
Treatise on Marine Ecology and Paleoecology. Geological Society of America, Boulder, Colorado.

Heese, R.; Allee, W. C.; and Schmidt, K. P. (1951). 136
Ecological Animal Geography. John Wiley & Sons, New York.

Heldridge, L. R. (1967). 137
Life Zone Ecology. Tropical Research Center, San Jose, Costa Rica.

Helfferich, C. *See:* McRoy, C. P.; and Helfferich, C. (1977).

Hill, M. N. (ed.). (1963). 138
The Sea. John Wiley & Sons, New York.

Himmelblau, D. M. *See:* O'Laoghaire, D. T.; and Himmelblau, D. M. (1974).

Hiorns, R. W. (eds.). *See:* Bartlett, M. S.; and Hiorns, R. W. (eds.). (1973).

Hodder, I.; and Orton, C. (1976). 139
Spatial Analysis in Archaeology. Cambridge University Press, New York.

Hodgson, J. M. (1978). 140
Soil Sampling and Soil Description. Clarendon Press, New York.

Holdgate, M. W. (eds.). *See:* Le Cren, E. D.; and Holdgate, M. W. (eds.). (1962).

Holdgate, M. W.; and White, G. F. (1977). 141
Environmental Issues - SCOPE 10. John Wiley & Sons, New York.

Holdgate, M. W.; and Woodman, M. J. (eds.). (1978). 142
The Breakdown and Restoration of Ecosystems. Plenum Press, New York.

Holt, S. J. *See:* Beverton, R. J. H.; and Holt, S. J. (1957).

Hoppensteadt, F. (1975). 143
Mathematical Theories of Populations: Demographics, Genetics, and Epidemics. SIAM Publishers, Philadelphia.

Horn, H. S. (1971). 144
The Adaptive Geometry of Trees. Monographs in Population Biology 3, Princeton University Press, Princeton, New Jersey.

Huffaker, C. B.; and Messenger, P. S. (1976). 145
Theory and Practice of Biological Control. Academic Press, New York.

Hughes, R. D. *See:* Clark, L. R.; Geier, P. W.; Hughes, R. D.; and Morris, R. F. (1967).

Hughes, R. E. *See:* Milner, C.; and Hughes, R. E. (1968).

Husch, B. (1963). 146
Forest Mensuration and Statistics. Roland Press, New York.

Hutchinson, G. E. (1978). 147
An Introduction to Population Ecology. Yale University Press, New Haven, Connecticut.

Inhaber, R. (1976). 148
Environmental Indices. John Wiley & Sons, New York.

Innis, G. S. (1977). 149
Ecological Studies - Volume 26, Grassland Simulation Model. Springer-Verlag, New York.

Innis, G.; and O'Neill, R. V. (eds.). (1979). 150
Systems Analysis of Ecosystems. Satellite Program in Statistical Ecology, International Cooperative Publishing House, Fairland, Maryland.

Iosifescu, M. *See:* Tautu, P.; and Iosifescu, M. (1968).

Iosifescu, M.; and Tautu, P. (1973). 151
Stochastic Processes and Applications in Biology and Medicine II: Models. Springer-Verlag, New York.

Ivlev, V. S. (1961). 152
Experimental Ecology of the Feeding of Fishes. Yale University Press, New Haven, Connecticut.

Jacquard, A. (1974). 153
The Genetic Structure of Populations. Springer-Verlag, New York.

Jardine, N.; and Sibson, R. (1971). 154
Mathematical Taxonomy. John Wiley & Sons, New York.

Jeffers, J. N. R. (ed.). (1972). 155
Mathematical Models in Ecology. Blackwell Scientific, Oxford.

Jeffers, J. N. R. (1959). 156
Experimental Design and Analysis in Forest Research. Almquist and Wiksell, Stockholm, Sweden.

Johnson, C. (1976). 157
Introduction to Natural Selection. University Park Press, Baltimore, Maryland.

Johnson, M. L. (eds.). *See:* Watts, J. A.; and Johnson, M. L. (eds.). (1977).

Johnson, P. L. (ed.). (1969). 158
Remote Sensing in Ecology. University of Georgia Press, Athens, Georgia.

Jorgensen, S. E. (ed.). (1979). 159
Handbook of Environmental Data and Ecological Parameters. Pergamon Press, New York.

Jorgensen, S. E. (ed.). (1979). 160
State-of-the-Art of Ecological Modelling. Pergamon Press, New York.

Keenan, P. C. *See:* Hauck, B.; and Keenan, P. C. (1977).

Keinath, T. M.; and Wanielista, M. 161
(eds.). (1975).
Mathematical Modeling of Water Pollution Control Processes. Ann Arbor Science Publishers, Ann Arbor, Michigan.

Kempthorne, O.; Bancroft, T. A.; Gowen, J. W.; and Lush, J. L. 162
(eds.). (1964).
Statistics and Mathematics in Biology. Hafner Publishing, New York.

Kerner, E. H. (1972). 163
Gibbs Ensemble: Biological Ensemble. Gordon and Breach, New York.

Kershaw, K. A. (1974). 164
Quantitative and Dynamic Plant Ecology, 2nd ed. American Elsevier, New York.

Keyfitz, N. (eds.). *See:* Smith, D.; and Keyfitz, N. (eds.). (1977).

Keyfitz, N. (1968). 165
Introduction to the Mathematics of Population. Addison-Wesley, Reading, Massachusetts.

Keyfitz, N.; and Flieger, W. (1971). 166
Population: Facts and Methods of Demography. W. H. Freeman, San Francisco.

Kimura, M. *See:* Crow, J. F.; and Kimura, M. (1970).

Kirkpatrick, J. (1977). 167
Mathematics for Water and Wastewater Treatment Plant Operators. Ann Arbor Science Publishers, Ann Arbor, Michigan.

Kojima, K. (ed.). (1970). 168
Mathematical Topics in Population Genetics. Springer-Verlag, New York.

Kormandy, E. J. (1969). 169
Concepts of Ecology. Prentice-Hall, Englewood Cliffs, New Jersey.

Kostitzin, V. A. (1939). 170
Mathematical Biology. (Transl. by T. H. Savory). George G. Harrap, London.

Kranz, J. (1974). 171
Epidemics of Plant Diseases, Mathematical Analysis and Modeling. Ecological Studies (13). Springer-Verlag, New York.

Krebs, C. J. (1972). 172
Ecology - The Experimental Analysis of Distribution and Abundance. Harper & Row, New York.

Kruskal, W. H. *See:* Tanur, J. M.; Mosteller, F.; Kruskal, W. H.; Link, R. F.; Pieters, R. S.; Rising, G. R.; and Lehmann, E. L. (1977).

Kruskal, W. H. *See:* Tanur, J. M.; Mosteller, F.; Kruskal, W. H.; Link, R. F.; Pieters, R. S.; and Rising, G. R. (1972).

Krutilla, J. V. (ed.). (1972). 173
Natural Environments: Studies in Theoretical and Applied Analysis. Johns Hopkins University Press, Baltimore.

Kuczynski, R. R. (1969). 174
The Measurement of Population Growth. Gordon and Breach, New York.

Lack, D. (1968). 175
Population Studies of Birds. Oxford University Press, Clarendon.

Lack, D. (1970). 176
The Natural Regulation of Animal Numbers. Oxford University Press, Clarendon.

Lawrence, P. A. (1976). 177
Insect Development. Halsted Press, New York.

Le Cren, E. D.; and Holdgate, M. W. 178
(eds.). (1962).
The Exploitation of Natural Animal Populations. Blackwell Scientific, Oxford.

Legay, J. M.; and Tomassone, R. 179
(eds.). (1978).
Biometrie et Ecologie. Societe Francaise de Biometrie, Paris.

Lehmann, E. L. *See:* Tanur, J. M.; Mosteller, F.; Kruskal, W. H.; Link, R. F.; Pieters, R. S.; Rising, G. R.; and Lehmann, E. L. (1977).

Lenihan, J.; and Fletcher, W. W. 180
(1978).
Measuring and Monitoring the Environment. Academic Press, New York.

Levin, S. A. (ed.). (1974). 181
Ecosystem Analysis and Prediction. SIAM Publishers, Philadelphia.

Levin, S. A. (ed.). (1978). 182
Studies in Mathematical Biology. Part II. Popu-

lations and Communities. Mathematical Association of America, Providence.

Levin, S. A. (ed.). (1978). 183
Studies in Mathematical Biology: Part I, Cellular Behaviour and the Development of Pattern. Mathematical Association of America.

Levins, R. (1968). 184
Evolution in Changing Environments. Monographs in Population Biology 2, Princeton University Press, Princeton, New Jersey.

Lewis, T.; and Taylor, L. R. (1967). 185
Introduction to Experimental Ecology. Academic Press, New York.

Lewontin, R. C. *See:* Simpson, G. G.; Roe, A.; and Lewontin, R. C. (1960).

Lieth, H. (ed.). (1974). 186
Phenology and Seasonality Modeling. Springer-Verlag, New York.

Link, R. F. *See:* Tanur, J. M.; Mosteller, F.; Kruskal, W. H.; Link, R. F.; Pieters, R. S.; and Rising, G. R. (1972).

Link, R. F. *See:* Tanur, J. M.; Mosteller, F.; Kruskal, W. H.; Link, R. F.; Pieters, R. S.; Rising, G. R.; and Lehmann, E. L. (1977).

Lotka, A. J. (1925). 187
Elements of Physical Biology. Williams and Wilkins, Baltimore, Maryland.

Lotka, A. J. (1956). 188
Elements of Mathematical Biology. Dover Publications, New York.

Lowe-McConnell, R. H. (eds.). *See:* Van Dobben, W. H.; and Lowe-McConnell, R. H. (eds.). (1975).

Luckmann, W. *See:* Metcalf, R.; and Luckmann, W. (1975).

Ludwig, D. (1974). 189
Stochastic Population Theories. Springer-Verlag, New York.

Ludwig, D.; and Cooke, K. L. (1975). 190
Epidemiology. SIAM Publishers, Philadelphia.

Lush, J. L. (eds.). *See:* Kempthorne, O.; Bancroft, T. A.; Gowen, J. W.; and Lush, J. L. (eds.). (1964).

Lynn, W. R. (eds.). *See:* Charnes, A.; and Lynn, W. R. (eds.). (1975).

Macfadyen, A. (ed.). (1975). 191
Advances in Ecological Research. Academic Press, New York.

MacArthur, R. H. (1966). 192
The Biology of Populations. John Wiley & Sons, New York.

MacArthur, R. H. (1972). 193
Geographical Ecology: Patterns in the Distribution of Species. Harper & Row, New York.

MacArthur, R. H.; and Wilson, E. O. 194
(1967).
The Theory of Island Biogeography. Monographs in Population Biology 1, Princeton University Press, Princeton, New Jersey.

MacDonald, N. (1978). 195
Time Lags in Biological Models. Springer-Verlag, New York.

MacFadyen, A. *See:* Petrusewicz, K.; and MacFadyen, A. (1970).

Maitra, S. C. *See:* Goel, N. S.; Maitra, S. C.; and Montroll, E. W. (1971).

Maki, A. W. *See:* Cairns, J. Jr.; Dickson, K. L.; and Maki, A. W. (1978).

Maloney, T. E. (eds.). *See:* Middlebrooks, E. J.; Falkenborg, D. H.; and Maloney, T. E. (eds.). (1977).

Marble, D. F. *See:* Berry, B. J. L.; and Marble, D. F. (1968).

Margalef, R. (1968). 196
Perspectives in Ecological Theory. The University of Chicago Press, Chicago.

Maruyama, T. (1977). 197
Stochastic Problems in Population Genetics. Springer-Verlag, New York.

Matern, B. (1970). 198
Spatial Variation. Meddelander Fran Statens Skogsforskningsinstitut.

Matis, J. H.; Patten, B. C.; and White, 199
G. C. (eds.). (1979).
Compartmental Analysis of Ecosystem Models. Satellite Program in Statistical Ecology, International Co-operative Publishing House, Fairland, Maryland.

May, R. M. (ed.). (1976). 200
Theoretical Ecology Principles and Applications. W. B. Saunders, Philadelphia.

May, R. M. (1973). 201
Stability and Complexity in Model Ecosystems. Monographs in Population Biology 6, Princeton University Press, Princeton, New Jersey.

Maynard Smith, J. (1974). 202
Mathematical Ideas in Biology. Cambridge University Press, New York.

Maynard Smith, J. (1974). 203
Models in Ecology. Cambridge University Press, New York.

McDave, I. N. *See:* Goldberg, E. D.; McDave, I. N.; O'Brien, J. J.; and Steele, J. H. (1977).

McMillan, M. *See:* Bromley, D. W.; McMillan, M.; Robertson, M.; and Schoeder, A. (1977).

McRoy, C. P.; and Helfferich, C. 204
(1977).
Seagrass Ecosystems. Marcel Dekker, New York.

Merriam, D. F. *See:* Harbaugh, J. W.; and Merriam, D. F. (1968).

Mesarovic, M. D. (ed.). (1968). 205
Systems Theory and Biology. Springer-Verlag, New York.

Messenger, P. S. *See:* Huffaker, C. B.; and Messenger, P. S. (1976).

Metcalf, R.; and Luckmann, W. (1975). 206
Introduction to Insect Pest Management. John Wiley & Sons, New York.

Middlebrooks, E. J.; Falkenborg, D. H.; and Maloney, T. E. (eds.). (1977). 207
Modeling the Eutrophication Process. Ann Arbor Science Publishers, Ann Arbor, Michigan.

Milner, C.; and Hughes, R. E. (1968). 208
Methods for the Measurement of the Primary Production of Grassland. IBP Handbook No. 6, Blackwell Scientific, Oxford.

Montroll, E. W. *See:* Goel, N. S.; Maitra, S. C.; and Montroll, E. W. (1971).

Moore, H. B. (1958). 209
Marine Ecology. John Wiley & Sons, New York.

Morowitz, H. J. (1968). 210
Energy Flow in Biology. Academic Press, New York.

Morresi, A. C. (eds.). *See:* Cheremisinoff, P. N.; and Morresi, A. C. (eds.). (1978).

Morresi, A. C. *See:* Cheremisinoff, P. N.; and Morresi, A. C. (1977).

Morris, R. F. *See:* Clark, L. R.; Geier, P. W.; Hughes, R. D.; and Morris, R. F. (1967).

Mosimann, J. E. (1968). 211
Elementary Probability for the Biological Sciences. Appleton-Century-Crofts.

Mosteller, F. *See:* Tanur, J. M.; Mosteller, F.; Kruskal, W. H.; Link, R. F.; Pieters, R. S.; and Rising, G. R. (1972).

Mosteller, F. *See:* Tanur, J. M.; Mosteller, F.; Kruskal, W. H.; Link, R. F.; Pieters, R. S.; Rising, G. R.; and Lehmann, E. L. (1977).

Mound, L. A.; and Waloff, N. (eds.). (1978). 212
Diversity of Insect Faunas. Blackwell Scientific, Oxford.

Mozley, A. (1960). 213
Consequences of Disturbances: The Pest Situation Examined. H. K. Lewis and Company, London.

Mueller-Dombois, D.; and Ellenberg, H. (1974). 214
Aims and Methods of Vegetation Ecology. John Wiley & Sons, New York.

Muirhead-Thomson, R. C. (1968). 215
Ecology of Insect Vector Populations. Academic Press, New York.

Munn, R. E. (1970). 216
Biometeorological Methods. Academic Press, New York.

Newbould, P. J. (1967). 217
Methods for Estimating the Primary Production of Forests. IBP Handbook No. 2, Blackwell Scientific, Oxford.

Nihoul, J. C. J. (ed.). (1975). 218
Modelling of Marine Systems. Elsevier, New York.

Nikolsky, G. V. (1969). 219
Theory of Fish Population Dynamics as the Background for Rational Exploitation and Management of Fishery Resources. Oliver and Boyd, Edinburgh.

O'Brien, J. J. *See:* Goldberg, E. D.; McDave, I. N.; O'Brien, J. J.; and Steele, J. H. (1977).

O'Laoghaire, D. T.; and Himmelblau, D. M. (1974). 220
Optimal Expansion of a Water Resources System. Academic Press, New York.

O'Neill, R. V. (eds.). *See:* Innis, G.; and O'Neill, R. V. (eds.). (1979).

O'Riordan, T. (1971). 221
Perspectives on Resource Management. Pion Limited, London.

Odum, E. P. (1971). 222
Fundamentals of Ecology. W. B. Saunders, Philadelphia.

Odum, H. J. (1971). 223
Environment, Power, and Society. John Wiley & Sons, New York.

Ord, J. K. (eds.). *See:* Cormack, R. M.; and Ord, J. K. (eds.). (1979).

Ord, J. K. *See:* Cliff, A. D.; and Ord, J. K. (1973).

Ord, J. K.; Patil, G. P.; and Taillie, C. (eds.). (1979). 224
Statistical Distributions in Ecological Work. Satellite Program in Statistical Ecology, International Co-operative Publishing House, Fairland, Maryland.

Orloci, L. (1978). 225
Multivariate Analysis in Vegetation Research. Dr. Junk Publishers, The Hague, Netherlands.

Orloci, L.; Rao, C. R.; and Stiteler, W. M. (eds.). (1979). 226
Multivariate Methods in Ecological Work. Satellite Program in Statistical Ecology, International Co-operative Publishing House, Fairland, Maryland.

Orton, C. *See:* Hodder, I.; and Orton, C. (1976).

Oster, G.; and Wilson, E. O. (1978). 227
Caste and Ecology in the Social Insects. Princeton University Press, Princeton, New Jersey.

Ott, W. R. (ed.). (1976). 228
Proceedings of the EPA Conference on Environmental Modeling and Simulation. U.S. Environmental Protection Agency, Washington, D.C.

Ott, W. R. (1978). 229
Environmental Indices: Theory and Practice. Ann Arbor Science Publishers, Ann Arbor, Michigan.

Pankhurst, R. J. (ed.). (1975). 230
Biological Identification with Computers. Academic Press, New York.

Pantell, R. H. (1976). 231
Techniques of Environmental Systems Analysis. John Wiley & Sons, New York.

Park, O. *See:* Allee, W. C.; Emerson, A. E.; Park, O.; Park, T.; and Schmidt, K. P. (1949).

Park, T. *See:* Allee, W. C.; Emerson, A. E.; Park, O.; Park, T.; and Schmidt, K. P. (1949).

Parkinson, D.; Gray, T. R. G.; and 232
Williams, S. T. (eds.). (1971). *Methods for Studying the Ecology of Soil Microorganisms*. IBP Handbook No. 19, Blackwell Scientific, Oxford.

Parsons, T. R.; Takahashi, M.; and 233
Hargrave, B. (1973). *Biological Oceanographic Processes*. Pergamon Press, New York.

Patil, G. P. (1970). *Random Counts in* 234
Biomedical and Social Sciences. The Pennsylvania State University Press, University Park, Pennsylvania.

Patil, G. P. (1970). *Random Counts in* 235
Models and Structures. The Pennsylvania State University Press, University Park, Pennsylvania.

Patil, G. P.; and Rosenzweig, M. 236
(eds.). (1979). *Contemporary Quantitative Ecology and Related Ecometrics*. Satellite Program in Statistical Ecology, International Co-operative Publishing House, Fairland, Maryland.

Patil, G. P.; Pielou, E. C.; and Waters, 237
W. E. (eds.). (1971). *Statistical Ecology, Volume 1, Spatial Patterns and Statistical Distributions*. The Pennsylvania State University Press, University Park, Pennsylvania.

Patil, G. P.; Pielou, E. C.; and Waters, 238
W. E. (eds.). (1971). *Statistical Ecology, Volume 2, Sampling and Modeling Biological Populations and Population Dynamics*. The Pennsylvania State University Press, University Park, Pennsylvania.

Patil, G. P.; Pielou, E. C.; and Waters, 239
W. E. (eds.). (1971). *Statistical Ecology, Volume 3, Many Species Populations, Ecosystems, and Systems Analysis*. The Pennsylvania State University Press, University Park, Pennsylvania.

Patil, G. P. *See:* Cairns, J. Jr.; Patil, G. P.; and Waters, W. E. (eds.). (1979).

Patil, G. P. *See:* Cormack, R. M.; Patil, G. P.; and Robson, D. S. (eds.). (1979).

Patil, G. P. *See:* Grassle, F.; Patil, G. P.; Smith, W.; and Taillie, C. (eds.). (1979).

Patil, G. P. *See:* Ord, J. K.; Patil, G. P.; and Taillie, C. (eds.). (1979).

Patten, B. C. (ed.). (1971). *Systems* 240
Analysis and Simulation in Ecology, Volume I. Academic Press, New York.

Patten, B. C. (ed.). (1972). *Systems* 241
Analysis and Simulation in Ecology, Volume II. Academic Press, New York.

Patten, B. C. (ed.). (1975). *Systems* 242
Analysis and Simulation in Ecology, Volume III. Academic Press, New York.

Patten, B. C. (ed.). (1976). *Systems* 243
Analysis and Simulation in Ecology, Volume IV. Academic Press, New York.

Patten, B. C. *See:* Matis, J. H.; Patten, B. C.; and White, G. C. (eds.). (1979).

Pavlidis, T. (1973). *Biological Oscilla-* 244
tors: Their Mathematical Analysis. Academic Press, New York.

Petrusewicz, K.; and MacFadyen, A. 245
(1970). *Productivity of Terrestrial Animals: Principles and Methods*. IBP Handbook No. 13, Blackwell Scientific, Oxford.

Phillipson, J. (ed.). (1969). *Methods of* 246
Study in Soil Ecology. UNESCO, Paris.

Pianka, E. R. (1974). *Evolutionary* 247
Ecology. Harper and Row, New York.

Pielou, E. C. (1969). *An Introduction to* 248
Mathematical Ecology. John Wiley & Sons, New York.

Pielou, E. C. (1974). *Population and* 249
Community Ecology - Principles and Methods. Gordon and Breach, New York.

Pielou, E. C. (1975). *Ecological Diver-* 250
sity. John Wiley & Sons, New York.

Pielou, E. C. (1977). *Mathematical* 251
Ecology. John Wiley & Sons, New York.

Pielou, E. C. *See:* Patil, G. P.; Pielou, E. C.; and Waters, W. E. (eds.). (1971).

Pieters, R. S. *See:* Tanur, J. M.; Mosteller, F.; Kruskal, W. H.; Link, R. F.; Pieters, R. S.; Rising, G. R.; and Lehmann, E. L. (1977).

Pieters, R. S. *See:* Tanur, J. M.; Mosteller, F.; Kruskal, W. H.; Link, R. F.; Pieters, R. S.; and Rising, G. R. (1972).

Pimentel, D. *See:* Smith, E. H.; and Pimentel, D. (1978).

Pirie, N. W. (eds.). *See:* Cragg, J. B.; and Pirie, N. W. (eds.). (1955).

Poole, R. W. (1974). *An Introduction to* 252
Quantitative Ecology. McGraw-Hill, New York.

Prasad, J. *See:* Griffith, A. L.; and Prasad, J. (1949).

Pratt, J. W. (ed.). (1974). *Statistical and* 253
Mathematical Aspects of Pollution Problems. Marcel Dekker, New York.

Prodan, M. (1968). *Forest Biometrics*. 254
Pergamon Press, New York.

Ra, B. S. *See:* Griffith, A. L.; and Ra, B. S. (1947).

Rao, C. R. (1970). *Advanced Statistical* 255
Methods in Biometric Research. Hafner Publishing, New York.

Rao, C. R. *See:* Orloci, L.; Rao, C. R.; and Stiteler, W. M. (eds.). (1979).

Raunkiaer, C. (1934). *The Life Forms* 256
of Plants and Statistical Plant Geography. Clarendon Press, Oxford.

Raup, D. M.; and Stanley, S. M. 257
(1978). *Principles of Paleontology.* W. H. Freeman, San Francisco.

Reddingius, J. (1971). *Gambling for* 258
Existence. E. J. Brill, Leiden, Netherlands.

Reichle, D. E. (ed.). (1970). *Analysis of* 259
Temperate Forest Ecosystems. (Ecological Studies - Analysis and Synthesis, Vol. 1). Springer-Verlag, New York.

Reichle, D. E. (ed.). (1970). *Studies in* 260
Ecology. Springer-Verlag, New York.

Reyment, R. A. (1971). *Introduction to* 261
Quantitative Paleoecology. American Elsevier, New York.

Reyment, R. A. *See:* Blackith, R. E.; and Reyment, R. A. (1971).

Ricciardi, L. M. (1977). *Diffusion Proc-* 262
esses and Related Topics in Biology. Springer-Verlag, New York.

Richter-Dyn, N. *See:* Goel, N. S.; and Richter-Dyn, N. (1974).

Ricker, W. E. (ed.). (1968). *Methods* 263
for Assessment of Fish Production in Fresh Waters. IBP Handbook No. 3, Blackwell Scientific, Oxford.

Ricker, W. E. (1948). *Methods of* 264
Estimating Vital Statistics of Fish Populations. Indiana University Publications, Science Series No. 15.

Ricker, W. E. (1975). *Computation and* 265
Interpretation of Biological Statistics of Fish Populations. Department of the Environment, Fisheries and Marine Service.

Ricklefs, R. E. (1973). *Ecology.* Chiron 266
Press, Portland, Oregon.

Rising, G. R. *See:* Tanur, J. M.; Mosteller, F.; Kruskal, W. H.; Link, R. F., Pieters, R. S.; and Rising, G. R. (1972).

Rising, G. R. *See:* Tanur, J. M.; Mosteller, F.; Kruskal, W. H.; Link, R. F.; Pieters, R. S.; Rising, G. R.; and Lehmann, E. L. (1977).

Roberts, F. S. (1975). *Discrete Mathe-* 267
matical Models with Applications to Social, Biological, and Environmental Problems. Prentice Hall, Englewood Cliffs, New Jersey.

Robertson, A. (eds.). *See:* Scavia, D.; and Robertson, A. (eds.). (1978).

Robertson, M. *See:* Bromley, D. W.; McMillan, M.; Robertson, M.; and Schoeder, A. (1977).

Robson, D. S. (eds.). *See:* Cormack, R. M.; Patil, G. P.; and Robson, D. S. (eds.). (1979).

Robson, D. S. *See:* Brownie, C.; Anderson, D. R.; Burnham, K. P.; and Robson, D. S. (1978).

Roe, A. *See:* Simpson, G. G.; Roe, A.; and Lewontin, R. C. (1960).

Roedel, P. M. (ed.). (1975). *Optimum* 268
Sustainable Yield as a Concept in Fisheries Management. The American Fisheries Society, Washington, D. C.

Rogers, A. (1968). *Matrix Analysis of* 269
Population Growth and Distribution. University of California Press, Berkeley.

Rogers, A. (1974). *Statistical Analysis of* 270
Spatial Dispersion. Pion Limited, London.

Rosen, R. (1970). *Dynamical System* 271
Theory in Biology, Volume I: Stability Theory and Its Applications. John Wiley & Sons, New York.

Rosenzweig, M. (eds.). *See:* Patil, G. P.; and Rosenzweig, M. (eds.). (1979).

Rounsfell, G. A.; and Everhart, W. H. 272
(1953). *Fishery Science: Its Methods and Applications.* John Wiley & Sons, New York.

Rutter, A. J. *See:* Wadsworth, R. M.; Chapas, L. C.; Rutter, A. J.; Solomon, M. E.; and Wilson, J. W. (eds.). (1967).

Scavia, D.; and Robertson, A. (eds.). 273
(1978). *Perspectives on Lake Ecosystem Modeling.* Ann Arbor Science Publishers, Ann Arbor, Michigan.

Schanda, E. (ed.). (1976). *Remote* 274
Sensing for Environmental Sciences. Springer-Verlag, New York.

Schmerl, R. B. *See:* Gates, D. M.; and Schmerl, R. B. (1975).

Schmidt, K. P. *See:* Allee, W. C.; Emerson, A. E.; Park, O.; Park, T.; and Schmidt, K. P. (1949).

Schmidt, K. P. *See:* Heese, R.; Allee, W. C.; and Schmidt, K. P. (1951).

Schoeder, A. *See:* Bromley, D. W.; McMillan, M.; Robertson, M.; and Schoeder, A. (1977).

Schultz, V.; Eberhardt, L. L.; 275
Thomas, J. M.; and Cochran, M. I. (1976). *A Bibliography of Quantitative Ecology.* Dowden Hutchinson & Ross, Stroudsburg, Pennsylvania.

Scudo, F. M.; and Ziegler, J. R. 276
(1978). *The Golden Age of Theoretical Ecology: 1923-1940.* Springer-Verlag, New York.

Seber, G. A. F. (1973). *The Estimation* 277
of Animal Abundance and Related Parameters. Charles Griffin, London.

Shimwell, D. W. (1972). *The Descrip-* 278
tion and Classification of Vegetation. University of Washington Press, Seattle, Washington.

Shugart, H. H. (ed.). (1978). *Time* 279
Series and Ecological Processes. SIAM Publishers, Philadelphia.

Sibson, R. *See:* Jardine, N.; and Sibson, R. (1971).

Simpson, G. G. (1978). *Concession to* 280
the Improbable. Yale University Press, New Haven, Connecticut.

Simpson, G. G.; Roe, A.; and 281
Lewontin, R. C. (1960). *Quantitative Zoology*. Harcourt, Brace & World, New York.

Singh, J. S.; and Gopal, B. (eds.). 282
(1978). *Glimpses of Ecology*. Prakash Publishers, Jaipur, India.

Sladen, B. K.; and Bang, F. B. (eds.). 283
(1969). *Biology of Populations - The Biological Basis of Public Health*. American Elsevier, New York.

Slobodkin, L. B. (1961). *Growth and* 284
Regulation of Animal Populations. Holt, Rinehart and Winston, New York.

Smith, D.; and Keyfitz, N. (eds.). 285
(1977). *Mathematical Demography*. Springer-Verlag, New York.

Smith, E. H.; and Pimentel, D. (1978). 286
Pest Control Strategies. Academic Press, New York.

Smith, H. H. (eds.). *See:* Woodwell, G. M.; and Smith, H. H. (eds.). (1969).

Smith, W. *See:* Grassle, F.; Patil, G. P.; Smith, W.; and Taillie, C. (eds.). (1979).

Sneath, P. H. A. *See:* Sokal, R. R.; and Sneath, P. H. A. (1963).

Sneath, P. H. A.; and Sokal, R. R. 287
(1973). *Numerical Taxonomy: The Principles and Practice of Numerical Classification*. W. H. Freeman, San Francisco.

Sokal, R. R. *See:* Sneath, P. H. A.; and Sokal, R. R. (1973).

Sokal, R. R.; and Sneath, P. H. A. 288
(1963). *Principles of Numerical Taxonomy*. W. H. Freeman, San Francisco.

Solomon, M. E. *See:* Wadsworth, R. M.; Chapas, L. C.; Rutter, A. J.; Solomon, M. E.; and Wilson, J. W. (eds.). (1967).

Southwick, C. H. (1972). *Ecology and* 289
the Quality of Our Environment. D. Van Nostrand, New York.

Southwood, T. R. E. (1966). *Ecological* 290
Methods. Chapman and Hall, London.

Stanley, S. M. *See:* Raup, D. M.; and Stanley, S. M. (1978).

Steele, J. H. (ed.). (1970). *Marine Food* 291
Chains. Oliver and Boyd, Edinburgh.

Steele, J. H. (ed.). (1978). *Spatial Pat-* 292
tern in Plankton Communities. Plenum Press, New York.

Steele, J. H. (1974). *The Structure of* 293
Marine Ecosystems. Blackwell Scientific, Oxford.

Steele, J. H. *See:* Goldberg, E. D.; McDave, I. N.; O'Brien, J. J.; and Steele, J. H. (1977).

Stiteler, W. M. (eds.). *See:* Orloci, L.; Rao, C. R.; and Stiteler, W. M. (eds.). (1979).

Taillie, C. (eds.). *See:* Grassle, F.; Patil, G. P.; Smith, W.; and Taillie, C. (eds.). (1979).

Taillie, C. (eds.). *See:* Ord, J. K.; Patil, G. P.; and Taillie, C. (eds.). (1979).

Takahashi, M. *See:* Parsons, T. R.; Takahashi, M.; and Hargrave, B. (1973).

Tanur, J. M.; Mosteller, F.; Kruskal, 294
W. H.; Link, R. F.; Pieters, R. S.; Rising, G. R.; and Lehmann, E. L. (1977). *Statistics: A Guide to Biological and Health Sciences*. Holden-Day, San Francisco.

Tanur, J. M.; Mosteller, F.; Kruskal, 295
W. H.; Link, R. F.; Pieters, R. S.; and Rising, G. R. (1972). *Statistics: A Guide to the Unknown*. Holden-Day, San Francisco.

Tautu, P. *See:* Iosifescu, M.; and Tautu, P. (1973).

Tautu, P.; and Iosifescu, M. (1968). 296
Stochastic Processes and Applications in Biology and Medicine II. Springer-Verlag, New York.

Taylor, L. R. *See:* Lewis, T.; and Taylor, L. R. (1967).

Thom, R. (1972). *Stabilite Structurelle* 297
et Morphogenese (English ed. 1975). Benjamin, New York.

Thomas, J. M. *See:* Schultz, V.; Eberhardt, L. L.; Thomas, J. M.; and Cochran, M. I. (1976).

Thomas, R. W. (1977). *An Introduction* 298
to Quadrat Analysis. Geological Abstracts Ltd., University of East Anglia, Norwich, NR4 7TJ.

Thomas, W. A. (ed.). (1975). *Indicators* 299
of Environmental Quality. Plenum Press, New York.

Thomas, W. A.; Goldstein, G.; and 300
Wilcox, W. H. (1976). *Biological Indicators of Environmental Quality*. Ann Arbor Science Publishers, Ann Arbor, Michigan.

Thornley, J. H. M. (1976). *Mathe-* 301
matical Models in Plant Pathology. Academic Press, New York.

Tomassone, R. (eds.). *See:* Legay, J. M.; and Tomassone, R. (eds.). (1978).

Tomovic, R. (1963). *Sensitivity Analysis* 302
of Dynamic Systems. McGraw-Hill, New York.

Tummala, R. L.; Haynes, D. L.; and 303
Croft, B. A. (eds.). (1976). *Modeling for Pest Management*. Michigan State University, East Lansing, Michigan.

Turk, A. *See:* Turk, J.; Wittes, J. T.; Wittes, R.; and Turk, A. (1975).

Turk, J.; Wittes, J. T.; Wittes, R.; and 304
Turk, A. (1975). *Ecosystems, Energy, Population.* W. B. Saunders, Philadelphia.

Udvardy, M. D. F. (1969). *Dynamic* 305
Zoogeography, with Special Reference to Land Animals. van Nostrand Reinhold, New York.

Underwood, E. E. (1970). *Quantitative* 306
Stereology. Addison-Wesley, Reading, Massachusetts.

Usher, M. B.; and Williamson, M. H. 307
(eds.). (1974). *Ecological Stability.* Chapman and Hall, London.

van den Driessche, P. (1974). 308
Mathematical Problems in Biology. Springer-Verlag, New York.

van Dobben, W. H.; and Lowe- 309
McConnell, R. H. (eds.). (1975). *Unifying Concepts in Ecology.* Dr. Junk Publishers, The Hague, Netherlands.

van Dyne, G. M. (ed.). (1969). *The* 310
Ecosystem Concept in Natural Resource Management. Academic Press, New York.

van Ryzin, J. (1977). *Classification and* 311
Clustering. Academic Press, New York.

Vann, E. (1972). *Fundamentals of Bio-* 312
statistics. D. C. Heath, Lexington, Massachusetts.

Varley, G. C.; Gradwell, G. R.; and 313
Hassell, M. P. (1973). *Insect Population Ecology, An Analytical Approach.* Blackwell Scientific, Oxford.

Vollenweider, R. A. (1969). *A Manual* 314
on Methods for Measuring Primary Production in Aquatic Environments. IBP Handbook No. 12, Blackwell Scientific, Oxford.

Volterra, V. (1931). *Lecons sur la* 315
theorie mathematique de la lutte pour la vie. Gauthiers-Villars, Paris.

Wadley, F. M. (1967). *Experimental* 316
Statistics in Entomology. Graduate School Press, U. S. Dept. of Agriculture, Washington, D. C.

Wadsworth, R. M.; Chapas, L. C.; 317
Rutter, A. J.; Solomon, M. E.; and Wilson, J. W. (eds.). (1967). *The Measurement of Environmental Factors in Terrestrial Ecology.* Blackwell Scientific, Oxford.

Waloff, N. (eds.). *See:* Mound, L. A.; and Waloff, N. (eds.). (1978).

Walsh, J. J. (eds.). *See:* Cushing, D. H.; and Walsh, J. J. (eds.). (1976).

Waltman, P. (1974). *Deterministic* 318
Threshold Models in the Theory of Epidemics. Springer-Verlag, New York.

Wanielista, M. (eds.). *See:* Keinath, T. M.; and Wanielista, M. (eds.). (1975).

Ward, D. V. (1978). *Biological Envi-* 319
ronmental Impact Studies: Theory and Methods. Academic Press, New York.

Waters, W. E. (eds.). *See:* Patil, G. P.; Pielou, E. C.; and Waters, W. E. (eds.). (1971).

Waters, W. E. (eds.). *See:* Cairns, J. Jr.; Patil, G. P.; and Waters, W. E. (eds.). (1979).

Watt, K. E. F. (ed.). (1966). *Systems* 320
Analysis in Ecology. Academic Press, New York.

Watt, K. E. F. (1968). *Ecology and Re-* 321
source Management. McGraw-Hill, New York.

Watt, K. E. F. (1973). *Principles of En-* 322
vironmental Science. McGraw-Hill, New York.

Watts, J. A.; and Johnson, M. L. 323
(eds.). (1977). *Abstracts - U.S. International Biological Program - Ecosystem Analysis Studies (Vol. 1, No. 4).* Oak Ridge National Laboratory, Tennessee.

Weatherley, A. H. (1972). *Growth and* 324
Ecology of Fish Populations. Academic Press, New York.

Weiss, P. A. (1971). *Hierarchically Or-* 325
ganized Systems in Theory and Practice. Hafner Publishing, New York.

Wesley, J. P. (1974). *Ecophysics: The* 326
Application of Physics to Ecology. Charles C. Thomas, Springfield, Illinois.

Westlake, G. G. (eds.). *See:* Cairns, J. Jr.; Dickson, K. L.; and Westlake, G. G. (eds.). (1975).

Weyl, P. K. (1970). *Oceanography: An* 327
Introduction to the Marine Environment. John Wiley & Sons, New York.

White, G. C. (eds.). *See:* Matis, J. H.; Patten, B. C.; and White, G. C. (eds.). (1979).

White, G. F. *See:* Holdgate, M. W.; and White, G. F. (1977).

Whittaker, R. H. (ed.). (1978). 328
Classification of Plant Communities. Dr. Junk Publishers, The Hague, Netherlands.

Whittaker, R. H. (ed.). (1978). *Or-* 329
dination of Plant Communities. Dr. Junk Publishers, The Hague, Netherlands.

Whittaker, R. H. (1975). *Communities* 330
and Ecosystems. MacMillan Publishing, New York.

Wicklund, E. C. *See:* Wilimovsky, N. J.; and Wicklund, E. C. (1963).

Wilcox, W. H. *See:* Thomas, W. A.; Goldstein, G.; and Wilcox, W. H. (1976).

Wilimovsky, N. J.; and Wicklund, E. 331
C. (1963). *Tables of the Incomplete Beta Function for the Calculation of Fish Population Yield.* University of British Columbia Press, Vancouver.

Williams, C. B. (1964). *Patterns in the* 332
Balance of Nature. Academic Press, New York.

Williams, G. C. (1975). *Sex and Evolu-* 333
tion. Princeton University Press, Princeton, New Jersey.

Williams, S. T. (eds.). *See:* Parkinson, D.; Gray, T. R. G.; and Williams, S. T. (eds.). (1971).

Williams, W. T. (ed.). (1976). *Pattern* 334
Analysis in Agricultural Science. CSIRO, Melbourne; and American Elsevier, New York.

Williamson, M. (1972). *The Analysis of* 335
Biological Populations. Edward Arnold, London.

Williamson, M. H. (eds.). *See:* Usher, M. B.; and Williamson, M. H. (eds.). (1974).

Willis, J. C. (1922). *Age and Area*. 336
Cambirdge University Press.

Wilson, E. O. *See:* MacArthur, R. H.; and Wilson, E. O. (1967).

Wilson, E. O. *See:* Oster, G.; and Wilson, E. O. (1978).

Wilson, E. O.; and Bossert, W. H. 337
(1971). *A Primer of Population Biology*. Sinauer Associates, Stamford, Connecticut.

Wilson, J. W. (eds.). *See:* Wadsworth, R. M.; Chapas, L. C.; Rutter, A. J.; Solomon, M. E.; and Wilson, J. W. (eds.). (1967).

Winberg, G. G. (ed.). (1971). *Methods* 338
for the Estimation of Production of Aquatic Animals. (Translated by A. Duncan). Academic Press, New York.

Wittes, J. T. *See:* Turk, J.; Wittes, J. T.; Wittes, R.; and Turk, A. (1975).

Wittes, R. *See:* Turk, J.; Wittes, J. T.; Wittes, R.; and Turk, A. (1975).

Wood, E. J. F. (1965). *Marine Micro-* 339
bial Ecology. Chapman and Hall, London.

Woodman, M. J. (eds.). *See:* Holdgate, M. W.; and Woodman, M. J. (eds.). (1978).

Woodwell, G. M.; and Smith, H. H. 340
(eds.). (1969). *Diversity and Stability in Ecological Systems*. Brookhaven National Laboratory, National Technical Information Service, Springfield, Virginia.

Young, I. G. *See:* Bibbero, R. J.; and Young, I. G. (1974).

Zaika, V. E. (1972). *Udel'Naya* 341
Produktsiya Vodnykh bespozvonochnykyh. izdatel'stvo 'naukova dumka', Kiev. (English translation by A. Mercado and edite Halsted Press, New York.

Ziegler, J. R. *See:* Scudo, F. M.; and Ziegler, J. R. (1978).

(Received June 1977. Revised July 1979)

AUTHOR INDEX

SUBJECT INDEX

INTERNATIONAL STATISTICAL ECOLOGY PROGRAM

The International Statistical Ecology Program (ISEP) consists of the activities of the Statistical Ecology Section of the International Association for Ecology and of the Liaison Committee on Statistical Ecology of the International Statistical Institute, the Biometric Society, and the International Association for Ecology. The ISEP is a non-profit program formulated to serve the needs of interdisciplinary research and training in the newly emerging fields of Statistical Ecology and Ecological Statistics.

SATELLITE PROGRAM IN STATISTICAL ECOLOGY

The Second International Congress of Ecology was held in Jerusalem during September 1978. In this connection, ISEP organized a Satellite Program in Statistical Ecology during 1977 and 1978. The emphasis was on research, review, and exposition concerned with the interface between quantitative ecology and relevant quantitative methods. Both theory and application of ecology and ecometrics received attention. The Satellite Program consisted of instructional coursework, seminar series, thematic research conferences, and collaborative research workshops.

Research papers and research-review-expositions were specially prepared for the program by concerned experts and expositors. These materials have been refereed and revised, and are now available in a series of ten edited volumes listed on page ii of this volume.

The Satellite Program takes as its theme the better melding of fundamental ecological concepts with rigorous empirical quantification. The overall result should be progress toward a stronger body of general ecologic and ecometric theory and practice.

FUTURE DIRECTIONS

The satellite-like-programs help create and sustain enthusiasm, inward strength, and working efficiency of those who desire to meet a contemporary social need in the form of some interdisciplinary work. It should be only proper and rewarding for everyone involved that such programs are planned from time to time.

Plans are being made for a satellite program in conjunction with the next Biennial Conference of the International Statistical Institute and the next International Congress of Ecology. Care should be exercised that the next program not become a mere replica of the present one, however successful it has been. Instead, the next program should be organized so that it helps further the evolution of statistical ecology as a productive field.

The next program is being discussed in terms of subject area groups. Each subject group is to have a coordinator assisted by small committees, such as a program committee, a research committee, an annual review committee, a journal committee, and an education committee. This approach is expected to respond to the need for a journal on statistical ecology, and also to the need of bringing out well planned annual review volumes. The education committee would formulate plans for timely modules and monographs. Interested readers may feel free to communicate their ideas and interests to those involved in planning the next program. The mailing address is: International Statistical Ecology Program, P. O. Box 218, State College, PA 16801, USA.